Springer Handbook of Enzymes Volume 28

Dietmar Schomburg and
Ida Schomburg (Eds.)

Springer Handbook of Enzymes

Volume 28

Class 2 · Transferases I

EC 2.1.1

coedited by Antje Chang

Second Edition

Springer

Professor Dietmar Schomburg
e-mail: d.schomburg@uni-koeln.de

Dr. Ida Schomburg
e-mail: i.schomburg@uni-koeln.de

Dr. Antje Chang
e-mail: a.chang@uni-koeln.de

University to Cologne
Institute for Biochemistry
Zülpicher Strasse 47
50674 Cologne
Germany

Library of Congress Control Number: 2006922083

ISBN-10 3-540-31917-4 2nd Edition Springer Berlin Heidelberg New York
ISBN-13 978-3-540-31917-7 2nd Edition Springer Berlin Heidelberg New York

The first edition was published as Volume 11 (ISBN 3-540-60295-X) of the "Enzyme Handbook".

Springer is a part of Springer Science+Business Media
springeronline.com

Printed in Germany

Cover design: Erich Kirchner, Heidelberg
Typesetting: medionet AG, Berlin

Printed on acid-free paper 2/3141m-5 4 3 2 1 0

Preface

Today, as the full information about the genome is becoming available for a rapidly increasing number of organisms and transcriptome and proteome analyses are beginning to provide us with a much wider image of protein regulation and function, it is obvious that there are limitations to our ability to access functional data for the gene products – the proteins and, in particular, for enzymes. Those data are inherently very difficult to collect, interpret and standardize as they are widely distributed among journals from different fields and are often subject to experimental conditions. Nevertheless a systematic collection is essential for our interpretation of genome information and more so for applications of this knowledge in the fields of medicine, agriculture, etc. Progress on enzyme immobilisation, enzyme production, enzyme inhibition, coenzyme regeneration and enzyme engineering has opened up fascinating new fields for the potential application of enzymes in a wide range of different areas.

The development of the enzyme data information system BRENDA was started in 1987 at the German National Research Centre for Biotechnology in Braunschweig (GBF) and is now continuing at the University at Cologne, Institute of Biochemistry. The present book "Springer Handbook of Enzymes" represents the printed version of this data bank. The information system has been developed into a full metabolic database.

The enzymes in this Handbook are arranged according to the Enzyme Commission list of enzymes. Some 4,000 "different" enzymes are covered. Frequently enzymes with very different properties are included under the same EC-number. Although we intend to give a representative overview on the characteristics and variability of each enzyme, the Handbook is not a compendium. The reader will have to go to the primary literature for more detailed information. Naturally it is not possible to cover all the numerous literature references for each enzyme (for some enzymes up to 40,000) if the data representation is to be concise as is intended.

It should be mentioned here that the data have been extracted from the literature and critically evaluated by qualified scientists. On the other hand, the original authors' nomenclature for enzyme forms and subunits is retained. In order to keep the tables concise, redundant information is avoided as far as possible (e.g. if K_m values are measured in the presence of an obvious cosubstrate, only the name of the cosubstrate is given in parentheses as a commentary without reference to its specific role).

The authors are grateful to the following biologists and chemists for invaluable help in the compilation of data: Cornelia Munaretto and Dr. Antje Chang.

Cologne
Spring 2006

Dietmar Schomburg, Ida Schomburg

List of Abbreviations

A	adenine
Ac	acetyl
ADP	adenosine 5'-diphosphate
Ala	alanine
All	allose
Alt	altrose
AMP	adenosine 5'-monophosphate
Ara	arabinose
Arg	arginine
Asn	asparagine
Asp	aspartic acid
ATP	adenosine 5'-triphosphate
Bicine	N,N'-bis(2-hydroxyethyl)glycine
C	cytosine
cal	calorie
CDP	cytidine 5'-diphosphate
CDTA	trans-1,2-diaminocyclohexane-N,N,N,N-tetraacetic acid
CMP	cytidine 5'-monophosphate
CoA	coenzyme A
CTP	cytidine 5'-triphosphate
Cys	cysteine
d	deoxy-
D-	(and L-) prefixes indicating configuration
DFP	diisopropyl fluorophosphate
DNA	deoxyribonucleic acid
DPN	diphosphopyridinium nucleotide (now NAD^+)
DTNB	5,5'-dithiobis(2-nitrobenzoate)
DTT	dithiothreitol (i.e. Cleland's reagent)
EC	number of enzyme in Enzyme Commission's system
E. coli	Escherichia coli
EDTA	ethylene diaminetetraacetate
EGTA	ethylene glycol bis(-aminoethyl ether) tetraacetate
ER	endoplasmic reticulum
Et	ethyl
EXAFS	extended X-ray absorption fine structure
FAD	flavin-adenine dinucleotide
FMN	flavin mononucleotide (riboflavin 5'-monophosphate)
Fru	fructose
Fuc	fucose
G	guanine
Gal	galactose

GDP	guanosine 5'-diphosphate
Glc	glucose
GlcN	glucosamine
GlcNAc	N-acetylglucosamine
Gln	glutamine
Glu	glutamic acid
Gly	glycine
GMP	guanosine 5'-monophosphate
GSH	glutathione
GSSG	oxidized glutathione
GTP	guanosine 5'-triphosphate
Gul	gulose
h	hour
H4	tetrahydro
HEPES	4-(2-hydroxyethyl)-1-piperazineethane sulfonic acid
His	histidine
HPLC	high performance liquid chromatography
Hyl	hydroxylysine
Hyp	hydroxyproline
IAA	iodoacetamide
IC 50	50% inhibitory concentration
Ig	immunoglobulin
Ile	isoleucine
Ido	idose
IDP	inosine 5'-diphosphate
IMP	inosine 5'-monophosphate
ITP	inosine 5'-triphosphate
K_m	Michaelis constant
L-	(and D-) prefixes indicating configuration
Leu	leucine
Lys	lysine
Lyx	lyxose
M	mol/l
mM	millimol/l
m-	*meta*-
Man	mannose
MES	2-(N-morpholino)ethane sulfonate
Met	methionine
min	minute
MOPS	3-(N-morpholino)propane sulfonate
Mur	muramic acid
MW	molecular weight
NAD^+	nicotinamide-adenine dinucleotide
NADH	reduced NAD
$NADP^+$	NAD phosphate
NADPH	reduced NADP
NAD(P)H	indicates either NADH or NADPH

NBS	N-bromosuccinimide
NDP	nucleoside 5'-diphosphate
NEM	N-ethylmaleimide
Neu	neuraminic acid
NMN	nicotinamide mononucleotide
NMP	nucleoside 5'-monophosphate
NTP	nucleoside 5'-triphosphate
o-	*ortho-*
Orn	ornithine
p-	*para-*
PBS	phosphate-buffered saline
PCMB	*p*-chloromercuribenzoate
PEP	phosphoenolpyruvate
pH	$-\log 10[H^+]$
Ph	phenyl
Phe	phenylalanine
PHMB	*p*-hydroxymercuribenzoate
PIXE	proton-induced X-ray emission
PMSF	phenylmethane-sulfonylfluoride
p-NPP	*p*-nitrophenyl phosphate
Pro	proline
Q_{10}	factor for the change in reaction rate for a 10°C temperature increase
Rha	rhamnose
Rib	ribose
RNA	ribonucleic acid
mRNA	messenger RNA
rRNA	ribosomal RNA
tRNA	transfer RNA
Sar	N-methylglycine (sarcosine)
SDS-PAGE	sodium dodecyl sulfate polyacrylamide gel electrophoresis
Ser	serine
T	thymine
t_H	time for half-completion of reaction
Tal	talose
TDP	thymidine 5'-diphosphate
TEA	triethanolamine
Thr	threonine
TLCK	N^{α}-*p*-tosyl-L-lysine chloromethyl ketone
T_m	melting temperature
TMP	thymidine 5'-monophosphate
Tos-	tosyl-(*p*-toluenesulfonyl-)
TPN	triphosphopyridinium nucleotide (now $NADP^+$)
Tris	tris(hydroxymethyl)-aminomethane
Trp	tryptophan
TTP	thymidine 5'-triphosphate
Tyr	tyrosine
U	uridine

U/mg	μmol/(mg*min)
UDP	uridine 5'-diphosphate
UMP	uridine 5'-monophosphate
UTP	uridine 5'-triphosphate
Val	valine
Xaa	symbol for an amino acid of unknown constitution in peptide formula
XAS	X-ray absorption spectroscopy
Xyl	xylose

List of Deleted and Transferred Enzymes

Since its foundation in 1956 the Nomenclature Committee of the International Union of Biochemistry and Molecular Biology (NC-IUBMB) has continually revised and updated the list of enzymes. Entries for new enzymes have been added, others have been deleted completely, or transferred to another EC-number in the original class or to different EC classes, catalyzing other types of chemical reactions. The old numbers have not been allotted to new enzymes; instead the place has been left vacant or cross-references given to the changes in nomenclature.

Deleted and Transferred Enzymes

For EC class 2.1.1 these changes are:

Recommended name	Old EC-number	Alteration
protein-arginine N-methyltransferase	2.1.1.23	transferred to EC 2.1.1.124, EC 2.1.1.125, EC 2.1.1.126
protein-gamma-glutamate O-methyltransferase	2.1.1.24	deleted, included in EC 2.1.1.77, EC 2.1.1.80, EC 2.1.1.100
tRNA (purine-2- or -6-)-methyltransferase	2.1.1.30	deleted
mRNA (adenosine-2'-O-)-methyltransferase	2.1.1.58	deleted, included in EC 2.1.1.57
site-specific DNA-methyl-transferase (cytosine-specific)	2.1.1.73	deleted, reaction is that of EC 2.1.1.37
nicotine N-methyltransferase	2.1.1.81	deleted, included in EC 2.1.1.49
myo-inositol 6-O-methyltransferase	2.1.1.134	transferred to EC 2.1.1.129
[methionine synthase]-cobalamin methyltransferase (cob(II)alamin reducing)	2.1.1.135	transferred to EC 1.16.1.8
methylarsonite methyltransferase	2.1.1.138	deleted, reaction due to EC 2.1.1.137

Index of Recommended Enzyme Names

Description of Data Fields

All information except the nomenclature of the enzymes (which is based on the recommendations of the Nomenclature Committee of IUBMB (**I**nternational **U**nion of **B**iochemistry and **M**olecular **B**iology) and IUPAC (**I**nternational **U**nion of **P**ure and **A**pplied **C**hemistry) is extracted from original literature (or reviews for very well characterized enzymes). The quality and reliability of the data depends on the method of determination, and for older literature on the techniques available at that time. This is especially true for the fields *Molecular Weight* and *Subunits*.
The general structure of the fields is: **Information – Organism – Commentary – Literature**
The information can be found in the form of numerical values (temperature, pH, K_m etc.) or as text (cofactors, inhibitors etc.).
Sometimes data are classified as *Additional Information*. Here you may find data that cannot be recalculated to the units required for a field or also general information being valid for all values. For example, for *Inhibitors*, *Additional Information* may contain a list of compounds that are not inhibitory.
The detailed structure and contents of each field is described below. If one of these fields is missing for a particular enzyme, this means that for this field, no data are available.

1 Nomenclature

EC number

The number is as given by the IUBMB, classes of enzymes and subclasses defined according to the reaction catalyzed.

Systematic name

This is the name as given by the IUBMB/IUPAC Nomenclature Committee

Recommended name

This is the name as given by the IUBMB/IUPAC Nomenclature Committee

Synonyms

Synonyms which are found in other databases or in the literature, abbreviations, names of commercially available products. If identical names are frequently used for different enzymes, these will be mentioned here, cross references are given. If another EC number has been included in this entry, it is mentioned here.

CAS registry number

The majority of enzymes have a single chemical abstract (CAS) number. Some have no number at all, some have two or more numbers. Sometimes two enzymes share a common number. When this occurs, it is mentioned in the commentary.

2 Source Organism

For listing organisms their systematic name is preferred. If these are not mentioned in the literature, the names from the respective literature are used. For example if an enzyme from yeast is described without being specified further, yeast will be the entry. This field defines the code numbers for the organisms in which the enzyme with the respective EC-number is found. These code numbers (form <_>) are displayed together with each entry in all fields of BRENDA where organism-specific information is given.

3 Reaction and Specificity

Catalyzed reaction

The reaction as defined by the IUBMB. The commentary gives information on the mechanism, the stereochemistry, or on thermodynamic data of the reaction.

Reaction type

According to the enzyme class a type can be attributed. These can be oxidation, reduction, elimination, addition, or a name (e.g. Knorr reaction)

Natural substrates and products

These are substrates and products which are metabolized in vivo. A natural substrate is only given if it is mentioned in the literature. The commentary gives information on the pathways for which this enzyme is important. If the enzyme is induced by a specific compound or growth conditions, this will be included in the commentary. In *Additional information* you will find comments on the metabolic role, sometimes only assumptions can be found in the references or the natural substrates are unknown.

In the listings, each natural substrate (indicated by a bold **S**) is followed by its respective product (indicated by a bold **P**). Products are given with organisms and references included only if the respective authors were able to demonstrate the formation of the specific product. If only the disappearance of the substrate was observed, the product is included without organisms of references. In cases with unclear product formation only a **?** as a dummy is given.

Substrates and products

All natural or synthetic substrates are listed (not in stoichiometric quantities). The commentary gives information on the reversibility of the reaction, on isomers accepted as substrates and it compares the efficiency of substrates. If a specific substrate is accepted by only one of several isozymes, this will be stated here.

The field *Additional Information* summarizes compounds that are not accepted as substrates or general comments which are valid for all substrates. In the listings, each substrate (indicated by a bold **S**) is followed by its respective product (indicated by a bold **P**). Products are given with organisms and references included if the respective authors demonstrated the formation of the specific product. If only the disappearance of the substrate was observed, the product will be included without organisms or references. In cases with unclear product formation only a **?** as a dummy is given.

Inhibitors

Compounds found to be inhibitory are listed. The commentary may explain experimental conditions, the concentration yielding a specific degree of inhibition or the inhibition constant. If a substance is activating at a specific concentration but inhibiting at a higher or lower value, the commentary will explain this.

Cofactors, prosthetic groups

This field contains cofactors which participate in the reaction but are not bound to the enzyme, and prosthetic groups being tightly bound. The commentary explains the function or, if known, the stereochemistry, or whether the cofactor can be replaced by a similar compound with higher or lower efficiency.

Activating Compounds

This field lists compounds with a positive effect on the activity. The enzyme may be inactive in the absence of certain compounds or may require activating molecules like sulfhydryl compounds, chelating agents, or lipids. If a substance is activating at a specific concentration but inhibiting at a higher or lower value, the commentary will explain this.

Metals, ions

This field lists all metals or ions that have activating effects. The commentary explains the role each of the cited metal has, being either bound e.g. as Fe-S centers or being required in solution. If an ion plays a dual role, activating at a certain concentration but inhibiting at a higher or lower concentration, this will be given in the commentary.

Turnover number (min^{-1})

The k_{cat} is given in the unit min^{-1}. The commentary lists the names of the substrates, sometimes with information on the reaction conditions or the type of reaction if the enzyme is capable of catalyzing different reactions with a single substrate. For cases where it is impossible to give the turnover number in the defined unit (e.g., substrates without a defined molecular weight, or an undefined amount of protein) this is summarized in *Additional Information*.

Specific activity (U/mg)

The unit is micromol/minute/milligram of protein. The commentary may contain information on specific assay conditions or if another than the natural substrate was used in the assay. Entries in *Additional Information* are included if the units of the activity are missing in the literature or are not calculable to the obligatory unit. Information on literature with a detailed description of the assay method may also be found.

K_m-Value (mM)

The unit is mM. Each value is connected to a substrate name. The commentary gives, if available, information on specific reaction condition, isozymes or presence of activators. The references for values which cannot be expressed in mM (e.g. for macromolecular, not precisely defined substrates) are given in *Additional Information.* In this field we also cite literature with detailed kinetic analyses.

K_i-Value (mM)

The unit of the inhibition constant is mM. Each value is connected to an inhibitor name. The commentary gives, if available, the type of inhibition (e.g. competitive, non-competitive) and the reaction conditions (pH-value and the temperature). Values which cannot be expressed in the requested unit and references for detailed inhibition studies are summerized under *Additional information.*

pH-Optimum

The value is given to one decimal place. The commentary may contain information on specific assay conditions, such as temperature, presence of activators or if this optimum is valid for only one of several isozymes. If the enzyme has a second optimum, this will be mentioned here.

pH-Range

Mostly given as a range e.g. 4.0–7.0 with an added commentary explaining the activity in this range. Sometimes, not a range but a single value indicating the upper or lower limit of enzyme activity is given. In this case, the commentary is obligatory.

Temperature optimum (°C)

Sometimes, if no temperature optimum is found in the literature, the temperature of the assay is given instead. This is always mentioned in the commentary.

Temperature range (°C)

This is the range over which the enzyme is active. The commentary may give the percentage of activity at the outer limits. Also commentaries on specific assay conditions, additives etc.

4 Enzyme Structure

Molecular weight

This field gives the molecular weight of the holoenzyme. For monomeric enzymes it is identical to the value given for subunits. As the accuracy depends on the method of determination this is given in the commentary if provided in the literature. Some enzymes are only active as multienzyme complexes for which the names and/or EC numbers of all participating enzymes are given in the commentary.

Subunits

The tertiary structure of the active species is described. The enzyme can be active as a monomer a dimer, trimer and so on. The stoichiometry of subunit composition is given. Some enzymes can be active in more than one state of complexation with differing effectivities. The analytical method is included.

Posttranslational modifications

The main entries in this field may be proteolytic modification, or side-chain modification, or no modification. The commentary will give details of the modifications e.g.:

- proteolytic modification <1> (<1>, propeptide Name) [1];
- side-chain modification <2> (<2>, N-glycosylated, 12% mannose) [2];
- no modification [3]

5 Isolation / Preparation / Mutation / Application

Source / tissue

For multicellular organisms, the tissue used for isolation of the enzyme or the tissue in which the enzyme is present is given. Cell-lines may also be a source of enzymes.

Localization

The subcellular localization is described. Typical entries are: cytoplasm, nucleus, extracellular, membrane.

Purification

The field consists of an organism and a reference. Only references with a detailed description of the purification procedure are cited.

Renaturation

Commentary on denaturant or renaturation procedure.

Crystallization

The literature is cited which describes the procedure of crystallization, or the X-ray structure.

Cloning
Lists of organisms and references, sometimes a commentary about expression or gene structure.

Engineering
The properties of modified proteins are described.

Application
Actual or possible applications in the fields of pharmacology, medicine, synthesis, analysis, agriculture, nutrition are described.

6 Stability

pH-Stability
This field can either give a range in which the enzyme is stable or a single value. In the latter case the commentary is obligatory and explains the conditions and stability at this value.

Temperature stability
This field can either give a range in which the enzyme is stable or a single value. In the latter case the commentary is obligatory and explains the conditions and stability at this value.

Oxidation stability
Stability in the presence of oxidizing agents, e.g. O_2, H_2O_2, especially important for enzymes which are only active under anaerobic conditions.

Organic solvent stability
The stability in the presence of organic solvents is described.

General stability information
This field summarizes general information on stability, e.g., increased stability of immobilized enzymes, stabilization by SH-reagents, detergents, glycerol or albumins etc.

Storage stability
Storage conditions and reported stability or loss of activity during storage.

References
Authors, Title, Journal, Volume, Pages, Year.

Nicotinamide N-methyltransferase 2.1.1.1

1 Nomenclature

EC number
2.1.1.1

Systematic name
S-adenosyl-L-methionine:nicotinamide N-methyltransferase

Recommended name
nicotinamide N-methyltransferase

Synonyms
NNMT
methyltransferase, nicotinamide

CAS registry number
9029-74-7

2 Source Organism

<1> *Rattus norvegicus* (bearing Walker 256 carcinosarcoma [7]) [1, 7]
<2> *Sus scrofa* [2]
<3> *Homo sapiens* (radioresistant bladder carcinoma cell line MGH-UI [10]) [3-5, 9, 10, 12]
<4> *Mus musculus* (bearing Ehrlich ascites tumor [7]; C57BL/6J [8]; bearing colon 26 adenocarcinoma [11]) [6-8, 11]

3 Reaction and Specificity

Catalyzed reaction
S-adenosyl-L-methionine + nicotinamide = S-adenosyl-L-homocysteine + 1-methylnicotinamide

Reaction type
methyl group transfer

Natural substrates and products
S L-methionine + nicotinamide <1> (Reversibility: ? <1> [1]) [1]
P L-homocysteine + 1-methylnicotinamide
S Additional information <2> (<2> enzyme may function in detoxificating of numerous alkaloids [2]) [2]
P ?

Substrates and products

- **S** S-adenosyl-L-methionine + 1,2,3,4-tetrahydroisoquinoline <2> (Reversibility: ? <2> [2]) [2]
- **P** S-adenosyl-L-homocysteine + ?
- **S** S-adenosyl-L-methionine + 3-acetylpyridine <2> (Reversibility: ? <2> [2]) [2]
- **P** S-adenosyl-L-homocysteine + ?
- **S** S-adenosyl-L-methionine + isoquinoline <2> (Reversibility: ? <2> [2]) [2]
- **P** S-adenosyl-L-homocysteine + ?
- **S** S-adenosyl-L-methionine + nicotinamide <1-3> (<1> L-methionine + nicotinamide [1]) (Reversibility: ? <1-3> [1-3]) [1-3]
- **P** S-adenosyl-L-homocysteine + 1-methylnicotinamide
- **S** S-adenosyl-L-methionine + quinoline <2> (Reversibility: ? <2> [2]) [2]
- **P** S-adenosyl-L-homocysteine + ?
- **S** S-adenosyl-L-methionine + thionicotinamide <2> (Reversibility: ? <2> [2]) [2]
- **P** S-adenosyl-L-homocysteine + 1-methylthionicotinamide

Inhibitors

N^1-methylnicotinamide <3> (<3> IC50 0.06 mM for human protein, 0.03 mM for recombinant protein [5]) [3, 5]

S-adenosyl-L-homocysteine <3> [3]

Additional information <3> (<3> not inhibitory: Ca^{2+}, SKF 525A, 3,4-dimethoxy-5-hydroxybenzoic acid [3]) [3]

Specific activity (U/mg)

Additional information <2-4> (<2> assay [2]; <3> activity in different individuals, correlation with phenotypic differences [4]; <4> activity in male animals of different inbred strains [8]) [2, 4, 8]

K_m-Value (mM)

0.00176 <3> (S-adenosyl-L-methionine) [3]
0.0018 <3> (S-adenosyl-L-methionine) [5]
0.0022 <3> (S-adenosyl-L-methionine, <3> recombinant protein [5]) [5]
0.0065 <4> (S-adenosyl-L-methionine) [8]
0.016 <2> (S-adenosyl-L-methionine) [2]
0.2 <2> (1,2,3,4-tetrahydroisoquinoline) [2]
0.347 <3> (nicotinamide) [3]
0.37 <4> (nicotinamide) [8]
0.38 <3> (nicotinamide, <3> recombinant protein [5]) [5]
0.43 <3> (nicotinamide) [5]
0.83 <2> (nicotinamide) [2]
1.3 <2> (quinoline) [2]
1.5 <2> (isoquinoline) [2]

pH-Optimum

7.2 <2> [2]
7.4 <4> [8]

pH-Range

5.7-9.8 <2> (<2> about 50% of maximum activity at pH 5.7 and 9.8 [2]) [2]

Temperature optimum (°C)

25 <2> (<2> assay at [2]) [2]

4 Enzyme Structure

Molecular weight

27000-30000 <2> (<2> gel filtration, SDS-PAGE [2]) [2]

Subunits

? <3> (<3> x * 29000, SDS-PAGE, recombinant protein [5]) [5]
monomer <2> (<2> 1 * 27000, SDS-PAGE [2]) [2]

5 Isolation/Preparation/Mutation/Application

Source/tissue

bladder <3> [10]
brain <3> (<3> lumbar cerebrospinal fluid [9]) [9, 12]
liver <1-4> [1-5, 7, 8, 11]

Localization

cytosol <3, 4> [4, 8]
soluble <1> [1]

Purification

<2> [2]

Cloning

<3> [5]
<4> [6]

Application

medicine <1, 3, 4> (<1,4> enzyme activity correlates with tumor growth and morphological changes resembling apoptosis [7]; <3> enzyme protein level is significantly elevated in younger patients with Parkinson's disease [9]; <3> possibly involved in response to radiation [10]; <4> levels of hepatic enzyme activity are closely associated with the degree of weight loss, use as marker of cancer cachexia [11]; <3> significantly increased levels of enzyme protein and activity in brain of patients with Parkinson's disease, study on brain tissue distribution [12]) [7, 9, 10, 11, 12]

6 Stability

General stability information

<2>, glycerol stabilizes [2]
<2>, thiols required for stability during isolation [2]

Storage stability

<2>, -20°C, 20% glycerol w/v, stable for 1 month, partially purified enzyme [2]
<2>, 4°C, 20% w/v glycerol, pH 7.6, 30% loss of activity, purified enzyme [2]
<2>, 4°C, 20% w/v glycerol, stable for 1 week, partially purified enzyme [2]

References

[1] Cantoni, G.L.: Methylation of nicotinamide with a soluble enzyme system from rat liver. J. Biol. Chem., **189**, 203-217 (1951)
[2] Alston, T.A.; Abeles, R.H.: Substrate specificity of nicotinamide methyltransferase isolated from porcine liver. Arch. Biochem. Biophys., **260**, 601-608 (1988)
[3] Rini, J.; Szumlanski, C.; Guerciolini, R.; Weinshilboum, R.M.: Human liver nicotinamide N-methyltransferase: ion-pairing radiochemical assay, biochemical properties and individual variation. Clin. Chim. Acta, **186**, 359-374 (1990)
[4] Smith, M.L.; Burnett, D.; Bennett, P.; Waring, R.H.; Brown, H.M.; Williams, A.C.; Ramsden, D.B.: A direct correlation between nicotinamide N-methyltransferase activity and protein levels in human liver cytosol. Biochim. Biophys. Acta, **1442**, 238-244 (1998)
[5] Aksoy, S.; Szumlanski, C.L.; Weinshilboum, R.M.: Human liver nicotinamide N-methyltransferase. cDNA cloning, expression, and biochemical characterization. J. Biol. Chem., **269**, 14835-14840 (1994)
[6] Yan, L.; Otterness, D.M.; Kozak, C.A.; Weinshilboum, R.M.: Mouse nicotinamide N-methyltransferase gene: molecular cloning, structural characterization, and chromosomal localization. DNA Cell Biol., **17**, 659-667 (1998)
[7] Fujimura, S.; Okamura, A.; Ohmura, Y.; Moriyama, Y.; Ohwada, H.; Horitsu, K.; Ohkubo, M.: Nicotinamide methyltransferase activity and cell-death in the liver of the mouse bearing Ehrlich ascites tumor. Adv. Exp. Med. Biol., **398**, 507-511 (1996)
[8] Scheller, T.; Orgacka, H.; Szumlanski, C.L.; Weinshilboum, R.M.: Mouse liver nicotinamide N-methyltransferase pharmacogenetics: biochemical properties and variation in activity among inbred strains. Pharmacogenetics, **6**, 43-53 (1996)
[9] Aoyama, K.; Matsubara, K.; Kondo, M.; Murakawa, Y.; Suno, M.; Yamashita, K.; Yamaguchi, S.; Kobayashi, S.: Nicotinamide-N-methyltransferase is higher in the lumbar cerebrospinal fluid of patients with Parkinson's disease. Neurosci. Lett., **298**, 78-80 (2001)

[10] Kassem, H.S.; Sangar, V.; Cowan, R.; Clarke, N.; Margison, G.P.: A potential role of heat shock proteins and nicotinamide N-methyl transferase in predicting response to radiation in bladder cancer. Int. J. Cancer, **101**, 454-460 (2002)
[11] Okamura, A.; Ohmura, Y.; Islam, M.M.; Tagawa, M.; Horitsu, K.; Moriyama, Y.; Fujimura, S.: Increased hepatic nicotinamide N-methyltransferase activity as a marker of cancer cachexia in mice bearing colon 26 adenocarcinoma. Jpn. J. Cancer Res., **89**, 649-656 (1998)
[12] Parsons, R.B.; Smith, M.L.; Williams, A.C.; Waring, R.H.; Ramsden, D.B.: Expression of nicotinamide N-methyltransferase (E.C. 2.1.1.1) in the Parkinsonian brain. J. Neuropathol. Exp. Neurol., **61**, 111-124 (2002)

Guanidinoacetate N-methyltransferase 2.1.1.2

1 Nomenclature

EC number
2.1.1.2

Systematic name
S-adenosyl-L-methionine:N-guanidinoacetate methyltransferase

Recommended name
guanidinoacetate N-methyltransferase

Synonyms
GA methylpherase
guanidinoacetate methyltransferase
guanidinoacetate transmethylase
guanidoacetate methyltransferase
methionine-guanidinoacetic transmethylase

CAS registry number
9029-75-8

2 Source Organism

<1> *Cavia porcellus* [1]
<2> *Oryctolagus cuniculus* [1]
<3> *Bos taurus* [1]
<4> *Sus scrofa* [1-3]
<5> *Rattus norvegicus* (expressed in E. coli [5,6,7]) [4-8, 11-13]
<6> *Mus musculus* [9]
<7> *Rattus norvegicus* [10]

3 Reaction and Specificity

Catalyzed reaction
S-adenosyl-L-methionine + guanidinoacetate = S-adenosyl-L-homocysteine + creatine

Reaction type
methyl group transfer

Natural substrates and products

S S-adenosyl-L-methionine + guanidinoacetate <1-7> (<1-4> biosynthesis of creatine [1]) (Reversibility: ? <1-7> [1-13]) [1-13]

P S-adenosyl-L-homocysteine + creatine <1-7> [1-13]

Substrates and products

S S-adenosyl-L-methionine + guanidinoacetate <1-7> (<1-4> S-adenosyl-L-methionine is the only possible methyl donor [1]) (Reversibility: ? <1-7> [1-13]) [1-13]

P S-adenosyl-L-homocysteine + creatine <1-7> [1-13]

Inhibitors

3-deaza-S-adenosyl-L-homocysteine <4> (<4> competitive with S-adenosylmethionine [2]) [2, 3]
DTNB <5> (<5> 2 mol DTNB per mol enzyme causes complete inactivation [6]) [6]
N-ethylmaleimide <5> (<5> irreversible [4]) [4]
S-adenosyl-L-homocysteine <4, 5> (<4,5> competitive with S-adenosylmethionine [2,8]) [2, 3, 8]
glutathione <5> (<5> oxidized form, S-adenosylmethionine protects against inactivation [4]; <5> readily reactivated by reduced glutathione, substrates do not protect against inactivation, guanidinoacetate together with sinefungin protects against inactivation, can be reactivated with thiol compounds [7]) [4, 7]
iodoacetate <5> [6]
p-chloromercuribenzoate <1-4> (<1-4> complete inhibition at 1 mM, reversible by addition of glutathione [1]) [1]
sinefungin <4, 5> (<4,5> competitive with S-adenosylmethionine [2,7,8]) [2, 7, 8]
Additional information <5> (<5> complete and irreversible inactivation by ultraviolet irradiation [8]) [8]

Cofactors/prosthetic groups

ATP <1-4> [1]

Activating compounds

2-mercaptoethanol <5> [4]
CN^- <1-4> [1]
cysteine <1-5> [1, 4]
dithiothreitol <5> [4]
glutathione <1-5> (<1-5> required for maximum activity, can be replaced by other reducing agents [1,4]) [1, 4]

Metals, ions

Mg^{2+} <1-4> [1]

Turnover number (min^{-1})

4.73 <5> (guanidinoacetate, <5> wild type enzyme [11]) [11]
4.86 <7> (guanidinoacetate, <7> wild type enzyme [10]) [10]

Additional information <5, 7> (<5,7> all mutant enzymes listed with turnover [10,11]) [10, 11]

Specific activity (U/mg)

0.022 <4> (<4> purified enzyme [1]) [1]
0.12 <5> (<5> native enzyme [5]) [5]
0.13 <5> (<5> recombinant enzyme [5]) [5]
0.16 <5> (<5> purified enzyme [4]) [4]
27.2 <4> (<4> purified enzyme [2]) [2]

K_m-Value (mM)

0.003 <5> (S-adenosyl-L-methionine, <5> native enzyme [5]) [5]
0.003 <5> (S-adenosyl-L-methionine, <5> recombinant enzyme [5]) [5]
0.005 <5> (S-adenosyl-L-methionine, <5> wild type enzyme [11]) [11]
0.007 <5, 7> (S-adenosyl-L-methionine, <7> wild type enzyme [10]) [4, 10]
0.011 <5, 7> (guanidinoacetate, <5,7> wild type enzyme [10,11]) [10, 11]
0.027 <4> (guanidinoacetate) [2]
0.031 <5> (guanidinoacetate, <5> native enzyme [5]) [5]
0.032 <5> (guanidinoacetate, <5> recombinant enzyme [5]) [5]
0.049 <4> (S-adenosyl-L-methionine) [2]
0.098 <5> (guanidinoacetate) [4]
Additional information <5, 7> (<5,7> all mutant enzymes listed with K_m [10,11]) [10, 11]

K_i-Value (mM)

0.0004 <5> (S-adenosyl-L-homocysteine) [8]
0.006 <5> (sinefungin) [8]
0.016 <4> (adenosylhomocysteine) [2]
0.018 <4> (sinefungin) [2]
0.039 <4> (3-deaza-S-adenosyl-L-homocysteine) [2]

pH-Optimum

7.5 <1-4> [1]
9 <5> [4]

4 Enzyme Structure

Molecular weight

25000 <5, 6> (<5> gel filtration [4]; <6> SDS-PAGE [9]) [4, 9]
26000 <5> (<5> SDS-PAGE, in presence and absence of SDS [4]; <5> SDS-PAGE, gel filtration [5]) [4, 5]
31000 <4> (<4> SDS-PAGE, native PAGE without SDS yields heterogeneity in molecular weight [2]) [2]
31500 <4> (<4> gel filtration [2]) [2]

Subunits

monomer <5> (<5> SDS-PAGE, gel filtration [4,5]) [4, 5]

5 Isolation/Preparation/Mutation/Application

Source/tissue

brain <5> (<5> all parts of brain, neurons and glia, very low levels in astrocytes [12]) [12]
caput epididymis <6> (<6> highest enzyme level [9]) [9]
heart <6> (<6> low enzyme level [9]) [9]
kidney <6> (<6> low enzyme level [9]) [9]
liver <1-6> (<6> higher levels in female than in male [9]) [1-7, 9]
muscle <2, 6> (<6> low enzyme level [9]) [1, 9]
ovary <6> [9]
oviduct <6> (<6> low enzyme level [9]) [9]
testis <6> (<6> mainly found in Sertoli cells of seminiferous tubules [9]) [9]
uterus <6> (<6> low enzyme level [9]) [9]
Additional information <6> (<6> absent in brain, small intestine, seminal vesicle, lung, vas deferens, cauda epididymis, coagulating gland and spleen [9]) [9]

Localization

nucleus <5> (<5> perinuclear localization [12]) [12]

Purification

<4> [1, 2]
<5> (inactivated during purification, can be reactivated by addition of a thiol [4]) [4, 11]
<7> [10]

Crystallization

<5> [13]

Cloning

<5> (in Escherichia coli [5,6]) [5-7]

Engineering

D129A <7> (<7> active enzyme [10]) [10]
D129N <7> (<7> active enzyme [10]) [10]
D134A <7> (<7> inactive [10]) [10]
D134E <7> (<7> lower activity than wild type enzyme [10]) [10]
D134N <7> (<7> much lower activity than wild type enzyme [10]) [10]
D92N <5> (<5> lower activity than wild type enzyme [11]) [11]
E89Q <5> (<5> lower activity than wild type enzyme [11]) [11]
G67A <5> (<5> inactive enzyme [11]) [11]
G69A <5> (<5> inactive enzyme [11]) [11]
T135A <7> (<7> lower activity than wild type enzyme [10]) [10]
W143F <5> (<5> lower activity than wild type enzyme [11]) [11]
W143L <5> (<5> lower activity than wild type enzyme [11]) [11]
Y133F <7> (<7> lower activity than wild type enzyme [10]) [10]
Y133V <5> (<5> lower activity than wild type enzyme [11]) [11]
Y136F <5> (<5> higher turnover than wild type enzyme [11]) [11]

Y136F <7> (<7> higher turnover than wild type enzyme [10]) [10]
Y136V <5> (<5> lower activity than wild type enzyme [11]) [11]

6 Stability

Temperature stability

4 <5> (<5> remains active for 24 h [4]) [4]

General stability information

<5>, inactivated during purification, can be reactivated by addition of a thiol [4]

Storage stability

<4>, -60°C, purified enzyme, stable for months [2]
<5>, -20°C, 10 mM phosphate buffer, pH 7.2, 1 mM EDTA, 0.1 M NaCl, concentrated by ultrafiltration [4]

References

[1] Cantoni, G.L.; Vignos, P.J.: Enzymatic mechanism of creatine synthesis. J. Biol. Chem., **209**, 647-659 (1954)

[2] Im, Y.S.; Chiang, P.K.; Cantoni, G.L.: Guanidoacetate methyltransferase. Purification and molecular properties. J. Biol. Chem., **254**, 11047-11050 (1979)

[3] Im, Y.S.; Cantoni, G.L.; Chiang, P.K.: A radioactive assay for guanidoacetate methyltransferase. Anal. Biochem., **95**, 87-88 (1979)

[4] Ogawa, H.; Ishiguro, Y.; Fujioka, M.: Guanidoacetate methyltransferase from rat liver: purification, properties, and evidence for the involvement of sulfhydryl groups for activity. Arch. Biochem. Biophys., **226**, 265-275 (1983)

[5] Ogawa, H.; Date, T.; Gomi, T.; Konishi, K.; Pitot, H. C.; Cantoni, G.L.; Fujioka, M.: Molecular cloning, sequence analysis, and expression in Escherichia coli of the cDNA for guanidinoacetate methyltransferase from rat liver. Proc. Natl. Acad. Sci. USA, **85**, 694-698 (1988)

[6] Fujioka, M.; Konishi, K.; Takata, Y.: Recombinant rat liver guanidinoacetate methyltransferase: reactivity and function of sulfhydryl groups. Biochemistry, **27**, 7658-7664 (1988)

[7] Konishi, K.; Fujioka, M.: Reversible inactivation of recombinant rat liver guanidinoacetate methyltransferase by glutathione disulfide. Arch. Biochem. Biophys., **289**, 90-96 (1991)

[8] Takata, Y.; Fujioka, M.: Identification of a tyrosine residue in rat guanidinoacetate methyltransferase that is photolabeled with S-adenosyl-L-methionine. Biochemistry, **31**, 4369-4374 (1992)

[9] Lee, H.; Ogawa, H.; Fujioka, M.; Gerton, G.L.: Guanidinoacetate methyltransferase in the mouse: extensive expression in Sertoli cells of testis and in microvilli of caput epididymis. Biol. Reprod., **50**, 152-162 (1994)

[10] Takata, Y.; Konishi, K.; Gomi, T.; Fujioka, M.: Rat guanidinoacetate methyltransferase. Effect of site-directed alteration of an aspartic acid residue that is conserved across most mammalian S-adenosylmethionine-dependent methyltransferases. J. Biol. Chem., **269**, 5537-5542 (1994)
[11] Hamahata, A.; Takata, Y.; Gomi, T.; Fujioka, M.: Probing the S-adenosylmethionine-binding site of rat guanidinoacetate methyltransferase: effect of site-directed mutagenesis of residues that are conserved across mammalian non-nucleic acid methyltransferases. Biochem. J., **317**, 141-145 (1996)
[12] Braissant, O.; Henry, H.; Loup, M.; Eilers, B.; Bachmann, C.: Endogenous synthesis and transport of creatine in the rat brain: an in situ hybridization study. Mol. Brain Res., **86**, 193-201 (2001)
[13] Komoto, J.; Huang, Y.; Takata, Y.; Yamada, T.; Konishi, K.; Ogawa, H.; Gomi, T.; Fujioka, M.; Takusagawa, F.: Crystal structure of guanidinoacetate methyltransferase from rat liver: A model structure of protein arginine methyltransferase. J. Mol. Biol., **320**, 223-235 (2002)

Thetin-homocysteine S-methyltransferase 2.1.1.3

1 Nomenclature

EC number
2.1.1.3

Systematic name
dimethylsulfonioacetate:L-homocysteine S-methyltransferase

Recommended name
thetin-homocysteine S-methyltransferase

Synonyms
dimethylthetin-homocysteine methyltransferase
methyltransferase, dimethylthetin-homocysteine
thetin-homocysteine methylpherase
Additional information <1> (<1> cf. EC 2.1.1.5, probably one enzyme [2]) [2]

CAS registry number
9029-76-9

2 Source Organism

<1> *Rattus norvegicus* [1, 2]

3 Reaction and Specificity

Catalyzed reaction
dimethylsulfonioacetate + L-homocysteine = S-methylthioglycolate + L-methionine (<1> mechanism [1])

Reaction type
methyl group transfer

Natural substrates and products
S dimethylthetin + L-homocysteine <1> (Reversibility: ? <1> [1]) [1]
P S-methylthioglycolate + L-methionine

Substrates and products
S dimethylthetin + L-homocysteine <1> (<1> i.e. dimethylsulfonioacetate, specific for L-homocysteine [1,2]) (Reversibility: ? <1> [1]) [1, 2]

P S-methylthioglycolate + L-methionine <1> (<1> i.e. methylthioacetate [1, 2]) [1, 2]

S Additional information <1> (<1> methyl group donors are ethylmethylthetin, dimethyl-α- and dimethyl-β-propiothetin, ethylmethyl-β-propiothetin, dimethyl-γ-butyrothetin, methionine methylsulfonium, trimethylsulfonium, ethyldimethylsulfonium or butyldimethylsulfonium, no acceptors are cysteamine, 2-mercaptoethanol, DL-homocystine, sodium mercaptoacetate [1]) [1]

P ?

Inhibitors

Ag^{2+} <1> [1]
Co^{2+} <1> [1]
Cu^{2+} <1> (<1> 0.01 M, strong [1]) [1]
Fe^{2+} <1> [1]
Hg^{2+} <1> (<1> 0.01 M, strong [1]) [1]
K^+ <1> [1]
Mn^{2+} <1> [1]
Rb^+ <1> [1]
Zn^{2+} <1> (<1> 0.01 M, strong [1]) [1]
glycine betaine <1> (<1> 0.1 M [1]) [1]
hydroxylamine <1> (<1> 0.01 M [1]) [1]
phenylhydrazine <1> (<1> 0.01 M [1]) [1]
sodium mercaptoacetate <1> [1]
Additional information <1> (<1> no inhibition by dimethylthetin, sulfocholine (up to 0.1 M), mercaptoethanol, cysteine, Na_2SO_4, Li^+, Cs^+, Ca^{2+}, Mg^{2+}, NaN_3, NaCN, NaF, EDTA, semicarbazide, iodoacetate [1]) [1]

Metals, ions

Additional information <1> (<1> no activation by metal ions [1]) [1]

K_m-Value (mM)

5.3 <1> (DL-homocysteine) [2]
10.2 <1> (thetin) [2]

pH-Optimum

7.4 <1> (<1> assay at [1,2]) [1, 2]

Temperature optimum (°C)

37 <1> (<1> assay at [1]) [1]

5 Isolation/Preparation/Mutation/Application

Source/tissue

liver <1> [1, 2]

Purification

<1> (partial [2]) [2]

6 Stability

Storage stability

<1>, 4°C, several weeks [2]

References

[1] Maw, G.A.: Thetin-homocysteine transmethylase: some further characteristics of the enzyme from rat liver. Biochem. J., **70**, 168-173 (1958)

[2] Maw, G.A.: Thetin-homocysteine transmethylase: a preliminary manometric study of the enzyme from rat liver. Biochem. J., **63**, 116-124 (1956)

Acetylserotonin O-methyltransferase 2.1.1.4

1 Nomenclature

EC number
2.1.1.4

Systematic name
S-adenosyl-L-methionine:N-acetylserotonin O-methyltransferase

Recommended name
acetylserotonin O-methyltransferase

Synonyms
ASMT
HIOMT
N-acetylserotonin O-methyltransferase
acetylserotonin methyltransferase
hydroxyindole O-methyltransferase
hydroxyindole methyltransferase
hydroxyindole-O-methyl transferase
methyltransferase, acetylserotonin

CAS registry number
9029-77-0

2 Source Organism

<1> *Bos taurus* [1, 3, 5, 7, 9, 11, 13]
<2> *Macaca mulatta* (monkey [1]) [1]
<3> *Felis catus* [1]
<4> *Rattus norvegicus* [2, 4, 5, 8, 12, 16, 18, 20]
<5> *Gallus gallus* [2, 5, 9]
<6> *frog* [2]
<7> *Diceros bicornis* (black rhinoceros [6]) [6]
<8> *Oncorhynchus tshawytscha* (Chinook salmon [10]) [10]
<9> *Ovis aries* [5]
<10> *Ovibos moschatus* (muskox [14]) [14]
<11> *Oncorhynchus mykiss* (rainbow trout [15]) [15]
<12> *Homo sapiens* (tissues and Y79 retinoblastoma cells [17]; Y79 retinoblastoma cells [19]) [17, 19]

3 Reaction and Specificity

Catalyzed reaction

S-adenosyl-L-methionine + N-acetylserotonin = S-adenosyl-L-homocysteine + melatonin (<1, 4, 5, 7, 9> reaction sequence is ordered bi-bi type [5, 6]; <1, 4, 5, 9, 10> addition of S-adenosyl-L-methionine occuring first followed by N-acetylserotonin [5,14])

Reaction type

O-methylation
methyl group transfer

Natural substrates and products

S N-acetylserotonin + S-adenosyl-L-methionine <1-5, 9> (<1-5,9> formation of the hormone melatonin [1,5]) (Reversibility: ? <1-5, 9> [1, 5]) [1, 5]

P S-adenosyl-L-homocysteine + N-acetyl-5-methoxytryptamine

Substrates and products

S 4-hydroxy-N-acetyltryptamine + S-adenosyl-L-methionine <1> (<1> 8% of the activity with N-acetylserotonin [1]) (Reversibility: ? <1> [1]) [1]

P S-adenosyl-L-homocysteine + N-acetyl-4-methoxytryptamine

S 5-hydroxyindoleacetic acid + S-adenosyl-L-methionine <1> (<1> 12% of the activity with N-acetylserotonin [1]) (Reversibility: ? <1> [1]) [1]

P S-adenosyl-L-homocysteine + 5-methoxyindoleacetic acid

S N-methylserotonin + S-adenosyl-L-methionine <1> (<1> 9% of the activity with N-acetylserotonin [1]) (Reversibility: ? <1> [1]) [1]

P S-adenosyl-L-homocysteine + N-methyl-5-methoxytryptamine

S S-adenosyl-L-methionine + N-acetylserotonin <1-10, 12> (Reversibility: ? <1-10, 12> [1-14, 17, 18]) [1-14, 17, 18]

P S-adenosyl-L-homocysteine + N-acetyl-5-methoxytryptamine

S bufotenine + S-adenosyl-L-methionine <1, 5> (<1> 14% of the activity with N-acetylserotonin [1]) (Reversibility: ? <1, 5> [1, 9]) [1, 9]

P S-adenosyl-L-homocysteine + N,N-dimethyl-5-hydroxytryptamine

S serotonin + S-adenosyl-L-methionine <1, 5> (<1> 7% of the activity with N-acetylserotonin [1]) (Reversibility: ? <1, 5> [1, 9]) [1, 9]

P S-adenosyl-L-homocysteine + 5-methoxytryptamine

S Additional information <1-5> (<1-3> absolute requirement for S-adenosylmethionine [1]; <1,5> bovine enzyme has high specificity towards N-acetylserotonin, chicken enzyme methylates N-acetylserotonin and to some extent serotonin and bufotenine [9]; <4> activity is controlled by the endogenous clock [16]) [1, 9, 16]

P ?

Inhibitors

Ca^{2+} <1> [7]
Cu^{2+} <1> [7]

L-glutamate <4> (<4> inhibition is reversible, dose-dependent, no inhibition of mRNA expression [20]) [20]
Mg^{2+} <1> [7]
N-acetylserotonin <7, 10, 12> (<1> activation [13]) [6, 14, 17]
S-adenosyl-L-homocysteine <1, 7, 10> [3, 6, 13, 14]
S-adenosyl-L-methionine <7> (<1> activation [13]) [6]
arginine vasotocin <4> [4]
cysteamine <4> (<4> rat enzyme inhibited to 50% at 0.128 mM, bovine and ovine enzyme not [4]) [4]
cysteic acid <1> [13]
glutathione <1> [13]
homocysteic acid <1> [13]
oxytocin <4> [4]
p-chloromercuribenzoate <1> [1]
pyridoxal 5'-phosphate <4> [12]
selenocysteamine <4> (<4> 50% inhibition at 0.013 mM [4]) [4]
vasopressin <4> [4]
Additional information <11> (<11> not inhibitory: dexamethasone [15]; <4> not inhibitory: (2S,2'R,3'R)-2-(2',3'-dicarboxycyclopropyl)glycine [20]) [15, 20]

Activating compounds

N-acetylserotonin <1> (activation, <1> [13]; inhibition, <7> [6]) [13]
S-adenosyl-L-methionine <1> (<1> activation [13]; <7> inhibition [6]) [13]
bicarbonate <1> (<1> activation [13]) [13]
oxaloacetate <1> (<1> activation [13]) [13]
Additional information <1, 12> (<1> citric acid cycle intermediates are good activators [13]; <12> retinoic acid stereoisomers induce an increase in activity and mRNA level [19]) [13, 19]

Specific activity (U/mg)

Additional information <1-3> (<1-3> enzyme assay [1]) [1]

K_m-Value (mM)

0.005 <8> (N-acetylserotonin) [10]
0.005 <4> (S-adenosyl-L-methionine) [12]
0.00856 <1> (S-adenosyl-L-methionine) [11]
0.015 <10> (N-acetylserotonin) [14]
0.0291 <1> (N-acetylserotonin) [11]
0.04 <4> (N-acetylserotonin) [18]
0.04 <4> (S-adenosyl-L-methionine) [18]
0.046 <1> (S-adenosyl-L-methionine) [1]
0.054 <1> (N-acetylserotonin) [1]
0.1 <4> (N-acetylserotonin) [12]
0.11 <8> (S-adenosyl-L-methionine) [10]

K_i-Value (mM)

0.0059 <4> (pyridoxal 5'-phosphate, <4> S-adenosylmethionine [12]) [12]
0.042 <10> (N-acetylserotonin) [14]

pH-Optimum

7.5-8.3 <1> [1]
7.6 <8> [10]
7.9-8.4 <1> [11]
8.2 <7, 10> [6, 14]

pH-Range

7.8-8.4 <7> [6]

Temperature optimum (°C)

37 <4-6> (<4-6> assay at [2]) [2]

4 Enzyme Structure

Molecular weight

76000-78000 <1> (<1> [5]) [5]

Subunits

? <1, 4, 12> (<4> x * 38000, SDS-PAGE [8]; <1> x * 38000, SDS-PAGE [11]; <12> x * 42000, SDS-PAGE [17]) [8, 11, 17]
dimer <1, 5> (<1,5> 2 * 39000, SDS-PAGE [9]) [9]
polymer <1> (<1> x * 21800 [13]) [13]

5 Isolation/Preparation/Mutation/Application

Source/tissue

brain <6> (<6> low activity [2]; <1-3> not [1]) [2]
diencephalon <8> (<8> dorsal portion [10]) [10]
ovary <4> [18]
pineal gland <1-12> (<4-6> not surrounding areas [2]) [1-6, 8-10, 12-15, 17, 20]
retina <4-6, 8, 9, 12> (<12> very low levels [17]) [2, 5, 10, 17]
Additional information <4-6> (<4-6> not: liver, kidney, heart, lung, adrenal gland, pancreas, salivary gland, skin [2]) [2]

Localization

cytosol <1, 4, 5, 9> [5]

Purification

<1> [1, 9, 11, 13]
<4> [5, 8]
<5> [9]

6 Stability

Storage stability

<1>, frozen, stable for several weeks [1]
<4>, 4°C, 96 h, inactivated [5]
<1, 9>, 4°C, 96 h, pineal enzyme stable [5]

References

[1] Axelrod, J.; Weissbach, H.: Purification and properties of hydroxyindole-O-methyl transferase. J. Biol. Chem., **236**, 211-213 (1961)

[2] Suzuki, O.; Yagi, K.: A nonisotopic assay for acetylserotonin methyltransferase. Anal. Biochem., **88**, 580-586 (1978)

[3] Kuwano, R.; Takahashi, Y.: Binding of S-adenosylhomocysteine to hydroxyindole O-methyltransferase. Biochim. Biophys. Acta, **787**, 1-7 (1984)

[4] Sugden, D.; Klein, D.C.: Inactivation of rat pineal hydroxyindole-O-methyltransferase by disulfide-containing compounds. J. Biol. Chem., **262**, 6489-6493 (1987)

[5] Sugden, D.; Cena, V.; Klein, D.C.: Hydroxyindole O-methyltransferase. Methods Enzymol., **142**, 590-596 (1987)

[6] Morton, D.J.; Kock, N.: Some properties of pineal gland hydroxyindole-O-methyltransferase from black rhinoceros (Diceros bicornis). J. Pineal Res., **8**, 35-40 (1990)

[7] Morton, D.J.: Mechanism of inhibition of bovine pineal gland hydroxyindole-O-methyltransferase (EC 2.1.1.4) by divalent cations. J. Pineal Res., **4**, 295-303 (1987)

[8] Sugden, D.; Voisin, P.; Klein, D.C.: Purification of rat pineal hydroxyindole-O-methyltransferase using S-adenosyl-L-homocysteine agarose chromatography. J. Pineal Res., **3**, 389-395 (1986)

[9] Nakane, M.; Yokoyama, E.; Deguchi, T.: Species heterogeneity of pineal hydroxyindole-O-methyltransferase. J. Neurochem., **40**, 790-796 (1983)

[10] Birks, E.K.; Ewing, R.D.: Characterization of hydroxyindole-O-methyltransferase (HIOMT) from the pineal gland of chinook salmon (Oncorhynchus tshawytscha). Gen. Comp. Endocrinol., **43**, 269-276 (1981)

[11] Kuwano, R.; Yoshida, Y.; Takahshim, Y.: Purification of bovine pineal hydroxyindole O-methyltransferase by immunoadsorption chromatography. J. Neurochem., **31**, 815-824 (1978)

[12] Nir, I.; Hirschmann, N.; Sulman, F.G.: Inhibition of pineal hydroxyindole-O-methyl transferase by pyridoxal-5-phosphate. Biochem. Pharmacol., **25**, 581-583 (1976)

[13] Karahasanoglu, A.M.; Ozand, P.T.: The purification and properties of bovine pineal S-adenosylmethionine: N-acetylserotonin O-methyl transferase. J. Neurochem., **19**, 411-430 (1972)

[14] Tedesco, S.C.; Moron, D.J.; Reiter, R.J.: Hydroxyindole-O-methyltransferase activity in the pineal gland of the muskox (Ovibos moschatus). J. Pineal Res., **16**, 121-126 (1994)
[15] Benyassi, A.; Schwartz, C.; Ducouret, B.; Falcon, J.: Glucocorticoid receptors and serotonin N-acetyltransferase activity in the fish pineal organ. NeuroReport, **12**, 889-892 (2001)
[16] Ribelayga, C.; Gauer, F.; Calgari, C.; Pevet, P.; Simonneaux, V.: Photoneural regulation of rat pineal hydroxyindole-O-methyltransferase (HIOMT) messenger ribonucleic acid expression: an analysis of its complex relationship with HIOMT activity. Endocrinology, **140**, 1375-1384 (1999)
[17] Bernard, M.; Donohue, S.J.; Klein, D.C.: Human hydroxyindole-O-methyltransferase in pineal gland, retina and Y79 retinoblastoma cells. Brain Res., **696**, 37-48 (1995)
[18] Itoh, M.T.; Ishizuka, B.; Kudo, Y.; Fusama, S.; Amemiya, A.; Sumi, Y.: Detection of melatonin and serotonin N-acetyltransferase and hydroxyindole-O-methyltransferase activities in rat ovary. Mol. Cell. Endocrinol., **136**, 7-13 (1997)
[19] Bernard, M.; Klein, D.C.: Retinoic acid increases hydroxyindole-O-methyltransferase activity and mRNA in human Y-79 retinoblastoma cells. J. Neurochem., **67**, 1032-1038 (1996)
[20] Ishio, S.; Yamada, H.; Craft, C.M.; Moriyama, Y.: Hydroxyindole-O-methyltransferase is another target for L-glutamate-evoked inhibition of melatonin synthesis in rat pinealocytes. Brain Res., **850**, 73-78 (1999)

Betaine-homocysteine S-methyltransferase 2.1.1.5

1 Nomenclature

EC number
2.1.1.5

Systematic name
trimethylammonioacetate:L-homocysteine S-methyltransferase

Recommended name
betaine-homocysteine S-methyltransferase

Synonyms
BHMT
betaine-homocysteine methyltransferase
betaine-homocysteine transmethylase
methyltransferase, betaine-homocysteine
Additional information (<1> probably identical with EC 2.1.1.3 [2])

CAS registry number
9029-78-1

2 Source Organism

<1> *Homo sapiens* [1, 2, 4, 7, 10, 11, 13, 15, 18-20]
<2> *Rattus norvegicus* (strain Sprague-Dawley [5]) [3, 5, 6, 8, 9, 19]
<3> *Equus caballus* [2]
<4> *Ovis aries* (lamb [1]) [1]
<5> *Gallus gallus* [1, 12]
<6> *Sus scrofa* [13, 19]
<7> *Macaca mulatta* (rhesus monkey [14]) [14]
<8> *Aphanothece halophytica* [16, 17]

3 Reaction and Specificity

Catalyzed reaction
trimethylammonioacetate + L-homocysteine = dimethylglycine + L-methionine (<2> ordered bi-bi mechanism with homocysteine the first substrate to be added and methionine the last product released [3, 8]; <1> mechanism [7])

Reaction type
methyl group transfer

Natural substrates and products

S N,N,N-trimethylglycine + L-homocysteine <2> (<2> i.e. betaine, involved in betaine metabolism [3]; <7> enzyme cristallin in lens [14]) (Reversibility: ? <2> [3]) [3, 14]

P N,N-dimethylglycine + L-methionine

Substrates and products

S N,N,N-trimethylglycine + L-homocysteine <1-3> (<1> i.e. betaine, methyl transfer directly from one substrate to the other [7]) (Reversibility: ? <1-3> [2-7]) [2-7]

P N,N-dimethylglycine + L-methionine <1-3> [2-7]

S dimethylacetothetin + L-homocysteine <6> (Reversibility: ? <6> [13]) [13]

P ? + L-methionine

S glycinebetaine + L-homocysteine <8> (Reversibility: ? <8> [16]) [16]

P ? + L-methionine

Inhibitors

3,3-dimethylbutyrate <1> (<1> competitive to betaine [2,4]) [2, 4]
D-cystine <2> [3]
L-cysteine <2> [3]
L-cystine <2> [3]
N,N-dimethylglycine <8> (<8> 70% inhibition at 5 mM [17]) [17]
NaCl <8> (<8> 100% inhibition above 200 mM [17]) [17]
S-(δ-carboxybutyl)-DL-homocysteine <1> (<1> strong [7]) [2, 7]
S-adenosyl-L-ethionine <2> (<2> irreversible, S-adenosyl-L-homocysteine and L-homocysteine prevent, but not DL-homocysteine, GSH, DTT or L-cysteine [6]) [6]
S-adenosyl-L-homocysteine <2> (<2> non-linear/competitive to homocysteine, mixed/non-competitive to betaine [8]) [8]
S-adenosyl-L-methionine <2> (<2> irreversible, S-adenosyl-L-homocysteine and L-homocysteine prevent, not DL-homocysteine, GSH, DTT or L-cysteine [6]; <1> no effect on recombinant enzyme [18]) [6]
betaine aldehyde <8> (<8> 100% inhibition at 2.5 mM [17]) [17]
butyrate <1> (<1> moderate, competitive to betaine [2]) [2, 7]
choline <8> (<8> 60% inhibition at 5 mM [17]) [17]
isovalerate <1> (<1> competitive to betaine [2,4]) [2, 4]
pinanyl N,N,N-trimethylaminomethane boronate <2> (<2> substrate analogue [5]) [5]
Additional information <1-2> (<1> no inhibition by sarcosine [2, 7]; <1> not inhibitory: S-(γ-carboxypropyl)-DL-homocysteine, S-(β-carboxyethyl)-DL-homocysteine [7]; <1> not inhibitory: methionine or ethionine [4]; <2> not inhibitory: taurine, cysteic acid, cysteine sulfinate, sulfate, glycine, L-serine, L-threonine [3]; <1,2> product inhibition [3, 4, 6]) [2-4, 6, 7]

Activating compounds

D-cysteine <2> (<2> increase of product formation [3]) [3]
L-cysteine <2> (<2> stimulation at low concentration [3]) [3]

cystathione <2> [3]
dithiothreitol <2> [3]
Additional information <2> (<2> taurine, cysteic acid, cysteine sulfinate, sulfate, glycine, L-serine, L-threonine do not affect activity [3]) [3]

Metals, ions

Zn^{2+} <1> (<1> removal results in loss of activity, restoration is possible [11]) [11]

Specific activity (U/mg)

0.000103 <4> [1]
0.000557 <1> [1]
0.00084 <5> [1]
0.0187 <2> [5]
0.314 <1> [2]

K_m-Value (mM)

0.002 <1> (betaine, <1> recombinant enzyme [11]) [11]
0.004 <1> (L-homocysteine, <1> recombinant enzyme [11]) [11]
0.012-0.021 <2> (L-homocysteine) [3]
0.023 <6> (betaine) [13]
0.032 <6> (L-homocysteine) [13]
0.039-0.056 <2> (betaine) [3]
0.1 <1> (betaine) [4]
0.12 <1, 2> (L-homocysteine, <2> betaine [5]) [4, 5]
0.155 <6> (dimethylacetothetin) [13]
1.3 <8> (L-homocysteine) [17]
4.3 <8> (glycinebetaine) [17]

K_i-Value (mM)

0.045 <2> (pinanyl N,N,N-trimethylaminomethane boronate) [5]

pH-Optimum

7.4 <1, 4, 5> (<1,4,5> assay at [1]) [1]
7.5 <8> [17]
8 <2> (<2> 0.05 M Tricine buffer [3]) [3]
Additional information <2> (<2> 3 active forms with pI: 7.0, 7.6 and 8.0 [5]; <2> Tris, TES and HEPES decrease activity [3]) [3, 5]

Temperature optimum (°C)

37 <1, 4, 5, 8> (<1,4,5> assay at [1]) [1, 17]

4 Enzyme Structure

Molecular weight

270000 <1, 2> (<1,2> gel filtration [2,4,5]) [2, 4, 5]
350000 <8> (<8> gel filtration [17]) [17]
Additional information <1> (<1> amino acid composition) [4]

Subunits

? <1> (<1> x * 52000, SDS-PAGE [11]) [11]
hexamer <1, 2> (<1> 6 * 45000, SDS-PAGE [2,4]; <2> 6 * 45000, SDS-PAGE [5]) [2, 4, 5]
octamer <8> (<8> 8 * 45000, SDS-PAGE [17]) [17]
tetramer <1> (<1> 4 * 45000, recombinant His-tagged enzyme, SDS-PAGE [18]) [18]

5 Isolation/Preparation/Mutation/Application

Source/tissue

kidney <1, 2, 3, 6> (<1,2,6> proximal tubes of cortex [19]) [2, 19]
lens <7> [14]
liver <1-6> [1-8, 11, 13, 15]
Additional information <1> (<1> no activity in phytohemagglutinin-stimulated human peripheral blood lymphocytes and cultured human skin fibroblasts [1]) [1]

Localization

cytoplasm <1, 2, 6> [4, 19]
nucleus <7> [14]

Purification

<1> (two forms of the enzyme [2,4]; partial [11]) [2, 4, 11]
<2> (FPLC-chromatofocusing, 3 active forms [5]) [5]
<6> [13]
<8> [17]

Crystallization

<2,1> (recombinant enzyme [9,10]) [9, 10]

Cloning

<1> [13]

Engineering

Cys217Ala <1> (<1> complete loss of activity, reduction in zinc binding, identification of zinc binding motif [15]) [15]
Cys299Ala <1> (<1> complete loss of activity, reduction in zinc binding, identification of zinc binding motif [15]) [15]
Cys300Ala <1> (<1> complete loss of activity, reduction in zinc binding, identification of zinc binding motif [15]) [15]
Additional information <1> (<1> random mutagenesis of zinc binding motif, Gly214 is essential [20]) [20]

Application

medicine <5> (<5> methionine-deficient diet produces large increase in enzyme activity [12]) [12]

6 Stability

pH-Stability

5.5 <1> (<1> above, 4 h stable at 37°C [4]) [4]

Temperature stability

Additional information <1> (<1> significant heat stability [4]) [4]

General stability information

<1>, divalent cations and 2-mercaptoethanol do not stabilize during purification [2, 4]
<1>, lyophilization inactivates [2]
<1>, substrates stabilize during purification [2, 4, 5]
<2>, glycerol does not stabilize [5]

Storage stability

<1>, -20°C, many months in 50% v/v glycerol [4]
<1>, -20°C, over a year in 50% v/v glycerol [2]
<2>, -20°C, 2 weeks in the presence of substrates [5]
<2>, -20°C, 40% loss of activity after 4 weeks in the presence of substrates [5]
<2>, -20°C, 70% loss of activity after 10 weeks in the presence of substrates [5]

References

[1] Wang, J.; Dudman, N.P.; Lynch, J.; Wilcken, D.E.L.: Betaine:homocysteine methyltransferase – a new assay for the liver enzyme and its absence from human skin fibroblasts and peripheral blood lymphocytes. Clin. Chim. Acta, **204**, 239-249 (1991)

[2] Skiba, W.E.; Wells, M.S.; Mangum, J.H.; Awad, W.M.: Betaine-homocysteine S-methyltransferase (human). Methods Enzymol., **143**, 384-388 (1987)

[3] Finkelstein, J.D.; Harris, B.J.; Kyle, W.E.: Methionine metabolism in mammals: kinetic study of betaine-homocysteine methyltransferase. Arch. Biochem. Biophys., **153**, 320-324 (1972)

[4] Skiba, W.E.; Taylor, M.P.; Wells, M.S.; Mangum, J.H.; Awad, W.M.: Human hepatic methionine biosynthesis. Purification and characterization of betaine:homocysteine S-methyltransferase. J. Biol. Chem., **257**, 14944-14948 (1982)

[5] Lee, K.H.; Cava, M.; Amiri, P.; Ottoboni, T.; Lindquist, R.N.: Betaine:homocysteine methyltransferase from rat liver: purification and inhibition by a boronic acid substrate analog. Arch. Biochem. Biophys., **292**, 77-86 (1992)

[6] Finkelstein, J.D.; Martin, J.J.: Inactivation of betaine-homocysteine methyltransferase by adenosylmethionine and adenosylethionine. Biochem. Biophys. Res. Commun., **118**, 14-19 (1984)

[7] Awad, W.M.; Whitney, P.L.; Skiba, W.E.; Mangum, J.H.; Wells, M.S.: Evidence for direct methyl transfer in betaine: homocysteine S-methyl-transferase. J. Biol. Chem., **258**, 12790-12792 (1983)

[8] Finkelstein, J.D.; Kyle, W.E.; Harris, B.J.: Methionine metabolism in mammals: regulatory effects of S-adenosylhomocysteine. Arch. Biochem. Biophys., **165**, 774-779 (1974)
[9] Gonzalez, B.; Pajares, M.A.; Too, H.P.; Garrido, F.; Blundell, T.L.; Sanz-Aparicio, J.: Crystallization and preliminary X-ray study of recombinant betaine-homocysteine S-methyltransferase from rat liver. Acta Crystallogr. Sect. D, **58**, 1507-1510 (2002)
[10] Bose, N.; Momany, C.: Crystallization and preliminary X-ray crystallographic studies of recombinant human betaine-homocysteine S-methyltransferase. Acta Crystallogr. Sect. D, **57**, 431-433 (2001)
[11] Millian, N.S.; Garrow, T.A.: Human betaine-homocysteine methyltransferase is a zinc metalloenzyme. Arch. Biochem. Biophys., **356**, 93-98 (1998)
[12] Emmert, J.L.; Garrow, T.A.; Baker, D.H.: Hepatic betaine-homocysteine methyltransferase activity in the chicken is influenced by dietary intake of sulfur amino acids, choline and betaine. J. Nutr., **126**, 2050-2058 (1996)
[13] Garrow, T.A.: Purification, kinetic properties, and cDNA cloning of mammalian betaine-homocysteine methyltransferase. J. Biol. Chem., **271**, 22831-22838 (1996)
[14] Rao, P.V.; Garrow, T.A.; John, F.; Garland, D.; Millian, N.S.; Zigler, J.S., Jr.: betaine-homocysteine methyltransferase is a developmentally regulated enzyme crystallin in rhesus monkey lens. J. Biol. Chem., **273**, 30669-30674 (1998)
[15] Breksa, A.P.; Garrow, T.A.: Recombinant human liver betaine-homocysteine S-methyltransferase: Identification of three cysteine residues critical for zinc binding. Biochemistry, **38**, 13991-13998 (1999)
[16] Incharoensakdi, A.; Waditee, R.: Degradation of glycinebetaine by betaine-homocysteine methyltransferase in Aphanothece halophytica: effect of salt downshock and starvation. Curr. Microbiol., **41**, 227-231 (2000)
[17] Waditee, R.; Incharoensakdi, A.: Purification and kinetic properties of betaine-homocysteine methyltransferase from Aphanothece halophytica. Curr. Microbiol., **43**, 107-111 (2001)
[18] Bose, N.; Greenspan, P.; Momany, C.: Expression of recombinant human betaine:homocysteine S-methyltransferase for X-ray crystallographic studies and further characterization of interaction with S-adenosylmethionine. Protein Expr. Purif., **25**, 73-80 (2002)
[19] Delgado-Reyes, C.V.; Wallig, M.A.; Garrow, T.A.: Immunohistochemical detection of betaine-homocysteine S-methyltransferase in human, pig, and rat liver and kidney. Arch. Biochem. Biophys., **393**, 184-186 (2001)
[20] Breksa, A.P.; Garrow, T.A.: Random mutagenesis of the zinc-binding motif of betaine-homocysteine methyltransferase reveals that Gly214 Is essential. Arch. Biochem. Biophys., **399**, 73-80 (2002)

Catechol O-methyltransferase 2.1.1.6

1 Nomenclature

EC number
2.1.1.6

Systematic name
S-adenosyl-L-methionine:catechol O-methyltransferase

Recommended name
catechol O-methyltransferase

Synonyms
COMT I
COMT II
MB-COMT <3, 7> (<3,7> membrane-bound form of catechol-O-methyltransferase [6]) [6]
S-COMT <3, 7> (<3,7> soluble form of catechol-O-methyltransferase [6]) [6]
catechol methyltransferase
catecholamine O-methyltransferase
methyltransferase, catechol

CAS registry number
9012-25-3

2 Source Organism

<1> *Rattus norvegicus* (gene expression in Escherichia coli [5]; particulate enzyme [9]; both S-COMT and MB-COMT [33]; S-COMT [39]) [1-3, 5, 8-11, 14-16, 21, 25, 27, 30, 31, 33, 35-37, 39, 43, 47, 49]
<2> *Bos taurus* [1]
<3> *Sus scrofa* [1, 6, 7, 12, 24, 48]
<4> *Mus musculus* [1]
<5> *Cavia porcellus* [1]
<6> *Candida tropicalis* [19, 20]
<7> *Homo sapiens* (2 forms which differ in nature of intramolecular disulfide bonds [21]; hepatoma cell line G2 [6]; 2 forms [2]; both S-COMT and MB-COMT [33]; wild type and variant isoform [44]) [1, 2, 4-6, 13, 17, 18, 21-23, 28, 29, 32, 33, 36, 38, 40-44, 46]
<8> *Oryctolagus cuniculus* [1, 2]
<9> *monkey* [22]
<10> *Clarias gariepinus* (African catfish [26]) [26]

<11> *Cricetulus griseus* (CHO cell line [34]) [34]
<12> *Streptomyces griseus* (S-COMT [45]) [45]
<13> *hamster* [48]

3 Reaction and Specificity

Catalyzed reaction

S-adenosyl-L-methionine + a catechol = S-adenosyl-L-homocysteine + a guaiacol (<1,7> ordered reaction mechanism with S-adenosyl-L-methionine as the leading substrate [4,18,28,30]; <1> mechanism [16]; <7> mechanism [40]; <1> study on mechanism [49])

Reaction type

O-methylation
methyl group transfer

Natural substrates and products

S S-adenosyl-L-methionine + a catechol <1, 7, 8> (Reversibility: ? <1, 7, 8> [2, 3, 9]) [2, 3, 9]
P ?
S homoprotocatechuic acid + S-adenosyl-L-methionine <6> (Reversibility: ? <6> [20]) [20]
P ?
S Additional information <1, 7, 8, 11> (<1> membrane protein and (or) lipid components may play an important role in catecholamine metabolism [9]; <1,7,8> physiological role is the inactivation of catecholamine hormones and neurotransmitters as well as detoxification of a variety of xenobiotic amines and drugs [2]; <1> primary role in extraneuronal inactivation of endogenous catecholamines and in the further metabolism of oxidized catecholamine metabolites [3]; <11> activity in cells expressing norepinephrine transporter [34]; <7> comparison of wild type and variant isoforms and implications for estrogen levels [44]) [2, 3, 9, 34, 44]
P ?

Substrates and products

S (+)-epicatechin + S-adenosyl-L-methionine <7> (Reversibility: ? <7> [41]) [41]
P S-adenosyl-L-homocysteine + ?
S (-)-epicatechin + S-adenosyl-L-methionine <7> (Reversibility: ? <7> [41]) [41]
P S-adenosyl-L-homocysteine + ?
S (-)-epigallocatechin + S-adenosyl-L-methionine <7> (Reversibility: ? <7> [41]) [41]
P S-adenosyl-L-homocysteine + ?
S 1-carboxy-1-methyl-6,7-dimethoxy-1,2,3,4-tetrahydroisoquinoline + S-adenosyl-L-methionine <1> (Reversibility: ? <1> [14]) [14]
P S-adenosyl-L-homocysteine + ?

S 1-methyl-6,7-dimethoxy-1,2,3,4-tetrahydroisoquinoline + S-adenosyl-L-methionine <1> (Reversibility: ? <1> [14]) [14]
P S-adenosyl-L-homocysteine + ?
S 2-hydroxyestradiol-17β + S-adenosyl-L-methionine <7> (Reversibility: ? <7> [17]) [17]
P 2-methoxyestradiol-17β + 2-hydroxyestradiol-17β methyl ether + S-adenosyl-L-homocysteine <7> [17]
S 3,4-dihydroxymandelic acid + S-adenosyl-L-methionine <1> (Reversibility: ? <1> [1]) [1]
P S-adenosyl-L-homocysteine + ?
S 3,4-dihydroxyphenylacetic acid + S-adenosyl-L-methionine <1, 3> (Reversibility: ? <1, 3> [1, 7]) [1, 7, 12]
P S-adenosyl-L-homocysteine + ?
S 4-nitrocatechol + S-adenosyl-L-methionine <12> (Reversibility: ? <12> [45]) [45]
P S-adenosyl-L-homocysteine + ?
S 6,7-dihydroxycoumarin + S-adenosyl-L-methionine <12> (Reversibility: ? <12> [45]) [45]
P S-adenosyl-L-homocysteine + ?
S DL-3,4-dihydroxyamphetamine + S-adenosyl-L-methionine <1> (Reversibility: ? <1> [1]) [1]
P S-adenosyl-L-homocysteine + ?
S DL-3,4-dihydroxybenzoic acid + S-adenosyl-L-methionine <1, 7> (Reversibility: ? <1, 7> [1, 40]) [1, 40]
P S-adenosyl-L-homocysteine + ?
S DL-3,4-dihydroxyephedrine + S-adenosyl-L-methionine <1> (Reversibility: ? <1> [1]) [1]
P S-adenosyl-L-homocysteine + ?
S adrenaline + S-adenosyl-L-methionine <1> (Reversibility: ? <1> [37, 39, 47]) [37, 39, 47]
P S-adenosyl-L-homocysteine + ?
S caffeic acid + S-adenosyl-L-methionine <6, 12> (Reversibility: ? <6, 12> [20, 45]) [20, 45]
P S-adenosyl-L-homocysteine + ?
S catechol + S-adenosyl-L-methionine <1> (Reversibility: ? <1> [1]) [1]
P S-adenosyl-L-homocysteine + ?
S catecholestradiol + S-adenosyl-L-methionine <10> (Reversibility: ? <10> [26]) [26]
P S-adenosyl-L-homocysteine + ?
S catecholestrone + S-adenosyl-L-methionine <10> (Reversibility: ? <10> [26]) [26]
P S-adenosyl-L-homocysteine + ?
S dobutamine + S-adenosyl-L-methionine <7> (Reversibility: ? <7> [38]) [38]
P S-adenosyl-L-homocysteine + ?
S dopa + S-adenosyl-L-methionine <1, 7> (Reversibility: ? <1, 7> [1, 40]) [1, 40]
P S-adenosyl-L-homocysteine + ?

S dopamine + S-adenosyl-L-methionine <1, 7> (Reversibility: ? <1, 7> [1, 18, 38, 40]) [1, 18, 38, 40]
P S-adenosyl-L-homocysteine + ?
S epinephrine + S-adenosyl-L-methionine <1-5, 7, 8> (Reversibility: ? <1-5, 8> [1]) [1]
P metanephrine + S-adenosyl-L-homocysteine <1-5, 7, 8> [1]
S epinine + S-adenosyl-L-methionine <1> (Reversibility: ? <1> [1, 16]) [1, 16]
P S-adenosyl-L-homocysteine + ?
S esculetin + S-adenosyl-L-methionine <6> (Reversibility: ? <6> [19, 20]) [19, 20]
P S-adenosyl-L-homocysteine + ?
S fisetin + S-adenosyl-L-methionine <3, 13> (Reversibility: ? <3, 13> [48]) [48]
P S-adenosyl-L-homocysteine + ?
S noradrenaline + S-adenosyl-L-methionine <7> (Reversibility: ? <7> [40]) [40]
P S-adenosyl-L-homocysteine + ?
S norepinephrine + S-adenosyl-L-methionine <1, 11> (Reversibility: ? <1, 11> [1, 8, 34]) [1, 8, 34]
P S-adenosyl-L-homocysteine + normetanephrine [34]
S *p*-nitrocatechol + S-adenosyl-L-methionine <1> (Reversibility: ? <1> [27]) [27]
P S-adenosyl-L-homocysteine + ?
S protocatechualdehyde + S-adenosyl-L-methionine <6> (Reversibility: ? <6> [20]) [20]
P S-adenosyl-L-homocysteine + ?
S protocatechuic acid + S-adenosyl-L-methionine <6> (Reversibility: ? <6> [20]) [20]
P S-adenosyl-L-homocysteine + ?
S quercetin + S-adenosyl-L-methionine <3, 13> (Reversibility: ? <3, 13> [48]) [48]
P S-adenosyl-L-homocysteine + ?
S Additional information <1, 3, 6-8> (<1> strict requirement for S-adenosyl-L-methionine as methyl donor and a catechol as acceptor substrate [3]; <1,6> not: monophenols [1, 3, 20]; <3> benzimidazole [7]) [1, 3, 7, 20]
P ?

Inhibitors

1-carboxysalsoline <1> (<1> i.e. 1-carboxy-1-methyl-6,7-dimethoxy-1,2,3,4-tetrahydroisoquinoline, competitive inhibition of 3,4-dihydroxybenzoic acid methylation [14]) [14]
2-hydroxyestradiol-17β 3-methyl ether <7> (<7> product inhibition [17]) [17]
2-iodosobenzoic acid <6> [19]
2-methoxyestradiol-17β <7> (<7> product inhibition [17]) [17]

3,5-dinitrocatechol <1> (<1> liver S-COMT, 50% inhibition at 74 nM [39]) [39]
3-carboxysalsolinol <1> [30]
4-nitrocatechol <6> [19]
5-substituted 3-hydroxy-4-methoxybenzaldehydes <1> [31]
5-substituted 3-hydroxy-4-methoxybenzoic acids <1> [31]
Ca^{2+} <1, 6, 7> [3, 19, 29]
Hg^{2+} <12> (<12> complete inhibition at 1 mM [45]) [45]
L-ascorbic acid <1, 6> [10, 19]
L-dopa <1, 6> [10, 19]
Mg^{2+} <1> (<1> inhibition above 2 mM [3]; <1,6,7,10> required for activity [3, 9, 17, 19, 26, 27, 29]) [3]
N-(3,4-dihydroxyphenyl)maleimide <3> (<3> irreversible [24]) [24]
N-(3,4-dihydroxyphenyl)succinimide <3> (<3> reversible [24]) [24]
N-ethylmaleimide <1, 12> (<1> 70-100% inhibition at 0.2 mM [9]; <12> complete inhibition at 1 mM [45]) [3, 9, 45]
OR-462 <1> (<1> disubstituted catechol [25]) [25]
RO-4-4602 <1> (<1> competitive [10]) [10]
Ro 41-0960 <11> [34]
S-adenosyl-L-homocysteine <1, 6, 7> (<7> 50% inhibition at about 0.005 mM [41]) [3, 8, 16-19, 40]
Triton X-100 <7> [29]
U-0521 <11> [34]
amino group reagents <3> [24]
analogs of S-adenosyl-L-homocysteine <1> (<1> overview: inhibition of the liver, heart and brain enzyme [8]) [8]
β-thujaplicin <1> (<1> 70-100% inhibition at 0.2 mM [9]) [9]
catechol <6> [19]
dobutamine <7> (<7> competitive to dopamine [38]) [38]
dopamine <7> (<7> competitive to dobutamine [38]) [38]
flavonoids <1> (<1> overview: relationship between structure and ability to inhibit [15]) [15]
gallic acid <1, 6> [10, 19]
gallic acid methylester <6> [19]
high ionic strength <1> [27]
homocysteine <12> [45]
iodoacetamide <6> (<6> slight [19]) [19]
iodoacetic acid <1-5, 7, 8> [1, 3, 27]
nordihydroguaiaretic acid <7> [29]
p-chloromercuribenzoate <1, 6, 12> (<12> complete inhibition at 1 mM [45]) [1, 3, 19, 45]
p-hydroxymercuribenzoate <1> (<1> 70-100% inhibition at 0.2 mM [9]) [9]
phenolic compounds <1> [1]
polyphenolic compounds <1> [1]
purpurogallin <6> [19]

pyrogallol <1, 6> (<1> 70-100% inhibition at 0.2 mM [9]) [9, 10, 19]
salsolidine <1> (<1> i.e. 1-methyl-6,7-dimethoxy-1,2,3,4-tetrahydroisoquinoline, competitive inhibition of 3,4-dihydroxybenzoic acid methylation [14]) [14, 30]
tropolone <1, 6, 7> (<1> 70-100% inhibition at 0.2 mM [9]) [1, 9, 10, 18, 19, 28, 29]
Additional information <1, 7> (<7> substrate inhibition is dependent on the concentration of S-adenosylmethionine and $MgCl_2$ [17]; <7> activity of the enzyme is strongly influenced by the nature of the buffer [23]; <1> inhibition by analogs of S-adenosyl-homcysteine [8]; <1> relationship between the structure of flavonoids and their inhibitory activity [15]) [8, 15, 17, 23]

Activating compounds

cysteine <7> (<7> 20 mM, activates [13]; <7> required [17]) [13, 17]
Additional information <7> (<7> activity of the enzyme is strongly influenced by the nature of the buffer [23]; <7> L-dopa induces expression and activity of enzyme [46]) [23]

Metals, ions

Cd^{2+} <1> (<1> can substitute for Mg^{2+} [1]) [1]
Co^{2+} <1> (<1> Co^{2+} and Mn^{2+} most effective in activation [1]) [1]
Fe^{2+} <1> (<1> can substitute for Mg^{2+} [1]) [1]
Mg^{2+} <1, 6, 7, 10> (<1> below 2 mM [3]; <1, 6, 7, 10, 12> required [3, 9, 17, 19, 26, 27, 29, 45]; <1> optimum range: 2-5 mM [9]; <10> optimum 10 mM [26]; <1> inhibition above 2 mM [3]) [1, 3, 9, 17, 19, 26, 27, 29]
Mn^{2+} <1, 6, 7> (<1> Co^{2+} and Mn^{2+} most effective in activation [1]) [1, 17, 19]
Ni^{2+} <1> (<1> can substitute for Mg^{2+} [1]) [1]
Zn^{2+} <1, 6> (<1,6> can substitute for Mg^{2+} [1,19]) [1, 19]
Additional information <7> (<7> activity of the enzyme is strongly influenced by the nature of the buffer [23]) [23]

Specific activity (U/mg)

2172 <1> [11]
Additional information <1, 3, 6, 7, 9> [2, 3, 6-9, 11, 12, 20, 22]

K_m-Value (mM)

0.0005 <10> (dopamine) [26]
0.0006 <10> (catecholestradiol) [26]
0.0012 <10> (catecholestrone) [26]
0.0014 <1> (adrenaline, <1> brain MB-COMT, native [37]) [37]
0.0022-0.0082 <7> ((-)-epicatechin) [41]
0.0025 <1> (adrenaline, <1> liver MB-COMT, native [37]) [37]
0.0031 <6, 7> (S-adenosyl-L-methionine, <6> plus esculetin [19]) [18, 19]
0.0033 <7> (dopamine) [18]
0.0038 <1> (*p*-nitrocatechol) [27]
0.0039-0.0067 <7> ((-)-epigallocatechin) [41]
0.0061 <3> (quercetin) [48]
0.0062 <6> (esculetin) [19]

0.0069 <13> (quercetin) [48]
0.0093 <1> (adrenaline, <1> liver MB-COMT, solubilized [37]) [37]
0.012 <1> (adrenaline, <1> brain MB-COMT, solubilized [37]) [37]
0.014 <7> (2-hydroxyestradiol-17β) [17]
0.027 <1> (S-adenosyl-L-methionine, <1> S-COMT, recombinant fusion protein [39]) [39]
0.04 <6> (S-adenosyl-L-methionine, <6> plus esculetin [19]) [19]
0.05 <7> (dobutamine) [38]
0.056 <3> (S-adenosyl-L-methionine, <3> plus 3,4-dihydroxyphenylacetic acid [7]) [7]
0.12 <1> (epinephrine) [1]
0.169 <1> (norepinephrine, <1> heart [8]) [8]
0.257 <1> (adrenaline, <1> liver S-COMT, native [37]) [37]
0.276 <1> (norepinephrine, <1> brain [8]) [8]
0.283 <1> (S-adenosyl-L-methionine) [27]
0.292 <1> (adrenaline, <1> brain S-COMT, native [37]) [37]
0.295 <1> (adrenaline, <1> liver S-COMT, detergent-treated [37]) [37]
0.428 <1> (adrenaline, <1> S-COMT, recombinant fusion protein [39]) [39]
0.44 <7> (dopamine) [38]
0.48 <1> (adrenaline, <1> brain S-COMT, detergent-treated [37]) [37]
0.5 <12> (6,7-dihydroxycoumarin) [45]
0.6 <12> (S-adenosyl-L-methionine) [45]
0.64 <3> (3,4-dihydroxyphenylacetic acid) [7]
1.856 <1> (norepinephrine, <1> liver [8]) [8]
Additional information <1, 3, 6, 7, 12> (<3> pH-dependence of K_m, methods for kinetic studies [12]; <1> overview: K_m values of liver, heart and brain enzymes with various substrates [8]; <7> kinetics and mechanism [40]; <12> assay method [45]; <1> assay conditions [47]) [8-10, 12, 15, 18, 19, 29, 40, 45, 47]

K_i-Value (mM)

0.001 <7> (S-adenosyl-L-homocysteine, <7> plus S-adenosylmethionine [18]) [18]
0.005 <7> (tropolone, <7> plus dopamine [18]) [18]
0.0053 <1> (RO-4-4602, <1> plus norepinephrine [10]) [10]
0.007 <1> (RO-4-4602, <1> plus dopamine [10]) [10]
0.007 <6> (S-adenosyl-L-homocysteine, <6> plus S-adenosyl-L-methionine [19]) [19]
0.009 <1> (RO-4-4602, <1> plus epinephrine [10]) [10]
0.022-0.04 <1> (S-adenosyl-L-homocysteine) [8]
0.039 <7> (S-adenosyl-L-homocysteine) [17]
0.07 <6> (purpurogallin) [19]
0.19 <1> (salsolidine) [14]
0.224 <12> (homocysteine) [45]
0.44 <1> (1-carboxysalsoline) [14]

pH-Optimum

6.8-8.4 <7> (<7> leads to maximal formation of 2-methoxyestradiol-17β [17]) [17]
7.1-7.4 <3> [7]
7.3-8.2 <1> [3]
7.5 <12> [45]
7.5-8.2 <1> [1]
7.6-7.8 <6> [20]
7.8 <1> [27]
8 <7> [29]
8-8.5 <1> [9]
9.2 <7> (<7> leads to maximal formation of 2- and 3-monomethyl ether [17]) [17]

pH-Range

6.8-8 <3> (<3> about 50% of maximum activity at pH 6.8 and pH 8 [7]) [7]

Temperature optimum (°C)

35 <12> [45]
37 <7> (<7> assay at [38]) [38]
42 <7> (<7> 2-hydroxyestradiol-17β [17]) [17]
50 <1> [27]

4 Enzyme Structure

Molecular weight

21000 <1> (<1> gel filtration [9]) [9]
23000 <1, 3, 7> (<1> gel filtration [8]; <7> gel filtration [4]; <3> gel filtration [7]; <7> gel filtration [23]) [4, 7, 8, 23]
24000 <1> (<1> COMT I, gel filtration, sedimentation data [2]) [2]
24000 <1> (<1> S-COMT, SDS-PAGE [35]) [35]
25000 <1> (<1> gel filtration [21]) [21]
25000-26000 <7> (<7> gel filtration, SDS-PAGE [13]) [13]
27500 <7> [28]
28000 <1> (<1> MB-COMT, SDS-PAGE [35]) [35]
32000 <6> (<6> gel filtration [20]) [20]
36000 <12> (<12> gel filtration [45]) [45]
47500 <1> (<1> COMT II, gel filtration, sedimentation data [2]) [2]

Subunits

? <1, 7> (<1> x * 23000, SDS-PAGE [3]; <7> x * 29000, hepatoma cell line G2, SDS-PAGE [6]) [3, 6]
monomer <1, 3, 7, 12> (<3> 1 * 23000, SDS-PAGE [7]; <7> 1 * 23000, SDS-PAGE [4,23]; <1> 1 * 25000, SDS-PAGE [21]; <1> 1 * 25700, liver, SDS-PAGE [27]; <12> 1 * 36000, SDS-PAGE, gel filtration [45]) [4, 7, 21, 23, 27, 45]

5 Isolation/Preparation/Mutation/Application

Source/tissue

Hep-G2 cell <7> [6]
blood <7> (<7> mononuclear cell [38]) [38]
brain <1, 7> [1, 2, 8, 18, 28, 30, 37]
duodenum <1> [25]
erythrocyte <7, 9> [22]
frontal lobe <7> [29]
heart <1> [1, 2, 8]
kidney <1, 13> [1, 2, 35, 48]
liver <1-5, 7, 8> [1-3, 5-12, 14-17, 21, 24, 25, 27, 31, 32, 35, 37, 48]
lung <1, 7, 8> [1, 2]
placenta <7> [2, 4, 5, 13, 23, 41]
skeletal muscle <1> [1]
small intestine <1> [1]
spleen <1> [1]
Additional information <1> (<1> distribution in various tissues [35]) [35]

Localization

cytoplasm <1, 7> (<1,7> S-COMT, recombinant enzyme [33]) [33]
endoplasmic reticulum <1, 7> (<1,7> MB-COMT, recombinant enzyme [33]) [33]
membrane <3, 7> (<7> integral membrane protein [29]; <3,7> exists in a soluble cytosolic and in a particulate membrane-bound form [6]) [6, 18, 29]
microsome <1, 3, 7> [6, 8, 9]
nucleus <1, 7> (<1,7> S-COMT, recombinant enzyme [33]) [33]
soluble <1, 3, 7> (<3,7> exists in a soluble cytosolic and in a particulate membrane-bound form [6]) [2, 6, 8, 13, 17, 21, 28]
Additional information <1, 7> (<1,7> MB-COMT, recombinant enzyme, not: plasma membrane [33]; <7> S-COMT highly expressed in liver and kidney, MB-COMT in brain [36]) [33, 36]

Purification

<1> (affinity chromatography [11]; microsomal [9]; recombinant enzyme [5]; soluble and microsomal enzyme [8]; overview on purification of native and recombinant protein [43]) [2, 3, 5, 8, 9, 11, 21, 27, 43]
<3> (soluble form [6]) [6, 12]
<6> (partial [20]) [20]
<7> (overview on purification of native and recombinant protein [43]) [4, 5, 13, 23, 28, 43]
<12> [45]

Crystallization

<7> (crystal structure [32]) [32]

Cloning

<1> (gene expressed in Escherichia coli [5]; S-COMT and MB-COMT are encoded by a single gene with two different sites of transcription initiation, overview on genetic polymorphism [36]; S-COMT [39]) [5, 36]
<7> (S-COMT and MB-COMT are encoded by a single gene with two different sites of transcription initiation, overview on genetic polymorphism [36]) [36]

Application

medicine <7> (<7> COMT activity may alter neurotransmitter deposition and transport [42]; <7> comparison of COMT isoforms and implications for breast cancer risk [44]) [42, 44]

6 Stability

Temperature stability

37 <1> (<1> 2 h, inactivation [1]) [1]

General stability information

<3>, complete loss of activity on freezing solid [7]

Storage stability

<1>, -10°C, 3 months, stable [1]
<1>, -20°C, or -60°C, 50 mM Tris-HCl, pH 7.5, 20 mM 2-mercaptoethanol, 20% v/v glycerol, 60-80% of activity retained after 6 months [2]
<1>, -4°C, stable for several weeks [3]
<3>, -5°C, 20% v/v glycerol, stable for 6 months [7]
<7>, -20°C, 100 mM potassium phosphate, pH 8.0, 1 mM EDTA, 1 mM 2-mercaptoethanol, 5 mM dithiothreitol, 40% v/v glycerol, 1 mg/ml bovine serum albumin, stable for 4 months [4]

References

[1] Axelrod, J.; Tomchick, R.: Enzymatic O-methylation of epinephrine and other catechols. J. Biol. Chem., **233**, 702-705 (1958)
[2] Huh, M.M.O.; Friedhoff, A.J.: Multiple molecular forms of catechol-O-methyltransferase. Evidence for two distinct forms, and their purification and physical characterization. J. Biol. Chem., **254**, 299-308 (1979)
[3] Borchardt, R.T.: Catechol O-methyltransferase. Methods Enzymol., **77**, 267-272 (1981)
[4] Bhaird, N.N.; Tipton, K.F.: Catechol-O-methyltransferase from human placenta: purification and some properties. Biochem. Soc. Trans., **19**, 20S (1991)
[5] Lundström, K.; Tilgmann, C.; Peränen, J.; Kalkkinen, N.; Ulmanen, I.: Expression of enzymatically active rat liver and human placental catechol-O-methyltransferase in Escherichia coli; purification and partial characterization of the enzyme. Biochim. Biophys. Acta, **1129**, 149-154 (1992)

[6] Bertocci, B.; Garotta, G.; Da Prada, M.; Lahm, H.W.; Zurcher, G.; Virgallita, G.; Miggiano, V.: Immunoaffinity purification and partial amino acid sequence analysis of catechol-O-methyltransferase from pig liver. Biochim. Biophys. Acta, **1080**, 103-109 (1991)
[7] Gulliver, P.A.; Tipton, K.F.: The purification and properties of pig-liver catechol-O-methyl transferase. Eur. J. Biochem., **88**, 439-444 (1978)
[8] Borchardt, R.T.; Cheng, C.F.: Purification and characterization of rat heart and brain catechol methyltransferase. Biochim. Biophys. Acta, **522**, 49-62 (1978)
[9] Tong, J.H.; DIorio, A.: Solubilization and partial purification of particulate catechol-O-methyltransferase from rat liver. Can. J. Biochem., **55**, 1108-1113 (1977)
[10] Buu, N.T.; Kuchel, O.; Parent, M.T.: Competitive inhibition of catechol O-methyltransferase by RO-4-4602. Can. J. Biochem., **55**, 771-773 (1977)
[11] Borchardt, R.T.; Cheng, C.F.; Thakker, D.R.: Purification of catechol-O-methyltransferase by affinity chromatography. Biochem. Biophys. Res. Commun., **63**, 69-77 (1975)
[12] Raxworthy, M.J.; Gulliver, P.A.: The effect of bulk hydrogen ion concentration upon the apparent kinetic parameters of purified pig liver catechol O-methyltransferase. Biochim. Biophys. Acta, **870**, 417-425 (1986)
[13] Tilgmann, C.; Kalkkinen, N.: Purification and partial sequence analysis of the soluble catechol-O-methyltransferase from human placenta: comparison to the rat liver enzyme. Biochem. Biophys. Res. Commun., **174**, 995-1002 (1991)
[14] Sanft, K.; Thomas, H.: Competitive inhibition of catechol-O-methyltransferase by tetrahydro-isochinoline alkaloids salsolidin and 1-carboxysalsolin. Z. Naturforsch. C, **44**, 173-176 (1989)
[15] Schwabe, K.P.; Flohe, L.: Catechol-O-methyltransferase, 3. Relationship between the structure of flavonoids and their ability to inhibit catechol-O-methyltransferase. Hoppe-Seyler's Z. Physiol. Chem., **353**, 476-482 (1972)
[16] Coward, J.K.; Slisz, E.P.; Wu, F. Y.H.: Kinetic studies on catechol O-methyltransferase. Product inhibition and the nature of the catechol binding site. Biochemistry, **12**, 2291-2297 (1973)
[17] Ball, P.; Knuppen, R.; Haupt, M.; Breuer, H.: Kinetic properties of a soluble catechol O-methyltransferase of human liver. Eur. J. Biochem., **26**, 560-569 (1972)
[18] Rivett, A.J.; Roth, J.A.: Kinetic studies on the O-methylation of dopamine by human brain membrane-bound catechol O-methyltransferase. Biochemistry, **21**, 1740-1742 (1982)
[19] Veser, J.: Kinetics and inhibition studies of catechol O-methyltransferase from the yeast Candida tropicalis. J. Bacteriol., **169**, 3696-3700 (1987)
[20] Veser, J.; Geywitz, P.; Thomas, H.: Purification and properties of a catechol methyltransferase of the yeast Candida tropicalis. Z. Naturforsch. C, **34**, 709-714 (1979)
[21] Tilgmann, C.; Kalkkinen, N.: Purification and partial characterization of rat liver soluble catechol-O-methyltransferase. FEBS Lett., **264**, 95-99 (1990)

[22] Lin, R.L.; Narasimhachari, N.: Specific tlc, gc, and gc-ms methods for kinetic studies with COMT. Anal. Biochem., **57**, 46-58 (1974)
[23] Nic A'Bhaird, N.; Tipton, K.F.: Behavior and properties of catechol-O-methyltransferase from human placenta. J. Neural Transm., **32**, 359-368 (1990)
[24] Piedrafita, F.J.; Elorriaga, C.; Fernandez-Alvarez, E.; Nieto, O.: Inhibition of catechol-O-methyltransferase by N-(3,4-dihydroxyphenyl) maleimide. J. Enzyme Inhib., **4**, 43-50 (1990)
[25] Schultz, E.; Nissinen, E.: Inhibition of rat liver and duodenum soluble catechol-O-methyltransferase by a tight-binding inhibitor OR-462. Biochem. Pharmacol., **38**, 3953-3956 (1989)
[26] Timmers, R.J.M.; Lambert, J.G.D.: Kinetic studies of catechol-O-methyltransferase from the brain of the African catfish, Clarias gariepinus. Comp. Biochem. Physiol. C Comp. Pharmacol. Toxicol., **91**, 399-402 (1988)
[27] Rhee, J.; Myung Un, C.: Rat liver catechol-O-methyltransferase: Purification and general properties. Hanguk Saenghwahakhoe Chi, **21**, 60-67 (1988)
[28] Jeffrey, D.R.; Roth, J.A.: Purification and kinetic mechanism of human brain soluble catechol-O-methyltransferase. J. Neurochem., **44**, 881-885 (1985)
[29] Jeffrey, D.R.; Roth, J.A.: Characterization of membrane-bound and soluble catechol-O-methyltransferase from human frontal cortex. J. Neurochem., **42**, 826-832 (1984)
[30] Tunnicliff, G.; Ngo, T.T.: Kinetics of rat brain soluble catechol-O-methyltransferase and its inhibition by substrate analogues. Int. J. Biochem., **15**, 733-738 (1983)
[31] Borchardt, R.T.; Huber, J.H.; Houston, M.: Catechol O-methyltransferase. 10. 5-Substituted 3-hydroxy-4-methoxybenzoic acids (isovanillic acids) and 5-substituted 3-hydroxy-4-methoxybenzaldehydes (isovanillins) as potential inhibitors. J. Med. Chem., **25**, 258-263 (1982)
[32] Vidgren, J.; Svensson, L.A.; Liljas, A.: Crystal structure of catechol O-methyltransferase. Nature, **368**, 354-358 (1994)
[33] Ulmanen, I.; Peranen, J.; Tenhunen, J.; Tilgmann, C.; Karhunen, T.; Panula, P.; Bernasconi, L.; Aubry, J.P.; Lundstrom, K.: Expression and intracellular localization of catechol O-methyltransferase in transfected mammalian cells. Eur. J. Biochem., **243**, 452-459 (1997)
[34] Percy, E.; Kaye, D.M.; Lambert, G.W.; Gruskin, S.; Esler, M.D.; Du, X.J.: Catechol-O-methyltransferase activity in CHO cells expressing norepinephrine transporter. Br. J. Pharmacol., **128**, 774-780 (1999)
[35] Karhunen, T.; Tilgmann, C.; Ulmanen, I.; Julkunen, I.; Panula, P.: Distribution of catechol-O-methyltransferase enzyme in rat tissues. J. Histochem. Cytochem., **42**, 1079-1090 (1994)
[36] Weinshilboum, R.M.; Otterness, D.M.; Szumlanski, C.L.: Methylation pharmacogenetics: catechol O-methyltransferase, thiopurine methyltransferase, and histamine N-methyltransferase. Annu. Rev. Pharmacol. Toxicol., **39**, 19-52 (1999)
[37] Bonifacio, M.J.; Vieira-Coelho, M.A.; Borges, N.; Soares-da-Silva, P.: Kinetics of rat brain and liver solubilized membrane-bound catechol-O-methyltransferase. Arch. Biochem. Biophys., **384**, 361-367 (2000)

[38] Yan, M.; Webster, L.T., Jr.; Blumer, J.L.: Kinetic interactions of dopamine and dobutamine with human catechol-O-methyltransferase and monoamine oxidase in vitro. J. Pharmacol. Exp. Ther., **301**, 315-321 (2002)
[39] Bonifacio, M.J.; Vieira-Coelho, M.A.; Soares-da-Silva, P.: Expression and characterization of rat soluble catechol-O-methyltransferase fusion protein. Protein Expr. Purif., **23**, 106-112 (2001)
[40] Lotta, T.; Vidgren, J.; Tilgmann, C.; Ulmanen, I.; Melen, K.; Julkunen, I.; Taskinen, J.: Kinetics of human soluble and membrane-bound catechol O-methyltransferase: A revised mechanism and description of the thermolabile variant of the enzyme. Biochemistry, **34**, 4202-4210 (1995)
[41] Zhu, B.T.; Patel, U.K.; Cai, M.X.; Conney, A.H.: O-Methylation of tea polyphenols catalyzed by human placental cytosolic catechol-O-methyltransferase. Drug Metab. Dispos., **28**, 1024-1030 (2000)
[42] Eshleman, A.J.; Stewart, E.; Evenson, A.K.; Mason, J.N.; Blakely, R.D.; Janowsky, A.; Neve, K.A.: Metabolism of catecholamines by catechol-O-methyltransferase in cells expressing recombinant catecholamine transporters. J. Neurochem., **69**, 1459-1466 (1997)
[43] Tilgmann, C.; Ulmanen, I.: Purification methods of mammalian catechol-O-methyltransferases. J. Chromatogr. B, **684**, 147-161 (1996)
[44] Dawling, S.; Roodi, N.; Mernaugh, R.L.; Wang, X.; Parl, F.F.: Catechol-O-methyltransferase (COMT)-mediated metabolism of catechol estrogens: comparison of wild-type and variant COMT isoforms. Cancer Res., **61**, 6716-6722 (2001)
[45] Dhar, K.; Rosazza, J.P.N.: Purification and characterization of Streptomyces griseus catechol O-methyltransferase. Appl. Environ. Microbiol., **66**, 4877-4882 (2000)
[46] Zhao, W.Q.; Latinwo, L.; Liu, X.X.; Lee, E.S.; Lamango, N.; Charlton, C.G.: L-Dopa upregulates the expression and activities of methionine adenosyl transferase and catechol-O-methyltransferase. Exp. Neurol., **171**, 127-138 (2001)
[47] Borges, N.; Vieira-Coelho, M.A.; Parada, A.; Soares-Da-Silva, P.: Rat liver catechol-O-methyltransferase kinetics and assay methodology. J. Enzyme Inhib., **13**, 473-483 (1998)
[48] Zhu, B.T.; Ezell, E.L.; Liehr, J.G.: Catechol-O-methyltransferase-catalyzed rapid O-methylation of mutagenic flavonoids. J. Biol. Chem., **269**, 292-299 (1994)
[49] Zheng, Y.; Bruice, T.C.: A theoretical examiniation of the factors controlling the catalytic efficiency of a transmethylation enzyme: Catechol O-methyltransferase. J. Am. Chem. Soc., **119**, 8137-8145 (1997)

Nicotinate N-methyltransferase 2.1.1.7

1 Nomenclature

EC number
2.1.1.7

Systematic name
S-adenosyl-L-methionine:nicotinate N-methyltransferase

Recommended name
nicotinate N-methyltransferase

Synonyms
furanocoumarin 8-O-methyltransferase
furanocoumarin 8-methyltransferase
methyltransferase, nicotinate
nicotinate methyltransferase
nicotinic acid methyltransferase

CAS registry number
9029-79-2

2 Source Organism

<1> *Pisum sativum* (pea [1]) [1]
<2> *Lemna paucicostata* (151, duckweed [2]) [2]
<3> *Glycine max* (soybean [3]) [3]

3 Reaction and Specificity

Catalyzed reaction
S-adenosyl-L-methionine + nicotinate = S-adenosyl-L-homocysteine + N-methylnicotinate

Reaction type
methyl group transfer

Natural substrates and products
S S-adenosyl-L-methionine + nicotinate <1-3> (<1> trigonelline biosynthesis [1]) (Reversibility: ? <1-3> [1-3]) [1-3]
P S-adenosyl-L-homocysteine + 1-methylnicotinate

Substrates and products

S S-adenosyl-L-methionine + nicotinate <1-3> (<3> high specificity for nicotinate, no other pyridine derivative is a suitable methyl group acceptor [3]) (Reversibility: ? <1-3> [1-3]) [1-3]

P S-adenosyl-L-homocysteine + 1-methylnicotinate <1-3> (<1> i.e. trigonelline [1]) [1-3]

Inhibitors

Hg^{2+} <3> [3]
Mg^{2+} <3> (<3> slight [3]) [3]
S-adenosyl-L-homocysteine <3> [3]
SH-blocking reagents <3> [3]
benzoic acid <2> [2]
heavy metal ions <2> [2]
isonicotinic acid <2> [2]
p-hydroxymercuribenzoate <3> [3]
picolinic acid <2> [2]
pyrazinamide <2> [2]
salicylic acid <2> [2]
trigonelline <2> (<2> i.e. N-methylnicotinic acid [2]) [2]

Metals, ions

Additional information <3> (<3> no requirement for divalent cations [3]) [3]

Specific activity (U/mg)

Additional information <3> [3]

K_m-Value (mM)

0.055 <2, 3> (S-adenosyl-L-methionine, <2> nicotinate [2]) [2, 3]
0.078 <3> (nicotinate) [3]
0.1 <2> (S-adenosyl-L-methionine) [2]
0.4 <1> (nicotinate) [1]

pH-Optimum

6-7 <1> [1]
7 <2> [2]
8 <3> [3]

pH-Range

5-9 <3> (<3> pH 5.0: about 50% of maximum activity, pH 9.0: nearly inactive [3]) [3]
5.2-8.5 <1> (<1> about 55% of maximum activity at pH 5.2 and 8.5 [1]) [1]

Temperature optimum (°C)

30 <2> [2]
35-40 <3> [3]

4 Enzyme Structure

Molecular weight

90000 <3> (<3> ultracentrifugation, gel filtration [3]) [3]

5 Isolation/Preparation/Mutation/Application

Source/tissue

cell suspension culture <3> (<3> heterotrophic [3]) [3]
seedling <1> [1]

Purification

<3> [3]

6 Stability

Oxidation stability

<3>, rigorous exclusion of O_2 by saturating the buffer with N_2 allows partial stabilization during purification [3]

General stability information

<3>, no stabilization by glycerol, 10-40% v/v, or nicotinic acid [3]
<3>, sulfhydryl group protecting agents required for stabilization during purification [3]
<3>, total loss of activity by freezing at -20°C [3]

Storage stability

<3>, 4°C, 24 h, N_2-atmosphere [3]

References

[1] Joshi, J.G.; Handler, P.: Biosynthesis of trigonelline. J. Biol. Chem., **235**, 2981-2983 (1960)

[2] Taguchi, H.; Nishitani, H.; Okumura, K.; Shimabayashi, Y.; Iwai, K.: Biosynthesis and metabolism of trigonelline in Lemna paucicostata 151. Agric. Biol. Chem., **53**, 2867-2871 (1989)

[3] Upmeier, B.; Gross, W.; Köster, S.; Barz, W.: Purification and properties of S-adenosyl-L-methionine:nicotinic acid-N-methyltransferase from cell suspension cultures of Glycine max L. Arch. Biochem. Biophys., **262**, 445-454 (1988)

Histamine N-methyltransferase 2.1.1.8

1 Nomenclature

EC number
2.1.1.8

Systematic name
S-adenosyl-L-methionine:histamine N-tele-methyltransferase

Recommended name
histamine N-methyltransferase

Synonyms
S-adenosylmethionine-histamine N-methyltransferase
histamine 1-methyltransferase
histamine methyltransferase
histamine-methylating enzyme
imidazole N-methyltransferase
imidazole methyltransferase
imidazolemethyltransferase
methyltransferase, histamine

CAS registry number
9029-80-5

2 Source Organism

<1> *Cavia porcellus* [1, 2, 10, 15, 16, 18, 20, 22, 24]
<2> *Sus scrofa* [2, 3, 14, 28]
<3> *Rattus norvegicus* [1, 5-7, 11, 13, 17, 19, 20, 22]
<4> *Felis catus* [1]
<5> *Oryctolagus cuniculus* [1]
<6> *Mus musculus* [1, 9, 22]
<7> *Homo sapiens* (two polymorphic forms of enzyme [21,25,27]) [4, 9, 21, 23, 25-27]
<8> *Bos taurus* [8, 12]

3 Reaction and Specificity

Catalyzed reaction

S-adenosyl-L-methionine + histamine = S-adenosyl-L-homocysteine + N^{τ}-methylhistamine (<7> sequential reaction mechanism [4]; <7> ordered sequential bi bi reaction mechanism with S-adenosylmethionine as the first substrate and histamine as the second substrate, 1-methylhistamine is the first product to leave and S-adenosylhomocysteine the second [4]; <1> 2 half-reactions in a ping pong mechanism with the intermediate formation of a methylated enzyme [18]; <8> ordered steady state mechanism with S-adenosylmethionine being the first substrate to bind to the enzyme and N-methylhistamine being the first product to dissociate [12])

Reaction type

N-methylation
methyl group transfer

Natural substrates and products

S Additional information <1, 3-7> (<1, 3-6> major pathway of histamine metabolism in mammals [1]; <3, 7> major enzyme for histamine inactivation in mammalian tissues [5, 23]; <7> possible role in modulation of histamine mediated reactions in skin [4]) [1, 4, 5]
P ?

Substrates and products

S S-adenosyl-L-methionine + (+/-)β-methylhistamine <1> (Reversibility: ? <1> [16]) [16]
P S-adenosyl-L-homocysteine + ?
S S-adenosyl-L-methionine + (R)-α-methylhistamine <1, 2> (Reversibility: ? <1, 2> [14, 16]) [14, 16]
P S-adenosyl-L-homocysteine + ?
S S-adenosyl-L-methionine + (S)-α-methylhistamine <1, 2> (Reversibility: ? <1, 2> [14, 16]) [14, 16]
P S-adenosyl-L-homocysteine + ?
S S-adenosyl-L-methionine + histamine <1-8> (<1,2,8> highly specific for histamine, [2, 12, 18]) (Reversibility: ? <1-8> [1-20]) [1-20]
P S-adenosyl-L-homocysteine + N^{τ}-methylhistamine <1-6> [1, 2]
S S-adenosyl-L-methionine + homohistamine <1> (Reversibility: ? <1> [16]) [16]
P S-adenosyl-L-homocysteine + ?

Inhibitors

(R)-chloroquine <7> (<7> 50% inhibition at 0.018 mM, liver enzyme, 50% inhibition at 0.0022 mM, brain enzyme [26]) [26]
(S)-chloroquine <7> (<7> 50% inhibition at 0.005 mM, liver enzyme, 50% inhibition at 0.007 mM, brain enzyme [26]) [26]
1-methylhistamine <2, 7> (<2> strong [3]) [3, 4]
2-bromolysergic acid diethylamide <1> [1, 2]
2-methylhistamine <2> (<2> strong [3]) [3]

3-methylhistamine <2> (<2> strong [3]) [3]
5-methylhistamine <2> (<2> weak [3]) [3]
N-ethylmaleimide <3> [11]
N^{α}-methylhistamine <2> (<2> weak [3]) [3]
S-adenosylhomocysteine <1, 3, 7> [4, 6, 15, 19, 20]
SKF 91488 <1, 3, 7> [16, 17, 23]
amodiaquine <3, 6, 8> [8, 9, 17]
biogenic amines <3> [13]
bromo-lysergic acid diethylamide <1> (<1> 55% inhibition at 0.005 mM [1]) [1]
bufotenine <1> [1]
chlorpromazine <1-6> (<1> 60% inhibition at 0.005 mM [1]) [1, 2]
d-chlorpheniramine <3> [17]
dimaprit <3> [17]
ethylamine <3> [17]
histamine <3, 7> (<7> 0.025-0.1 mM [4]; <8> not [12]) [4, 6]
impromidine <3> (<3> competitive, reversible [17]) [17]
iodoacetamide <3> (<3> activation [11]; <3> complete inhibition at 0.01 mM [6]) [6]
methylhistamine <1-3> (<1> 40% inhibition at 0.1 mM [1]) [1, 2, 6, 18]
methylthioadenosine <3> [6]
metoprine <8> [8]
p-chloromercuribenzoate <1, 3> (<1> 90% inhibition at 0.01 mM [1]) [1, 2, 11]
p-hydroxymercuribenzoate <3> (<3> complete inhibition at 0.1 mM [6]) [6]
serotonin <1-6> (<1> 33% inhibition at 0.005 mM [1]) [1, 2]
tryptamine <3> (<3> and the hydroxyl derivative [13]) [13]
tubocurare <1> [1]
tyramine <3> (<3> and the hydroxyl derivative [13]) [13]
Additional information <1, 7> (<7> not inhibitory: S-adenosyl-L-methionine [4]; <1> overview: adenosine analogs bearing a lipophilic side chain [10]) [4, 10]

Activating compounds

5-methylhistamine <2> (<2> 0.001-0.5 mM, activation [3]) [3]
N',N'-dimethylhistamine <2> (<2> 0.001-0.5 mM, activation [3]) [3]
N^{α}-methylhistamine <2> (<2> 0.001-0.5 mM, activation [3]) [3]
α,N^{α}-dimethylhistamine <2> (<2> activation [14]) [14]
iodoacetamide <3> (<3> activation [11]; <3> inhibition [6]) [11]
iodoacetic acid <3> (<3> activation [11]) [11]
succinic anhydride <3> (<3> activation [11]) [11]

Specific activity (U/mg)

0.004 <3> [11]
0.517 <8> [8]
10.4 <2> [2]
25.4 <1> [1]
Additional information <3> (<3> HPLC assay method [7]) [6, 7]

K_m-Value (mM)
0.0017 <6> (S-adenosyl-L-methionine) [9]
0.0018 <7> (S-adenosyl-L-methionine) [4]
0.0042 <7> (histamine) [4]
0.006 <1> (S-adenosyl-L-methionine) [15]
0.0061 <3> (histamine) [6]
0.0071 <3> (S-adenosyl-L-methionine) [6]
0.012 <3> (histamine) [11]
0.014 <1> (histamine) [15]
0.026 <6> (histamine) [9]
0.035 <1> (histamine) [1, 2]
0.04 <3> (S-adenosyl-L-methionine) [11]
0.043 <1> (histamine) [18]
0.38 <1> (S-adenosyl-L-methionine) [1, 2]
Additional information <1, 3, 6, 7> [4, 9, 15, 16, 18, 19]

K_i-Value (mM)
0.00005 <3> (amodiaquin) [17]
0.0007 <3> (d-chlorpheniramine) [17]
0.001 <3> (SKF 91488) [17]
0.008 <3> (dimaprit) [17]

pH-Optimum
7.2-7.4 <1, 2, 6> [1, 2]
7.5 <1> (<1> 2 optima at pH 7.5 and pH 9.0, brain enzyme [15,20]) [15, 20]
8-8.3 <3> [5]
8.3 <2> (<2> N-methylhistamine [3]) [3]
8.5-9 <3> (<3> kidney enzyme [20]) [20]
9 <1, 3> (<3> kidney [20]; <1> 2 optima: pH 7.5 and pH 9.0, brain enzyme [15]) [15, 20]
9-9.4 <3> [6]

pH-Range
6.5-10.5 <2> (<2> about 50% of maximum activity at pH 6.5 and pH 10.5, N-methylhistamine [3]) [3]
7.3-10.3 <3> (<3> about 50% of maximum activity at pH 7.3 and pH 10.3 [6]) [6]

Temperature optimum (°C)
37 <1-6> (<1-6> assay at [1,2,5,6]) [1, 2, 5, 6]

4 Enzyme Structure

Molecular weight
29000 <1> (<1> gel filtration, SDS-PAGE [15]) [15]
31500 <3> (<3> SDS-PAGE, gel filtration [19]) [19]
34800 <8> (<8> gel filtration [12]) [12]
36000 <3> (<3> gel filtration [6]) [6]

Subunits

? <3, 7, 8> (<8> x * 33200, SDS-PAGE [8]; <3> x * 33400, SDS-PAGE [5]; <3> x * 35000, SDS-PAGE [6]; <6> x * 32000, SDS-PAGE [22]; <1> x * 33000, SDS-PAGE [22]; <7> x * 33000, deduced from DNA sequence [23]) [5, 6, 8, 22, 23]
monomer <1, 3> (<1> 1 * 29000, SDS-PAGE [15]; <3> 1 * 31500, SDS-PAGE [19]) [15, 19]
Additional information <7> (<7> two polymorphic forms, Thr105 corresponds to high activity, Ile105 corresponds to low activity phenotype [21,25]; <7> theoretical 3D model of isoforms, Ile105 energetically destabilizes [27]) [21, 25, 27]

5 Isolation/Preparation/Mutation/Application

Source/tissue

abdomen <1, 3-6> (<1,3-6> abdominal skin [1]) [1]
adrenal gland <1, 4, 5> [1]
brain <1, 3-8> [1, 5, 8, 10, 12, 15, 18, 20, 22, 26]
duodenum <1, 3> [5, 24]
erythrocyte <7> [9]
gall bladder <1> (<1> columnar epithelium [24]) [24]
gastrointestinal tract <1> [24]
heart <1, 3-6> [1]
ileum <1, 3-5> [1, 5]
jejunum <1, 3> [5, 24]
kidney <1, 3-7> [1, 5-7, 9, 13, 17, 19, 20, 22]
liver <1, 2, 4-6, 7> (<3> not in rat [1, 2]) [1, 2, 22, 23, 26]
lung <1, 4-6> (<3> not in rat [1, 2]) [1, 22]
mucosa <2> (antrum mucosa) [3]
mucosa <2> (fundic mucosa) [14]
pancreas <4> [1]
pituitary gland <3> (<3> posterior [5]) [5]
renal artery <3> [5]
sciatic nerve <1, 4, 5> [1]
skin <7> [4]
spleen <1, 2, 4-6> (<3> not in rat [1, 2]) [1, 2, 22]
stomach <1, 4-6> [1, 22]
thigh muscle <1, 3-6> [1]
trachea <1, 4, 7> [1, 23, 24]
Additional information <1, 3> (<3> tissue distribution [5]; <1> inhibitory activity of adenosine analogs [10]; <1> tissue distribution [24]) [5, 10]

Localization

soluble <1, 3-6> [1]

Purification

<1> (recombinant protein [22]) [1, 15, 18, 20, 22]
<2> [2]

<3> (renal enzyme [5,6]) [5, 6, 11, 19, 20]
<8> [8, 12]

Cloning
<2> (kidney, intestine [28]) [28]
<7> [23]

6 Stability

Temperature stability
37 <3> (<3> 90 min, 1 mM dithiothreitol, stable, 40% loss of activity without addition of dithiothreitol [6]) [6]

General stability information
<3>, dithiothreitol, 1 mM, stabilizes [6]

Storage stability
<3>, -80°C, 10 mM potassium phosphate buffer, pH 7.4, 10% glycerol, 1 mM dithiothreitol, 8 months, 16% loss of activity [6]
<3>, -80°C, 5 months [19]
<1, 3>, -80°C, 3 months [15, 20]

References

[1] Brown, D.D.; Tomchick, R.; Axelrod, J.: The distribution and properties of a histamine-methylating enzyme. J. Biol. Chem., **234**, 2948-2950 (1959)
[2] Axelrod, J.: Histamine N-methyltransferase (pig liver). Methods Enzymol., **17**, 766-769 (1971)
[3] Barth, H.; Lorenz, W.; Niemeyer, I.: Inhibition and activation of histamine methyltransferase by methylated histamines. Hoppe-Seyler's Z. Physiol. Chem., **354**, 1021-1026 (1973)
[4] Francis, D.M.; Thompson, M.F.; Greaves, M.W.: The kinetic properties and reaction mechanism of histamine methyltransferase from human skin. Biochem. J., **187**, 819-828 (1980)
[5] Bowsher, R.R.; Verburg, K.M.; Henry, D.P.: Rat histamine N-methyltransferase. Quantification, tissue distribution, purification, and immunologic properties. J. Biol. Chem., **258**, 12215-12220 (1983)
[6] Harvima, R.J.; Kajander, E.O.; Harvima, I.T.; Fraki, J.E.: Purification and partial characterization of rat kidney histamine-N-methyltransferase. Biochim. Biophys. Acta, **841**, 42-49 (1985)
[7] Fukuda, H.; Yamatodani, A.; Imamura, I.; Maeyama, K.; Watanabe, T.; Wada, H.: High-performance liquid chromatographic determination of histamine N-methyltransferase activity. J. Chromatogr., **567**, 459-464 (1991)
[8] Nishibori, M.; Oishi, R.; Itoh, Y.; Saeki, K.: Purification and partial characterization of histamine N-methyltransferase from bovine brain. Neurochem. Int., **19**, 135-141 (1991)

[9] Scott, M.C.; Guerciolini, R.; Szumlanski, C.; Weinshilboum, R.M.: Mouse kidney histamine N-methyltransferase: assay conditions, biochemical properties and strain variation. Agents Actions, **32**, 194-202 (1991)
[10] Crooks, P.A.; Hassan, S.F.; Benghiat, E.; Hemrick-Luecke, S.K.; Fuller, R.W.: 5'-Thioadenosine derivatives as potent and selective inhibitors of histamine N-methyltransferase. Drug Metab. Drug Interact., **7**, 111-141 (1989)
[11] Rhim, H.; Choi, M.U.: Purification and general characterization of rat brain histamine N-methyltransferase. Hanguk Saenghwahakhoe Chi, **22**, 455-461 (1989)
[12] Gitomer, W.L.; Tipton, K.F.: Purification and kinetic properties of ox brain histamine N-methyltransferase. Biochem. J., **233**, 669-676 (1986)
[13] Fuhr, N.; Kownatzki, E.: Inhibition of rat kidney histamine-N-methyltransferase by biogenic amines. Pharmacology, **32**, 114-120 (1986)
[14] Barth, H.; Schunack, W.; Crombach, M.; Lorenz, W.: Gastric histamine methyltransferase: different methylation rates for enantiomers of side-chain methylated histamine analogues using a highly purified enzyme preparation. Agents Actions, **14**, 346-350 (1984)
[15] Matuszewska, B.; Borchardt, R.T.: Guinea pig brain histamine N-methyltransferase: purification and partial characterization. J. Neurochem., **41**, 113-118 (1983)
[16] Dent, C.; Nilam, F.; Smith, I.R.: Sidechain-modified histamine analogues as substrates for histamine N-methyltransferase. Biochem. Pharmacol., **31**, 2297-2300 (1982)
[17] Beaven, M.A.; Roderick, N.B.: Impromidine, a potent inhibitor of histamine methyltransferase (HMT) and diamine oxidase (DAO). Biochem. Pharmacol., **29**, 2897-2900 (1980)
[18] Thithapandha, A.; Cohn, V.H.: Brain histamine N-methyltransferase purification, mechanism of action, and inhibition by drugs. Biochem. Pharmacol., **27**, 263-271 (1978)
[19] Matuszewska, B.; Borchardt, R.T.: Rat kidney histamine N-methyltransferase: purification and partial characterization. Prep. Biochem., **15**, 145-157 (1985)
[20] Borchardt, R.T.; Matuszewska, B.: S-adenosylmethionine dependent transmethylation of histamine: Purification and partial characterization of guinea pig brain and rat kidney histamine N-methyltransferase. Adv. Biosci., **51**, 163-172 (1985)
[21] Horton, J.R.; Sawada, K.; Nishibori, M.; Zhang, X.; Cheng, X.: Two polymorphic forms of human histamine methyltransferase. Structural, thermal, and kinetic comparisons. Structure, **9**, 837-849 (2001)
[22] Takemura, M.; Imamura, I.; Mizuguchi, H.; Fukui, H.; Yamatodani, A.: Tissue distribution of histamine N-methyltransferase-like immunoreactivity in rodents. Life Sci., **54**, 1059-1071 (1994)
[23] Yamauchi, K.; Sekizawa, K.; Suzuki, H.; Nakazawa, H.; Ohkawara, Y.; Katayose, D.; Ohtsu, H.; Tamura, G.; Shibahara, S.; et al.: Structure and function of human histamine N-methyltransferase: critical enzyme in histamine metabolism in airway. Am. J. Physiol., **267**, L342-L349 (1994)

[24] Tahara, A.; Nishibori, M.; Ohtsuka, A.; Sawada, K.; Sakiyama, J.; Saeki, K.: Immunohistochemical localization of histamine N-methyltransferase in guinea pig tissues. J. Histochem. Cytochem., **48**, 943-954 (2000)
[25] Preuss, C.V.; Wood, T.C.; Szumlanski, C.L.; Raftogianis, R.B.; Otterness, D.M.; Girard, B.; Scott, M.C.; Weinshilboum, R.M.: Human histamine N-methyltransferase pharmacogenetics: common genetic polymorphisms that alter activity. Mol. Pharmacol., **53**, 708-717 (1998)
[26] Donatelli, P.; Marchi, G.; Giuliani, L.; Gustafsson, L.L.; Pacifici, G.M.: Stereoselective inhibition by chloroquine of histamine N-methyltransferase in the human liver and brain. Eur. J. Clin. Pharmacol., **47**, 345-349 (1994)
[27] Pang, Y.P.; Zheng, X.E.; Weinshilboum, R.M.: Theoretical 3D model of histamine N-methyltransferase: Insights into the effects of a genetic polymorphism on enzymatic activity and thermal stability. Biochem. Biophys. Res. Commun., **287**, 204-208 (2001)
[28] Schwelberger, H.G.; Drasche, A.: Structure and expression of porcine histamine N-methyltransferase. Inflamm. Res., **51**, S85-S86 (2002)

Thiol S-methyltransferase 2.1.1.9

1 Nomenclature

EC number
2.1.1.9

Systematic name
S-adenosyl-L-methionine:thiol S-methyltransferase

Recommended name
thiol S-methyltransferase

Synonyms
S-methyltransferase
TMT
methyltransferase, thiol
thiol methyltransferase

CAS registry number
9029-81-6

2 Source Organism

<1> *Rattus norvegicus* [1-4, 9, 12]
<2> *Mus musculus* (high-affinity form and low-affinity form [7]) [3, 7]
<3> *Homo sapiens* (high-affinity form and low-affinity form [6]) [6, 10, 13]
<4> *Oryctolagus cuniculus* [3, 5]
<5> *Cavia porcellus* [3]
<6> *Bos taurus* [3]
<7> *Ovis aries* [3]
<8> *Gallus gallus* (low activity [3]) [3]
<9> *Euglena gracilis* [8]
<10> *Brassica oleracea* (cabbage, several functional isoforms [11]) [11]
<11> *Brassica oleracea* [14]
<12> *Brassica oleracea* [14]

3 Reaction and Specificity

Catalyzed reaction
S-adenosyl-L-methionine + a thiol = S-adenosyl-L-homocysteine + a thioether (<1> mechanism [2])

Reaction type

S-methylation
methyl group transfer

Natural substrates and products

S Additional information <1, 9> (<1> role in detoxification of hydrogen sulfide that is generated by the colonic flora [4]; <9> detoxification of xenobiotic thiols [8]) [4, 8]
P ?

Substrates and products

S 1-methylimidazole-2-thiol + S-adenosyl-L-methionine <1> (Reversibility: ? <1> [2, 4]) [2, 4]
P 1-methyl-2-(methylthio)-1H-imidazole + S-adenosyl-L-homocysteine
S 2,3-dimercaptopropanol + S-adenosyl-L-methionine <1> (Reversibility: ? <1> [2, 4]) [2, 4]
P ? + S-adenosyl-L-homocysteine
S 2-benzimidazole thiol + S-adenosyl-L-methionine <1> (Reversibility: ? <1> [2, 4]) [2, 4]
P ? + S-adenosyl-L-homocysteine
S 2-mercaptobenzothiazole + S-adenosyl-L-methionine <1> (Reversibility: ? <1> [2]) [2]
P S-methyl-2-mercaptobenzothiazole + S-adenosyl-L-homocysteine <1> [2]
S 2-mercaptoethanol + S-adenosyl-L-methionine <1, 3> (Reversibility: ? <1, 3> [1-3, 6, 10]) [1-3, 6, 10]
P S-methyl-2-mercaptoethanol + S-adenosyl-L-homocysteine <1> (<1> i.e. 2-hydroxy-S-methyl-1-thioethane [3]) [3]
S 2-mercaptopropionic acid + S-adenosyl-L-methionine <1, 2> (Reversibility: ? <1, 2> [2-4]) [2-4]
P S-methyl-2-mercaptopropionic acid + S-adenosyl-L-homocysteine
S 2-thioacetanilide + S-adenosyl-L-methionine <1> (Reversibility: ? <1> [2, 4]) [2, 4]
P S-methyl-2-thioacetanilide + S-adenosyl-L-homocysteine
S 3-mercaptopropionic acid methyl ester + S-adenosyl-L-methionine <1> (Reversibility: ? <1> [2]) [2]
P S-methyl-3-mercaptopropionic acid methyl ester + S-adenosyl-L-homocysteine
S 4,4'-thiobisbenzenethiol + S-adenosyl-L-methionine <10> (Reversibility: ? <10> [11]) [11]
P ? + S-adenosyl-L-homocysteine
S 4-chlorothiophenol + S-adenosyl-L-methionine <1> (Reversibility: ? <1> [2, 4]) [2, 4]
P 4-chloro-S-methylphenylmercaptan + S-adenosyl-L-homocysteine
S 4-nitrothiophenol + S-adenosyl-L-methionine <1> (Reversibility: ? <1> [2, 4]) [2, 4]
P 4-nitro-S-methylphenylmercaptan + S-adenosyl-L-homocysteine
S 6-propyl-2-thiouracil + S-adenosyl-L-methionine <1> (Reversibility: ? <1> [2, 4]) [2, 4]

P 6-propyl-S-methyl-2-thiouracil + S-adenosyl-L-homocysteine

S 6-thiopurine + S-adenosyl-L-methionine <1> (Reversibility: ? <1> [2]) [2]

P S-methyl-6-thiopurine + S-adenosyl-L-homocysteine <1> [2]

S H_2S + S-adenosyl-L-methionine <1, 10> (Reversibility: ? <1, 10> [9, 11]) [9, 11]

P methanethiol + S-adenosyl-L-homocysteine <1> [9]

S L-cysteine methyl ester + S-adenosyl-L-methionine <1> (Reversibility: ? <1> [2, 4]) [2, 4]

P S-methyl-L-cysteine methyl ester + S-adenosyl-L-homocysteine

S N-acetyl-L-cysteine + S-adenosyl-L-methionine <1> (Reversibility: ? <1> [2, 4]) [2, 4]

P N-acetyl-S-methyl-L-cysteine + S-adenosyl-L-homocysteine

S O-methylmercaptoethanol + S-adenosyl-L-methionine <1, 2, 4-8> (Reversibility: ? <1, 2, 4-8> [3]) [3]

P 2-methoxy-S-methyl-1-thioethane + S-adenosyl-L-homocysteine

S S-3'-deoxyadenosyl-L-methionine + 2-mercaptoethanol <1> (Reversibility: ? <1> [1]) [1]

P S-3'-deoxyadenosyl-L-homocysteine + S-methyl-2-mercaptoethanol

S S-3-deazaadenosyl-L-methionine + 2-mercaptoethanol <1> (Reversibility: ? <1> [1]) [1]

P S-3-deazaadenosyl-L-homocysteine + S-methyl-2-mercaptoethanol

S S-8-azaadenosyl-L-methionine + 2-mercaptoethanol <1> (Reversibility: ? <1> [1]) [1]

P S-8-azaadenosyl-L-homocysteine + S-methyl-2-mercaptoethanol

S S-N^6-methyladenosyl-L-methionine + 2-mercaptoethanol <1> (Reversibility: ? <1> [1]) [1]

P S-N^6-methyladenosyl-L-homocysteine + S-methyl-2-mercaptoethanol

S S-adenosyl-L-methionine + a thiol <1> (<1> high specificity for S-adenosyl-L-methionine [1]) (Reversibility: ? <1> [1]) [1]

P S-adenosyl-L-homocysteine + a thioether

S S-aristeromycinyl-L-methionine + 2-mercaptoethanol <1> (Reversibility: ? <1> [1]) [1]

P S-aristeromycinyl-L-homocysteine + S-methyl-2-mercaptoethanol

S S-tubercidinyl-L-methionine + 2-mercaptoethanol <1> (Reversibility: ? <1> [1]) [1]

P S-tubercidinyl-L-homocysteine + S-methyl-2-mercaptoethanol

S benzylmercaptan + S-adenosyl-L-methionine <1> (Reversibility: ? <1> [2]) [2]

P benzylmethylsulfide + S-adenosyl-L-homocysteine <1> [2]

S diethylthiocarbamoyl sulfide + S-adenosyl-L-methionine <1> (Reversibility: ? <1> [2, 4]) [2, 4]

P ? + S-adenosyl-L-homocysteine

S dithiothreitol + S-adenosyl-L-methionine <3> (Reversibility: ? <3> [10]) [10]

P ? + S-adenosyl-L-homocysteine

S mercaptoacetic acid + S-adenosyl-L-methionine <1> (<1> i.e. thioglycolic acid [1]) (Reversibility: ? <1> [1-4]) [1-4]
P S-methylmercaptoacetic acid + S-adenosyl-L-homocysteine
S methylmercaptan + S-adenosyl-L-methionine <1> (<1> i.e. methanethiol [3]) (Reversibility: ? <1> [3]) [3]
P dimethylsulfide + S-adenosyl-L-homocysteine
S pentachlorobenzenethiol + S-adenosyl-L-methionine <9> (Reversibility: ? <9> [8]) [8]
P 1,2,3,4,5-pentachloro-6-(methylthio)benzene + S-adenosyl-L-homocysteine
S phenyl sulfide + S-adenosyl-L-methionine <1> (Reversibility: ? <1> [2, 4]) [2, 4]
P ? + S-adenosyl-L-homocysteine
S thiocyanate + S-adenosyl-L-methionine <10> (Reversibility: ? <10> [11]) [11]
P ? + S-adenosyl-L-homocysteine
S thiophenol + S-adenosyl-L-methionine <1, 10> (Reversibility: ? <1, 10> [2, 11]) [2, 11]
P S-methylphenylmercaptan + S-adenosyl-L-homocysteine <1> [2]
S Additional information <1, 2, 9> (<1,2> not: glutathione [1,3]; <1,2,9> not: cysteine [1,3,8]; <1> not: homocysteine [3]) [1, 3, 8]
P ?

Inhibitors

N-ethylmaleimide <1> (<1> more than 50% inhibition at 0.2 mM [1]) [1, 4]
S-3-deazaadenosyl-L-homocysteine <1> [1]
S-adenosyl-L-homocysteine <1> [1, 2]
S-tubercidinyl-L-homocysteine <1> [1]
SKF 525A <2, 3> [7, 10]
cystine <10> (<10> 55% inhibition at 10 mM [11]) [11]
iodoacetamide <10> (<10> 55% inhibition at 10 mM [11]) [11]
p-chloromercuribenzoate <1> (<1> more than 50% inhibition at 0.1 mM [1]) [1, 4]
p-hydroxymercuribenzoic acid <10> (<10> 95% inhibition at 10 mM, reversed in presence of 20 mM Cys [11]) [11]
Additional information <1> (<1> overview: inhibition by analogs of S-adenosyl-L-homocysteine and analogs of S-adenosyl-L-methionine [1]) [1]

Metals, ions

Additional information <1, 3> (<1> no stimulation by: Na^+, K^+, Mn^{2+} [1]; <1,3> no stimulation by: Mg^{2+}, [1, 10]; <3> no stimulation by: Ca^{2+} [10]) [1, 10]

Specific activity (U/mg)

0.01 <1> [2]
Additional information <4> (<4> gas chromatographic method for quantitative assay [5]) [5]

K_m-Value (mM)

0.00054 <1> (4-chlorothiophenol) [2, 4]
0.0011 <1> (phenyl sulfide) [2, 4]
0.0028 <1> (4-nitrothiophenol) [2, 4]
0.012 <1> (diethylthiocarbamoyl sulfide) [2, 4]
0.043 <1> (2-thioacetanilide) [2, 4]
0.06 <1> (S-aristeromycinyl-L-methionine) [1]
0.064 <1> (H_2S) [9]
0.082 <9> (pentachlorobenzenetriol) [8]
0.106 <1> (S-adenosyl-L-methionine) [1]
0.11 <1> (2-benzimidazole thiol) [2, 4]
0.12 <1> (S-3'-deoxyadenosyl-L-methionine) [1]
0.14 <9> (S-adenosyl-L-methionine) [8]
0.154 <1> (S-tubercidinyl-L-methionine) [1]
0.19 <1> (thioglycolic acid) [2, 4]
0.21 <1> (L-cysteine methyl ester) [2, 4]
0.224 <1> (S-3-deazaadenosyl-L-methionine) [1]
0.24 <1> (methanethiol) [9]
0.4 <1> (N-acetyl-L-cysteine) [2, 4]
0.719 <1> (S-8-azaadenosyl-L-methionine) [1]
1 <1> (6-propyl-2-thiouracil) [2, 4]
1.017 <1> (S-N^6-methyladenosyl-L-methionine) [1]
1.2 <1> (2-mercaptopropionic acid) [2, 4]
1.4 <1> (1-methylimidazol-2-thiol) [2, 4]
1.6 <1> (2,3-dimercaptopropanol) [2, 4]
2.37 <1> (2-mercaptoethanol) [1]
4.7 <1> (3-mercaptopropionic acid methyl ester) [2, 4]
8.1 <1> (2-mercaptoethanol) [2, 4]
Additional information <1-3, 9> (<2,3> K_m of high-affinity form and low-affinity form [6, 7]) [2, 4, 6-10]

pH-Optimum

6.5 <9> [8]
7.2-7.6 <3> (<3> high-affinity form [6]) [6]
7.2-8.1 <2> (<2> high-affinity form [7]) [7]
7.5 <1> [2, 4]
7.8-9 <1> (<1> S-methylation of 2-mercaptoethanol [1]) [1]
8 <10> (<10> form V [11]) [11]
8.8 <3> (<3> low-affinity form [6]) [6]
9 <3> [10]
9.2 <2> (<2> low-affinity form [7]) [7]
9.5 <1> (<1> reaction rate increases with pH to at least pH 9.5 [3]) [3]

pH-Range

5.2-6.4 <10> (<10> form II [11]) [11]
5.8-7.8 <10> (<10> form III [11]) [11]
6.6-8.6 <1> (<1> about 50% of maximum activity at pH 6.6 and pH 8.6 [2,4]) [2, 4]

Temperature optimum (°C)

37 <1, 2, 4-8> (<1,2,4-8> assay at [1-4]) [1-4]

4 Enzyme Structure

Molecular weight

21000 <9> (<9> gel filtration [8]) [8]
28000 <1> (<1> gel filtration [2]) [2]

Subunits

? <11, 12> (<11> x * 30000, SDS-PAGE, recombinant enzyme [14]; <12> x * 30000, SDS-PAGE, recombinant enzyme [14]) [14]
monomer <1> (<1> 1 * 27500, SDS-PAGE [2]; <1> 1 * 28000, SDS-PAGE [4]) [2, 4]

5 Isolation/Preparation/Mutation/Application

Source/tissue

erythrocyte <3> [10, 13]
intestine <1> (<1> large bowel mucosa [9]) [4, 9]
kidney <1> [4]
liver <1-8> [1-7]
spleen <1> [4]

Localization

cytoplasm <9> [8]
membrane <3> [10]
microsome <1-8> [1, 3, 4, 6, 7]
soluble <1> (<1> in extracts of liver, kidney, intestine and spleen [4]) [4]

Purification

<1> (partial [1]) [1, 2, 4]
<9> (partial [8]) [8]

Cloning

<11, 12> [14]

Application

medicine <1, 3> (<1> enzyme activity in cholestatic liver [12]; <3> enzyme activity in inflammatory bowel disease [13]) [12, 13]

6 Stability

pH-Stability

7.9 <1> (<1> 30°C, absence of nucleotide substrate, half-life: less than 1 min [1]) [1]

General stability information

<1>, freezing destroys activity [2]

Storage stability

<1>, 4°C, 140 mM NaCl, 10 mM Tris-chloride, pH 7.4, half-life: 4 days, best conditions of storage [1]

References

[1] Borchardt, R.T.; Cheng, C.F.: Purification and characterization of rat liver microsomal thiol methyltransferase. Biochim. Biophys. Acta, **522**, 340-353 (1978)

[2] Weisiger, R.A.; Jakoby, W.B.: Thiol S-methyltransferase from rat liver. Arch. Biochem. Biophys., **196**, 631-637 (1979)

[3] Bremer, J.; Greenberg, D.M.: Enzymic methylation of foreign sulfhydryl compounds. Biochim. Biophys. Acta, **46**, 217-224 (1961)

[4] Weisiger, R.A.; Jakoby, W.B.: Thiol S-methyltransferase. Methods Enzymol., **77**, 257-262 (1981)

[5] Holloway, C.J.; Tegtmeier, F.; Brunner, G.; Trautschold, I.: Gas chromatographic method for the quantitative assay of alkane thiol S-methyltransferase. J. Chromatogr., **202**, 423-429 (1980)

[6] Glauser, T.A.; Kerremans, A.L.; Weinshilboum, R.M.: Human hepatic microsomal thiol methyltransferase. Assay conditions, biochemical properties, and correlation studies. Drug Metab. Dispos., **20**, 247-255 (1992)

[7] Otterness, D.M.; Keith, R.A.; Kerremans, A.L.; Weinshilboum, R.M.: Mouse liver thiol methyltransferase. Assay conditions, biochemical properties, and strain variation. Drug Metab. Dispos., **14**, 680-688 (1986)

[8] Drotar, A.M.; Fall, R.: Methylation of xenobiotic thiols by Euglena gracilis: Characterization of a cytoplasmic thiol methyltransferase. Plant Cell Physiol., **26**, 847-854 (1985)

[9] Weisiger, R.A.; Pinkus, L.M.; Jakoby, W.B.: Thiol S-methyltransferase: Suggested role in detoxication of intestinal hydrogen sulfide. Biochem. Pharmacol., **29**, 2885-2887 (1980)

[10] Weinshilboum, R.M.; Sladek, S.; Klumpp, S.: Human erythrocyte thiol methyltransferase: Radiochemical microassay and biochemical properties. Clin. Chim. Acta, **97**, 59-71 (1979)

[11] Attieh, J.; Sparace, S.A.; Saini, H.S.: Purification and properties of multiple isoforms of a novel thiol methyltransferase involved in the production of volatile sulfur compounds from Brassica oleracea. Arch. Biochem. Biophys., **380**, 257-266 (2000)

[12] Kim, Y.H.; Joo, I.: Arylamine N-methyltransferase and thiol methyltransferase activities in cholestatic rat liver induced by common bile duct ligation. Exp. Mol. Med., **33**, 23-28 (2001)
[13] Roediger, W.E.; Babidge, W.J.: Thiol methyltransferase activity in inflammatory bowel disease. Gut, **47**, 206-210 (2000)
[14] Attieh, J.; Djiana, R.; Koonjul, P.; Etienne, C.; Sparace, S.A.; Saini, H.S.: Cloning and functional expression of two plant thiol methyltransferases: a new class of enzymes involved in the biosynthesis of sulfur volatiles. Plant Mol. Biol., **50**, 511-521 (2002)

Homocysteine S-methyltransferase 2.1.1.10

1 Nomenclature

EC number
2.1.1.10

Systematic name
S-adenosyl-L-methionine:L-homocysteine S-methyltransferase

Recommended name
homocysteine S-methyltransferase

Synonyms
L-homocysteine S-methyltransferase
S-adenosyl-L-methionine:L-homocysteine methyltransferase
S-adenosylmethionine homocysteine transmethylase
S-adenosylmethionine-homocysteine transmethylase
S-adenosylmethionine:homocysteine methyltransferase
S-methylmethionine homocysteine transmethylase
adenosylmethionine transmethylase
adenosylmethionine:homocysteine methyltransferase
homocysteine methylase
homocysteine methyltransferase
homocysteine transmethylase
methylmethionine:homocysteine methyltransferase

CAS registry number
9012-40-2

2 Source Organism

<1> *Escherichia coli* (strain Texas M [1]; strains K12, B, W [2]) [1-3]
<2> *Equus caballus* [2]
<3> *Aerobacter aerogenes* (strain NRRL 199 [2]) [2-4]
<4> *Saccharomyces cerevisiae* [2-5, 7, 8]
<5> *Torulopsis utilis* (strain ATCC 9950 [2]) [2]
<6> *Rattus norvegicus* [3, 4]
<7> *Arabidopsis sp.* (sensitive to methionine inhibition [6]) [6]
<8> *Arabidopsis sp.* (insensitive to methionine inhibition [6]) [6]

3 Reaction and Specificity

Catalyzed reaction

S-adenosyl-L-methionine + L-homocysteine = S-adenosyl-L-homocysteine + L-methionine

Reaction type

methyl group transfer

Natural substrates and products

S S-adenosyl-L-methionine + L-homocysteine <1-8> (<1> biosynthesis of methionine [1-8]) (Reversibility: ? <1-8> [1-8]) [1-8]
P S-adenosyl-L-homocysteine + L-methionine <1-8> [1-8]

Substrates and products

S S-adenosyl-D-methionine + L-homocysteine <3, 4> (<4> low activity [5]) (Reversibility: ? <3, 4> [2, 5]) [2, 5]
P ?
S S-adenosyl-L-ethionine + L-homocysteine <4> (Reversibility: ? <4> [4]) [4]
P S-adenosyl-L-homocysteine + L-ethionine <4> [4]
S S-adenosyl-L-methionine + D-cysteine <7> (<7> very poor substrate [6]) (Reversibility: ? <7> [6]) [6]
P ?
S S-adenosyl-L-methionine + D-homocysteine <4, 7, 8> (Reversibility: ? <4, 7, 8> [4, 6]) [4, 6]
P ?
S S-adenosyl-L-methionine + L-cysteine <7> (<7> very poor substrate [6]) (Reversibility: ? <7> [6]) [6]
P ?
S S-adenosyl-L-methionine + L-homocysteine <1-8> (Reversibility: ? <1-8> [1-8]) [1-8]
P S-adenosyl-L-homocysteine + L-methionine <1-8> [1-8]
S S-adenosyl-L-methionine + L-homocysteine thiolactone <4> (Reversibility: ? <4> [4]) [4]
P ?
S S-methyl-L-methionine + L-homocysteine <1, 3, 4, 6> (<1, 3, 4, 6> S-methyl-L-methionine more effective methyl donor than S-adenosyl-L-methionine [3]) (Reversibility: ? <1, 3, 4, 6> [2]) [2, 3]
P ?

Inhibitors

EDTA <4, 7> (<7> 26% inhibition with 1 mM [6]; <4> complete inhibition, reversible by addition of excess of Zn^{2+} [8]) [6, 8]
L-methionine <1, 4, 7> (<1,4> product inhibition [1,4]; <4> no inhibition with D-isomer [4]) [1, 4, 6]
N-hydroxymethyl-DL-methionine <4> [4]
S-adenosyl-L-homocysteine <1, 4> (<1,4> product inhibition [1,4]) [1, 4]

Additional information <7, 8> (<7,8> two isoforms of Arabidopsis enzyme with isoform 1 found to be inhibited by methionine and EDTA while isoform 2 is not [6]) [6]

Activating compounds

2-mercaptoethanol <4> (<4> increases activity at low substrate concentrations [4]) [4]
cystathione <1> [1]
cysteine <1> [1]
homocysteine <1> (<1> induces reaction [1]) [1]
homoserine <1> [1]
Additional information <3-5> (<3-5> not inducible by substrates [2]) [2]

Metals, ions

Cd^{2+} <4> (<4> activating [4]; <4> can replace Zn^{2+} [8]) [4, 8]
Zn^{2+} <4> (<4> activating [4]; <4> required for activity [8]) [4, 8]
Additional information <7, 8> (<7,8> zinc-binding motif identified [6]) [6]

Specific activity (U/mg)

0.0008 <6> (<6> kidney, S-methyl-L-methionine as methyl donor [3]) [3]
0.0022 <6> (<6> liver, S-adenosyl-L-methionine as methyl donor [3]) [3]
0.0087 <6> (<6> liver, S-methyl-L-methionine as methyl donor [3]) [3]
0.011 <8> (<8> recombinant enzyme with D-homocysteine [6]) [6]
0.012 <1> (<1> S-methyl-L-methionine as methyl donor [3]) [3]
0.013 <8> (<8> recombinant enzyme with L-homocysteine [6]) [6]
0.014 <4> (<4> S-methyl-L-methionine as methyl donor [3]) [3]
0.017 <3> (<3> S-methyl-L-methionine as methyl donor [3]) [3]
0.021 <7> (<7> recombinant enzyme with D-homocysteine [6]) [6]
0.115 <7> (<7> recombinant enzyme with L-homocysteine [6]) [6]
1.37 <4> (<4> purified enzyme, S-adenosyl-L-methionine as methyl donor [4]) [4]

K_m-Value (mM)

0.035 <8> (L-homocysteine) [6]
0.077 <7> (L-homocysteine) [6]
0.225 <8> (S-adenosyl-L-methionine) [6]
1.95 <7> (S-adenosyl-L-methionine) [6]

pH-Optimum

6.5-7 <3-5> [2]
7.5 <7, 8> [6]
7.5-8 <4> [4]

Temperature optimum (°C)

30-35 <3-5> [2]

4 Enzyme Structure

Molecular weight

30000 <4> (<4> from sedimentation coefficient [4]) [4]
36000 <7, 8> (<7,8> gel filtration [6]; <7> calculated from DNA-sequence [6]) [6]
36400 <8> (<8> calculated from DNA sequence [6]) [6]

Subunits

monomer <7, 8> (<7,8> gel filtration data combined with calculated MW from DNA-sequence [6]) [6]

5 Isolation/Preparation/Mutation/Application

Source/tissue

kidney <6> (<6> low activity [3]) [3]
liver <2, 6> (<6> high activity [3]) [2, 3]
Additional information <6> (<6> no activity detected in brain and heart [3]) [3]

Purification

<4> [4]

Cloning

<4> (in Escherichia coli [7]) [7]
<7> (in Escherichia coli and yeast [6]) [6]
<8> (in Escherichia coli and yeast [6]) [6]

Application

medicine <4> (<4> determination of total homocysteine in plasma and serum [8]) [8]

6 Stability

Storage stability

<7, 8>, -80°C, after freezing in liquid N_2, no loss of activity [6]

References

[1] Balish, E.; Shapiro, S.K.: Methionine biosynthesis in Escherichia coli: induction and repression of methylmethionine(or adenosylmethionine):homocysteine methyltransferase. Arch. Biochem. Biophys., **119**, 62-68 (1967)
[2] Shapiro, S.K.: Adenosylmethionine-homocysteine transmethylase. Biochim. Biophys. Acta, **29**, 405-409 (1958)

[3] Shapiro, S.K.; Yphantis, D.A.: Assay of S-methylmethionine and S-adenosylmethionine homocysteine transmethylases. Biochim. Biophys. Acta, **36**, 241-244 (1959)
[4] Shapiro, S.K.: S-Adenosylmethionine:L-homocysteine S-methyltransferase (Saccharomyces cerevisiae). Methods Enzymol., **17B**, 400-405 (1971)
[5] Nakamura, K.D.; Schlenk, F.: The activity of adenosyl-D-methionine and adenosyl-2-methylmethionine in transmethylations. Arch. Biochem. Biophys., **177**, 170-175 (1976)
[6] Ranocha, P.; Bourgis, F.; Ziemak, M.J.; Rhodes, D.; Gage, D.A.; Hanson, A.D.: Characterization and functional expression of cDNAs encoding methionine-sensitive and -insensitive homocysteine S-methyltransferases from Arabidopsis. J. Biol. Chem., **275**, 15962-15968 (2000)
[7] Thomas, D.; Becker, A.; Surdin-Kerjan, Y.: Reverse methionine biosynthesis from S-adenosylmethionine in eukaryotic cells. J. Biol. Chem., **275**, 40718-40724 (2000)
[8] Matsuyama, N.; Yamaguchi, M.; Toyosato, M.; Takayama, M.; Mizuno, K.: New enzymatic colorimetric assay for total homocysteine. Clin. Chem., **47**, 2155-2157 (2001)

Magnesium protoporphyrin IX methyltransferase 2.1.1.11

1 Nomenclature

EC number
2.1.1.11

Systematic name
S-adenosyl-L-methionine:magnesium-protoporphyrin-IX O-methyltransferase

Recommended name
magnesium protoporphyrin IX methyltransferase

Synonyms
(-)-S-adenosyl-L-methionine:magnesium-protoporphyrin IX methyltransferase
MgPIXMT protein <5> [8]
S-adenosyl-L-methionine:Mg protoporphyrin methyltransferase
magnesium protoporphyrin IX methyltransferase
magnesium protoporphyrin O-methyltransferase
methyltransferase, magnesium protoporphyrin

CAS registry number
9029-82-7

2 Source Organism

<1> *Rhodopseudomonas sphaeroides* [1, 3, 9]
<2> *Triticum aestivum* [2, 5, 7]
<3> *Euglena gracilis* [4]
<4> *Hordeum vulgare* [6, 11, 12]
<5> *Arabidopsis thaliana* [8]
<6> *Spinacia oleracea* [8]
<7> *Rhodobacter capsulatus* [10]
<8> *Synechocystis sp.* (PCC6803 [13]) [13]

3 Reaction and Specificity

Catalyzed reaction
S-adenosyl-L-methionine + magnesium protoporphyrin IX = S-adenosyl-L-homocysteine + magnesium protoporphyrin IX 13-monomethyl ester (<2>,

ping-pong mechanism [2]; <1>, equilibrium-ordered bi bi mechanism [3]; <3>, random bi-bi mechanism with 2 dead-end ternary complexes [4]; <8>, ternary complex is likely to be formed either by a random or ordered addition of substrates [13])

Reaction type

methyl group transfer

Natural substrates and products

S Additional information <1, 2, 4> (<2>, chlorophyll-forming enzyme [7]; <1>, enzyme is involved in bacteriochlorophyll biosynthesis [9]; <4>, enzyme is involved in chlorophyll biosynthesis [12]) [7, 9, 12]
P ?

Substrates and products

S S-adenosyl-L-methionine + calcium protoporphyrin <1> (Reversibility: ? <1> [1]) [1]
P S-adenosyl-L-homocysteine + calcium protoporphyrin monomethyl ester <1> [1]
S S-adenosyl-L-methionine + magnesium deuteroporphyrin <1> (Reversibility: ? <1> [1]) [1]
P S-adenosyl-L-homocysteine + magnesium deuteroporphyrin monomethyl ester <1> [1]
S S-adenosyl-L-methionine + magnesium mesoporphyrin <1-8> (Reversibility: ? <1-8> [1-13]) [1-13]
P S-adenosyl-L-homocysteine + magnesium mesoporphyrin monomethyl ester <1> [1]
S S-adenosyl-L-methionine + magnesium protoporphyrin <1-8> (<4>, transfer of methyl group from S-adenosylmethionine to the propanoic acid side chain on the third pyrrole ring of magnesium protoporphyrin IX [12]) (Reversibility: ? <1-8> [1-13]) [1-13]
P S-adenosyl-L-homocysteine + magnesium protoporphyrin monomethyl ester <1> [1-13]
S S-adenosyl-L-methionine + protoporphyrin <1> (Reversibility: ? <1> [1]) [1]
P S-adenosyl-L-homocysteine + protoporphyrin monomethyl ester <1> [1]
S S-adenosyl-L-methionine + zinc protoporphyrin <1> (Reversibility: ? <1> [1]) [1]
P S-adenosyl-L-homocysteine + zinc protoporphyrin monomethyl ester <1> [1]

Inhibitors

1,10-phenanthroline <1> [1]
8-hydroxyquinoline <1> [1]
$AgNO_3$ <1> [1]
$MnCl_2$ <1> [1]
PCMB <1> [1]

S-adenosyl-L-homocysteine <1, 3, 4> (<1>, non-competitive inhibitor of Mg protoporphyrin [3]; <3,4>, competitive with S-adenosyl-L-methionine [4,6]) [1, 3, 4, 6]
S-adenosylethionine <3, 4> (<3,4>, competitive with S-adenosyl-L-methionine [4,6]) [4, 6]
$ZnSO_4$ <1> [1]
chlorophyllide a <2> [7]
diphenylthiocarbazone <1> [1]
ferric protoporphyrin <1> [1]
ferrous protoporphyrin <1> [1]
hemin <4> (<4>, noncompetitive with respect to magnesium protoporphyrin [6]) [6]
iodoacetic acid <1> [1]
magnesium protoporphyrin monomethyl ester <3> (<3>, noncompetitive inhibitor of S-adenosyl-L-methionine [4]) [4]
magnesium protoporphyrin IX monomethyl ester <2> [7]
manganic protoporphyrin <1> [1]
manganous protoporphyrin <1> [1]
pheophorbide a <2> (<2>, stronger inhibitor than the corresponding Mg-chelate [7]) [7]
protochlorophyllide a <2> [7]
protopheophorbide <2> (<2>, stronger inhibitor than the corresponding Mg-chelate [7]) [7]
sinefungin <4> (<4>, selective inhibitor and potentially usefull tool for chlorophyll biosynthetic studies [12]) [12]
sodium azide <1> [1]
sodium diethyldithiocarbamate <1> [1]

Activating compounds
thiol compounds <4> (<4>, stimulate [6]) [6]

K_m-Value (mM)
0.036 <2> (magnesium protoporphyrin IX, <2>, pH 7.7, 25°C [5]) [5]
0.038 <8> (S-adenosyl-L-methionine) [13]
0.039 <4> (S-adenosyl-L-methionine) [6]
0.048 <2> (S-adenosyl-L-methionine, <2>, pH 7.7, 25°C [5]) [5]
0.048 <4> (magnesium protoporphyrin IX) [6]

pH-Optimum
7.7 <2> (<2>, phosphate buffer [5]) [5]
8.4 <1> [1]

pH-Range
6.7-8.8 <1> (<1>, pH 6.7: about 60% of maximal activity, pH 8.8: about 85% of maximal activity [1]) [1]

Temperature optimum (°C)
37 <2> [5]

4 Enzyme Structure

Subunits

? <1> (<1>, x * 25681, calculation from nucleotide sequence [9]) [9]

Posttranslational modification

proteolytic modification <5> (<5>, the full-length polypeptide exhibits a chloroplast transit peptide, is processed during import into the chloroplast [8]) [8]

5 Isolation/Preparation/Mutation/Application

Source/tissue

leaf <4> (<4>, etiolated [11]; <4>, greening [12]) [11, 12]
seedling <2> (<2>, etiolated [5]) [5]
Additional information <1> (<1>, no activity can be detected in extracts of Rhodopseudomonas sphaeroides grown under high aeration in the dark [1]) [1]

Localization

chloroplast envelope <5> (<5>, dual localization in chloroplast envelope membranes as well as in thylakoids [8]) [8]
chromatophore <1> (<1>, firmly bound [1]) [1, 3]
etioplast prolamellar body <4> (<4>, containing photoactive protochlorophyllide [11]) [11]
membrane <5> (<5>, embedded within one leaflet of the membrane [8]) [8]
ribosome <2> (<2>, enzyme is synthesized in the 80S cytoplasmic ribosomes [7]) [7]
thylakoid <4, 5> (<5>, dual localization in chloroplast envelope membranes as well as in thylakoids [8]) [8, 11]

Purification

<1> [3]
<3> [4]
<4> [6]
<8> [13]

Cloning

<1> (overexpression in Escherichia coli [9]) [9]
<5> (expression in Escherichia coli [8]) [8]
<7> (expression in Escherichia coli [10]) [10]

References

[1] Gibson, K.D.; Neuberger, A.; Tait, G.H.: Studies on the biosynthesis of porphyrin and bacteriochlorophyll by Rhodopseudomonas spheroides. Biochem. J., **88**, 325-334 (1963)

[2] Yee, W.C.; Eglsaer, S.J.; Richards, W.R.: Confirmation of a ping-pong mechanism for S-adenosyl-L-methionine:magnesium protoporphyrin methyltransferase of etiolated wheat by an exchange reaction. Biochem. Biophys. Res. Commun., **162**, 483-490 (1989)
[3] Hinchigeri, S.B.; Richards, W.R.: The purification and reaction mechanism of S-adenosyl-L-methionine:magnesium protoporphyrin methyltransferase from Rhodopseudomonas sphaeroides. Photosynthetica, **16**, 554-560 (1982)
[4] Hinchigeri, S.B.; Nelson, D.W.; Richards, W.R.: The reaction mechanism of S-adenosyl-L-methionine:magnesium protoporphyrin methyltransferase from Euglena gracilis. Photosynthetica, **18**, 168-178 (1984)
[5] Ellsworth, R.K.; Dullaghan, J.P.: Activity and properties of (-)-S-adenosyl-L-methionine:magnesium-protoporphyrin IX methyltransferase in crude homogenates from wheat seedlings. Biochim. Biophys. Acta, **268**, 327-333 (1972)
[6] Hinchigeri, S.B.; Chan, J.C.S.; Richards, W.R.: Purification of S-adenosyl-L-methionine:magnesium protoporphyrin methyltransferase by affinity chromatography. Photosynthetica, **15**, 351-359 (1981)
[7] Ellsworth, R.K.; St. Pierre, M.E.: Biosynthesis and inhibition of (-)-S-adenosyl-L-methionine:magnesium protoporphyrin methyltransferase of wheat. Photosynthetica, **10**, 291-301 (1976)
[8] Block, M.A.; Tewari, A.K.; Albrieux, C.; Marechal, E.; Joyard, J.: The plant S-adenosyl-L-methionine:Mg-protoporphyrin IX methyltransferase is located in both envelope and thylakoid chloroplast membranes. Eur. J. Biochem., **269**, 240-248 (2002)
[9] Gibson, L.C.D.; Hunter, C.N.: The bacteriochlorophyll biosynthesis gene, bchM, of Rhodobacter sphaeroides encodes S-adenosyl-L-methionine:Mg protoporphyrin IX methyltransferase. FEBS Lett., **352**, 127-130 (1994)
[10] Bollivar, D.W.; Jiang, Z.Y.; Bauer, C.E.; Beale, S.I.: Heterologous expression of the bchM gene product from Rhodobacter capsulatus and demonstration that it encodes S-adenosyl-l-methionine:Mg-protoporphyrin IX methyltransferase. J. Bacteriol., **176**, 5290-5296 (1994)
[11] Averina, N.; Rassadina, V.; Fradkin, L.: Native state, energetic interaction of chlorophyll precursors and intraplastid location of S-adenosyl-L-methionine: Mg-protoporphyrin IX methyltransferase in etiolated leaves. Indian J. Exp. Biol., **40**, 192-201 (2002)
[12] Vothknecht, U.C.; Willows, R.D.; Kannangara, C.G.: Sinefungin inhibits chlorophyll synthesis by blocking the S-adenosylmethionine:Mg-protoporphyrin IX O-methyltransferase in greening barley leaves. Plant Physiol. Biochem., **33**, 759-763 (1995)
[13] Shepherd, M.; Reid, J.D.; Hunter, C.N.: Purification and kinetic characterisation of the magnesium protoporphyrin IX methyltransferase from Synechocystis PCC6803. Biochem. J., **18**, 1-33 (2002)

Methionine S-methyltransferase 2.1.1.12

1 Nomenclature

EC number
2.1.1.12

Systematic name
S-adenosyl-L-methionine:L-methionine S-methyltransferase

Recommended name
methionine S-methyltransferase

Synonyms
MMT
S-adenosyl methionine:methionine methyl transferase
S-adenosylmethionine transmethylase
S-adenosylmethionine-methionine methyltransferase
methionine methyltransferase
methyltransferase, methionine S-

CAS registry number
9027-77-4

2 Source Organism

<1> *Triticum aestivum* [1]
<2> *Lemna paucicostata* [2]
<3> *Glycine max* [2]
<4> *Daucus carota* [2]
<5> *Wollastonia biflora* [3]
<6> *Zea mays* [3]
<7> *Trifolium repens* [3]
<8> *Lactuca sativa* [3]
<9> *Brassica oleracea* [3]
<10> *Hordeum vulgare* [4]
<11> *Arabidopsis sp.* [5]
<12> *Brassica juncea* [5]
<13> *Hordeum vulgare* [6]

3 Reaction and Specificity

Catalyzed reaction

S-adenosyl-L-methionine + L-methionine = S-adenosyl-L-homocysteine + S-methyl-L-methionine (<5>, ordered bi bi mechanism with S-adenosylmethionine the first reactant to bind and S-adenosylhomocysteine the last product to be released [3])

Reaction type

methyl group transfer

Natural substrates and products

S S-adenosyl-L-methionine + L-methionine <2> (<2>, enzyme is involved in S-methylmethionine cycle [2]) (Reversibility: ? <2> [2]) [2]
P S-adenosyl-L-homocysteine + S-methyl-L-methionine
S S-adenosyl-L-methionine + selenomethionine <11, 12> (<11,12>, key enzyme in volatilization of selenium [5]) (Reversibility: ? <11, 12> [5]) [5]
P S-adenosyl-L-homocysteine + selenium-methyl-L-seleniummethionine <11, 12> [5]

Substrates and products

S S-adenosyl-L-methionine + L-methionine <1-13> (Reversibility: ? <1-13> [1-6]) [1-6]
P S-adenosyl-L-homocysteine + S-methyl-L-methionine <1> [1]
S S-adenosyl-L-methionine + selenomethionine <11, 12> (Reversibility: ? <11, 12> [5]) [5]
P S-adenosyl-L-homocysteine + selenium-methyl-L-seleniummethionine <11, 12> [5]

Inhibitors

3-(methylthio)propanol <5> [3]
3-(methylthio)propionate <5> [3]
3-(methylthio)propionate methyl ester <5> [3]
3-(methylthio)propylamine <5> [3]
3-dimethylsulfoniopropionaldehyde <5> [3]
D-Met <5> [3]
DL-2-hydroxy-4-(methylthio)butyrate <5> [3]
Gly-Met <5> [3]
L-Met ethyl ester <5> [3]
L-Met methyl ester <5> [3]
L-Met *p*-nitroanilide <5> [3]
L-ethionine <5> [3]
L-methioninamide <5> [3]
L-methioninol <5> [3]
N-2,4-dinitrophenyl-DL-Met <5> [3]
N-formyl-L-Met <5> [3]
NEM <5> (<5>, 0.5 mM, 99% inhibition [3]) [3]
S-adenosyl-L-homocysteine <5> [3]
S-methyl-L-methionine <5> [3]

p-hydroxymercuribenzoate <5> (<5>, 0.1 mM; 97% inhibition [3]) [3]
seleno-L-methionine <5> [3]

Metals, ions
Zn^{2+} <1> (<1>, enzyme requires 0.1 mM Zn^{2+} or 1 mM Mn^{2+} [1]) [1]

Specific activity (U/mg)
0.506 <10> [4]
Additional information <5, 13> (<13>, quick, simple, and specific method for measuring enzyme activity [6]) [3, 6]

K_m-Value (mM)
0.034 <5> (S-adenosyl-L-methionine) [3]
0.14 <5> (L-Met) [3]

K_i-Value (mM)
0.013 <5> (S-adenosyl-L-homocysteine) [3]
0.226 <5> (S-methyl-L-methionine) [3]

pH-Optimum
6.5 <1> [1]
7.2 <5> [3]

pH-Range
5.7-7.3 <1> (<1>, about 60% of activity maximum at pH 5.7 and pH 7.3 [1]) [1]
5.8-9 <5> (<5>, about 50-55% of optimal activity at pH 5.8 and 9.0, only 1% of optimal activity at pH 5.0 [3]) [3]

4 Enzyme Structure

Molecular weight
450000 <5> (<5>, gel filtration [3]) [3]

Subunits
? <10> (<10>, x * 115000, SDS-PAGE [4]) [4]
tetramer <5> (<5>, 4 * 115000, SDS-PAGE [3]) [3]

5 Isolation/Preparation/Mutation/Application

Source/tissue
aleurone layer <10> [4]
germ <1> [1]
kernel <10> (<10>, activity does not change significantly during the first 2 d of germination, increases during the 3rd and 4th day [4]) [4]
leaf <5, 12> [3, 5]
plant <2> [2]
root <12> [5]

scutellum <10> [4]
shoot <10> [4]
suspension culture <3, 4> [2]
Additional information <10> (<10>, no activity in the root or endosperm [4]) [4]

Purification
<1> (partial [1]) [1]
<5> [3]
<10> [4]
<13> [6]

Cloning
<11> (expression in Escherichia coli [5]) [5]

6 Stability

Temperature stability
4 <5> (<5>, complete inactivation after 1 week [3]) [3]
25 <5> (<5>, complete inactivation after 24 h [3]) [3]

Storage stability
<5>, -20°C, complete inactivation after flash freezing in liquid N_2 [3]

References

[1] Karr, D.; Tweto, J.; Albersheim, P.: S-Adenosyl methionine:methionine methyl transferase from wheat germ. Arch. Biochem. Biophys., **121**, 732-738 (1967)
[2] Mudd, S.H.; Datko, A.H.: The S-methylmethionine cycle in Lemna paucicostata. Plant Physiol., **93**, 623-630 (1990)
[3] James, F.; Nolte, K.D.; Hanson, A.D.: Purification and properties of S-adenosyl-L-methionine:L-methionine S-methyltransferase from Wollastonia biflora leaves. J. Biol. Chem., **270**, 22344-22350 (1995)
[4] Pimenta, M.J.; Kaneta, T.; Larondelle, Y.; Dohmae, N.; Kamiya, Y.: S-Adenosyl-L-methionine:L-methionine S-methyltransferase from germinating barley. Purification and localization. Plant Physiol., **118**, 431-438 (1998)
[5] Tagmount, A.; Berken, A.; Terry, N.: An essential role of S-adenosyl-L-methionine:L-methionine S-methyltransferase in selenium volatilization by plants. Methylation of selenomethionine to selenium-methyl-L-selenium-methionine, the precursor of volatile selenium. Plant Physiol., **130**, 847-856 (2002)
[6] Pimenta, M.J.; Vandercammen, A.V.; Dufour, J.P.; Larondelle, Y.: Determination of S-adenosyl-L-methionine:L-methionine S-methyltransferase activity by selective adsorption of [methyl-^{3}H]S-adenosylmethionine onto activated charcoal. Anal. Biochem., **225**, 167-169 (1995)

Methionine synthase 2.1.1.13

1 Nomenclature

EC number
2.1.1.13

Systematic name
5-methyltetrahydrofolate:L-homocysteine S-methyltransferase

Recommended name
methionine synthase

Synonyms
5-methyltetrahydrofolate-homocysteine S-methyltransferase
5-methyltetrahydrofolate-homocysteine transmethylase
B_{12} N^5-methyltetrahydrofolate homocysteine methyltransferase
N-methyltetrahydrofolate:L-homocysteine methyltransferase
N^5-methyltetrahydrofolate methyltransferase
N^5-methyltetrahydrofolate-homocysteine cobalamin methyltransferase
N^5-methyltetrahydrofolic-homocysteine vitamin B_{12} transmethylase
methionine synthase
methionine synthetase
methyltetrahydrofolate-homocysteine vitamin B_{12} methyltransferase
methyltransferase, methyltetrahydrofolate-homocysteine
tetrahydrofolate methyltransferase
tetrahydropteroylglutamate methyltransferase
tetrahydropteroylglutamic methyltransferase
vitamin B_{12} methyltransferase

CAS registry number
9033-23-2

2 Source Organism

<1> *Cercopithecus aethiops* (African Green monkey, Cos-7, i.e. SV40-transformed kidney cells [41]) [41]
<2> *mackerel* [44]
<3> *Bos taurus* [6, 12, 23, 44]
<4> *Escherichia coli* (B [1,7,10,15,17,24,28]; K12 [21,48]) [1, 3-5, 7, 8, 10-13, 15, 17, 19, 21, 24, 25, 28, 29, 31, 33-35, 37, 38, 40, 42, 46-48]
<5> *Sus scrofa* [2, 9, 12, 16, 32, 44]
<6> *Gallus gallus* [12, 44]

<7> *Rattus norvegicus* (HTC and Phi-1 tumour cells [43]) [12, 22, 24, 25, 30, 36, 39, 43, 44]
<8> *Homo sapiens* (HeLa cells, HepG2 cells and transformed kidney cells [41]) [12, 19, 24, 25, 26, 28, 41]
<9> *Phaseolus vulgaris* [14]
<10> *Aerobacter aerogenes* [12]
<11> *Salmonella typhimurium* [12]
<12> *Clostridium pyrenoidosa* [12]
<13> *Rhodopseudomonas sphaeroides* [12]
<14> *Ochromonas malhamensis* [12]
<15> *Euglena gracilis* [18, 45]
<16> *Chromatium sp.* (D [20]) [20]
<17> *Rhodospirillum rubrum* [20]
<18> *Propionibacterium shermanii* [25]
<19> *Oryctolagus cuniculus* [27]
<20> *Limanda ferruginea* (yellowtail) [44]

3 Reaction and Specificity

Catalyzed reaction

5-methyltetrahydrofolate + L-homocysteine = tetrahydrofolate + L-methionine (<3-8, 10-14> mechanism [7, 11, 12, 17, 24, 32, 33, 39, 48]; <4> review, detailed mechanism [29]; <5> ordered sequential mechanism [32]; <4> remethylation of inactivated enzyme and interaction with flavodoxin [34, 35]; <4> enzyme exists in two different conformations that interconvert in the cob(II)alamin oxidation state, control of reactivity [37])

Reaction type

methyl group transfer

Natural substrates and products

S 5-methyltetrahydrofolate + L-homocysteine <5> (<5> methionine biosynthesis [2,16]) [2, 16]
P tetrahydrofolate + L-methionine

Substrates and products

S 5-methyl-5,6,7,8-tetrahydropteroylheptaglutamate + homocysteine <3> (Reversibility: ? <3> [23]) [23]
P ?
S 5-methyl-5,6,7,8-tetrahydropteroylpentaglutamate + L-homocysteine <3, 7> (Reversibility: ? <3, 7> [22, 23]) [22, 23]
P ?
S 5-methyltetrahydrofolate + 2-mercaptoethanol <4> (Reversibility: ? <4> [12]) [12]
P S-methylmercaptoethanol + tetrahydrofolate <4> [12]
S N^5-methyltetrahydrofolate + L-homocysteine <3, 4, 7, 15> (Reversibility: ? <3, 4, 7, 15> [6-8, 17, 18, 21-23]) [6-8, 17, 18, 21-23]

P tetrahydrofolate + L-methionine <3> [6]
S N^5-methyltetrahydropteroylheptaglutamate + L-homocysteine <3> (Reversibility: ? <3> [23]) [23]
P tetrahydropteroylheptaglutamate + L-methionine
S N^5-methyltetrahydropteroylmonoglutamate + L-homocysteine <1-8, 10-14> (Reversibility: ? <1-8, 10-14> [1-3, 12, 14, 16, 22, 23]) [1-3, 12, 14, 16, 22, 23]
P tetrahydropteroylmonoglutamate + L-methionine <5> [2]
S N^5-methyltetrahydropteroylpentaglutamate + L-homocysteine <3, 7> (Reversibility: ? <3, 7> [22, 23]) [22, 23]
P tetrahydropteroylpentaglutamate + L-methionine
S N^5-methyltetrahydropteroyltriglutamate + L-homocysteine <1-8, 10-14> (Reversibility: ? <1-8, 10-14> [1, 2, 9, 12, 16, 23]) [1, 2, 9, 12, 16, 23]
P tetrahydropteroyltriglutamate + L-methionine <5> [2]
S S-adenosyl-L-methionine + L-homocysteine <4> (Reversibility: ? <4> [12]) [12]
P methionine + S-adenosyl-L-homocysteine <4> [12]
S S-adenosyl-L-methionine + L-selenohomocysteine <4> (Reversibility: ? <4> [46]) [46]
P selenomethionine + S-adenosyl-L-homocysteine <4> [46]
S S-adenosyl-L-methionine + tetrahydrofolate <4> (Reversibility: ? <4> [12]) [12]
P 5-methyltetrahydrofolate + S-adenosyl-L-homocysteine <4> [12]
S Additional information <4> (<4> S-adenosylhomocysteine replaces homocysteine as methyl acceptor to a limited extent [1]; <4> binding of the folate substrate [7]) [1, 7]
P ?

Cofactors/prosthetic groups

ATP <4> (<4> required [1]) [1]
FAD <5> (<5> enzyme requires reduced FAD and S-adenosyl-L-methionine [2]) [2]
S-adenosyl-L-methionine <3-8, 10-17> (<5> bacterial enzyme requires reduced FAD and S-adenosyl-L-methionine [2]; <3-8, 10-15, 17> required [6, 11, 12, 16, 18, 20, 32]; <5> partially stimulates [9]; <4> role of S-adenosyl-L-methionine [12]) [2, 6, 9, 12, 16, 18, 20]
cobinamide <4, 7, 8> (<4,7,8> more active than cyanocobalamin with rat liver enzyme, but inhibitory with E. coli enzyme [25]) [25]
vitamin B_{12} <3-8, 10-19> (<8> i.e. cobalamin, 1 mol per mol of enzyme [26]; <7,8> methylcobalamin tightly bound and required [24]; <4> 1 mol per mol of protein, MW 186000 [21]; <3-8,10-17,19> enzyme contains a derivative of vitamin B_{12} as prosthetic group [1-5,7-10,12,16-21,24-27]; <4,7,8> mechanism of conversion of apo- to holomethionine synthase by various forms of cobalamin [24]) [1-5, 7-10, 12, 16-21, 24-27]
Additional information <3-8, 10-15> (<3-8,10-15> absolute requirement for a reducing system [9,11,12,18]; <4> enzyme requires another protein fraction, ATP, Mg^{2+}, FAD and a NADH-generating system [1,3]) [1, 3, 9, 12, 18]

Activating compounds

K^+ <4> (<4> at 67 mM 1.4 fold activity [8]) [8]
Li^+ <4> (<4> at 67 mM 1.3 fold activity [8]) [8]
NH_4^+ <4> (<4> at 67 mM 1.5 fold activity [8]) [8]
Na^+ <4> (<4> at 67 mM 1.5 fold activity [8]) [8]
S-adenosylmethionine <4> (<4> at about 0.0001 mM [8]) [8]
TrisHCl <4> (<4> at 67 mM 1.2 fold activity [8]) [8]
cadaverine <7> (<7> up to 40% stimulation at 1 mm [36]) [36]
cyanocobalamin <4, 7, 8> (<4,7,8> supports methyltransferase in extracts of animal tissues and E. coli [25]) [25]
putrescine <7> (<7> up to 50% stimulation at 1 mm [36]) [36]
spermidine <7> (<7> up to 2.5 fold stimulation at 1 mm [36]) [36]
spermine <7> (<7> up to 4 fold stimulation at 1 mm [36]) [36]
vitamin B_{12} <1, 8> (<1,8> 2 to 14 fold stimulation after addition to media due to posttranscriptional regulation [41]) [41]

Metals, ions

Co^{2+} <5> [32]
Cu <4> (<4> 1:1 stoichiometry with Co [11]) [11]
Mg^{2+} <4> (<4> required [1,3]; <4,9> activates [8,14]) [1, 3, 8, 14]
Zn^{2+} <4> (<4> 0.9 equiv. of zinc per mole, removal results in inactive enzyme [38]; <4> characterization of zinc site [47]) [38, 47]
Additional information <5> (<5> no other metals except for Co [32]) [32]

Turnover number (min^{-1})

1.3 <4> (S-adenosyl-L-methionine; <4> with tetrahydrofolate as cosubstrate [12]) [12]
2.2 <4> (S-adenosyl-L-methionine; <4> with L-homocysteine as cosubstrate [12]) [12]
500-780 <4> (S-methyltetrahydrofolate) [12]
Additional information <8> [26]

Specific activity (U/mg)

3.8 <4> [21]
9.3-11.7 <4> [11]
Additional information <2-14, 16, 17, 20> (<4, 5, 8, 9> assay method [12, 14, 16, 17, 19, 31]; <2, 3, 5-7, 20> assay using HPLC with fluorescence detection [44]) [2, 6, 9, 12, 14, 16, 17, 19, 20, 31, 44]

K_m-Value (mM)

0.00065 <4> (S-adenosyl-L-methionine) [3]
0.0016 <4> (S-adenosyl-L-methionine) [12, 17]
0.0026 <7> (L-homocysteine) [39]
0.004 <7> (5-methyl-5,6,7,8-tetrahydropteroylpentaglutamate) [22]
0.013 <7> (5-methyl-5,6,7,8-tetrahydropteroylmonoglutamate) [22]
0.016 <4> (L-homocysteine) [12, 17]
0.017 <4> (L-selenohomocysteine) [46]
0.0223 <3> (5-methyl-5,6,7,8-tetrahydropteroylheptaglutamate) [23]
0.0244 <3> (5-methyl-5,6,7,8-tetrahydropteroyltriglutamate) [23]

0.0277 <3> (5-methyl-5,6,7,8-tetrahydropteroylpentaglutamate) [23]
0.03 <4> (methyltetrahydrofolate) [12]
0.035 <4> (S-methyltetrahydrofolate) [7]
0.06 <4> (methyltetrahydrofolate) [17]
0.069 <4> (L-homocysteine) [46]
0.0731 <3> (5-methyl-5,6,7,8-tetrahydropteroylmonoglutamate) [23]
0.089 <7> (N^5-methyltetrahydrofolate) [39]
0.43 <9> (L-homocysteine) [14]
2.4 <9> (5-methyltetrahydropteroylglutamate) [14]
Additional information <4> [12]

pH-Optimum
6.5 <7, 9> [14, 22]
6.5-6.6 <17> [20]
6.5-7.2 <3> [6]
7 <5> [2]
7.5-7.8 <16> [20]

pH-Range
6-7.5 <17> (<17> pH 6.0: about 75% of maximum activity, pH 7.5: about 35% of activity maximum) [20]
6.2-8.2 <5> (<5> pH about 50% of maximum activity at pH 6.2 and 8.2 [2]) [2]
6.5-8.3 <16> (<16> pH 6.5: about 30% of maximum activity, pH 8.3: about 85% of maximum activity [20]) [20]

Temperature optimum (°C)
31 <9> [14]
39 <9> [14]
45 <4, 3> [5, 6]

Temperature range (°C)
10-39 <9> (<9> 10°C: about 10% of maximum activity, 39°C: maximum activity [14]) [14]
15-31 <9> (<9> 15°C: about 10% of maximum activity, 31°C: maximum activity [14]) [14]

4 Enzyme Structure

Molecular weight
40000 <9> (<9> gel filtration [14]) [14]
140000 <4> (<4> sucrose density gradient centrifugation [17]) [17]
140000-150000 <4> (<4> gel filtration, sedimentation coefficient in sucrose gradients [12]) [12]
150000 <4> (<4> gel filtration [10]) [10]
151000-155000 <5> (<5> gel filtration [32]) [32]
153000 <4> (<4> gel filtration [11]) [11]
160000 <8> (<8> gel filtration [26]) [26]

186000 <4> (<4> gel filtration [21]) [21]
200000 <4> (<4> gel filtration [28]) [28]

Subunits

? <4, 7, 8> (<8> x * 90000 + x * 45000 + x * 35000, SDS-PAGE of reduced enzyme [26]; <4> x * 49500, 2 different types of subunits, SDS-PAGE [28]; <7> x * 140000, SDS-PAGE, recombinant enzyme [39]) [26, 28, 39]
monomer <4, 5> (<5> 1 * 151000, SDS-PAGE [32]; <4> x * 133000, SDS-PAGE [11]) [11, 32]

5 Isolation/Preparation/Mutation/Application

Source/tissue

adrenal gland <3> [6]
brain <3, 8> [6, 12, 23]
heart <3> [6]
kidney <1, 3, 5, 8> [6, 9, 12, 25, 41]
liver <3, 5-8, 19> [2, 6, 12, 16, 22, 25, 27]
lymphocyte <8> [24, 25]
pancreas <3, 8> [6, 24, 25]
placenta <8> [28, 26]
spinal cord <3> [6]
spleen <3> [6]
thyroid gland <3> [6]
tissue culture <3, 5, 7, 8, 12> (<3, 5, 7, 12> a variety of mammalian cell lines in tissue culture [12]; <8> leukemia cells [19]) [12, 19]
Additional information <3> (<3> distribution in tissues [6]) [6]

Localization

chloroplast <15> (<15> 18% of total activity [45]) [45]
cytoplasm <7> [12]
cytosol <15> (<15> 69% of total activity [45]) [45]
mitochondrion <7, 15> (<15> 10% of total activity [45]) [12, 45]

Purification

<3> [6]
<4> (K12 [21]; B [12, 17]; affinity chromatography [13]; preparation of different cobalamin forms [31]) [5, 11, 12, 13, 17, 21, 25, 28, 31]
<5> [2, 9, 16, 32]
<7> [22]
<8> [17, 24, 26]
<9> [14]
<15> (partial [18]) [18]
<16> (partial [20]) [20]
<17> (partial [20]) [20]

Renaturation

<4> (incubation of resolved apoenzyme with methyl-B_{12} results in spontaneous formation of holoenzyme [12]) [12]

Crystallization

<4> (cobalamin-binding domain [5]) [5]

Engineering

Cys310Ser <4> (<4> contains less than 0.05 equiv. of zinc, no proton release upon incubation with L-homocysteine, no enzymativc activity of holoenzyme [38]) [38]
Cys311Ser <4> (<4> contains less than 0.05 equiv. of zinc, no proton release upon incubation with L-homocysteine, no enzymativc activity of holoenzyme [38]) [38]
Additional information <4> (<4> truncated MetH(2-649) protein, contains 0.9 equiv. of zinc, still proton release upon incubation with L-homocysteine [38]; <4> truncated MetH(2-649) protein catalyzes partial enzymic reactions, truncated MetH(2-649) + Cys310Ala or + Cys311Ala indicate modular structure of enzyme comprising of four separate regions [40]) [38, 40]

Application

analysis <2, 3, 5-7, 20> (<2, 3, 5-7, 20> assay using HPLC with fluorescence detection [44]) [44]
medicine <7> (<7> B_{12}-deficient diet leads to extremely low activity of enzyme possibly due to effects of coenzyme stabilization [30]; <7> comparison of enzyme in methionine dependent HTC cells and methionine independent Phi-1 cells [43]) [30, 43]

6 Stability

pH-Stability

5 <5> (<5> unstable [16]; <5> precipitation at pH 5 inactivates [2]) [2, 16]
7 <9> (<9> stability optimum, 0°C, $t_{1/2}$: 6 days) [14]

Temperature stability

37 <5> (<5> 2 h, complete loss of activity [2,16]) [2, 16]

General stability information

<5>, acetone inactivates [2, 16]
<5>, freezing and thawing, 40% loss of activity [2, 16]
<5>, L-homocysteine, 0.01 M, stabilizes [9]
<5>, 2-mercaptoethanol, 0.2 M, inactivation after 2 h at 24°C or during freezing and thawing, no effect at 0°C [2, 16]
<5>, precipitation at pH 5 inactivates [2]
<7>, apoenzyme is very unstable, may be stabilized by forming a complex with methylcobalamin [30]
<8>, rapid loss of activity in frozen state [26]

Storage stability

<4>, -20°C, 15-20 mg/ml protein, stable for at least 1 month [17]
<4>, -80°C, methylated enzyme [5]
<5>, -20°C, ammonium sulfate paste containing homocysteine [9]
<5>, 0°C, glutathione, N_2-atmosphere, stable for at least a week [2, 16]
<8>, 4°C, 20% loss of activity after 2 weeks [26]
<9>, 0°C, pH 7, half-life 6 days [14]
<9>, freeze-dried, 20% loss of activity after 3 months [14]

References

[1] Guest, J.R.; Friedman, S.; Foster, M.A.; Tejerina, G.; Woods, D.D.: Transfer of the methyl group from N^5-methyltetrahydrofolates to homocysteine in Escherichia coli. Biochem. J., **92**, 497-504 (1964)
[2] Loughlin, R.E.; Elford, H.L.; Buchanan, J.M.: Enzymatic synthesis of the methyl group of methionine. J. Biol. Chem., **239**, 2888-2895 (1964)
[3] Foster, M.A.; Dilworth, M.J.; Woods, D.D.: Cobalamin and the synthesis of methionine by Escherichia coli. Nature, **201**, 39-42 (1964)
[4] Gonzalez, J.C.; Banerjee, R.V.; Huang, S.; Sumner, J. S.; Matthews, R.G.: Comparison of cobalamin-independent and cobalamin-dependent methionine synthases from Escherichia coli: two solutions to the same chemical problem. Biochemistry, **31**, 6045-6056 (1992)
[5] Luschinsky, C.L.; Drummond, J.T.; Matthews, R.G.; Ludwig, M.L.: Crystallization and preliminary X-ray diffraction studies of the cobalamin-binding domain of methionine synthase from Escherichia coli. J. Mol. Biol., **225**, 557-560 (1992)
[6] Magnum, J.H.; Steuart, B.W.; North, J.A.: The isolation of N^5 -methyltetrahydrofolate-homocysteine transmethylase from bovine brain. Arch. Biochem. Biophys., **148**, 63-69 (1972)
[7] Taylor, R.T.; Hanna, M.L.: Escherichia coli B N^5 -methyltetrahydrofolate-homocysteine cobalamin methyltransferase: binding of the folate substrate to the enzyme. Arch. Biochem. Biophys., **151**, 401-413 (1972)
[8] Rudiger, H.: On the cation sensibility of the vitamin B_{12}-dependent methionine synthetase (5-methyltetrahydrofolate-homocysteine-methyltransferase from Escherichia coli). FEBS Lett., **35**, 295-298 (1973)
[9] Mangum, J.H.; North, J.A.: Isolation of a cobalamin containing 5-methyltetrahydrofolate-homocysteine transmethylase from mammalian kidney. Biochemistry, **10**, 3765-3769 (1971)
[10] Taylor, R.T.: Escherichia coli B N^5 -methyltetrahydrofolate-homocysteine cobalamin methyltransferase: gel-filtration behavior of apoenzyme and holoenzymes. Biochim. Biophys. Acta, **242**, 355-364 (1971)
[11] Frasca, V.; Banerjee, R.V.; Dunham, W.R.; Sands, R.; Matthews, R.G.: Cobalamin-dependent methionine synthase from Escherichia coli B: Electron paramagnetic resonance spectra of the inactive form and the active methylated form of the enzyme. Biochemistry, **27**, 8458-8465 (1988)

[12] Taylor, R.T.; Weissbach, H.: N^5-Methyltetrahydrofolate-homocysteine methyltransferases. The Enzymes, 3rd Ed. (Boyer, P.D., ed.), **12**, 121-165 (1973)
[13] Sato, K.; Hiei, E.; Shimizu, S.: Affinity chromatography of N^5-methyltetrahydrofolate-homocysteine methyltransferase on a cobalamin-Sepharose. FEBS Lett., **85**, 73-76 (1978)
[14] Grabau, W.; Rudiger, H.: Purification and properties of the tetrahydropteroylglutamate methyltransferase from green beans (Phaseolus vulgaris). Hoppe-Seyler's Z. Physiol. Chem., **358**, 695-698 (1977)
[15] Kageyama, M.; Perlman, D.: Inhibition of N^5-methyltetrahodrofolate-homocysteine transmethylase by a vitamin B_{12}-antimetabolite. Biochem. Biophys. Res. Commun., **76**, 420-423 (1977)
[16] Buchanan, J.M.: Methionine biosynthesis (hog liver). Methods Enzymol., **17B**, 371-378 (1971)
[17] Taylor, R.T.; Weissbach, H.: N^5-Methytetrahydrofolate-homocysteine (vitamin B_{12}) methyltransferase (Escherichia coli B). Methods Enzymol., **17B**, 379-388 (1971)
[18] Lucchini, G.; Piazza, I.; Biachetti, R.: B_{12}-dependent methyl-H4-Folate homocysteine methyltransferase in Euglena gracilis. Plant Sci. Lett., **36**, 177-179 (1984)
[19] Sauer, H.: Methionine synthase. Methods Enzym. Anal., 3rd Ed. (Bergmeyer, H.U., ed.), **3**, 304-311 (1983)
[20] Ohmori, H.; Fukui, S.: Vitamin B_{12}-dependent methionine synthetase in photosynthetic bacteria: Partial purification and properties. Agric. Biol. Chem., **38**, 1317-1324 (1974)
[21] Fujii, K.; Huennekens, F.M.: Activation of methionine synthetase by a reduced triphosphopyridine nucleotide-dependent flavoprotein system. J. Biol. Chem., **249**, 6745-6753 (1974)
[22] Cheng, F.W.; Shane, B.; Stokstad, E.L.R.: Pentaglutamate derivatives of folate as substrates for rat liver tetrahydropteroylglutamate methyltransferase and 5,10-methylenetetrahydrofolate reductase. Can. J. Biochem., **53**, 1020-1027 (1975)
[23] Coward, J.K.; Chello, P.L.; Cashmore, A.R.; Parameswaran, K.N.; DeAngelis, L.M.; Bertiono, J.R.: 5-methyl-5,6,7,8-tetrahydropteroyl oligo-γ-L-glutamates: synthesis and kinetic studies with methionine synthetase from bovine brain. Biochemistry, **14**, 1548-1552 (1975)
[24] Kolhouse, J.F.; Utley, C.; Stabler, S.P.; Allen, R.H.: Mechanism of conversion of human apo- to holomethionine synthase by various forms of cobalamin. J. Biol. Chem., **266**, 23010-23015 (1991)
[25] Beck, W.S.; Thorndike, J.: Differences in coenzyme specificity of the N^5-methyltetrahydrofolate-homocysteine methyltransferases of various species: implications for corrin binding loci. Biochem. Biophys. Res. Commun., **120**, 74-80 (1984)
[26] Utley, C.S.; Marcell, P.D.; Allen, R.H.; Antony, A.C.; Kolhouse, J.F.: Isolation and characterization of methionine synthetase from human placenta. J. Biol. Chem., **260**, 13656-13665 (1985)

[27] Kolhouse, J.F.; Allen, R.H.: Recognition of two intracellular cobalamin binding proteins and their identification as methylmalonyl-CoA mutase and methionine synthetase. Proc. Natl. Acad. Sci. USA, **74**, 921-925 (1977)
[28] Paessens, A.; Rudiger, H.: Purification and chemical characterization of the vitamin-B_{12}-dependent 5-methyltetrahydrofolate: homocysteine methyltransferase from Escherichia coli B. Eur. J. Biochem., **112**, 47-51 (1980)
[29] Matthews, R.G.: Cobalamin-dependent methyltransferases. Acc. Chem. Res., **34**, 681-689 (2001)
[30] Yamada, K.; Kawata, T.; Wada, M.; Isshiki, T.; Onoda, J.; Kawanishi, T.; Kunou, A.; Tadokoro, T.; Tobimatsu, T.; Maekawa, A.; Toraya, T.: Extremely low activity of methionine synthase in vitamin B-12-deficient rats may be related to effects on coenzyme stabilization rather than to changes in coenzyme induction. J. Nutr., **130**, 1894-1900 (2000)
[31] Jarrett, J.T.; Goulding, C.W.; Fluhr, K.; Huang, S.; Matthews, R.G.: Purification and assay of cobalamin-dependent methionine synthase from Escherichia coli. Methods Enzymol., **281**, 196-213 (1997)
[32] Chen, Z.; Crippen, K.; Gulati, S.; Banerjee, R.: Purification and kinetic mechanism of a mammalian methionine synthase from pig liver. J. Biol. Chem., **269**, 27193-27197 (1994)
[33] Bandarian, V.; Matthews, R.G.: Quantitation of rate enhancements attained by the binding of cobalamin to methionine synthase. Biochemistry, **40**, 5056-5064 (2001)
[34] Hall, D.A.; Jordan-Starck, T.C.; Loo, R.O.; Ludwig, M.L.; Matthews, R.G.: Interaction of flavodoxin with cobalamin-dependent methionine synthase. Biochemistry, **39**, 10711-10719 (2000)
[35] Hall, D.A.; Vander Kooi, C.W.; Stasik, C.N.; Stevens, S.Y.; Zuiderweg, E.R.P.; Matthews, R.G.: Mapping the interactions between flavodoxin and its physiological partners flavodoxin reductase and cobalamin-dependent methionine synthase. Proc. Natl. Acad. Sci. USA, **98**, 9521-9526 (2001)
[36] Kenyon, S.H.; Nicolaou, A.; Ast, T.; Gibbons, W.A.: Stimulation in vitro of vitamin B_{12}-dependent methionine synthase by polyamines. Biochem. J., **316**, 661-665 (1996)
[37] Jarrett, J.T.; Huang, S.; Matthews, R.G.: Methionine synthase exists in two distinct conformations that differ in reactivity toward methyltetrahydrofolate, adenosylmethionine, and flavodoxin. Biochemistry, **37**, 5372-5382 (1998)
[38] Goulding, C.W.; Matthews, R.G.: Cobalamin-dependent methionine synthase from Escherichia coli: Involvement of zinc in homocysteine activation. Biochemistry, **36**, 15749-15757 (1997)
[39] Yamada, K.; Yamada, S.; Tobimatsu, T.; Torava, T.: Heterologous high level expression, purification, and enzymological properties of recombinant rat cobalamin-dependent methionine synthase. J. Biol. Chem., **274**, 35571-35576 (1999)
[40] Goulding, C.W.; Postigo, D.; Matthews, R.G.: Cobalamin-dependent methionine synthase is a modular protein with distinct regions for binding homocysteine, methyltetrahydrofolate, cobalamin, and adenosylmethionine. Biochemistry, **36**, 8082-8091 (1997)

[41] Gulati, S.; Brody, L.C.; Banerjee, R.: Posttranscriptional regulation of mammalian methionine synthase by B_{12}. Biochem. Biophys. Res. Commun., **259**, 436-442 (1999)
[42] Drummond, J.T.; Matthews, R.G.: Nitrous oxide degradation by cobalamin-dependent methionine synthase: characterization of the reactants and products in the inactivation reaction. Biochemistry, **33**, 3732-3741 (1994)
[43] Kenyon, S.H.; Waterfield, C.J.; Timbrell, J.A.; Nicolaou, A.: Methionine synthase activity and sulphur amino acid levels in the rat liver tumour cells HTC and Phi-1. Biochem. Pharmacol., **63**, 381-391 (2002)
[44] Huang, L.; Zhang, J.; Hayakawa, T.; Tsuge, H.: Assays of methylenetetrahydrofolate reductase and methionine synthase activities by monitoring 5-methyltetrahydrofolate and tetrahydrofolate using high-performance liquid chromatography with fluorescence detection. Anal. Biochem., **299**, 253-259 (2001)
[45] Isegawa, Y.; Watanabe, F.; Kitaoka, S.; Nakano, Y.: Subcellular distribution of cobalamin-dependent methionine synthase in Euglena gracilis Z. Phytochemistry, **35**, 59-61 (1994)
[46] Zhou, Z.S.; Smith, A.E.; Matthews, R.G.: L-Selenohomocysteine: One-step synthesis from L-selenomethionine and kinetic analysis as substrate for methionine synthases. Bioorg. Med. Chem. Lett., **10**, 2471-2475 (2000)
[47] Peariso, K.; Zhou, Z.S.; Smith, A.E.; Matthews, R.G.; Penner-Hahn, J.E.: characterization of the zinc sites in cobalamin-independent and cobalamin-dependent methionine synthase using zinc and selenium X-ray absorption spectroscopy. Biochemistry, **40**, 987-993 (2001)
[48] Bannerjee, R.V.; Harder, S.R.; Ragsdale, S.W.; Matthews, R.G.: Mechanism of reductive activation of cobalamin-dependent methionine synthase: An electron paramagnetic resonance spectroelectrochemical study. Biochemistry, **29**, 1129-1135 (1990)

5-Methyltetrahydropteroyltriglutamate-homocysteine S-methyltransferase 2.1.1.14

1 Nomenclature

EC number
2.1.1.14

Systematic name
5-methyltetrahydropteroyltri-L-glutamate:L-homocysteine S-methyltransferase

Recommended name
5-methyltetrahydropteroyltriglutamate homocysteine S-methyltransferase

Synonyms
homocysteine methylase
methyltetrahydropteroylpolyglutamate:homocysteine methyltransferase
methyltransferase, tetrahydropteroylglutamate-homocysteine transmethylase
tetrahydropteroyltriglutamate methyltransferase

CAS registry number
9068-29-5

2 Source Organism

<1> *Aerobacter aerogenes* [5]
<2> *Salmonella typhimurium* [5]
<3> *Saccharomyces cerevisiae* [5, 7]
<4> *Neurospora crassa* [5, 9]
<5> *Chlorella pyrenoidosa* [5]
<6> *Bacillus subtilis* [6]
<7> *Escherichia coli* (K12 [2]) [2-5, 8, 11, 14, 15]
<8> *Phaseolus sp.* (bean [1]) [1]
<9> *Hordeum vulgare* (barley [1]) [1]
<10> *Spinacia oleracea* (spinach [1]) [1]
<11> *Pisum sativum* (pea, 2 enzyme forms: a 5-methyltetrahydropteroylglutamate dependent enzyme and a S-methylmethionine dependent enzyme [1]) [1]
<12> *Daucus carota* (carrot [1]) [1]
<13> *Catharanthus roseus* [10, 16]
<14> *Solanum tuberosum* (potato [12]) [12]
<15> *Phaseolus vulgaris* (green string bean) [13]

3 Reaction and Specificity

Catalyzed reaction

5-methyltetrahydropteroyltri-L-glutamate + L-homocysteine = tetrahydropteroyltri-L-glutamate + L-methionine (<7-12> final step in methionine synthesis [1,8]; <7> mechanism [14,15])

Reaction type

S-methylation
methyl group transfer

Natural substrates and products

S Additional information <7-12> (<7-12> final step in methionine synthesis [1,8]) [1, 8]
P ?

Substrates and products

S 5-methyltetrahydropteroyl-γ-glutamyl-γ-glutamylglutamate + L-homocysteine <7-13, 15> (Reversibility: ? <7-13, 15> [1-3, 10, 13, 16]) [1-3, 10, 13, 16]
P tetrahydropteroyl-γ-glutamyl-γ-glutamylglutamate + L-methionine <8-12> [1]
S 5-methyltetrahydropteroylmonoglutamate + L-homocysteine <15> (Reversibility: ? <15> [13]) [13]
P tetrahydropteroylmonoglutamate + L-methionine
S S-adenosylmethionine + L-selenohomocysteine <7> (Reversibility: ? <7> [14]) [14]
P L-selenomethionine + S-adenosylhomocysteine <7> (<7> reaction catalyzed both by vitamin B_{12} dependent and independent enzyme [14]) [14]
S S-adenosylmethionine + homocysteine <11> (Reversibility: ? <11> [1]) [1]
P methionine + S-adenosylhomocysteine <11> [1]
S S-methylmethionine + homocysteine <11> (Reversibility: ? <11> [1]) [1]
P methionine <11> [1]
S Additional information <1-5, 7-12> (<8-12> most active with the polyglutamate forms of N^5-methyltetrahydropteroylglutamate [1]; <7> 5-methyltetrahydropteroyl-α-glutamate or 5-methyl-tetrahydropteroyl-α-glutamylglutamate cannot replace the triglutamate folate derivative as methyl donor [2]; <1-5,7> monoglutamate analog not used [3, 5]; <8-12> enzymes of higher plants can use the monoglutamate [1]) [1-3, 5]
P ?

Inhibitors

5-methyltetrahydropteroyl-α-glutamate <7> [2]
EDTA <1-5, 7> (<7> in absence of Mg^{2+} [2]) [2, 5, 7]
Mn^{2+} <7> (<7> above 1 mM [3]) [3]
chloromethyl ketones <7> [4]
high ionic strength <7> [2]

iodoacetamide <7> [4]
methionine <8-12> [1]
pteroyl-α-glutamylglutamic acid <7> [2]
pteroyl-γ-glutamyl-γ-glutamylglutamic acid <7> [2]

Cofactors/prosthetic groups

Additional information <1-13, 15> (<1-13, 15> vitamin-B_{12} independent enzyme [1-10, 13]; <8-13, 15> no requirement for S-adenosylmethionine [1-10, 13]; <8-12, 15> no requirement for a reducing system [1, 13]; <13> no requirement for phosphate or magnesium [10]) [1-10, 13]

Metals, ions

Ca^{2+} <1, 2, 4, 5> (<1, 2, 4, 5> can replace Mg^{2+} [5]) [5]
Mg^{2+} <1, 2, 4, 5, 7-12> (<1, 2, 4, 5, 7-12> stimulates [1, 2, 5]; <7> at low phosphate concentration, 0.1 mM, the Mg^{2+}-stimulation is 4.5fold, smaller stimulation at higher concentrations of phosphate [2]; <7> required [3]; <3,13> no stimulation [5,16]) [1-3, 5]
Mn^{2+} <7> (<7> below 1 mM more effective than Mg^{2+}, above 1 mM inhibitory [3]) [3]
Zn^{2+} <7, 13> (<7> 1 equivalent per subunit [11]; <7> characterization of zinc site [15]) [11, 15, 16]
phosphate <1-5, 7-13> [1-3, 5, 7, 16]

Specific activity (U/mg)

0.24 <3> [7]
2.5 <7> [2]
999 <4> (<4> specific activity after growth in different media [9]) [9]

K_m-Value (mM)

0.000026 <11> (5-methyltetrahydropteroylglutamate) [1]
0.0022 <11> (S-methylmethionine) [1]
0.004 <11> (S-adenosylmethionine) [1]
0.0044 <7> (5-methyltetrahydropteroyltriglutamate) [2]
0.01 <13> (L-homocysteine, <13> or less [10]) [10]
0.016 <7> (selenohomocysteine) [14]
0.022 <3> (L-homocysteine) [7]
0.028 <13> (5-methyltetrahydropteroyl-γ-glutamyl-γ-glutamylglutamate) [10]
0.041 <8> (homocysteine, <8> plus S-methylmethionine [1]) [1]
0.055 <8> (S-methylmethionine) [1]
0.069 <7> (homocysteine) [14]
0.08 <13> (5-methyltetrahydropteroyl-γ-glutamyl-γ-glutamylglutamate) [16]
0.35 <13> (5-methyltetrahydropteroylmonoglutamate) [16]
0.4 <3> (5-methyltetrahydropteroyltriglutamate) [7]
0.6 <11> (homocysteine, <11> plus 5-methyltetrahydropteroylglutamate [1]) [1]

pH-Optimum

6.5 <11> (5-methyltetrahydropteroylglutamate) [1]
6.6-7.6 <3> [5, 7]

7.3 <11> (S-methylmethionine) [1]
7.5-7.8 <7> [2, 5]
7.9 <8> (S-methylmethionine) [1]
8.1 <7> [3]

pH-Range
6-8.5 <7> [2]
6-8.7 <7> (<7> pH 6.0: about 70% of maximum activity, pH 8.7: about 90% of maximum activity [3]) [3]

Temperature optimum (°C)
57 <3> [7]

4 Enzyme Structure

Molecular weight
75000 <3> (<3> gel filtration, sedimentation data [5,7]) [5, 7]
84000 <7> (<7> equilibrium ultracentrifugation [2,5]) [2, 5]
85000 <13> (<13> gel filtration, recombinant enzyme [16]) [16]

Subunits
? <13> (<13> x * 85000, deduced from DNA sequence, SDS-PAGE of recombinant protein [10]) [10]
? <7> (<7> x * 50800, equilibrium sedimentation of carboxymethylated enzyme in 5 M guanidine-HCl [2,5]) [2, 5]
monomer <13> (<13> 1 * 85000, SDS-PAGE, recombinant enzyme [16]) [16]

5 Isolation/Preparation/Mutation/Application

Source/tissue
cell culture <13> [16]
cotyledon <13> [16]
flower <14> [12]
hypocotyl <13> [16]
root <8-12> [1]
seedling <8-12, 13> (<13> low activity [16]) [1, 16]

Localization
chloroplast <8-12> [1]
cytosol <13, 14> [10, 12]
soluble <7> [2]

Purification
<3> [7]
<7> (rapid one-step purification [4]) [2, 4]
<8> [1]

Crystallization

<7> [2]

Cloning

<7> [4, 8]
<13> [10]
<14> [12]

Engineering

C726S <7> (<7> mutant does not contain zinc, no activity, probably due to lack of zinc [11]) [11]

6 Stability

Storage stability

<3>, -20°C, stable [7]

References

[1] Madison, J.T.: Sulphur metabolism. F. Enzymes involved in the synthesis of methionine. Methods Plant Biochem., **3**, 361-369 (1990)
[2] Whitefield, C.D.; Steers, E.J.; Weissbach, H.: Purification and properties of 5-methyltetrahydropteroyltriglutamate-homocysteine transmethylase. J. Biol. Chem., **245**, 390-401 (1970)
[3] Guest, J.R.; Friedman, S.; Foster, M.A.; Tejerina, G.; Woods, D.D.: Transfer of the methyl group from N^5-methyltetrahydrofolates to homocysteine in Escherichia coli. Biochem. J., **92**, 497-504 (1964)
[4] Gonzalez, J.C.; Banerjee, R.V.; Huang, S.; Sumner, J. S.; Matthews, R.G.: Comparison of cobalamin-independent and cobalamin-dependent methionine synthases from Escherichia coli: two solutions to the same chemical problem. Biochemistry, **31**, 6045-6056 (1992)
[5] Taylor, R.T.; Weissbach, H.: N^5-Methyltetrahydrofolate-homocysteine methyltransferases. The Enzymes, 3rd Ed. (Boyer, P.D., ed.), **9**, 121-165 (1973)
[6] Salem, A.R.; Pattison, J.R.; Foster, M.A.: Folic acid and the methylation of homocysteine by Bacillus subtilis. Biochem. J., **126**, 993-1004 (1972)
[7] Burton, E.; Sakami, W.: Methionine synthesis (Yeast). Methods Enzymol., **17B**, 388-392 (1971)
[8] Chu, J.; Shoeman, R.; Hart, J.; Coleman, T.; Mazaitis, A.; Kelker, N.; Brot, N.; Weissbach, H.: Cloning and expression of the metE gene in Escherichia coli. Arch. Biochem. Biophys., **239**, 467-474 (1985)
[9] Burton, E.G.; Metzenberg, R.L.: Regulation of methionine biosythesis in Neurospora crassa. Arch. Biochem. Biophys., **168**, 219-229 (1975)
[10] Eichel, J.; Gonzalez, J.C.; Hotze, M.; Matthews, R.G.; Schroeder, J.: Vitamin-B_{12}-independent methionine synthase from a higher plant (Catharanthus

roseus). Molecular characterization, regulation, heterologous expression, and enzyme properties. Eur. J. Biochem., **230**, 1053-1058 (1995)
[11] Gonzalez, J.C.; Peariso, K.; Penner-Hahn, J.E.; Matthews, R.G.: Cobalamin-independent methionine synthase from Escherichia coli: A zinc metalloenzyme. Biochemistry, **35**, 12228-12234 (1996)
[12] Zeh, M.; Leggewie, G.; Hoefgen, R.; Hesse, H.: Cloning and characterization of a cDNA encoding a cobalamin-independent methionine synthase from potato (Solanum tuberosum L.). Plant Mol. Biol., **48**, 255-265 (2002)
[13] Burton, E.G.; Sakami, W.: The formation of methionine from the monoglutamate form of methyltetrahydrofolate by higher plants. Biochem. Biophys. Res. Commun., **36**, 228-234 (1969)
[14] Zhou, Z.S.; Smith, A.E.; Matthews, R.G.: L-Selenohomocysteine: one-step synthesis from L-selenomethionine and kinetic analysis as substrate for methionine synthases. Bioorg. Med. Chem. Lett., **10**, 2471-2475 (2000)
[15] Peariso, K.; Zhou, Z.S.; Smith, A.E.; Matthews, R.G.; Penner-Hahn, J.E.: Characterization of the zinc sites in cobalamin-independent and cobalamin-dependent methionine synthase using zinc and selenium X-ray absorption spectroscopy. Biochemistry, **40**, 987-993 (2001)
[16] Eckermann, C.; Eichel, J.; Schroder, J.: Plant methionine synthase: new insights into properties and expression. Biol. Chem., **381**, 695-703 (2000)

Fatty-acid O-methyltransferase 2.1.1.15

1 Nomenclature

EC number
2.1.1.15

Systematic name
S-adenosyl-L-methionine:fatty-acid O-methyltransferase

Recommended name
fatty-acid O-methyltransferase

Synonyms
fatty acid O-methyltransferase
fatty acid methyltransferase

CAS registry number
37256-89-6

2 Source Organism

<1> *Mycobacterium phlei* [1, 2]
<2> *Mycobacterium sp.* [2]
<3> *Rattus norvegius* (insulinoma cell line RINm5F [3]; Sprangue Dawley rats [4]) [3, 4]

3 Reaction and Specificity

Catalyzed reaction
S-adenosyl-L-methionine + a fatty acid = S-adenosyl-L-homocysteine + a fatty acid methyl ester

Reaction type
methyl group transfer

Natural substrates and products
S S-adenosyl-L-methionine + a fatty acid <1-3> (<1> synthesis of 10-methylstearic acid [1]; <1, 2> involved in biosynthesis of branched chain fatty acids, mycolic acid, menaquinones and in esterification of fatty acid carboxyl group [2]; <3> involved in biosynthesis of phosphatidylcholine [3]) (Reversibility: ? <1-3> [1-4]) [1-4]
P S-adenosyl-L-homocysteine + a fatty acid methyl ester <1-3> [1-4]

Substrates and products

S S-adenosyl-L-methionine + a fatty acid <1-3> (<1> oleic acid is the most effective fatty acid acceptor, traces of activity with phosphatidylethanolamine and phosphatidylglycerol [1]) (Reversibility: ? <1-3> [1-4]) [1-4]
P S-adenosyl-L-homocysteine + a fatty acid methyl ester <1-3> [1-4]
S S-adenosyl-L-methionine + abietic acid <1, 2> (Reversibility: ? <1, 2> [2]) [2]
P S-adenosyl-L-homocysteine + methyl abietate <1, 2> [2]
S S-adenosyl-L-methionine + oleic acid <1> (<1> oleic acid is the most effective fatty acid acceptor, traces of activity with phosphatidylethanolamine and phosphatidylglycerol [1]) (Reversibility: ? <1> [1]) [1]
P S-adenosyl-L-homocysteine + oleic acid methyl ester <1> [1]
S S-adenosyl-L-methionine + palmitic acid <1, 2> (Reversibility: ? <1, 2> [2]) [2]
P S-adenosyl-L-homocysteine + methyl palmitate <1, 2> [2]
S S-adenosyl-L-methionine + phosphatidylethanolamine <3> (Reversibility: ? <3> [3]) [3]
P S-adenosyl-L-homocysteine + phospahtidyl-N-methylethanolamine <3> [3]
S Additional information <1> (<1> traces of activity with methionine, serine and S-adenosyl-L-methionine as methyl donor [1]) [1]
P ?

Inhibitors

Ca^{2+} <1> (<1> slight inhibition [1]; <3> no effect below 0.5 mM [3]) [1, 3]
N-(6-aminohexyl)-5-chloro-1-naphthalene-sulfonamide <3> (<3> W7, very strong inhibition [3]) [3]
S-adenosylhomocysteine <1, 3> [1, 3, 4]
trifluoperazine <3> (<3> very strong inhibition [3]) [3]
unsaturated fatty acids <1> [1]
Additional information <3> (<3> Ca^{2+}, EGTA and calmodulin have no effect [3]) [3]

Specific activity (U/mg)

0.000018 <3> [3]

K_m-Value (mM)

0.025 <1> (S-adenosylmethionine) [1]
1.3 <1> (oleic acid) [1]

pH-Optimum

8-9 <1> [1]

pH-Range

7-10 <1> [1]

5 Isolation/Preparation/Mutation/Application

Source/tissue

retina <3> (<3> membrane [3]) [3]

Purification

<2> [2]

6 Stability

Temperature stability

50 <1> (<1> 20% loss of activity within 5 min [1]) [1]
100 <1> (<1> 90% loss of activity within 10 min [1]) [1]

References

[1] Akamatsu, Y.; Law, J.H.: The enzymatic synthesis of fatty acid methyl esters by carboxyl group alkylation. J. Biol. Chem., **245**, 709-713 (1970)

[2] Orpiszewski, J.; Hebda, C.; Szykula, J.; Powls, R.; Clasper, S.; Rees, H.H.: Multiple forms of O-methyltransferase involved in the microbial conversion of abietic acid into methyl abietate by Mycobacterium sp. FEMS Microbiol. Lett., **82**, 233-236 (1991)

[3] Safayhi, H.; Anazodo, M.I.; Ammon, H.P.T.: Calmodulin- and calcium-insensitive fatty acid methyltransferase from RINm5F cells. Inhibition by trifluoperazine and W7. Int. J. Biochem., **23**, 769-772 (1991)

[4] Sastry, B.V.R.; Vidaver, P.S.; Janson, V.E.; Franks, J.J.: S-Adenosyl-L-methionine-mediated enzymic methylations in the rat retinal membranes. J. Ocul. Pharm., **10**, 253-263 (1994)

Methylene-fatty-acyl-phospholipid synthase 2.1.1.16

1 Nomenclature

EC number
2.1.1.16

Systematic name
S-adenosyl-L-methionine:unsaturated-phospholipid methyltransferase (methenylating)

Recommended name
methylene-fatty-acyl-phospholipid synthase

Synonyms
methyltransferase, unsaturated phospholipid
unsaturated phospholipid methyltransferase
unsaturated-phospholipid methyltransferase

CAS registry number
37256-90-9

2 Source Organism

<1> *Mycobacterium phlei* [1]

3 Reaction and Specificity

Catalyzed reaction
S-adenosyl-L-methionine + phospholipid olefinic fatty acid = S-adenosyl-L-homocysteine + phospholipid methylene fatty acid

Reaction type
methyl group transfer

Natural substrates and products
S S-adenosyl-L-methionine + phospholipid olefinic fatty acid <1> (<1>, enzymatic synthesis of tuberculostearic acid i.e. 10-methylstearic acid [1]) (Reversibility: ? <1> [1]) [1]
P ?

Substrates and products

- **S** S-adenosyl-L-methionine + phosphatidylethanolamine <1> (<1>, alkylenation at the 10-carbon to give a methylene group [1]) (Reversibility: ? <1> [1]) [1]
- **P** S-adenosyl-L-homocysteine + ?
- **S** S-adenosyl-L-methionine + phosphatidylglycerol <1> (<1>, alkylenation at the 10-carbon to give a methylene group [1]) (Reversibility: ? <1> [1]) [1]
- **P** S-adenosyl-L-homocysteine + ?
- **S** S-adenosyl-L-methionine + phosphatidylinositol <1> (<1>, alkylenation at the 10-carbon to give a methylene group [1]) (Reversibility: ? <1> [1]) [1]
- **P** S-adenosyl-L-homocysteine + ?
- **S** S-adenosyl-L-methionine + phospholipid olefinic fatty acid <1> (Reversibility: ? <1> [1]) [1]
- **P** S-adenosyl-L-homocysteine

References

[1] Akamatsu, Y.; Law, J.H.: Enzymatic alkylenation of phospholipid fatty acid chains by extracts of Mycobacterium phlei. J. Biol. Chem., **245**, 701-708 (1970)

Phosphatidylethanolamine N-methyltransferase 2.1.1.17

1 Nomenclature

EC number
2.1.1.17

Systematic name
S-adenosyl-L-methionine:phosphatidylethanolamine N-methyltransferase

Recommended name
phosphatidylethanolamine N-methyltransferase

Synonyms
LMTase
PE N-MTase
PEMT
lipid methyl transferase
methyltransferase, phosphatidylethanolamine
phosphatidylethanolamine methyltransferase
phosphatidylethanolamine-N-methylase
phosphatidylethanolamine-S-adenosylmethionine methyltransferase

CAS registry number
37256-91-0

2 Source Organism

<1> *Rattus norvegicus* (2 forms of enzyme: PEMT1 and PEMT2 [16]) [1, 2, 4, 5, 6, 7, 8, 9, 10, 11, 16, 17, 18, 20, 21, 23, 25, 26, 28, 29, 32]
<2> *Saccharomyces cerevisiae* [3]
<3> *Zymomomas mobilis* [12, 13]
<4> *Mus musculus* [14]
<5> *Canis sp.* [15]
<6> *Homo sapiens* [19, 27]
<7> *Cricetulus griseus* [22, 24]
<8> *Bos taurus* [30, 31]

3 Reaction and Specificity

Catalyzed reaction

S-adenosyl-L-methionine + phosphatidylethanolamine = S-adenosyl-L-homocysteine + phosphatidyl-N-methylethanolamine

Reaction type

methyl group transfer

Natural substrates and products

S S-adenosyl-L-methionine + phosphatidyl-N,N-dimethylethanolamine <1, 3> (<1>, the enzyme catalyzes the stepwise methylation of phosphatidylethanolamine to phosphatidylcholine, in addition to the main pathway that synthesizes phosphatidylcholine directly from choline [32]) (Reversibility: ? <1, 3> [12, 32]) [12, 32]

P S-adenosyl-L-homocysteine + phosphatidylcholine <1, 3> [12, 32]

S S-adenosyl-L-methionine + phosphatidyl-N-methylethanolamine <1, 3> (<1>, the enzyme catalyzes the stepwise methylation of phosphatidylethanolamine to phosphatidylcholine, in addition to the main pathway that synthesizes phosphatidylcholine directly from choline [32]) (Reversibility: ? <1, 3> [12, 32]) [12, 32]

P S-adenosyl-L-homocysteine + phosphatidyl-N,N-dimethylethanolamine <1, 3> [12, 32]

S S-adenosyl-L-methionine + phosphatidylethanolamine <1, 3, 6, 7> (<3>, the first-step methylation from phosphatidylethanolamine to phosphatidylcholine is rate-limiting [12, 13]; <1>, methylation of phosphatidylethanolamine may play a role in very low density lipoprotein secretion [18]; <1>, possible role of PEMT2 in the suppression of liver proliferation, and perhaps in hepatic carcinogenesis [20]; <1>, inactivation of enzyme form PEMT2 may have a role in the regulation of non-neoplastic growth of liver [21]; <1>, activation of phosphatidylethanolamine methylation at the level of gene expression may be the mechanism by which choline-deficient rats compensate for the lack of dietary choline [23]; <7>, expression of PEMT2 in CHO cells does not down-regulate, but rather enhance the expression of CTP-phosphocholine cytidylyltransferase [24]; <1>, enzyme form PEMT2 may have a role in the regulation of in vivo hepatoma and hepatocyte cell division as well as hepatocyte cell death by apoptosis [26]; <6>, a decrease in enzyme activity is observed in the frontal cortex of the brain affected with Alzheimer‘s disease [27]; <1>, enzyme form PEMT2 plays an important role in the regulation of hepatocyte cell division and as a liver tumor suppressor [29]; <1>, the enzyme catalyzes the stepwise methylation of phosphatidylethanolamine to phosphatidylcholine, in addition to the main pathway that synthesizes phosphatidylcholine directly from choline. Loss of PEMT function may contribute to malignant transformation of hepatocytes [32]) (Reversibility: ? <1, 3, 6, 7> [12, 18, 20, 21, 23, 24]) [12, 13, 18, 20, 21, 23, 24, 26, 27, 29, 32]

P S-adenosyl-L-homocysteine + phosphatidyl-N-methylethanolamine <1> [32]

Substrates and products

S S-adenosyl-L-methionine + phosphatidyl-N-methylethanolamine <1, 8> (<1>, ordered bi-bi mechanism [1, 8]; <1>, phospholipid substrates and products are the first to bind and the last to dissociate from the active site [8]; <1, 2, 4>, the transfer of a methyl group to phosphatidyl-N-methylethanolamine and phosphatidyl-N,N-dimethylethanolamine is catalyzed by a second enzyme [3, 9, 10, 14]) (Reversibility: ? <1, 8> [1, 4, 6, 8, 11, 31, 32]) [1, 4, 6, 8, 11, 31, 32]

P S-adenosyl-L-homocysteine + phosphatidyl-N,N-dimethylethanolamine <1> [2]

S S-adenosyl-L-methionine + phosphatidyl-N-methylethanolamine <1, 8> (<1>, ordered bi-bi mechanism [1, 8]; <1, 2, 4>, the transfer of a methyl group to phosphatidyl-N-methylethanolamine and phosphatidyl-N,N-dimethylethanolamine is catalyzed by a second enzyme [3, 9, 10, 14]; <1>, phospholipid substrates and products are the first to bind and the last to dissociate from the active site [8]) (Reversibility: ? <1, 8> [1, 4, 6, 8, 11, 31, 32]) [1, 4, 6, 8, 11, 31, 32]

P S-adenosyl-L-homocysteine + phosphatidylcholine <1> [2]

S S-adenosyl-L-methionine + phosphatidylethanolamine <1-8> (<1>, transfer of three methyl groups to the amino head group of phosphatidylethanolamine [1]; <1, 2, 4>, the transfer of a methyl group to monomethyl-N-phosphatidylcholine and dimethyl-N,N-phosphatidylcholine is catalyzed by a second enzyme [3, 9, 10, 14]) (Reversibility: ? <1-8> [1-32]) [1-32]

P S-adenosyl-L-homocysteine + phosphatidyl-N-methylethanolamine <1> [2]

Inhibitors

$HgCl_2$ <5> [15]
$HgCl_2$ <2> (<2>, 1 mM, about 90% inhibition [3]) [3]
NEM <5> [15]
PCMB <5> [15]
PCMB <2> (<2>, 1 mM, about 90% inhibition [3]) [3]
S-adenosylhomocysteine <1, 2, 3, 4> (<3>, competitive with respect to S-adenosylmethionine [12,13]) [1, 3, 12, 13, 14]
Triton X-100 <6> [27]
bezafibrate <1> (<1>, potent inhibitor of methylation of phosphoethanolamine [18]) [18]
clofibric acid <1> (<1>, potent inhibitor of methylation of phosphoethanolamine [18]) [18]
Additional information <1> (<1>, phosphatidylethanolamine, monomethylphosphatidylethanolamine and dimethyl-phosphatidylethanolamine compete for a common active site [1,8]) [1, 8]

Activating compounds

1-methyl-4-phenylpyridinium <1> (<1>, enhances activity. S-Adenosylmethionine hypermethylation can be involved in the etiology of Parkinson's disease and an increase of phospholipid methylation could be one of the me-

chanisms by which 1-methyl-4-phenyl-pyridinium causes parkinsonism [25]) [25]
L-Cys <3> (<3>, enhances activity, optimal concentration 10 mM [12, 13]) [12, 13]
Triton X-100 <3> (<3>, enhances activity, optimal concentration 0.04% [12,13]) [12, 13]
Tween 20 <6> (<6>, stimulates [27]) [27]
Additional information <8> (<8>, light-activation is mediated by transducin [31]) [31]

Metals, ions

Ca^{2+} <4> (<4>, 1 mM, increases activity [14]) [14]
Mg^{2+} <1, 4, 5, 8> (<4>, 10 mM, increases activity [14]; <5>, activates [15]; <1>, 3-5 fold activation of formation of phosphatidylcholine and phosphatidyl N,N-dimethylethanolamine. Synthesis of phosphatidyl N-monomethylethanolamine is not affected [28]; <8>, required, maximal activity at 10 mM [30]) [14, 15, 28, 30]

Specific activity (U/mg)

0.273 <1> [4]
0.63 <1> (<1>, activity with phosphatidylethanolamine [1]) [1]
3.75 <1> (<1>, activity with phosphatidyl-N-monomethylethanolamine [1]) [1]
8.59 <1> (<1>, activity with phosphatidyl-N,N-dimethyl-ethanolamine [1]) [1]
Additional information <1, 3> [2, 12, 13]

K_m-Value (mM)

0.0019 <3> (S-adenosyl-L-methionine, <3>, reaction with phosphatidyl-N-monomethylethanolamine [12,13]) [12, 13]
0.0036 <3> (S-adenosyl-L-methionine, <3>, reaction with phosphatidylethanolamine [12,13]) [12, 13]
0.0039 <3> (S-adenosyl-L-methionine, <3>, phosphatidyl-N,N-dimethylethanolamine [12,13]) [12, 13]
0.0067 <1> (S-adenosyl-L-methionine) [9]
0.0136 <1> (S-adenosyl-L-methionine, <1>, reaction with phosphatidyl-N,N-dimethylethanolamine [8]) [8]
0.016 <1> (S-adenosyl-L-methionine, <1>, reaction with phosphatidyl-N,N-dimethylethanolamine [1]) [1]
0.0175 <3> (phosphatidyl-N-monomethylethanolamine) [12, 13]
0.02 <4> (S-adenosyl-L-methionine) [14]
0.022 <1> (S-adenosyl-L-methionine, <1>, reaction with phosphatidyl-N-monomethylethanolamine [1]) [1]
0.0296 <1> (S-adenosyl-L-methionine, <1>, reaction with phosphatidylethanolamine [8]) [8]
0.0336 <3> (phosphatidylethanolamine) [12, 13]
0.037 <8> (S-adenosyl-L-methionine, <8>, at pH 8.5 [30]) [30]
0.038 <8> (S-adenosyl-L-methionine, <8>, at pH 10 [30]) [30]

0.0397 <1> (S-adenosyl-L-methionine, <1>, reaction with phosphatidyl-N-monomethylethanolamine [8]) [8]
0.0405 <3> (phosphatidyl-N-monomethylethanolamine) [12, 13]
0.057 <2> (phosphatidylethanolamine) [3]
0.08 <1> (S-adenosyl-L-methionine, <1>, with phosphatidyl-N-monomethylethanolamine as substrate [16]) [16]
0.08 <1> (phosphatidyl-N-monomethylethanolamine) [1]
0.11 <1, 2> (S-adenosyl-L-methionine) [3, 10]
0.45 <1> (phosphatidyl-N,N-dimethylethanolamine) [1]
0.74 <1> (phosphatidyl-N-monomethylethanolamine) [8]
2.12 <1> (phosphatidyl-N,N-dimethylethanolamine) [8]
5 <1> (phosphatidylethanolamine) [8]
Additional information <1> [28]

K_i-Value (mM)

0.0048 <3> (S-adenosyl-L-homocysteine, <3>, towards phosphatidyl-N,N-dimethylethanolamine [12,13]) [12, 13]
0.0049 <1> (S-adenosyl-L-homocysteine, <1>, reaction with phosphatidyl-N-monomethylethanolamine [1]) [1]
0.0067 <1> (S-adenosyl-L-homocysteine, <1>, reaction with phosphatidyl-N,N-dimethylethanolamine [1]) [1]
0.0087 <3> (S-adenosyl-L-homocysteine, <3>, towards phosphatidyl-N-monomethylethanolamine [12,13]) [12, 13]
0.0118 <3> (S-adenosyl-L-homocysteine, <3>, towards phosphatidylethanolamine [12,13]) [12, 13]
0.012 <2> (S-adenosyl-L-homocysteine, <2>, inhibition with respect to S-adenosylmethionine [3]) [3]
0.0739 <1> (S-adenosyl-L-homocysteine, <1>, reaction with phosphatidyl-N-dimethylethanolamine [1]) [1]
0.138 <1> (S-adenosyl-L-homocysteine, <1>, reaction with phosphatidyl-N-monomethylethanolamine [8]) [8]
0.17 <2> (S-adenosyl-L-homocysteine, <2>, inhibition with respect to phosphatidylethanolamine [3]) [3]
0.2775 <1> (S-adenosyl-L-homocysteine, <1>, reaction with phosphatidylethanolamine [8]) [8]

pH-Optimum

6.5 <1> (<1>, and a second optimum at pH 9.5 [9]) [9]
8 <1, 8> (<8>, at low concentrations of S-adenosylmethionine, 0.01 mM [30]) [10, 30]
8.5 <3> [12, 13]
9 <2, 4, 6> [3, 14, 27]
9.5 <1> (<1>, and a second optimum at pH 6.5 [9]) [9]
10 <1, 8> (<1>, methylation of phosphatidylethanolamine, phosphatidyl-N-monomethylethanolamine and phosphatidyl-N,N-dimethylethanolamine [11]; <8>, at high concentrations of S-adenosylmethionine, 0.2 mM [30]) [11, 28, 30]

Additional information <1> (<1>, activity increases in nearly a linear fashion between pH 6.0 and 9.5 [1]) [1]

Temperature optimum (°C)

40 <2> [3]

4 Enzyme Structure

Molecular weight

120000 <3> (<3>, gel filtration [12,13]) [12, 13]
200000 <1> (<1>, gel filtration [5]) [5]

Subunits

? <1> (<1>, 18300, SDS-PAGE [11]) [11]
dimer <1> (<1>, 1 * 25000, enzyme can exist as monomer and as dimer, SDS-PAGE [4]) [4]
monomer <1> (<1>, 1 * 18300, SDS-PAGE [1]; <1>,1 * 25000, enzyme can exist as monomer and as dimer, SDS-PAGE [4]) [1, 4]
tetramer <1> (<1>, $\alpha x \beta y$, x * 50000 + y * 50000, SDS-PAGE [5]) [5]
trimer <3> (<3>, 3 * 42000, SDS-PAGE [12,13]) [12, 13]
Additional information <1> (<1>, the catalytic subunit of 15000-18000 Da exists in the native form as a complex of approximately 200000 Da in association with other membrane-bound proteins and lipids [6]) [6]

Posttranslational modification

phosphoprotein <1> (<1>, the dimeric enzyme form can be phosphorylated by cyclic AMP-dependent protein kinase [4]; the enzyme can be converted by cyclic AMP-dependent protein kinase from a low activity dephosphorylated enzyme form to a high activity phosphorylated form [7]) [4, 7]

5 Isolation/Preparation/Mutation/Application

Source/tissue

Yoshida ascites hepatoma cell <1> [26]
brain <1, 6> (<6>, frontal and occipital cortex [27]; <6>, a decrease in enzyme activity is observed in the frontal cortex of the brain affected with Alzheimer‘s disease [27]) [25, 27, 28]
colon <1> [10]
eye <8> [31]
hepatocyte <1> (<1>, high level of activity is expressed in defined extranuclear regions [17]) [17]
hepatoma cell <1> (<1>, PEMT2 protein decreases in preneoplastic nodules and virtually disappears in hepatocellular carcinoma [32]) [32]
hepatoma cell <1> (<1>, enzyme form PEMT2 is absent in hepatoma cells growing exponentially in vivo. Both PEMT2 and mRNA appears when cells enter the stationary phase of tumor growth [26]) [26]

liver <1, 6> (<1>, in prenatal livers enzyme form PEMT2 is completely absent, high level of PEMT2 appears at birth and is sustained throughout adult life [17]; <1>, normal liver, nodules, advanced nodules and hepatomas. Enzyme form PEMT2 is diminished during liver carcinogenesis [20]) [1, 2, 4, 5, 6, 7, 8, 11, 16, 17, 19, 20, 21, 23, 25, 26, 28, 29, 32]
lung <5> [15]
ovary cell line <7> (<7>, ovary cell line MT58, with impaired CDP-choline pathway caused by a temperature-sensitive mutation in CTP:phosphocholine cytidyltransferase. PEMT2 expression fails to rescue the mutant cell line at 40°C because it does not maintain phosphatidylcholine levels required for cellular replication [22]; <7>, cells express only a trace amount of phosphatidylethanolamine N-methyltransferase. Transfection of CHO cells, ovary cell line MT58, with PEMT2 cDNA behind the cytomegalovirus promoter result in a series of cell lines that overexpress PEMT2 [24]) [22, 24]
pituitary gland <1> [9]
retina <8> (<8>, rod outer segment [30]) [30]
thymocyte <4> [14]

Localization
brush border <1> [10]
cytosol <1> [28]
endoplasmic reticulum <1, 6> (<1>, enzyme form PEMT1 [16]) [1, 2, 6, 16, 19]
membrane <1, 2, 8> (<2>, associated to [3]; <1>, loosely bound to membrane [9]; <1>, enzyme form PEMT2 [17]; <1,6> mitochondria-associated membrane [16,19,29]; <6>, topographical model: 4 transmembrane regions span the membrane such that both the N- and C-termini of the enzyme are localized external to the ER. Two hydrophilic connecting loops protrude into the luminal face of the microsomes whereas a corresponding loop protrudes on the cytosolic side remains proximate to the membrane [19]; <8>, intrinsic membrane protein [30]) [3, 9, 11, 16, 19, 19, 30]
microsome <1, 5> [1, 4, 5, 11, 15]
mitochondrion <1, 6> (<1,6>, mitochondria associated membrane [16,19,29]) [16, 19, 29]
plasma membrane <4> [14]

Purification
<1> [1, 2, 4, 5, 6, 11, 16]
<3> [12, 13]
<4> (partial [14]) [14]
<5> [15]

Cloning
<1> (PEMT gene encodes enzyme form PEMT1 and PEMT2 [16]) [16]

6 Stability

Temperature stability

50 <5> (<5>, 30 min, complete loss of activity [15]) [15]

General stability information

<1>, 20% loss of activity after trypsin treatment, 0.001 mg trypsin per 0.05 mg membrane protein [10]

Storage stability

<1>, 2°C, 30% of the activity is lost after 12 h [2]

References

[1] Ridgway, N.D.; Vance, D.E.: Phosphatidylethanolamine N-methyltransferase from rat liver. Methods Enzymol., **209**, 366-374 (1992)
[2] Vance, D.E.; Schneider, W.J.: Conversion of phosphatidylethanolamine to phosphatidylcholine. Methods Enzymol., **71**, 581-588 (1981)
[3] Gaynor, P.M.; Carman, G.M.: Phosphatidylethanolamine methyltransferase and phospholipid methyltransferase activities from Saccharomyces cerevisiae. Enzymological and kinetic properties. Biochim. Biophys. Acta, **1045**, 156-163 (1990)
[4] Pajares, M.A.; Villalba, M.; Mato, J.M.: Purification of phospholipid methyltransferase from rat liver microsomal fraction. Biochem. J., **237**, 699-705 (1986)
[5] McBride, K.; Furness, L.M.; Gibbons, W.A.: Lipid methyltransferase: purification and structure. Biochem. Soc. Trans., **15**, 924-925 (1987)
[6] Fonteh, A.N.; Gibbons, W.A.: Solubilization and purification of lipid methyltransferase. Biochem. Soc. Trans., **18**, 266-267 (1990)
[7] Varela, I.; Merida, I.; Pajares, M.; Villalba, M.; Mato, J.M.: Activation of partially purified rat liver lipid methyltransferase by phosphorylation. Biochem. Biophys. Res. Commun., **122**, 1065-1070 (1984)
[8] Ridgway, N.D.; Vance, D.E.: Kinetic mechanism of phosphatidylethanolamine N-methyltransferase. J. Biol. Chem., **263**, 16864-16871 (1988)
[9] Prasad, C.; Edwards, R.M.: Synthesis of phosphatidylcholine from phosphatidylethanolamine by at least two methyltransferases in rat pituitary extracts. J. Biol. Chem., **256**, 13000-13003 (1981)
[10] Dudeja, P.K.; Foster, E.S.; Brasitus, T.A.: Synthesis of phosphatidylcholine by two distinct methyltransferases in rat colonic brush-border membranes: evidence for extrinsic and intrinsic membrane activities. Biochim. Biophys. Acta, **875**, 493-500 (1986)
[11] Ridgway, N.D.; Vance, D.E.: Purification of phosphatidylethanolamine N-methyltransferase from rat liver. J. Biol. Chem., **262**, 17231-17239 (1987)
[12] Tahara, Y.; Ogawa, Y.; Sakakibara, T.; Yamada, Y.: Purification and characterization of phosphatidylethanolamine N-methyltransferase from Zymomonas mobilis. Agric. Biol. Chem., **51**, 1425-1430 (1987)

[13] Tahara, Y.; Ogawa, Y.; Sakakibara, T.; Yamada, Y.: Phosphatidylethanolamine N-methyltransferase from Zymomonas mobilis: purification and characterization. Agric. Biol. Chem., **50**, 257-259 (1986)
[14] Makishima, F.; Toyoshima, S.; Osawa, T.: Partial purification and characterization of phospholipid N-methyltransferases from murine thymocyte microsomes. Arch. Biochem. Biophys., **238**, 315-324 (1985)
[15] Morgan, T.E.: Isolation and characterization of lipid N-methyltransferase from dog lung. Biochim. Biophys. Acta, **178**, 21-34 (1969)
[16] Vance, D.E.; Walkey, C.J.; Cui, Z.: Phosphatidylethanolamine N-methyltransferase from liver. Biochim. Biophys. Acta, **1348**, 142-150 (1997)
[17] Cui, Z.; Shen, Y.J.; Vance, D.E.: Inverse correlation between expression of phosphatidylethanolamine N-methyltransferase-2 and growth rate of perinatal rat livers. Biochim. Biophys. Acta, **1346**, 10-16 (1997)
[18] Nishimaki-Mogami, T.; Suzuki, K.; Takahashi, A.: The role of phosphatidylethanolamine methylation in the secretion of very low density lipoproteins by cultured rat hepatocytes: rapid inhibition of phosphatidylethanolamine methylation by bezafibrate increases the density of apolipoprotein B48-containing lipoproteins. Biochim. Biophys. Acta, **1304**, 21-31 (1996)
[19] Shields, D.J.; Lehner, R.; Agellon, L.B.; Vance, D.E.: Membrane topography of human phosphatidylethanolamine N-methyltransferase. J. Biol. Chem., **278**, 2956-2962 (2003)
[20] Tessitore, L.; Dianzani, I.; Cui, Z.; Vance, D.E.: Diminished expression of phosphatidylethanolamine N-methyltransferase 2 during hepatocarcinogenesis. Biochem. J., **337**, 23-27 (1999)
[21] Tessitore, L.; Cui, Z.; Vance, D.E.: Transient inactivation of phosphatidylethanolamine N-methyltransferase-2 and activation of cytidine triphosphate:phosphocholine cytidylyltransferase during non-neoplastic liver growth. Biochem. J., **322**, 151-154 (1997)
[22] Waite, K.A.; Vance, D.E.: Why expression of phosphatidylethanolamine N-methyltransferase does not rescue Chinese hamster ovary cells that have an impaired CDP-choline pathway. J. Biol. Chem., **275**, 21197-21202 (2000)
[23] Cui, Z.; Vance, D.E.: Expression of phosphatidylethanolamine N-methyltransferase-2 is markedly enhanced in long term choline-deficient rats. J. Biol. Chem., **271**, 2839-2843 (1996)
[24] Lee, M.W.; Bakovic, M.; Vance, D.E.: Overexpression of phosphatidylethanolamine N-methyltransferase 2 in CHO-K1 cells does not attenuate the activity of the CDP-choline pathway for phosphatidylcholine biosynthesis. Biochem. J., **320**, 905-910 (1996)
[25] Lee, E.S.Y.; Charlton, C.G.: 1-Methyl-4-phenyl-pyridinium increases S-adenosyl-L-methionine dependent phospholipid methylation. Pharmacol. Biochem. Behav., **70**, 105-114 (2001)
[26] Tessitore, L.; Sesca, E.; Bosco, M.; Vance, D.E.: Expression of phosphatidylethanolamine N-methyltransferase in Yoshida ascites hepatoma cells and the livers of host rats. Carcinogenesis, **20**, 561-567 (1999)
[27] Guan, Z.Z.; Wang, Y.N.; Xiao, K.Q.; Hu, P.S.; Liu, J.L.: Activity of phosphatidylethanolamine-N-methyltransferase in brain affected by Alzheimer's disease. Neurochem. Int., **34**, 41-47 (1999)

[28] Kim, Y.J.; Park, H.S.; Choi, M.U.: Detection of cytosolic phosphatidylethanolamine N-methyltransferase in rat brain. J. Biochem. Mol. Biol., **31**, 170-176 (1998)
[29] Vance, D.E.: Phosphatidylethanolamine N-methyltransferase: Unexpected findings from curiosity-driven research. Eur. J. Med. Res., **1**, 182-188 (1996)
[30] Roque, M.E.; Giusto, N.M.: Phosphatidylethanolamine N-methyltransferase activity in isolated rod outer segments from bovine retina. Exp. Eye Res., **60**, 631-643 (1995)
[31] Roque, M.E.; Salvador, G.A.; Giusto, N.M.: Light activation of phosphatidylethanolamine N-methyltransferase in rod outer segments and its modulation by association states of transducin. Exp. Eye Res., **69**, 555-562 (1999)
[32] Tessitore, L.; Sesca, E.; Vance, D.E.: Inactivation of phosphatidylethanolamine N-methyltransferase-2 in aflatoxin-induced liver cancer and partial reversion of the neoplastic phenotype by PEMT transfection of hepatoma cells. Int. J. Cancer, **86**, 362-367 (2000)

Polysaccharide O-methyltransferase 2.1.1.18

1 Nomenclature

EC number
2.1.1.18

Systematic name
S-adenosyl-L-methionine:1,4-α-D-glucan 6-O-methyltransferase

Recommended name
polysaccharide O-methyltransferase

Synonyms
acylpolysaccharide 6-methyltransferase
methyltransferase, acylpolysacharide 6-O
methyltransferase, polysaccharide
polysaccharide methyltransferase

CAS registry number
37205-56-4

2 Source Organism

<1> *Mycobacterium phlei* [1, 2]

3 Reaction and Specificity

Catalyzed reaction
S-adenosyl-L-methionine + 1,4-α-D-glucooligosaccharide = S-adenosyl-L-homocysteine + oligosaccharide containing 6-methyl-D-glucose units

Reaction type
methyl group transfer

Natural substrates and products
S S-adenosyl-L-methionine + 1,4-α-D-glucooligosaccharide <1> (<1>, the enzyme is involved in biosynthesis of lipopolysaccharide [1]; <1>, the enzyme may function in the biosynthesis of the 10 α(1-4)-linked 6-O-methyl-D-glucose residues of mycobacterial polysaccharide [2]) (Reversibility: ? <1> [1, 2]) [1, 2]
P S-adenosyl-L-homocysteine + oligosaccharide containing 6-methyl-D-glucose units <1> [1]

Substrates and products

S S-adenosyl-L-methionine + 1,4-α-D-glucooligosaccharide <1> (<1>, D-glucooligosaccharides with α-(1-4)-linkages prepared by partial acetolysis of amylose [1]) (Reversibility: ? <1> [1, 2]) [1, 2]

P S-adenosyl-L-homocysteine + oligosaccharide containing 6-methyl-D-glucose units <1> [1]

Activating compounds

NH_4^+ <1> (<1>, activates [1]) [1]

Metals, ions

Mg^{2+} <1> (<1>, maximal activation at 7 mM [1]) [1]

K_m-Value (mM)

0.1 <1> (S-adenosyl-L-methionine) [1]

pH-Optimum

7 <1> [1]

5 Isolation/Preparation/Mutation/Application

Localization

soluble <1> [1]

References

[1] Ferguson, J.A.; Ballou, C.E.: Biosynthesis of a mycobacterial lipopolysaccharide. Properties of the polysaccharide methyltransferase. J. Biol. Chem., **245**, 4213-4223 (1970)

[2] Grellert, E.; Ballou, C.E.: Biosynthesis of a mycobacterial lipopolysaccharide. Evidence for an acylpolysaccharide methyltransferase. J. Biol. Chem., **247**, 3236-3241 (1972)

Trimethylsulfonium-tetrahydrofolate N-methyltransferase 2.1.1.19

1 Nomenclature

EC number
2.1.1.19

Systematic name
trimethylsulfonium:tetrahydrofolate N-methyltransferase

Recommended name
trimethylsulfonium-tetrahydrofolate N-methyltransferase

Synonyms
trimethylsulfonium-tetrahydrofolate methyltransferase

CAS registry number
37256-92-1

2 Source Organism

<1> *Pseudomonas sp.* (<1> atypical pseudomonad with two lateral, subpolar flagellae [1]) [1]

3 Reaction and Specificity

Catalyzed reaction
trimethylsulfonium + tetrahydrofolate = dimethylsulfide + 5-methyltetrahydrofolate

Reaction type
methyl group transfer

Substrates and products
S trimethylsulfonium + tetrahydrofolate <1> (Reversibility: ? <1> [1]) [1]
P dimethylsulfide + 5-methyltetrahydrofolate <1> [1]
S Additional information <1> (<1> S-adenosylmethionine and dimethyl-β-propiothetin are not accepted as methyl donors [1]) [1]
P ?

Inhibitors
2-mercaptoethanol <1> (<1> inhibitory in excess [1]) [1]

Cofactors/prosthetic groups
tetrahydrofolate <1> [1]

Activating compounds
2-mercaptoethanol <1> (<1> reducing agent required for activity [1]) [1]
ascorbate <1> (<1> can replace 2-mercaptoethanol, best reducing agent for reaction assay [1]) [1]
homocysteine <1> (<1> can replace 2-mercaptoethanol [1]) [1]

Specific activity (U/mg)
0.007 <1> (<1> purified enzyme [1]) [1]

K_m-Value (mM)
1 <1> (tetrahydrofolate) [1]
5.8 <1> (trimethylsulfonium) [1]

pH-Optimum
7.8 <1> (<1> broad optimum [1]) [1]

4 Enzyme Structure

Molecular weight
100000 <1> (<1> sucrose density gradient centrifugation) [1]

5 Isolation/Preparation/Mutation/Application

Purification
<1> [1]

6 Stability

Storage stability
<1>, -20°C, 80% activity retained after 4 months [1]

References

[1] Wagner, C.; Lusty, S.M.; Kung, H.; Rogers, N.L.: Preparation and properties of trimethylsulfonium-tetrahydrofolate methyltransferase. J. Biol. Chem., **242**, 1287-1293 (1967)

Glycine N-methyltransferase 2.1.1.20

1 Nomenclature

EC number
2.1.1.20

Systematic name
S-adenosyl-L-methionine:glycine N-methyltransferase

Recommended name
glycine N-methyltransferase

Synonyms
GNMT
S-adenosyl-L-methionine:glycine methyltransferase
glycine methyltransferase
methyltransferase, glycine

CAS registry number
37228-72-1

2 Source Organism

<1> *Oryctolagus cuniculus* [1, 15]
<2> *Rattus norvegicus* [2-15, 17, 18, 20, 21]
<3> *Mus musculus* [15, 19]
<4> *Homo sapiens* [15, 16]

3 Reaction and Specificity

Catalyzed reaction
S-adenosyl-L-methionine + glycine = S-adenosyl-L-homocysteine + sarcosine (<2> mechanism [9,10]; <2> enzyme is also a polycyclic aromatic hydrocarbon binding protein [11]; <2> homodimer of enzyme acts as polycyclic aromatic hydrocarbon binding protein involved in cytochrome P-450IA1 expression [12,14]; <2> N-terminal acetylation or deprotonation is required for cooperative reaction mechanism [13]; <1-4> overview mechanism [15])

Reaction type
methyl group transfer

Natural substrates and products

S S-adenosyl-L-methionine + glycine <2> (<2> regulates the ratio of S-adenosylmethionine to S-adenosylhomocysteine [5]) [5]

P S-adenosyl-L-homocysteine + sarcosine

Substrates and products

S S-adenosyl-L-methionine + glycine <1-4> (<1,2> strict specificity for glycine as methyl acceptor [1,2]) (Reversibility: ? <1-3> [1-7, 15, 16]) [1-7, 15, 16]

P S-adenosyl-L-homocysteine + sarcosine <1, 2> [1-7]

Inhibitors

5'-[*p*-(fluorosulfonyl)benzoyl]adenosine <2> [4, 6]
5,5'-dithiobis(2-nitrobenzoate) <2> [6]
5-methyltetrahydrofolate hexaglutamate <2> [5]
5-methyltetrahydrofolate pentaglutamate <2> [5, 7]
5-methyltetrahydrofolate triglutamate <2> [5]
5-methyltetrahydrofolic acid <2> [5]
5-methyltetrahydropteroylpentaglutamate <1-4> [2, 5, 15, 16]
S-adenosyl-L-homocysteine <1> [1]
acetate <1> [1]
folinic acid <2> [5]
iodoacetate <2> [5]
methotrexate <2> [5]
p-chloromercuribenzoate <1> [1]
sarcosine <1> (<1> weak [1]) [1]
thioglycolic acid <1> (<1> weak [1]) [1]

Activating compounds

phosphate <2> (<2> in vitro phosphorylation increases activity, about 0.55 mol of phosphate present per mol of N-methyltransferase subunit [2]) [2]

Specific activity (U/mg)

0.132 <2> [5]
0.7 <1> [1]

K_m-Value (mM)

0.03 <2> (S-adenosyl-L-methionine) [3]
0.1 <1> (S-adenosyl-L-methionine) [1]
0.13 <2> (glycine) [3]
2.2 <1> (glycine) [1]

pH-Optimum

8.4-8.8 <1> [1]
9 <2> [3]

Temperature optimum (°C)

37 <1> (<1> assay at [1]) [1]

4 Enzyme Structure

Molecular weight

123500 <1> (<1> sedimentation equilibrium centrifugation [1]) [1]
130000-132000 <2> (<2> sedimenation equilibrium centrifugation, gel filtration [3]) [3]
135000 <2> (<2> gel filtration [5]) [5]

Subunits

? <2> (x * 34000, rat, SDS-PAGE) [5]
tetramer <2> (<2> 4 * 31500, identical subunits, SDS-PAGE [3]; <2> 4 * 32500, deduced from cDNA, crystal structure [8]) [3, 8]
trimer or tetramer <1> (<1> 3 or 4 * 27000-33000, nonidentical subunits, SDS-PAGE [1]) [1]

Posttranslational modification

glycoprotein <1> (<1> 4 residues of sialic acid and 2 residues of hexose per mol of protein [1]) [1]

5 Isolation/Preparation/Mutation/Application

Source/tissue

jejunum <2> [20]
liver <1-4> [1-4, 6-9, 11, 13-16, 20]
pancreas <2> [5, 20]
Additional information <1-4> (<1-4> overview on tissue distribution [15]; <2> tissue distribution [20]) [15]

Localization

cytosol <2> [2]
nucleus <2> [14]
Additional information <1-4> (<1-4> overview on subcellular localization [15]) [15]

Purification

<1> [1]
<2> (overview [15]) [3, 5, 15]

Crystallization

<2> (including R175K mutant [10]) [8-10, 15]

Cloning

<3> [19]
<1-4> (overview [15]) [15]

Engineering

R175K <2> (<2> crystal structure: N-terminal domains of subunits have moved out of the active sites of adjacent subunits [10]) [10]
Additional information <4> (<4> patients with mutations Leu49Pro and His176Asn suffer from mild liver disease [16]) [16]

Application

medicine <2, 4> (<4> patients with mutations Leu49Pro and His176Asn suffer from mild liver disease [16]; <2> vitamin A may induce enzyme, induction by all-trans-retinoic acid can lead to impairment of essential transmethylation processes [17]; <2> activation and induction of enzyme by retinoids are tissue- and gender-specific [21]; <2> feeding with excess methionine leads to upregulation of hepatic enzyme [18]) [16, 17, 18]

References

[1] Heady, J.E.; Kerr, S.J.: Purification and characterization of glycine N-methyltransferase. J. Biol. Chem., **248**, 69-72 (1973)

[2] Wagner, C.; Decha-Umphai, W.; Corbin, J.: Phosphorylation modulates the activity of glycine N-methyltransferase, a folate binding protein. In vitro phosphorylation is inhibited by the natural folate ligand. J. Biol. Chem., **264**, 9638-9642 (1989)

[3] Ogawa, H.; Fujioka, M.: Purification and properties of glycine N-methyltransferase from rat liver. J. Biol. Chem., **257**, 3447-3452 (1982)

[4] Fujioka, M.; Ishiguro, Y.: Reaction of rat liver glycine methyltransferase with 5-*p*-fluorosulfonylbenzoyladenosine. J. Biol. Chem., **261**, 6346-6351 (1986)

[5] Yeo, E.J.; Wagner, C.: Purification and properties of pancreatic glycine N-methyltransferase. J. Biol. Chem., **267**, 24669-24674 (1992)

[6] Fujioka, M.; Takata, Y.; Konishi, K.; Ogawa, H.: Function and reactivity of sulfhydryl groups of rat liver glycine methyltransferase. Biochemistry, **26**, 5696-5702 (1987)

[7] Wagner, C.; Briggs, W.T.; Cook, R.J.: Inhibition of glycine N-methyltransferase activity by folate derivatives: implications for regulation of methyl group metabolism. Biochem. Biophys. Res. Commun., **127**, 746-752 (1985)

[8] Pattanayek, R.; Newcomer, M.E.; Wagner, C.: Crystal structure of apo-glycine N-methyltransferase (GNMT). Protein Sci., **7**, 1326-1331 (1998)

[9] Fu, Z.; Hu, Y.; Konishi, K.; Takata, Y.; Ogawa, H.; Gomi, T.; Fujioka, M.; Takusagawa, F.: Crystal structure of glycine N-methyltransferase from rat liver. Biochemistry, **35**, 11985-11993 (1996)

[10] Huang, Y.; Komoto, J.; Konishi, K.; Takata, Y.; Ogawa, H.; Gomi, T.; Fujioka, M.; Takusagawa, F.: Mechanisms for auto-inhibition and forced product release in glycine N-methyltransferase: Crystal structures of wild-type, mutant R175K and S-adenosylhomocysteine-bound R175K enzymes. J. Mol. Biol., **298**, 149-162 (2000)

[11] Bhat, R.; Bresnick, E.: Glycine N-methyltransferase is an example of functional diversity. Role as a polycyclic aromatic hydrocarbon-binding receptor. J. Biol. Chem., **272**, 21221-21226 (1997)

[12] Bhat, R.; Wagner, C.; Bresnick, E.: The homodimeric form of glycine N-methyltransferase acts as a polycyclic aromatic hydrocarbon-binding receptor. Biochemistry, **36**, 9906-9910 (1997)

[13] Ogawa, H.; Gomi, T.; Imamura, T.; Kobayashi, M.; Huh, N.H.: Rat liver 4 S-benzo[a]pyrene-binding protein is distinct from glycine N-methyltransferase. Biochem. Biophys. Res. Commun., **233**, 300-304 (1997)

[14] Raha, A.; Wagner, C.; MacDonald, R.G.; Bresnick, E.: Rat liver cytosolic 4 S polycyclic aromatic hydrocarbon-binding protein is glycine N-methyltransferase. J. Biol. Chem., **269**, 5750-5756 (1994)

[15] Ogawa, H.; Gomi, T.; Takusagawa, F.; Fujioka, M.: Structure, function and physiological role of glycine N-methyltransferase. Int. J. Biochem. Cell Biol., **30**, 13-26 (1998)

[16] Luka, Z.; Cerone, R.; Phillips, J.A.; Mudd, S.H.; Wagner, C.: Mutations in human glycine N-methyltransferase give insights into its role in methionine metabolism. Hum. Genet., **110**, 68-74 (2002)

[17] Rowling, M.J.; McMullen, M.H.; Schalinske, K.L.: Vitamin A and its derivatives induce hepatic glycine N-methyltransferase and hypomethylation of DNA in rats. J. Nutr., **132**, 365-369 (2002)

[18] Rowling, M.J.; McMullen, M.H.; Chipman, D.C.; Schalinske, K.L.: Hepatic glycine N-methyltransferase is up-regulated by excess dietary methionine in rats. J. Nutr., **132**, 2545-2550 (2002)

[19] Aida, K.; Tawata, M.; Negishi, M.; Onaya, T.: Mouse glycine N-methyltransferase is sexually dimorphic and regulated by growth hormone. Horm. Metab. Res., **29**, 646-649 (1997)

[20] Yeo, E.J.; Wagner, C.: Tissue distribution of glycine N-methyltransferase, a major folate-binding protein of liver. Proc. Natl. Acad. Sci. USA, **91**, 210-214 (1994)

[21] McMullen, M.H.; Rowling, M.J.; Ozias, M.K.; Schalinske, K.L.: Activation and induction of glycine N-methyltransferase by retinoids are tissue- and gender-specific. Arch. Biochem. Biophys., **401**, 73-80 (2002)

Methylamine-glutamate N-methyltransferase 2.1.1.21

1 Nomenclature

EC number
2.1.1.21

Systematic name
methylamine:L-glutamate N-methyltransferase

Recommended name
methylamine-glutamate N-methyltransferase

Synonyms
N-methylglutamate synthase
methylamine-glutamate methyltransferase

CAS registry number
9045-32-3

2 Source Organism

<1> *Pseudomonas sp.* (strain AM 1, no activity when grown without methylamine in growth medium [1]; strain ATCC 23819 [2]; strain AM 1 [3]) [1-5]

3 Reaction and Specificity

Catalyzed reaction
methylamine + L-glutamate = NH_3 + N-methyl-L-glutamate

Reaction type
alkyl group transfer
methyl group transfer

Natural substrates and products
S methylamine + L-glutamate <1> (<1> involved in one carbon metabolism [1,4]) (Reversibility: r <1> [1-5]) [1-5]
P NH_3 + N-methyl-L-glutamate <1> [1-5]

Substrates and products
S N-methyl-L-glutamate + ammonia <1> (<1> at 54% the rate of the reaction with cysteamine [4]) (Reversibility: ? <1> [4]) [4]
P ?

S N-methyl-L-glutamate + butylamine <1> (<1> at 7% the rate of the reaction with ammonia [4]) (Reversibility: ? <1> [4]) [4]
P ?
S N-methyl-L-glutamate + ethanolamine <1> (<1> at 27% the rate of the reaction with ammonia [4]) (Reversibility: ? <1> [4]) [4]
P ?
S N-methyl-L-glutamate + ethylamine <1> (at 32% the rate of the reaction with ammonia [4]) (Reversibility: ? <1> [4]) [4]
P ?
S N-methyl-L-glutamate + methylamine <1> (<1> at 70% the rate of the reaction with ammonia [4]) (Reversibility: ? <1> [4]) [4]
P ?
S N-methyl-L-glutamate + propylamine <1> (<1> at 18% the rate of the reaction with ammonia [4]) (Reversibility: ? <1> [4]) [4]
P ?
S butylamine + L-glutamate <1> (<1> at 5% the rate of the reaction with methylamine [4]) (Reversibility: ? <1> [4]) [4]
P NH_3 + N-butyl-L-glutamate
S ethanolamine + L-glutamate <1> (<1> at 18% the rate of the reaction with methylamine [4]) (Reversibility: ? <1> [4]) [4]
P ?
S ethylamine + L-glutamate <1> (<1> at 16% the rate of reaction with methylamine [4]) (Reversibility: ? <1> [4]) [4]
P NH_3 + N-ethyl-L-glutamate
S methylamine + L-glutamate <1> (<1> ping pong bi bi kinetic pattern, glutaryl enzyme intermediate, no inversion during reaction [2,5]; no activity with D-glutamate [3]) (Reversibility: r <1> [1-5]) [1-5]
P NH_3 + N-methyl-L-glutamate <1> [1-5]
S methylamine + L-glutamate-γ-methylester <1> (Reversibility: r <1> [1]) [1]
P NH_3 + N-methyl-L-glutamate-γ-methyl ester
S methylamine + L-glutamine <1> (Reversibility: r <1> [1]) [1]
P ?
S propylamine + L-glutamate <1> (<1> at 8% the rate of the reaction with methylamine [4]) (Reversibility: ? <1> [4]) [4]
P NH_3 + N-propyl-L-glutamate

Inhibitors

N-ethylmaleimide <1> [1]
N-methylglutamate <1> (<1> weak product inhibition [2]) [2]
ammonia <1> (<1> competitive with respect to methylamine [1,2]) [1, 2]
arsenite <1> [1]
hydroxylamine <1> (<1> reversible [1]) [1]
iodoacetate <1> [1]
p-chloromercuribenzoate <1> [1]

Cofactors/prosthetic groups

5-deazaFMN <1> (<1> 1% of activity compared to FMN reconstituted enzyme [3,5]) [3, 5]
FMN <1> [2-5]

Turnover number (min^{-1})

0.17 <1> (L-glutamate, <1> 5-deazaFMN reconstituted enzyme [3]) [3]
20.8 <1> (L-glutamate, <1> FMN reconstituted enzyme [3]) [3]

Specific activity (U/mg)

0.138 <1> (<1> purified enzyme [4]) [4]
3.8 <1> (<1> 5-deazaFMN reconstituted enzyme [3]) [3]
79.4 <1> (<1> FMN reconstituted enzyme [3]) [3]
115 <1> (<1> native enzyme [3]) [3]

K_i-Value (mM)

49 <1> (ammonia) [1]

pH-Optimum

8.3 <1> [1]
9 <1> [4]

pH-Range

7-9 <1> [1]

4 Enzyme Structure

Molecular weight

30000-35000 <1> (<1> SDS-PAGE [4]) [4]
350000-390000 <1> (<1> density gradient centrifugation [4]) [4]

Subunits

multimer <1> (<1> SDS-PAGE, density gradient centrifugation [4]) [4]

5 Isolation/Preparation/Mutation/Application

Purification

<1> [3, 4]

References

[1] Shaw, W.V.; Tsai, L.; Stadtman, E.R.: The enzymatic synthesis of N-methylglutamic acid. J. Biol. Chem., **241**, 935-945 (1966)
[2] Pollock, R.J.; Hersh, L.B.: N-methylglutamate synthetase. The use of flavin mononucleotide in oxidative catalysis. J. Biol. Chem., **248**, 6724-6733 (1973)

[3] Schuman Jorns, M.; Hersh, L.B.: N-Methylglutamate synthetase. Substrate-flavin hydrogen transfer reactions probed with deazaflavin mononucleotide. J. Biol. Chem., **250**, 3620-3628 (1975)
[4] Hersh, L.B.: N-Methyl-L-glutamate synthase. Methods Enzymol., **113**, 36-42 (1985)
[5] Bornemann, S.: Flavoenzymes that catalyse reactions with no net redox change. Nat. Prod. Rep., **19**, 761-772 (2002)

Carnosine N-methyltransferase 2.1.1.22

1 Nomenclature

EC number
2.1.1.22

Systematic name
S-adenosyl-L-methionine:carnosine N-methyltransferase

Recommended name
carnosine N-methyltransferase

CAS registry number
37256-93-2

2 Source Organism

<1> *Gallus gallus* (White Rock chick) [1]
<2> *Guinea pig* [1]
<3> *Rattus norvegicus* [1]
<4> *Felis catus* [1]
<5> *Oryctolagus cuniculus* [1, 2]

3 Reaction and Specificity

Catalyzed reaction
S-adenosyl-L-methionine + carnosine = S-adenosyl-L-homocysteine + anserine

Reaction type
methyl group transfer

Natural substrates and products
S S-adenosyl-L-methionine + carnosine <1-5> (<1-5> biosynthesis of anserine [1]) (Reversibility: ? <1-5> [1]) [1, 2]
P S-adenosyl-L-homocysteine + anserine <1-5> [1]
S actin peptide H + S-adenosyl-L-methionine <5> (<5> synthetic peptide based on amino acid sequence of actin [2]) [2]
P actin peptide H methylated at N^1-position of histidine + S-adenosyl-L-homocysteine <5> [2]

Substrates and products

- **S** S-adenosyl-L-methionine + carnosine <1-5> (Reversibility: ? <1-5> [1]) [1]
- **P** S-adenosyl-L-homocysteine + anserine <1-5> [1]
- **S** S-adenosyl-L-methionine + histidine <2, 5> (Reversibility: ? <2, 5> [1, 2]) [1, 2]
- **P** S-adenosyl-L-homocysteine + methylhistidine <2, 5> [1, 2]
- **S** S-adenosyl-L-methionine + homocarnosine <2, 3> (Reversibility: ? <2, 3> [1]) [1]
- **P** S-adenosyl-L-homocysteine + homoanserine <2, 3> [1]
- **S** actin peptide H + S-adenosyl-L-methionine <5> (<5> synthetic peptide based on amino acid sequence of actin [2]) [2]
- **P** actin peptide H methylated at N^1-position of histidine + S-adenosyl-L-homocysteine <5> [2]

Metals, ions

Co^{2+} <1> (<1> inhibitory, reversible by addition of EDTA [1]) [1]
Mn^{2+} <1> (<1> inhibitory, reversible by addition of EDTA [1]) [1]
Zn^{2+} <1> (<1> inhibitory, reversible by addition of EDTA [1]) [1]

Specific activity (U/mg)

0.0005 <1> (<1> purified enzyme [1]) [1]

K_m-Value (mM)

0.09 <1> (S-adenosyl-L-methionine) [1]
1.5 <5> (actin peptide H) [2]
4 <1> (carnosine) [1]

pH-Optimum

7.5-8.5 <1> [1]

pH-Range

6-9 <1> [1]

4 Enzyme Structure

Molecular weight

85000 <5> (<5> gel filtration, sucrose gradient density centrifugation [2]) [2]

5 Isolation/Preparation/Mutation/Application

Source/tissue

brain <1-3> (<1-3> low activity [1]) [1]
heart <2, 3> (<2,3> low activity [1]) [1]
kidney <1-5> (<1,2,4,5> much lower activity than in muscle [1]; <3> about the same activity as in muscle [1]) [1]

liver <1, 3> (<1> same activity than in muscle [1]; <3> 20% of activity in muscle [1]) [1]
lung <2, 3> [1]
muscle <1-5> [1, 2]

Localization
soluble <1> [1]

Purification
<1> (partial [1]) [1]

6 Stability

Temperature stability
50 <1> (<1> no loss of activity within 5 min in presence of carnosine; 30% loss of activity when heated for 10 min in presence of carnosine, 70% loss of activity within 3 min in absence of carnosine [1]) [1]

General stability information
<1>, stable on freezing and thawing [1]

Storage stability
<1>, -20°C, very stable [1]

References

[1] McManus, R.: Enzymatic synthesis of anserine in skeletal muscle by N-methylation of carnosine. J. Biol. Chem., **237**, 1207-1211 (1962)
[2] Raghavan, M.; Lindberg, U.; Schutt, C.: The use of alternative substrates in the characterization of actin-methylating and carnosine-methylating enzymes. Eur. J. Biochem., **210**, 311-318 (1992)

Protein-arginine N-methyltransferase 2.1.1.23

1 Nomenclature

EC number

2.1.1.23 (transferred to EC 2.1.1.124, EC 2.1.1.125, EC 2.1.1.126)

Recommended name

protein-arginine N-methyltransferase

Protein-γ-glutamate O-methyltransferase 2.1.1.24

1 Nomenclature

EC number

2.1.1.24 (deleted, included in EC 2.1.1.77, EC 2.1.1.80, EC 2.1.1.100)

Recommended name

protein-γ-glutamate O-methyltransferase

Phenol O-methyltransferase 2.1.1.25

1 Nomenclature

EC number
2.1.1.25

Systematic name
S-adenosyl-L-methionine:phenol O-methyltransferase

Recommended name
phenol O-methyltransferase

Synonyms
PMT
methyltransferase, phenol

CAS registry number
37256-94-3

2 Source Organism

<1> *Oryctolagus cuniculus* [1]
<2> *Cavia porcellus* [1]
<3> *Mus musculus* [1]
<4> *Rattus norvegicus* [1]
<5> *Homo sapiens* [2, 3]

3 Reaction and Specificity

Catalyzed reaction
S-adenosyl-L-methionine + phenol = S-adenosyl-L-homocysteine + anisole

Reaction type
methyl group transfer

Substrates and products
S S-adenosyl-L-methionine + acetaminophen <5> (Reversibility: ? <5> [3]) [3]
P S-adenosyl-L-homocysteine + N-(4-methoxyphenyl)-acetamide
S S-adenosyl-L-methionine + *m*-bromophenol <1> (Reversibility: ? <1> [1]) [1]
P S-adenosyl-L-homocysteine + *m*-bromoanisole

S S-adenosyl-L-methionine + *m*-chlorophenol <1> (Reversibility: ? <1> [1]) [1]
P S-adenosyl-L-homocysteine + *m*-chloroanisole
S S-adenosyl-L-methionine + *m*-methyoxyphenol <1> (Reversibility: ? <1> [1]) [1]
P S-adenosyl-L-homocysteine + 1,3-dimethoxybenzene
S S-adenosyl-L-methionine + *m*-nitrophenol <1> (Reversibility: ? <1> [1]) [1]
P S-adenosyl-L-homocysteine + *m*-nitroanisole
S S-adenosyl-L-methionine + *o*-chlorophenol <1> (Reversibility: ? <1> [1]) [1]
P S-adenosyl-L-homocysteine + *o*-chloroanisole
S S-adenosyl-L-methionine + *o*-methylphenol <1> (Reversibility: ? <1> [1]) [1]
P S-adenosyl-L-homocysteine + *o*-methylanisole
S S-adenosyl-L-methionine + *o*-methyoxyphenyol <1> (Reversibility: ? <1> [1]) [1]
P S-adenosyl-L-homocysteine + veratrole
S S-adenosyl-L-methionine + *p*-chlorophenol <1> (Reversibility: ? <1> [1]) [1]
P S-adenosyl-L-homocysteine + *p*-chloroanisole <1> [1]
S S-adenosyl-L-methionine + *p*-ethylphenol <1> (Reversibility: ? <1> [1]) [1]
P S-adenosyl-L-homocysteine + *p*-ethylanisole
S S-adenosyl-L-methionine + *p*-hydroxyacetanilide <1> (Reversibility: ? <1> [1]) [1]
P S-adenosyl-L-homocysteine + *p*-methoxyacetanilide
S S-adenosyl-L-methionine + *p*-methoxyphenol <1> (Reversibility: ? <1> [1]) [1]
P S-adenosyl-L-homocysteine + 1,4-dimethoxybenzene
S S-adenosyl-L-methionine + *p*-methylphenol <1> (Reversibility: ? <1> [1]) [1]
P S-adenosyl-L-homocysteine + *p*-methylanisole
S S-adenosyl-L-methionine + phenol <1, 2, 3, 5> (Reversibility: ? <1, 2, 3, 5> [1, 3]) [1, 3]
P S-adenosyl-L-homocysteine + anisole <1> [1]

Inhibitors

Mg^{2+} <2> (<2>, 10 mM, 61% inhibition [1]) [1]
NEM <2> (<2>, 0.1 mM, 90% inhibition [1]) [1]
PCMB <2> (<2>, 0.01 mM, 85% inhibition [1]) [1]
SKF 525 <2> (<2>, 0.5 mM, 56% inhibition [1]) [1]

K_m-Value (mM)

0.0048 <5> (S-adenosyl-L-methionine) [3]
2 <5> (phenol) [3]
21.2 <5> (acetaminophen) [3]

pH-Optimum

7.6-8.3 <2> [1]
9 <5> [3]

5 Isolation/Preparation/Mutation/Application

Source/tissue

adrenal gland <2> (<2>, low activity [1]) [1]
erythrocyte <5> (<5>, activity is lower in membranes from uremic patients on maintenance hemodialysis than in blood from randomly selected subjects [2]) [2, 3]
gut <2> (<2>, low activity [1]) [1]
kidney <2> [1]
liver <1, 2> [1]
lung <2> [1]
spleen <2> (<2>, low activity [1]) [1]
testis <2> (<2>, low activity [1]) [1]

Localization

membrane <5> (<5>, from erythrocyte [2]) [2]
microsome <2> [1]

References

[1] Axelrod, J.; Daly, J.: Phenol-O-methyltransferase. Biochim. Biophys. Acta, **159**, 472-478 (1968)
[2] Pazmino, P.; Rogoff, F.; Weinshilboum, R.M.: Inhibition of human erythrocyte phenol-O-methyltransferase in uremia. Clin. Pharmacol. Ther., **26**, 464-472 (1979)
[3] Pazmino, P.A.; Weinshilboum, R.M.: Human erythrocyte phenol O-methyltransferase: radiochemical microassay and biochemical properties. Clin. Chim. Acta, **89**, 317-329 (1978)

Iodophenol O-methyltransferase 2.1.1.26

1 Nomenclature

EC number
2.1.1.26

Systematic name
S-adenosyl-L-methionine:2-iodophenol O-methyltransferase

Recommended name
iodophenol O-methyltransferase

Synonyms
DIB O-methyltransferase (3,5-diiodo-4-hydroxy-benzoic acid)
iodophenol methyltransferase

CAS registry number
37256-95-4

2 Source Organism

<-1> no activity in *Cavia porcellus* (liver [1]) [1]
<1> *Rattus norvegicus* (Sprague-Dawley rat [1]) [1]
<2> *Oryctolagus cuniculus* [1]
<3> *pigeon* [1]

3 Reaction and Specificity

Catalyzed reaction
S-adenosyl-L-methionine + 2-iodophenol = S-adenosyl-L-homocysteine + 2-iodophenol methyl ether

Reaction type
methyl group transfer

Substrates and products
S 3,5,3',5'-tetraiodothyroacetic acid + S-adenosyl-L-methionine <1> (Reversibility: ? <1> [1]) [1]
P [4-(4-methoxy-3,5-diiodophenoxy)-3,5-diiodophenyl]-acetic acid + S-adenosyl-L-homocysteine
S 3,5,3'-triiodo-L-thyronine + S-adenosyl-L-methionine <1> (<1> poor substrate [1]) (Reversibility: ? <1> [1]) [1]

P 3,5,3'-triiodo-O-methyl-threonine + S-adenosyl-L-homocysteine
S 3,5,3'-triiodothyroacetic acid + S-adenosyl-L-methionine <1> (<1> poor substrate [1]) (Reversibility: ? <1> [1]) [1]
P [3,5-diiodo-4-(3-iodo-4-methoxyphenoxy)phenyl]-acetic acid + S-adenosyl-L-homocysteine
S 3,5-diiodo-4-hydroxybenzoic acid + S-adenosyl-L-methionine <1-3> (<1-3> DIB is the most effective substrate [1]) (Reversibility: ? <1-3> [1]) [1]
P 3,5-diiodo-4-methoxybenzoic acid + S-adenosyl-L-homocysteine <1-3> [1]
S L-thyroxine + S-adenosyl-L-methionine <1> (<1> poor substrate [1]) (Reversibility: ? <1> [1]) [1]
P 3,5,3'5'-tetraiodo-O'-methyl-L-thyronine + S-adenosyl-L-homocysteine

Inhibitors
3,5-diiodo-4-hydroxybenzoic acid <1> (<1> substrate inhibition above 0.2 mM [1]) [1]
Additional information <1> (<1> activity is suppressed by extract from microsomal fraction [1]) [1]

Activating compounds
ATP <1> (<1> required [1]) [1]

Metals, ions
Mg^{2+} <1> (<1> required [1]) [1]

Specific activity (U/mg)
0.0002 <1> (<1> purified enzyme with 3,5-diiodo-4-hydroxybenzoic acid as substrate [1]) [1]

K_m-Value (mM)
0.0015 <1> (S-adenosyl-L-methionine) [1]
0.0034 <1> (3,5-diiodo-4-hydroxybenzoic acid) [1]

pH-Optimum
6 <1> [1]

pH-Range
5.2-8 <1> [1]

5 Isolation/Preparation/Mutation/Application

Source/tissue
kidney <1, 2> [1]
liver <1-3> [1]
Additional information <2, 3> (<2,3> no activity found in muscle [1]) [1]

Localization
soluble <1> [1]

Purification

<1> (partial [1]) [1]

References

[1] Tonita, K.; Cha, M.C.; Lardy, H.A.: Enzymic O-methylation of iodinated phenols and thyroide hormones. J. Biol. Chem., **239**, 1202-1207 (1964)

Tyramine N-methyltransferase 2.1.1.27

1 Nomenclature

EC number
2.1.1.27

Systematic name
S-adenosyl-L-methionine:tyramine N-methyltransferase

Recommended name
tyramine N-methyltransferase

Synonyms
DIB O-methyltransferase (3,5-diiodo-4-hydroxy-benzoic acid)
S-adenosyl-methionine:tyramine N-methyltransferase
methyltransferase, tyramine N-
tyramine methylpherase

CAS registry number
37256-96-5

2 Source Organism

<1> *Hordeum vulgare* (L. var. Betzes, barley [1]) [1]

3 Reaction and Specificity

Catalyzed reaction
S-adenosyl-L-methionine + tyramine = S-adenosyl-L-homocysteine + N-methyltyramine

Natural substrates and products
S Additional information <1> (<1> involved in hordenine biosynthesis [1]) [1]
P ?

Substrates and products
S S-adenosyl-L-methionine + α-methyltyramine <1> (<1> i.e. paredrine, 53% of activity with tyramine [1]) (Reversibility: ? <1> [1]) [1]
P ?
S S-adenosyl-L-methionine + tyramine <1> (<1> conversion rate is higher with (-)-L-diastereoisomer of methionine [1]) (Reversibility: ? <1> [1]) [1]

P N-methyltyramine + S-adenosyl-L-homocysteine <1> [1]
S β-phenethanolamine + S-adenosyl-L-methionine <1> (<1> 23% of activity with tyramine [1]) (Reversibility: ? <1> [1]) [1]
P ?
S *p*-chloroethylamine + S-adenosyl-L-methionine <1> (<1> 29% of activity with tyramine [1]) (Reversibility: ? <1> [1]) [1]
P ?

Inhibitors
N-methyltyramine <1> [1]
homotyramine <1> [1]
p-chloromercuribenzoate <1> (<1> strong [1]) [1]
Additional information <1> (<1> EDTA is not inhibitory [1]) [1]

Cofactors/prosthetic groups
Additional information <1> (<1> no cofactor requirement [1]) [1]

K_m-Value (mM)
0.02 <1> (tyramine) [1]
0.03 <1> (S-adenosyl-L-methionine) [1]

pH-Optimum
8.4 <1> [1]

Temperature optimum (°C)
26 <1> (<1> assay at [1]) [1]

5 Isolation/Preparation/Mutation/Application

Source/tissue
root <1> (<1> of seedling [1]) [1]

Purification
<1> (partial [1]) [1]

6 Stability

Temperature stability
37 <1> (<1> heat-labile: 70% loss of activity after 10 min [1]) [1]

Oxidation stability
<1>, sulfhydryl compound necessary for stabilization [1]

Storage stability
<1>, -18°C, addition of sucrose to a final concentration of 0.2 M, several weeks [1]
<1>, 4°C, pH 7.0 and 5 mM 2-mercaptoethanol, several days [1]

References

[1] Mann, J.D.; Mudd, S.H.: Alkaloids and plant metabolism. J. Biol. Chem., **238**, 381-385 (1963)

Phenylethanolamine N-methyltransferase 2.1.1.28

1 Nomenclature

EC number
2.1.1.28

Systematic name
S-adenosyl-L-methionine:phenylethanolamine N-methyltransferase

Recommended name
phenylethanolamine N-methyltransferase

Synonyms
NMT
PNMT
methyltransferase, noradrenaline N-
noradrenalin N-methyltransferase
noradrenalin methyltransferase
noradrenaline N-methyltransferase
norepinephrin N-methyltransferase
norepinephrine N-methyltransferase
norepinephrine methyltransferase
phenethanolamine N-methyltransferase
phenethanolamine methyltransferase

CAS registry number
9037-68-7

2 Source Organism

<1> *Macaca mulatta* [1]
<2> *Rattus norvegicus* [1, 4, 23]
<3> *Bos taurus* [1, 2, 4, 8, 9, 13, 14, 18, 19, 26, 27, 28, 29]
<4> *Oryctolagus cuniculus* [1, 3, 5, 6, 7, 10, 11, 12, 17]
<5> *Cavia porcellus* [1]
<6> *Felis catus* [1]
<7> *Ovis aries* [15]
<8> *Columba livia* [16]
<9> *Homo sapiens* [20, 22, 24]
<10> *Mus musculus* [21]
<11> *Heteropneustes fossilis* [25]

3 Reaction and Specificity

Catalyzed reaction

S-adenosyl-L-methionine + phenylethanolamine = S-adenosyl-L-homocysteine + N-methylphenylethanolamine (<3>, random binding mechanism with a kinetically preferred order of binding where S-adenosyl-L-methionine is the preferred first substrate [2]; <4>, ordered sequential reaction [7])

Reaction type

methyl group transfer

Natural substrates and products

S 9-methylnorharman + S-adenosyl-L-methionine <9> (<9>, PNMT catalyzes the 2N-methylation of β-carbolines, forming 2N-methylated β-carbolinium cations, which are structural and functional analogs of the Parkinsonian-inducing toxin MPP+ [24]) (Reversibility: ? <9> [24]) [24]
P ? + S-adenosyl-L-homocysteine
S Additional information <2, 3, 4, 7> (<4>, terminal enzyme in the biosynthetic pathway for epinephrine [10]; <7>, terminal enzyme in epinephrine biosynthesis [15]; <2>, final enzyme in the pathway for epinephrine biosynthesis [23]; <3>, enzyme is involved in biosynthesis of epinephrine [27]) [10, 15, 23, 27]
P ?

Substrates and products

S 2-hydroxyphenylethanolamine + S-adenosyl-L-methionine <4> (<4>, i.e. *o*-octopamine [10]) (Reversibility: ? <4> [10]) [10]
P ? + S-adenosyl-L-homocysteine
S 3,4-dichlorophenylethanolamine + S-adenosyl-L-methionine <1, 4> (Reversibility: ? <1,4> [1,10]) [1, 10]
P N-methyl-3,4-dichlorophenylethanolamine + S-adenosyl-L-homocysteine
S 3,4-dichlorophenylethylenediamine + S-adenosyl-L-methionine <1> (Reversibility: ? <1> [1]) [1]
P ? + S-adenosyl-L-homocysteine
S 3,4-dihydroxynorephedrine + S-adenosyl-L-methionine <1> (<1>, 18% of the activity with normetanephrine [1]) (Reversibility: ? <1> [1]) [1]
P 3,4-dihydroxyephedrine + S-adenosyl-L-homocysteine
S 3-chloro-4-hydroxyphenylethanolamine + S-adenosyl-L-methionine <1> (Reversibility: ? <1> [1]) [1]
P ? + S-adenosyl-L-homocysteine
S 3-hydroxyphenylethanolamine + S-adenosyl-L-methionine <4> (<4>, i.e. *m*-octopamine [10]) (Reversibility: ? <4> [10]) [10, 11]
P N-methyl-*m*-octopamine + S-adenosyl-L-homocysteine <4> [11]
S 4-hydroxynorephedrine + S-adenosyl-L-methionine <1> (<1>, 18% of the activity with normetanephrine [1]) (Reversibility: ? <1> [1]) [1]
P 4-hydroxyephedrine + S-adenosyl-L-homocysteine
S 4-hydroxyphenylethanolamine + S-adenosyl-L-methionine <4> (<4>, i.e. octopamine [10]) (Reversibility: ? <4> [10]) [10]

P ? + S-adenosyl-L-homocysteine
S 9-methylnorharman + S-adenosyl-L-methionine <9> (<9>, 2N-methylation [24]) (Reversibility: ? <9> [24]) [24]
P ? + S-adenosyl-L-homocysteine
S D-norepinephrine + S-adenosyl-L-methionine <1, 4> (<1>, 15% of the activity with normetanephrine [1]) (Reversibility: ? <1, 4> [1, 3]) [1, 3]
P epinephrine + S-adenosyl-L-homocysteine
S DL-*m*-hydroxyphenylethanolamine + S-adenosyl-L-methionine <1> (<1>, 60% of the activity with normetanephrine [1]) (Reversibility: ? <1> [1]) [1]
P neosynephrine + S-adenosyl-L-homocysteine
S DL-neo-synephrine + S-adenosyl-L-methionine <1> (<1>, 20% of the activity with normetanephrine [1]) (Reversibility: ? <1> [1]) [1]
P 3-hydroxy-α-[(dimethylamine)methyl]benzenemethanol + S-adenosyl-L-homocysteine
S DL-octopamine + S-adenosyl-L-methionine <1> (<1>, 18% of the activity with normetanephrine [1]) (Reversibility: ? <1> [1]) [1]
P ? + S-adenosyl-L-homocysteine
S L-norephedrine + S-adenosyl-L-methionine <1> (<1>, 14% of the activity with normetanephrine [1]) (Reversibility: ? <1> [1]) [1]
P ephedrine + S-adenosyl-L-homocysteine
S L-norepinephrine + S-adenosyl-L-methionine <1, 4, 8> (<1>, 21% of the activity with normetanephrine [1]) (Reversibility: ? <1, 4, 8> [1, 3, 7, 16]) [1, 3, 7, 16]
P epinephrine + S-adenosyl-L-homocysteine
S dopamine + S-adenosyl-L-methionine <3> (Reversibility: ? <3> [18]) [18]
P ? + S-adenosyl-L-homocysteine
S epinephrine + S-adenosyl-L-methionine <3> (Reversibility: ? <3> [18]) [18]
P ? + D-adenosylhomocysteine
S norepinephrine + S-adenosyl-L-methionine <3, 4, 9> (Reversibility: ? <3, 4, 9> [5, 6, 7, 10, 11, 12, 17, 22]) [5, 6, 7, 10, 11, 12, 17, 18, 22]
P epinephrine + S-adenosyl-L-homocysteine <4, 9> [10, 22]
S normetanephrine + S-adenosyl-L-methionine <1, 3> (<4>, i.e. 3-methoxy-4-hydroxyphenylethanolamine [10]) (Reversibility: ? <1, 3> [1, 2, 10]) [1, 2, 10]
P metanephrine + S-adenosyl-L-homocysteine
S phenylethanolamine + S-adenosyl-L-methionine <1, 3, 4> (<1>, 72% of the activity with normetanephrine [1]) (Reversibility: ? <1, 3, 4> [1, 3, 8, 10, 18]) [1, 3, 8, 10, 18]
P N-methylphenylethanolamine + S-adenosyl-L-homocysteine

Inhibitors

(R)-3-methyl-1,2,3,4-tetrahydroisoquinoline <3> [29]
(S)-3-methyl-1,2,3,4-tetrahydroisoquinoline <3> [29]
1,2,3,4-tetrahydroisoquinoline-7-carboxylic acid <3> [27]
2,3,4,5-tetrahydro-1H-2-benzazepine <3> [29]
2,3,4,5-tetrahydro-5H-1,4-benzodiazepine <3> [29]

2,3,4,5-tetrahydro-5H-1,4-benzothiazepine <3> [29]
2,3,4,5-tetrahydro-5H-1,4-benzoxazepine <3> [29]
2,5-dimethyl-1-aminobenzamidazole <2, 3> [4]
2-aminotetralin <4> (<4>, most effective inhibitor [17]) [17]
2-chlorophenylethanolamine <4> [10]
2-fluorophenylethanolamine <4> [10]
3,4-dichloroamphetamine <4> [17]
3,4-dichlorophenylethanolamine <4> (<4>, substrate inhibition [3]) [3, 10]
3,4-dichlorophenylethylenediamine <4> (<4>, substrate inhibition [3]) [3]
3,4-dihydroxyphenylethanolamine <4> [10]
3-bromophenylethanolamine <4> [10]
3-fluoromethyl-7-(N-benzylaminosulfonyl)-1,2,3,4-tetrahydrosioquinoline <3> [26]
3-fluoromethyl-7-(N-methylaminosulfonyl)-1,2,3,4-tetrahydroisoquinoline <3> [26]
3-fluoromethyl-7-[N-(4-chlorophenyl)aminosulfonyl]-1,2,3,4-tetrahydroisoquinoline <3> [26]
3-fluoromethyl-7-aminosulfonyl-1,2,3,4-tetrahydroisoquinoline <3> [26]
3-fluoromethyl-7-azido-1,2,3,4-tetrahydrosioquinoline <3> [26]
3-fluoromethyl-7-bromo-1,2,3,4-tetrahydroisoquinoline <3> [26]
3-fluoromethyl-7-cyano-1,2,3,4-tetrahydroisoquinoline <3> [26]
3-fluoromethyl-7-iodo-1,2,3,4-tetrahydroisoquinoline <3> [26]
3-fluoromethyl-7-isothiocyanato-1,2,3,4-tetrahydroisoquinoline <3> [26]
3-fluoromethyl-7-methanesulfonyl-1,2,3,4-tetrahydroisoquinoline <3> [26]
3-fluoromethyl-7-nitro-1,2,3,4-tetrahydroisoquinoline <3> [26]
3-fluoromethyl-7-trifluoromethyl-1,2,3,4-tetrahydroisoquinoline <3> [26]
3-trifluoromethyl-1,2,3,4-tetrahydroisoquinoline <3> [28]
3-trifluoromethyl-7-bromo-1,2,3,4-tetrahydroisoquinoline <3> [28]
3-trifluoromethyl-7-cyano-1,2,3,4-tetrahydroisoquinoline <3> [28]
3-trifluoromethyl-7-nitro-1,2,3,4-tetrahydroisoquinoline <3> [28]
4-bromophenylethanolamine <4> [10]
4-fluorophenylethanolamine <4> [10]
4-hydroxyphenylethanolamine <4> [10]
5,6-dichloro-2-aminotetralin <4> (<4>, 0.00093 mM, 50% inhibition, competitive with norepinephrine as substrate [17]) [17]
7-acetamido-1,2,3,4-tetrahydroisoquinoline <3> [27]
7-allylsulfonyl-1,2,3,4-tetrahydrosioquinoline <3> [27]
7-aminocarbonyl-1,2,3,4-tetrahydroisoquinoline <3> [27]
7-aminomethyl-1,2,3,4-tetrahydroisoquinoline dihydrochloride <3> [27]
7-benzoyl-1,2,3,4-tetrahydroisoquinoline <3> [27]
7-benzyl-1,2,3,4-tetrahydroisoquinoline <3> [27]
7-bromo-N-triphenylmethyl-1,2,3,4-tetrahydroisoquinoline <3> [27]
7-hydroxymethyl-1,2,3,4-tetrahydroisoquinoline oxalate <3> [27]
7-iodo-1,2,3,4-tetrahydroisoquinoline <3> [27]
7-methoxycarbonyl-1,2,3,4-tetrahydroisoquinoline <3> [27]
7-methylsulfinyl-1,2,3,4-tetrahydroisoquinoline <3> [27]
7-methylsulfonyl-1,2,3,4-tetrahydroisoquinoline <3> [27]

7-methylsulfonyl-3-trifluoromethyl-1,2,3,4-tetrahydroisoquinoline <3> [28]
7-methylthio-1,2,3,4-tetrahydroisoquinoline <3> [27]
7-phenylsulfonyl-1,2,3,4-tetrahydroisoquinoline <3> [27]
7-trichloromethylsulfonyl-1,2,3,4-tetrahydroisoquinoline <3> [27]
7-trifluoroacetyl-1,2,3,4-tetrahydroisoquinoline <3> [27]
D-norepinephrine <4> (<4>, substrate inhibition [3]) [3]
L-norepinephrine <4> (<4>, substrate inhibition [3]) [3]
LY134046 <9> (<9>, selective inhibitor, IC50: 0.0019 mM [24]) [24]
PCMB <1, 2, 3> (<1>, 0.01 mM, complete inhibition [1]) [1, 4]
S-adenosyl-L-methionine <4> (<4>, substrate inhibition [3]; <4>, 4 distinct substrate inhibition patterns: enzyme form E1 is not subject to substrate inhibition, isoenzyme E-2 is inhibited by S-adenosyl-L-methionine and norepinephrine, isoenzyme E-3 and E-4 are inhibited by norepinephrine only, isoenzyme E5 is inhibited only by S-adenosyl-L-methionine [6]) [3, 6]
S-adenosylhomocysteine <3, 4> (<3>, competitive [14]) [7, 10, 14]
dopamine <3> (<3>, competitive inhibition with respect to phenylethanolamine [18]) [18]
epinephrine <3> (<3>, noncompetitive with normetanephrine as substrate [18]) [18]
epinephrine <2, 3, 4> (<4>, isoenzyme E2, noncompetitive inhibitor of S-adenosyl-L-methionine [7]) [4, 7, 10]
human cerebrospinal fluid <3> [19]
methanol <3> (<3>, at nonsaturating concentrations of phenylethanolamine, methanol inhibits in a competitive fashion. With respect to S-adenosyl-L-methionine inhibition by methanol is noncompetitive [8]) [8]
norepinephrine <2> (<2>, at high concentrations [4]) [4]
norepinephrine <3, 4> (<4>, 4 distinct substrate inhibition patterns: enzyme form E1 is not subject to substrate inhibition, isoenzyme E-2 is inhibited by S-adenosyl-L-methionine and norepinephrine, isoenzyme E-3 and E-4 are inhibited by norepinephrine only, isoenzyme E5 is inhibited only by S-adenosyl-L-methionine [6]; <3>, noncompetitive inhibition with phenylethanolamine as substrate, competitive inhibition with normetanephrine as substrate [18]) [6, 18]
octopamine <2> (<2>, at high concentrations [4]) [4]
p-hydroxymercuribenzoate <3> (<3>, 0.00001 mM, 75% loss of activity [2]) [2]
phenylethanolamine <2> (<2>, at high concentrations [4]) [4]
phenylethylamine <3> (<3>, competitive inhibition with respect to phenylethanolamine [18]) [18]
sinefungin <4> [12]
tyramine <2> (<2>, at high concentrations [4]) [4]
Additional information <3> (<3>, 7-substitited-1,2,3,4-tetrahydroisoquinolines are potent inhinbitors, unfortunately, most of these compounds also exhibit strong affinity for the α_2-adrenoceptor [27]) [27]

Specific activity (U/mg)

0.00658 <3> [9]
9.42 <4> (<4>, enzyme form E-5 from young rabbit [5]) [5]
11.5 <4> (<4>, enzyme form E-5 from adult [5]) [5]
35.9 <4> (<4>, enzyme form E-1 from young rabbit [5]) [5]
51.6 <4> (<4>, enzyme form E-1 from adult [5]) [5]
53.5 <4> (<4>, enzyme form E-4 from young rabbit [5]) [5]
89.4 <4> (<4>, enzyme form E-4 from adult [5]) [5]
175 <4> (<4>, enzyme form E-3 from young rabbit [5]) [5]
188 <4> (<4>, enzyme form E-2 from adult [5]) [5]
Additional information <2, 4> [3, 4]

K_m-Value (mM)

0.0008 <4> (3,4-dichlorophenylethanolamine) [10]
0.00132 <4> (3,4-dichlorophenylethanolamine, <4>, enzyme form H-2 [3]) [3]
0.00159 <4> (3,4-dichlorophenylethanolamine, <4>, enzyme form H-1 [3]) [3]
0.0024 <4> (3-chloro-4-hydroxyphenylethanolamine) [10]
0.00543 <4> (S-adenosyl-L-methionine, <4>, enzyme form H-1 [3]) [3]
0.012 <4> (3,4-dichlorophenylethylenediamine, <4>, enzyme form H-2 [3]) [3]
0.013 <4> (3,4-dihydroxyphenylethanolamine) [10]
0.0131 <4> (3,4-dichlorophenylethylenediamine, <4>, enzyme form H-1 [3]) [3]
0.0149 <4> (S-adenosyl-L-methionine, <4>, enzyme form H-2 [3]) [3]
0.015 <4> (L-norepinephrine) [11]
0.017 <4> (4-hydroxyphenylethanolamine) [10]
0.0184 <9> (norepinephrine, <9>, wild-type enzyme [22]) [22]
0.0194 <9> (S-adenosyl-L-methionine, <9>, wild-type enzyme [22]) [22]
0.024 <8> (S-adenosyl-L-methionine) [16]
0.0297 <4> (D-norepinephrine, <4>, enzyme form H-1 [3]) [3]
0.0299 <4> (L-norepinephrine, <4>, enzyme form H-2 [3]) [3]
0.0428 <4> (L-norepinephrine, <4>, enzyme form H-1 [3]) [3]
0.05 <1> (normetanephrine) [1]
0.077 <4> (phenylethanolamine) [10]
0.089 <4> (3-hydroxyphenylethanolamine) [10]
0.089 <4> (*m*-octopamine) [11]
0.101 <8> (L-norepinephrine) [16]
0.109 <4> (D-norepinephrine, <4>, enzyme form H-2 [3]) [3]
0.226 <4> (phenylethanolamine, <4>, enzyme form H-2 [3]) [3]
0.244 <4> (phenylethanolamine, <4>, enzyme form H-1 [3]) [3]
0.366 <4> (3-methoxy-4-hydroxy-phenylethanolamine) [10]
0.676 <4> (2-hydroxyphenylethanolamine) [10]
Additional information <9> (<9>, K_m-value of mutant enzymes [22]) [22]

K_i-Value (mM)

0.00018 <3> (7-trifluoromethyl-1,2,3,4-tetrahydroisoquinoline) [27]
0.00021 <3> (3-fluoromethyl-7-iodo-1,2,3,4-tetrahydroisoquinoline) [26]
0.00027 <4> (5,6-dichloro-2-aminotetralin) [17]
0.00029 <3> (7-bromo-N-triphenylmethyl-1,2,3,4-tetrahydroisoquinoline) [27]
0.00032 <3> (3-fluoromethyl-7-trifluoromethyl-1,2,3,4-tetrahydroisoquinoline) [26]
0.00037 <3> (7-iodo-1,2,3,4-tetrahydroisoquinoline) [27]
0.00052 <3> (3-trifluoromethyl-7-bromo-1,2,3,4-tetrahydroisoquinoline) [28]
0.00057 <3> (3-fluoromethyl-7-isothiocyanato-1,2,3,4-tetrahydroisoquinoline) [26]
0.00061 <3> (7-methylthio-1,2,3,4-tetrahydroisoquinoline) [27]
0.00064 <3> (3-fluoromethyl-7-bromo-1,2,3,4-tetrahydroisoquinoline) [26]
0.00066 <3> (3-fluoromethyl-7-aminosulfonyl-1,2,3,4-tetrahydroisoquinoline) [26]
0.00074 <3> (3-fluoromethyl-7-[N-(4-chlorophenyl)aminosulfonyl]-1,2,3,4-tetrahydroisoquinoline) [26]
0.00101 <3> ((S)-3-methyl-1,2,3,4-tetrahydroisoquinoline) [29]
0.0011 <3> (3-fluoromethyl-7-cyano-1,2,3,4-tetrahydroisoquinoline) [26]
0.0013 <3> (7-methylsulfonyl-1,2,3,4-tetrahydroisoquinoline) [27]
0.0013 <3> (7-trichloromethylsulfonyl-1,2,3,4-tetrahydroisoquinoline) [27]
0.0016 <3> (3-fluoromethyl-7-methanesulfonyl-1,2,3,4-tetrahydroisoquinoline) [26]
0.0017 <3> (3-fluoromethyl-7-azido-1,2,3,4-tetrahydrosioquinoline) [26]
0.002 <3> (3-trifluoromethyl-7-nitro-1,2,3,4-tetrahydroisoquinoline) [28]
0.0024 <3> (3-fluoromethyl-7-(N-methylaminosulfonyl)-1,2,3,4-tetrahydroisoquinoline) [26]
0.0029 <3> (7-benzoyl-1,2,3,4-tetrahydroisoquinoline) [27]
0.00334 <3> (2,3,4,5-tetrahydro-1H-2-benzazepine) [29]
0.0041 <3> (2,3,4,5-tetrahydro-5H-1,4-benzothiazepine) [29]
0.0057 <3> (7-trifluoroacetyl-1,2,3,4-tetrahydroisoquinoline) [27]
0.0059 <3> (norepinephrine, <3>, noncompetitive with phenylethanolamine [18]) [18]
0.0065 <3> (3-fluoromethyl-7-(N-benzylaminosulfonyl)-1,2,3,4-tetrahydroisoquinoline) [26]
0.0067 <3> (7-methoxycarbonyl-1,2,3,4-tetrahydroisoquinoline) [27]
0.0098 <3> (7-allylsulfonyl-1,2,3,4-tetrahydrosioquinoline) [27]
0.011 <3> (7-hydroxymethyl-1,2,3,4-tetrahydroisoquinoline oxalate) [27]
0.0124 <4> (S-adenosyl-L-methionine, <4>, enzyme form H-2, substrate inhibition [3]) [3]
0.013 <3> (3-trifluoromethyl-7-cyano-1,2,3,4-tetrahydroisoquinoline) [28]
0.014 <3> (7-phenylsulfonyl-1,2,3,4-tetrahydroisoquinoline) [27]
0.015 <3> (3-trifluoromethyl-1,2,3,4-tetrahydroisoquinoline) [28]
0.0212 <3> (2,3,4,5-tetrahydro-5H-1,4-benzoxazepine) [29]
0.028 <3> (2,3,4,5-tetrahydro-5H-1,4-benzodiazepine) [29]

0.036 <3> (7-methylsulfonyl-3-trifluoromethyl-1,2,3,4-tetrahydroisoquinoline) [28]
0.0383 <3> ((R)-3-methyl-1,2,3,4-tetrahydroisoquinoline) [29]
0.045 <3> (7-methylsulfinyl-1,2,3,4-tetrahydroisoquinoline) [27]
0.054 <4> (epinephrine) [10]
0.061 <3> (7-benzyl-1,2,3,4-tetrahydroisoquinoline) [27]
0.064 <3> (7-aminocarbonyl-1,2,3,4-tetrahydroisoquinoline) [27]
0.0894 <4> (S-adenosyl-L-methionine, <4>, enzyme form H-1, substrate inhibition [3]) [3]
0.109 <4> (D-norepinephrine, <4>, enzyme form H-2, substrate inhibition [3]) [3]
0.165 <4> (L-norepinephrine, <4>, enzyme form H-1, substrate inhibition [3]) [3]
0.19 <3> (7-acetamido-1,2,3,4-tetrahydroisoquinoline) [27]
0.47 <3> (1,2,3,4-tetrahydroisoquinoline-7-carboxylic acid) [27]
0.508 <4> (3,4-dichlorophenylethanolamine, <4>, enzyme form H-2, substrate inhibition [3]) [3]
0.54 <3> (3-fluoromethyl-7-nitro-1,2,3,4-tetrahydroisoquinoline) [26]
0.631 <4> (L-norepinephrine, <4>, enzyme form H-2, substrate inhibition [3]) [3]
0.909 <4> (3,4-dichlorophenylethanolamine, <4>, enzyme form H-1, substrate inhibition [3]) [3]
1.138 <4> (3,4-dichlorophenylethylenediamine, <4>, enzyme form H-2, substrate inhibition [3]) [3]
1.522 <4> (3,4-dichlorophenylethylenediamine, <4>, enzyme form H-1, substrate inhibition [3]) [3]
1.9 <3> (7-aminomethyl-1,2,3,4-tetrahydroisoquinoline dihydrochloride) [27]
3.6 <3> (phenylethylamine, <3>, competitive with phenylethanolamine [18]) [18]
8 <3> (dopamine, <3>, competitive with phenylethanolamine [18]) [18]
8.298 <4> (D-norepinephrine, <4>, enzyme form H-1, substrate inhibition [3]) [3]
Additional information <4> [6, 7]

pH-Optimum

7.5-8.2 <1> (<1>, phosphate buffer [1]) [1]
7.9 <3> (<3>, Tris or borate buffer [2]) [2]
8 <2> [4]
8-9 <1> (<1>, Tris buffer [1]) [1]

4 Enzyme Structure

Molecular weight

34000 <3> (<3>, gel filtration [9]) [9]
37000 <4> (<4>, gel filtration [3]) [3]
38000 <3> (<3>, sucrose density gradient centrifugation [2]) [2]

Subunits

monomer <4> (<4>, 1 * 35000, enzyme form E-3, SDS-PAGE [5]; <4>, 1 * 40000, enzyme form E-1, SDS-PAGE [5]) [5]

Posttranslational modification

no glycoprotein <3> (<3>, enzyme contains no carbohydrate [14]) [14]

5 Isolation/Preparation/Mutation/Application

Source/tissue

adrenal gland <1, 2, 3, 4> (<1>, adrenal medulla has considerably more activity than the cortex [1]) [1, 3, 4, 5, 6, 7, 8, 10, 12, 13, 17, 18, 26, 28, 29]
adrenal medulla <1, 3, 9> [1, 2, 9, 14, 18, 24, 27]
brain <8> [16]
brain stem <10> [21]
brown adipose tissue <7> (<7>, low activity [15]) [15]
embryo <2> (<2>, heart [23]) [23]
heart <1, 2, 4> (<1,4>, weak activity [1]; <2>, transient developmental burst in enzyme expression in the embryonic rat heart from E9.5 to E13.0 [23]) [1, 23]
lung <7> [15]
myometrium <7> (<7>, low activity [15]) [15]
thymus <10> [21]
Additional information <10, 11> (<10>, barely detectable in the spleen [21]; <11>, PNMT activity is low in the resting phase and increases to the peak in the prespawning and spawning phases. Maintenance of the fish under long photoperiods and high temperature stimulates activity [25]) [21, 25]

Purification

<1> [1]
<2> [4]
<3> (at least 4 active charged isoenzymes [14]) [2, 9, 14, 18, 19]
<4> (enzyme form H-1 and H-2 [3]; 5 charge isoenzymes: E-1, E-2, E-3, E-4, E-5 [5]) [3, 5, 10]

Crystallization

<9> (crystallized in complex with an inhibitor and the cofactor product S-adenosyl-L-homocysteine using hanging-drop technique with PEG 6000 and lithium chloride as precipitant [20]) [20]

Cloning

<3> (expressed in a mouse C127 cell line [13]) [13]
<9> (expression of wild-type enzyme and mutant enzymes in Escherichia coli [22]) [22]

Engineering

C131S <9> (<9>, mutant enzyme shows similar K_M-values and maximal velocity to thoses of the wild-type enzyme [22]) [22]
C139S <9> (<9>, mutant enzyme shows similar K_M-values and maximal velocity to thoses of the wild-type enzyme [22]) [22]
C183S <9> (<9>, mutant enzyme shows markedly reduced enzyme activity with less than 3% of the maximal activity of the wild-type enzyme, and ca. sixfold increased apparent K_M-value for both substrates [22]) [22]
C48S <9> (<9>, mutant enzyme shows similar K_M-values and maximal velocity to thoses of the wild-type enzyme [22]) [22]
C60S <9> (<9>, mutant enzyme shows similar K_M-values and maximal velocity to thoses of the wild-type enzyme [22]) [22]
C91S <9> (<9>, mutant enzyme shows similar K_M-values and maximal velocity to thoses of the wild-type enzyme [22]) [22]

Application

analysis <2> (<2>, enzyme serves as a marker for tissues and cells producing epinephrine [23]) [23]

6 Stability

Temperature stability

39 <3> (<3>, stable for at least 50 min [2]) [2]
45 <3> (<3>, activity is rapidly destroyed [2]) [2]
60 <2> (<2>, 5 min, littel activity remains [4]) [4]
Additional information <3> (<3>, presence of 2.5 mM normetanephrine has no effect on the heat denaturation, 1 mM S-adenosyl-L-methionine decreases the rate of inactivation [2]) [2]

Storage stability

<1>, -10°C, stable for at least 1 month [1]

References

[1] Axelrod, J.: Purification and properties of phenylethanolamine-N-methyltransferase. J. Biol. Chem., **237**, 1657-1660 (1962)
[2] Connett, R.J.; Kirshner, N.: Purification and properties of bovine phenylethanolamine N-methyltransferase. J. Biol. Chem., **245**, 329-334 (1970)
[3] Lee, H.-S.; Schulz, A.R.; Fuller, R.W.: Isolation of multiple forms of rabbit-adrenal norepinephrine N-methyltransferase. Arch. Biochem. Biophys., **163**, 735-741 (1974)

[4] Pohorecky, L.A.; Baliga, B.S.: Purification and properties of rat adrenal phenylethanolamine-N-methyl transferase. Arch. Biochem. Biophys., **156**, 703-711 (1973)
[5] Lee, H.-S.; Schulz, A.R.; Fuller, R.W.: Isolation and purification of rabbit adrenal norephinephrine N-methyl transferase isozymes. Arch. Biochem. Biophys., **185**, 222-227 (1978)
[6] Lee, H.-S.; Schulz, A.R.; Fuller, R.W.: The interaction of rabbit adrenal norepinephrine N-methyl transferase isozymes with substrates. Arch. Biochem. Biophys., **185**, 228-238 (1978)
[7] Lee, H.-S.; Schulz, A.R.; Fuller, R.W.: Product inhibition studies and the reaction sequence of rabbit adrenal norepinephrine N-methyl transferase isozymes. Arch. Biochem. Biophys., **185**, 239-250 (1978)
[8] Harralson, J.D.; Brown, F.C.: Phenylethanolamine-N-methyltransferase:alcohol and the mechanism of action. Arch. Biochem. Biophys., **221**, 354-360 (1983)
[9] Hurst, J.H.; Guchhait, R.B.; Billingsley, M.L.; Stolk, J.M.; Lovenberg, W.: Phenylethanolamine N-methyltransferase: notes on its purification from bovine adrenal medulla and separation from protein carboxymethyltransferase. Biochem. Biophys. Res. Commun., **112**, 1061-1068 (1983)
[10] Fuller, R.W.: Norepinephrine N-methyltransferase from rabbit adrenal glands. Methods Enzymol., **142**, 655-660 (1987)
[11] Fuller, R.W.; Hemrick, S.K.: *m*-Octopamine as a substrate for norepinephrine N-methyltransferase. IRCS Med. Sci. Libr. Compend., **8**, 284 (1980)
[12] Fuller, R.W.; Nagarajan, R.: Inhibition of methyltransferases by some new analogs of S-adenosylhomocysteine. Biochem. Pharmacol., **27**, 1981-1983 (1978)
[13] Park, D.H.; Wessel, T.; Baker, H.; Joh, T.H.; Samanta, H.: Characterization of recombinant bovine phenylethanolamine N-methyltransferase expressed in a mouse C127 cell line. Mol. Brain Res., **10**, 213-218 (1991)
[14] Wong, D.L.; Yamasaki, L.; Ciaranello, R.D.: Characterization of the isozymes of bovine adrenal medullary phenylethanolamine N-methyltransferase. Brain Res., **410**, 32-44 (1987)
[15] Padbury, J.F.; Lam, R.W.; Hobel, C.J.; Fisher, D.A.: Identification and partial purification of phenylethanolamine N-methyl transferase in the developing ovine lung. Pediatr. Res., **17**, 362-367 (1983)
[16] Fuller, R.W.; Hemrick, S.K.: Properties of norepinephrine N-methyltransferase from pigeon brain. Pharmacol. Biochem. Behav., **11**, 563-566 (1979)
[17] Fuller, R.W.; Molloy, B.B.: Inhibition in vitro of norepinephrine N-methyltransferase by 2-aminotetralins, analogs of phenylethylamines with rigid conformation. Biochem. Pharmacol., **26**, 446-447 (1977)
[18] Hoffman, A.R.; Ciaranello, R.D.; Axelrod, J.: Substrate and inhibitor kinetics of bovine phenylethanolamine-N-methyltransferase. Biochem. Pharmacol., **24**, 544-546 (1975)
[19] Cubeddu, L.; Gimenez, M.; Goldstein, D.J.: Inhibition of phenylethanolamine-N-methyltransferase by human cerebrospinal fluid - a sodium chloride effect. Biochem. Pharmacol., **25**, 2537-2538 (1976)

[20] Begun, J.; McLeish, M.J.; Caine, J.M.; Palant, E.; Grunewald, G.L.; Martin, J.L.: Crystallization of PNMT, the adrenaline-synthesizing enzyme, is critically dependent on a high protein concentration. Acta Crystallogr. Sect. D, **D58**, 314-315 (2002)

[21] Warthan, M.D.; Freeman, J.G.; Loesser, K.E.; Lewis, C.W.; Hong, M.; Conway, C.M.; Stewart, J.K.: Phenylethanolamine N-methyltransferase expression in mouse thymus and spleen. Brain Behav. Immun., **16**, 493-499 (2002)

[22] Kaneda, N.; Hikita, K.; Naruse, Y.; Fukuo, T.; Matsubara, K.; Nagatsu, T.: Identification of the essential cysteinyl residue located in the active site of human phenylethanolamine N-methyltransferase. Biochem. Biophys. Res. Commun., **249**, 405-409 (1998)

[23] Ebert, S.N.; Baden, J.M.; Mathers, L.H.; Siddall, B.J.; Wong, D.L.: Expression of phenylethanolamine N-methyltransferase in the embryonic rat heart. J. Mol. Cell. Cardiol., **28**, 1653-1658 (1996)

[24] Gearhart, D.A.; Neafsey, E.J.; Collins, M.A.: Phenylethanolamine N-methyltransferase has β-carboline 2N-methyltransferase activity: hypothetical relevance to Parkinson's disease. Neurochem. Int., **40**, 611-620 (2002)

[25] Senthilkumaran, B.; Joy, K.P.: Changes in hypothalamic catecholamines, dopamine-β-hydroxylase, and phenylethanolamine-N-methyltransferase in the catfish Heteropneustes fossilis in relation to season, raised photoperiod and temperature, ovariectomy, and estradiol-17β replacement. Gen. Comp. Endocrinol., **97**, 121-134 (1995)

[26] Grunewald, G.L.; Caldwell, T.M.; Li, Q.; Slavica, M.; Criscione, K.R.; Borchardt, R.T.; Wang, W.: Synthesis and biochemical evaluation of 3-fluoromethyl-1,2,3,4-tetrahydroisoquinolines as selective inhibitors of phenylethanolamine N-methyltransferase versus the α(2)-adrenoceptor. J. Med. Chem., **42**, 3588-3601 (1999)

[27] Grunewald, G.L.; Dahanukar, V.H.; Jalluri, R.K.; Criscione, K.R.: Synthesis, biochemical evaluation, and classical and three-dimensional quantitative structure-activity relationship studies of 7-substituted-1,2,3,4-tetrahydroisoquinolines and their relative affinities toward phenylethanolamine N-methyltransferase and the α_2-adrenoceptor. J. Med. Chem., **42**, 118-134 (1999)

[28] Grunewald, G.L.; Caldwell, T.M.; Li, Q.; Criscione, K.R.: Synthesis and evaluation of 3-trifluoromethyl-7-substituted-1,2,3,4-tetrahydroisoquinolines as selective inhibitors of phenylethanolamine N-methyltransferase versus the α(2)-adrenoceptor. J. Med. Chem., **42**, 3315-3323 (1999)

[29] Grunewald, G.L.; Dahanukar, V.H.; Ching, P.; Criscione, K.R.: Effect of ring size or an additional heteroatom on the potency and selectivity of bicyclic benzylamine-type inhibitors of phenylethanolamine N-methyltransferase. J. Med. Chem., **39**, 3539-3546 (1996)

tRNA (cytosine-5-)-methyltransferase 2.1.1.29

1 Nomenclature

EC number
2.1.1.29

Systematic name
S-adenosyl-L-methionine:tRNA (cytosine-5-)-methyltransferase

Recommended name
tRNA (cytosine-5-)-methyltransferase

Synonyms
methyltransferase, transfer ribonucleate cytosine 5-
tRNA cytosine 5-methyltransferase
tRNA:m5C-methyltransferase <8> [13]
transfer RNA cytosine 5-methyltransferase
transfer ribonucleate cytosine 5-methyltransferase

CAS registry number
37256-97-6

2 Source Organism

<1> *Escherichia coli* (strain W [1, 3]) [1, 3]
<2> *Homo sapiens* (HeLa-S3-cells [2, 11]) [2, 11]
<3> *Rattus norvegicus* (Sprague-Dawley strain and leukemic strain Fischer RL [8]) [4, 6, 8]
<4> *Mus musculus* (strain BALB/c [5]) [5, 7]
<5> *Tetrahymena pyriformis* [9]
<6> *Gallus gallus* [10]
<7> *Xenopus laevis* [12]
<8> *Saccharomyces cerevisiae* [13, 14]

3 Reaction and Specificity

Catalyzed reaction
S-adenosyl-L-methionine + tRNA = S-adenosyl-L-homocysteine + tRNA containing 5-methylcytosine (<8> phylogenetic tree [13]; <3, 7, 8> methylation site [4, 12, 14]; <3> sequential mechanism [6])

Reaction type

methyl group transfer

Natural substrates and products

S S-adenosyl-L-methionine + tRNA <7> (<7> methylation specifically at C49 of tRNA [12]) (Reversibility: ? <7> [12]) [12]

P S-adenosyl-L-homocysteine + tRNA containing 5-methylcytosine <7> [12]

Substrates and products

S S-adenosyl-L-methionine + tRNA <1-8> (<8> methylation of cytosines in yeast tRNA is not site specific [13]; <1> S-RNA [3]; <1> specific for S-adenosyl-L-methionine, cannot be replaced by methionine plus ATP, formaldehyde plus tetrahydrofolic acid, methylcobalamin and methyltetrahydrofolic acid [3]; <2> no strict specificity for tRNA [2,11]; <2,4> methyl group acceptors are tRNAs from E. coli, unfractionated tRNA, $tRNA^{Phe}$ [2,5,11]; <3> $tRNA_{f}^{Met}$, best substrate [4]; <1-4> tRNA2Glu [1-6]; <2> yeast: $tRNA^{Phe}$ [2,11]; <3> yeast: $tRNA^{Val}$ [4]; <1,2> rRNA from E. coli, RNA from tobacco mosaic virus [2,3]; <2> brome mosaic virus [2]; <1> turnip yellow mosaic virus, bacteriophage f2 [3]; <2> acceptors are poly-A,C,G,U, poly-A,C, poly-C,U, poly-A,C,U, poly-A,C,G, poly-C,G,U [2]) (Reversibility: ir <1> [3]; ? <1-8> [1,2,4-14]) [1-14]

P S-adenosyl-L-homocysteine + tRNA containing 5-methylcytosine <1-8> [1-14]

S Additional information <1, 2> (<1,2> poly-C are no methyl group acceptors [2,3]; <2> poly-dC,dG, DNA from Micrococcus luteus, calf thymus and vaccinia virus [2]; <1> CMP [3]; <2> enzyme preferentially methylates C-residues within specific RNAse T1-resistant oligonucleotides in E. coli or yeast $tRNA^{Phe}$ [2]) [2, 3]

P ?

Inhibitors

5'-methylthioadenosine <3> (<3> competitive to substrate tRNA, non-competitive to S-adenosyl-L-methionine [6]) [6]

Co^{2+} <4> [7]

Cu^{2+} <4> [7]

E. coli rRNA <3> (<3> dead-end inhibition, competitive to substrate tRNA, non-competitive to S-adenosyl-L-methionine [6]) [6]

Mg^{2+} <2, 4> (<2> strong, 5mM [2,11]; <4> only in presence of KCl [7]) [2, 5, 7, 11]

Mn^{2+} <1> (<1> slight inhibition [3]) [3]

NaCl <2> (<2> 50 mM [2]) [2, 11]

Ni^{2+} <1> [3]

PCMB <1> [3]

S-adenosyl-L-homocysteine <1-3> (<1> product inhibition, reversible by S-adenosyl-L-methionine [3]; <3> competitive to S-adenosyl-L-methionine [6]) [2, 3, 6]

Zn^{2+} <1> [3]

ammonium acetate <2> (<2> 120 mM [2]) [2]
ammonium sulfate <2> (<2> 50 mM [2]) [2, 11]
cytosine-methylated tRNA <3> (<3> product inhibition, competitive to substrate tRNA, non-competitive to S-adenosyl-L-methionine [6]) [6]
putrescine <4> (<4> only in presence of KCl, S-adenosyl-L-methionine or 20% v/v glycerol protects, activation in absence of monovalent cations [7]) [7]
ribonuclease <1> [3]
sodium phosphate <2> (<2> 50 mM [2]) [2]
spermidine <4> (<4> only in presence of KCl [7]) [7]
spermine <1, 4> (<4> strong, in presence of KCl [7]) [3, 7]
spleen phosphodiesterase <1> [3]
Additional information <1, 4> (<1> no inhibition by snake venom phosphodiesterase, adenosine, L-homocysteine [3]; <4> EDTA [7]) [3, 7]

Activating compounds

putrescine <4> (<4> activation, in the absence of monovalent cations [7]) [7]
Additional information <1, 4> (<1> no activation by 2-mercaptoethanol [3]; <4> no activation by Triton X-100 [5]; <4> no activation by cadaverine [7]) [3, 5, 7]

Metals, ions

KCl <4> (<4> activation, maximal at 0.25 M [7]) [7]
Mg^{2+} <4> (<4> activation, in absence of monovalent cations [7]; <1> no activation [3]; <2> inhibition at 5 mM [2]) [7]
Additional information <4> (<4> no activation by Mn^{2+}, Ca^{2+} [7]) [7]

Specific activity (U/mg)

0.0000725 <2> [2]
0.0039 <1> [3]
7.3 <1> (<1> purified enzyme [1]) [1]

K_m-Value (mM)

0.004 <3> (S-adenosyl-L-methionine) [6]
0.013 <1> (S-adenosyl-L-methionine) [3]
0.5 <1> (methyl deficient tRNA) [3]
Additional information <4> (<4> K_m-values at various KCl-concentrations with different tRNA substrates [7]) [7]

pH-Optimum

7.3 <2> [2, 11]
7.8 <4> (<4> assay at [7]) [7]
8 <8> (<8> assay at [13]) [13]
8-9 <1> [3]
8.3 <4> (<4> assay at [5]) [5]

Temperature optimum (°C)

30 <8> (<8> assay at [13]) [13]
37 <1, 4> (<1,4> assay at [1,5,7]) [1, 5, 7]

4 Enzyme Structure

Molecular weight

72000 <2> (<2> HeLa-S3-cells, sucrose gradient sedimentation [2]) [2, 11]
110000-140000 <4> (<4> gel filtration, sucrose density gradient centrifugation [7]) [7]

Subunits

? <8> (<8> x * 82000, recombinant His-tagged protein, SDS-PAGE [13]) [13]

5 Isolation/Preparation/Mutation/Application

Source/tissue

cell suspension culture <2> [2]
hepatoma cell <3> (<3> plasacytoma [4]) [4]
intracisternal A particle <4> [5]
liver <3, 6> (<3> hepatoma [4]) [4, 6, 8, 10, 11]
plastocytoma cell <3> (<3> hepatoma [4]) [4]
spleen <3> (<3> leukemic rat [8]) [8]
tumor <4> (<4> plasmacytoma MOPC-104E [5]; <4> MOPC173 [7]) [5, 7]

Localization

cytosol <1, 2, 4> [1, 3, 5, 11]

Purification

<1> [1, 3]
<2> [2, 11]
<4> (partial [7]) [7]
<8> (recombinant His-tagged from E. coli [13]) [13]

Cloning

<8> (yeast mutants [14]; hyperexpression of the His-tagged enzyme in Escherichia coli BL21(DE3), from ORF YBL024w of Saccharomyces cerevisiae, single copy gene [13]) [13, 14]

Engineering

C424A <8> (<8> hemagglutinin-tagged, altered substrate binding, evaluation of mehylation site [14]) [14]
C424S <8> (<8> hemagglutinin-tagged, altered substrate binding, evaluation of mehylation site [14]) [14]
C478A <8> (<8> hemagglutinin-tagged, altered substrate binding, evaluation of mehylation site [14]) [14]
Additional information <8> (<8> disrupted YBL024 strain shows no remaining activity [13]) [13]

6 Stability

Temperature stability

37 <4> (<4> at least 3 h stable [7]) [7]
47 <4> (<4> 2-step-denaturation pattern: $t_{1/2}$: 2-4 min and 15-25 min [7]) [7]

General stability information

<4>, dilution inactivates, non-specific proteins protect [7]
<4>, glycerol, 50% v/v, stabilizes [7]
<4>, purification leads to severe, irreversible loss of activity [7]

Storage stability

<1>, 0°C, at least 2 months stable, about 70% loss of activity after 6 months [1]
<2>, -70°C, more than 1 year [2, 11]
<2>, 0°C, 5 to 6 months [2, 11]
<4>, -20°C, 50% glycerol, more than 2 months [7]
<4>, 4°C, $t_{1/2}$: 2 weeks [7]

References

[1] Hurwitz, J.; Gold, M.; Anders, M.: The enzymatic methylation of ribonucleic acid and deoxyribonucleic acid. III. Purification of soluble ribonucleic acid-methylating enzymes. J. Biol. Chem., **239**, 3462-3473 (1964)
[2] Keith, J.M.; Winters, E.M.; Moss, B.: Purification and characterization of a HeLa cell transfer RNA(cytosine-5-)-methyltransferase. J. Biol. Chem., **255**, 4636-4644 (1980)
[3] Hurwitz, J.; Gold, M.; Anders, M.: The enzymatic methylation of ribonucleic acid and deoxyribonucleic acid. IV. The properties of the soluble ribonucleic acid-methylating enzymes. J. Biol. Chem., **239**, 3474-3482 (1964)
[4] Shershneva, L.P.; Venkstern, T.V.; Bayev, A.A.: A study of tRNA methylase action. FEBS Lett., **29**, 132-134 (1973)
[5] Chepelinsky, A.B.; Gantt, R.; Wivel, N.: Presence of RNA methylases in intracisternal A particles purified from a mouse plasma cell tumor. Eur. J. Biochem., **103**, 339-347 (1980)
[6] Kahle, P.; Kröger, H.: Kinetic mechanism of transfer ribonucleic acid (cytosine-5)-methylase from rat liver. Biochem. Soc. Trans., **3**, 908-909 (1975)
[7] Nau, F.; Pham-Coeur-Joly, G.; Dubert, J.-M.: A study of some molecular and kinetic properties of two tRNA methyltransferases from mouse plasmocytoma. Eur. J. Biochem., **130**, 261-268 (1983)
[8] Kraus, J.; Staehelin, M.: N^2-guanine specific transfer RNA methyltransferase I from rat liver and leukemic rat spleen. Nucleic Acids Res., **1**, 1455-1478 (1974)
[9] Reinhart, M.P.; Lewis, J.M.; Leboy, P.S.: A single tRNA (guanine)-methyltransferase from Tetrahymena with both mono- and di-methylating activity. Nucleic Acids Res., **14**, 1131-1148 (1986)

[10] Kehr, S.J.: tRNA methyltransferases. Methods Enzymol., **29**, 716-726 (1974)
[11] Söll, D.; Kline, L.K.: RNA methylation. The Enzymes, 3rd Ed., **15B**, 557-566 (1982)
[12] Brule, H.; Grosjean, H.; Giege, R.; Florentz, C.: A pseudoknotted tRNA variant is a substrate for tRNA (cytosine-5)-methyltransferase from Xenopus laevis. Biochimie, **80**, 977-985 (1998)
[13] Motorin, Y.; Grosjean, H.: Multisite-specific tRNA:m5C-methyltransferase (Trm4) in yeast Saccharomyces cerevisiae: identification of the gene and substrate specificity of the enzyme. RNA, **5**, 1105-1118 (1999)
[14] King, M.Y.; Redman, K.L.: RNA methyltransferases utilize two cysteine residues in the formation of 5-methylcytosine. Biochemistry, **41**, 11218-11225 (2002)
[15] Reid, R.; Greene, P.J.; Santi, D.V.: Exposition of a family of RNA m5C methyltransferases from searching genomic and proteomic sequences. Nucleic Acids Res., **27**, 3138-3145 (1999)

tRNA (purine-2- or -6-)-methyltransferase 2.1.1.30

1 Nomenclature

EC number
2.1.1.30 (deleted)

Recommended name
tRNA (purine-2- or -6-)-methyltransferase

tRNA (guanine-N^1-)-methyltransferase 2.1.1.31

1 Nomenclature

EC number
2.1.1.31

Systematic name
S-adenosyl-L-methionine:tRNA (guanine-N^1-)-methyltransferase

Recommended name
tRNA (guanine-N^1-)-methyltransferase

Synonyms
1MGT
m1GT
methyltransferase, transfer ribonucleate guanine 1-
tRNA guanine 1-methyltransferase
tRNA(m1G37)methyltransferase
tRNA-m1G methyltransferase
transfer ribonucleate guanine 1-methyltransferase

CAS registry number
37256-99-8

2 Source Organism

<1> *Escherichia coli* (strain W, 2 isoenzymes [1,2]; K12 overproducing strain C600 [9]; strain MJ109, His-tagged enzyme [12,13]) [1, 2, 5, 9-13]
<2> *Salmonella typhimurium* (strains TR1034, GT1381, GT2439, GT857, GT1380, GT1480, GT1482, GT2919, GT2920, GT2921, GT2922, GT3022, GT3023, GT3024, GT3025, GT3026, GT3027, GT3028, GT3029 [11]; syn. Salmonella enterica serovar Typhimurium, strain LT2 and its derivatives GT467, GT477, GT686, GT688, GT907, GT2677, GT2678, GT2685, GT2690, GT2732, GT2750, GT2752, GT2756, GT2772, GT2774, GT2780, GT2784, GT3162, GT3164, GT3645, GT3647, GT3649, GT3651, GT5473, GT5475, GT5477, GT5479 [14]) [4, 11, 14]
<3> *Rattus norvegicus* (adult male Sprague-Dawley rats [6]) [6]
<4> *Saccharomyces cerevisiae* (strain D38, 2 isoenzymes [7,8]; strain D84 [3]) [3, 7, 8]
<5> *Serratia marcescens* (possesses trmD gene [14]) [14]
<6> *Haemophilus influenzae* (possesses trmD gene [14]) [14]
<7> *Bacillus subtilis* (possesses trmD gene [14]) [14]

<8> *Acinetobacter calcoaceticus* (possesses trmD gene [14]) [14]
<9> *Synechocystis sp.* (possesses trmD gene [14]) [14]
<10> *Helicobacter pylori* (possesses trmD gene [14]) [14]
<11> *Borrelia burgdorferi* (possesses trmD gene [14]) [14]
<12> *Mycobacterium leprae* (possesses trmD gene [14]) [14]
<13> *Mycobacterium tuberculosis* (possesses trmD gene [14]) [14]
<14> *Mycoplasma pneumoniae* (possesses trmD gene [14]) [14]
<15> *Mycoplasma genitalium* (possesses trmD gene [14]) [14]

3 Reaction and Specificity

Catalyzed reaction
S-adenosyl-L-methionine + tRNA = S-adenosyl-L-homocysteine + tRNA containing N^1-methylguanine (<1> mechanism [5]; <1> random-sequential mechanism [12])

Reaction type
methyl group transfer

Natural substrates and products

S S-adenosyl-L-methionine + tRNA <1-3> (<1> post-transcriptional tRNA-modifying enzyme [5]; <3> involved in specific tRNA methylation [6]; <1> very low amount of enzyme in cell [9]; <1,2> tRNA modifying enzyme during maturation [9-14]; <2> conformation of the anticodon loop may be a major determining element for the formation of m1G37 in vivo, mutants of tRNAGGGPro may no longer be a substrate in vivo for enzyme [11]) (Reversibility: ? <1-3> [5, 6, 9-14]) [5, 6, 9-14]

P S-adenosyl-L-homocysteine + tRNA containing N^1-methylguanine <1-3> [5, 6, 9-14]

Substrates and products

S S-adenosyl-L-methionine + GpG <1> (<1> minimal substrate that can be methylated, forming m1G, low efficiency, poor binding affinity [12]) (Reversibility: ? <1> [12]) [12]

P S-adenosyl-L-homocysteine + N^1-methylated GpG <1> (<1> only methylation of the 3′ G residue [12]) [12]

S S-adenosyl-L-methionine + polyG <1> (<1> minimal substrate that can be methylated, forming m1G, low efficiency, poor binding affinity, but higher than of GpG [12]) (Reversibility: ? <1> [12]) [12]

P S-adenosyl-L-homocysteine + polyG containing N^1-methylguanine <1> [12]

S S-adenosyl-L-methionine + tRNA <1-4> (<1> highly specific for tRNA and S-adenosyl-L-methionine, cannot be replaced by methionine plus ATP, formaldehyde plus tetrahydrofolic acid, methylcobalamin and methyltetrahydrofolic acid [2]; <1> high specificity for tRNA species and the specific guanosine base to be methylated, gross tRNA structure is important for full enzyme activity, stacking in the stem structure is important for methylation activity, importance of the presence of the anticodon G adjacent to the G being methylated for enzyme recognition, methyl group acceptor: $tRNA_1^{Leu}$ [5]; <1> methyl group acceptor: isoenzymes I and II: methyl-deficient tRNA from E. coli, tRNA from Corynebacterium diphtheriae, isoenzyme II: tRNA from Clostridium pasteurianum [1]; <1,4> methyl group acceptor: methyl-deficient tRNA from E. coli [2,3]; <3> methyl group acceptor: bulk E. coli K12 tRNA, methylates only $tRNA_f^{Met}$ with G7,U8,G9 sequence and a short variable loop [6]; <4> methyl group acceptor: isoenzyme I: bulk E. coli K12 tRNA, isoenzymes I and II: E. coli W6 methyl-deficient tRNA, 2 isoenzymes with different site specificities: isoenzyme I may be specific for guanine at position 9 and isoenzyme II for guanine at position 43 [7]; <4> methyl group acceptor: isoenzyme I: E. coli normal and methyl-deficient $tRNA_f^{Met}$, specific for site G9, isoenzyme II: E. coli methyl-deficient $tRNA_1^{Leu}$, specific for site G37 [8]; <1> methyl group acceptor: bulk methyl-deficient tRNA from E. coli K12 strain IB1111 at position G37, next to the anticodon, and from E. coli W6, 12% of the bulk tRNA is methylated at G1 [9]; <1> extend of methylation is directly proportional to the amount of tRNA added, isoenzyme I: 0.04% and isoenzyme II: 0.034% of nucleotide residues methylated, isoenzyme I and II may have common methylation sites [2]; <1> enzyme requires entire tRNA structure and G residue of the anticodon at position 36 for optimal catalytic activity, tRNA recognition may require specific base contacts as well as phosphate backbone structures embodied in the tRNA structure, properly stacked bases rather than primary sequence in other positions of the stem-loop structure are important for recognition, GpG sequence at position 36/37 is specifically bound, perhaps in the active site, and involved in stabilizing the enzyme-tRNA complex, methyl group transfer only to G37 within 7 of the 46 tRNA species of E. coli, e.g. $tRNA_1^{Leu}$ [10]; <1> S-adenosyl-L-methionine is not required for tRNA binding [10,12]; <2> extreme dependence of enzyme on an almost perfect 3-dimensional structure of the tRNA, enzyme requires more than a simple mini-helix containing the target nucleotide, base-stacking and dinucleotide G36-G37 is important for tRNA recognition, activity is dependent on an intact acceptor stem [11]; <1,2> methyl group acceptors at position G37 next to the 3'end of the anticodon: $tRNA_1^{Pro}$, $tRNA_2^{Pro}$, $tRNA_3^{Pro}$, $tRNA_1^{Leu}$, $tRNA_2^{Leu}$, $tRNA_3^{Leu}$ and $tRNA_3^{Arg}$ [11]; <1> tRNA recognition by core tertiary structures of tRNA, G37pG36 and correct position of G37 in space relative to important contacts elsewhere in the molecule, acceptor stem structure, but not specific base sequence, methyl group acceptors: E. coli $tRNA_1^{Leu}$, E. coli $tRNA^{His}$ and yeast $tRNA^{Asp}$ substituted with G at positions 36 and 37 are excellent substrates, modifies a subset of tRNAs which all recognize only codons beginning with C, such as those for Leu, Pro and Arg [12]; <1> methyl group acceptor: E. coli $tRNA_1^{Leu}$, G37 and G36 are crucial for catalysis and substrate binding and may undergo structural perturbation in the presence of S-adenosyl-L-methionine [13]; <1> methylation of N^1-position of G37 in only 7 isoacceptors of E. coli, all recognize codon beginning with C and have G36, e.g. $tRNA_1^{Leu}$, enzyme requires complete tRNA structure for catalytic activity, the core tRNA structure, a canonical anticodon stem and loop structure and an acceptor stem, massive contacts of enzyme with the phosphate backbone in the anticodon stem and limited contacts in the core region, G36pG37 is crucial for tRNA binding and catalytic activity, conformational changes in 1MGT and tRNA occur upon complex formation [13]; <2> formation of m1G37 in seven tRNA species, e.g. tRNACGGPro and tRNAGGGPro [14]) (Reversibility: ir <1> [2]; ? <2-4> [1, 3-14]) [1-14]

P S-adenosyl-L-homocysteine + tRNA containing N^1-methylguanine <1-4> (<3> formation of 1-methylguanine only at position 9 [6]) [1-4, 6-14]

S Additional information <1-4> (<1> not as methyl group acceptors: methyl-deficient DNA from E. coli, polyadenylic acid, polycytidylic acid, polyuridylic acid, polyinosinic acid, copolymers of adenylic acid and uridylic acid and of guanylic acid and cytidylic acid, tobacco mosaic viral RNA, turnip yellow mosaic viral RNA, RNA from bacteriophage f2, CMP, AMP, UMP, GMP, rRNA [2]; <3> methylation occurs on polynucleotide level, no substrates: $tRNA^{Phe}$, $tRNA^{Val}$, $tRNA^{Arg}$, $tRNA_2^{Glu}$, $tRNA_3^{Ser}$, $tRNA_4^{Leu}$ [6]; <4> no substrates for isoenzyme I: $tRNA_1^{Ser}$, $tRNA_3^{Ser}$, $tRNA_1^{Leu}$, $tRNA_2^{Leu}$ [8]; <1> methylation of G1 only in position 37 of tRNA from prokaryotes, however, in tRNA from eukaryotes m1G may be found in both position 9 and 37 [9]; <2> many mutants of tRNAGGGPro are no longer substrates, e.g. mutations G43A in the anticodon stem, G72A, G68A and G7A in the acceptor stem reduce the level of m1G37 by more than 80%, base substitutions C11U, C23U and G15A in the D-stem, and a deletion in the extra arm severely reduce m1G37 level [11]; <1,2> methylation at G37 prevents translational frameshifting [12,14]) [2, 6, 8, 9, 11, 12, 14]

P Additional information <1> (<1> no methylation products are 6-hydroxy-2-methylguanine, 2-dimethylamino-6-hydroxypurine, 7-methylguanine [1]) [1]

Inhibitors

GpG <1> (<1> potent and specific inhibitor [10,12]) [10, 12]
K^+ <1> (<1> inhibits above 100 mM [9]) [9]
Mg^{2+} <1> (<1> inhibits at 2 mM [9]) [9]
NH_4^+ <1> (<1> inhibits above 150 mM [9]) [9]
Na^+ <1> (<1> inhibits above 100 mM [9]) [9]
S-adenosyl-L-homocysteine <1, 3> (<3> strong product inhibition [6]; <1> product inhibition, isoenzyme I is less sensitive than isoenzyme II, reversed by S-adenosylmethionine [2]; <1> competitive product inhibition [9]; <1> strong competitive product inhibition [12,13]) [2, 6, 9, 12, 13]
Zn^{2+} <1> [2]
iodine <1> (<1> partial inactivation [13]) [13]
mature $tRNA_1^{Leu}$ <1> (<1> fully modified $tRNA_1^{Leu}$ is an excellent inhibitor of $tRNA_1^{Leu}$ transcript methylation, 50% inhibition at 0.061 mM [12]) [12]
mature $tRNA_4^{Leu}$ <1> (<1> 50% inhibition at 0.054 mM [12]) [12]

mature $tRNA^{His}$ <1> (<1> 50% inhibition at 0.095 mM [12]) [12]
p-chloromercuribenzoate <1> (<1> 2mM, complete inhibition [9]) [9]
p-hydroxymercuribenzoate <1> [2]
polyG <1> (<1> potent and specific inhibitor [10,12]) [10, 12]
spleen phosphodiesterase <1> [2]
Additional information <1> (<1> not inhibited by Ni^{2+}, snake venom phosphodiesterase, L-homocysteine [2]; <1> not inhibited by polyU, GMP, ApG, GpU, GpC and GpA [12]; <1> enzyme is relatively resistant to Pb inactivation [13]) [2, 12]

Cofactors/prosthetic groups

S-adenosyl-L-methionine <1-4> [1-14]

Activating compounds

2-mercaptoethanol <1> (<1> activates [2]) [2]
putrescine <1, 3> (<3> activates at 40 mM [6]; <1> not required for activity, but stimulates 1.2fold at 6 mM [9]) [6, 9]
spermidine <1, 3> (<3> activates, at 2 mM 86% as effective as 40 mM putrescine [6]; <1> not required for activity, but stimulates 1.2fold at 2 mM [9]) [6, 9]
spermine <3, 4> (<3> poor activator, at 0.2 mM 27% as effective as 40 mM putrescine [6]; <4> greatly activates, isoenzyme I: optimum at 1 mM, isoenzyme II: optimum at 0.5 mM and around 3 mM [7]) [6, 7]
Additional information <1> (<1> not activated by spermine [2]) [2]

Metals, ions

Ca^{2+} <1> (<1> activates [2]) [2]
Mg^{2+} <1, 4> (<1> in presence of Mg^{2+} 4fold increase of activity [2]; <4> activates at 2-5 mM [7]) [2, 7]
Mn^{2+} <1> (<1> activates [2]) [2]
ammonium acetate <3> (<3> poor activator, at 0.3 M 22% as effective as 40 mM putrescine [6]) [6]
ammonium sulfate <3> (<3> activates, at 0.1 mM 40% as effective as 40 mM putrescine [6]) [6]
magnesium acetate <3> (<3> poor activator, at 3 mM 12% as effective as 40 mM putrescine [6]) [6]
Additional information <3> (<3> not activated by K^+ or Na^+ [6]) [6]

Turnover number (min^{-1})

10 <1> (tRNA) [9]

Specific activity (U/mg)

0.00347 <1> (<1> isoenzyme II [1]) [1]
0.0053 <1> (<1> isoenzyme I [1]) [1]
0.006 <3> (<3> above [6]) [6]
0.0124 <1> (<1> wild-type enzyme [12]) [12]
0.0152 <1> (<1> His-tagged enzyme [12,13]) [12, 13]
0.313 <1> [9]

K_m-Value (mM)

0.003 <3> (S-adenosyl-L-methionine, <3> bulk E. coli K12 tRNA [6]) [6]
0.0033-0.0036 <1> ($tRNA_1^{Leu}$ from E. coli) [12]
0.005 <1> (S-adenosyl-L-methionine) [9]
0.01 <1> ($tRNA^{Asp}$ from E. coli) [12]
0.02 <1> (methyl-deficient tRNA from E. coli strain IB1111, <1> value only a guideline [9]) [9]
0.024 <1> (S-adenosyl-L-methionine, <1> isoenzyme I [2]) [2]
0.067 <1> (S-adenosyl-L-methionine, <1> isoenzyme II [2]) [2]
0.67 <1> (methyl deficient tRNA from E. coli, <1> isoenzyme II [2]) [2]
0.83 <1> (methyl deficient tRNA from E. coli, <1> isoenzyme I [2]) [2]
1.1 <1> ($tRNA^{His}$ from E. coli) [12]
4 <1> ($tRNA^{Asp}$ $G^{36}pG^{37}$ from yeast) [12]
Additional information <1> (<1> values for E. coli and yeast tRNA derivatives [12]) [12]

K_i-Value (mM)

0.00011 <3> (S-adenosyl-L-homocysteine, <3> E. coli K12 tRNA [6]) [6]
0.006 <1> (S-adenosyl-L-homocysteine) [9]
2 <1> (polyG, <1> E. coli $tRNA_1^{Leu}$ [12]) [12]
6 <1> (GpG, <1> E. coli $tRNA_1^{Leu}$ [12]) [12]

pH-Optimum

8 <3> (<3> broad, TEA-HCl better than Tris or bicine buffer [6]) [6]
8-8.5 <1> (<1> isozyme I, in both Tris and triethanolamine buffers [2]; <1> broad range around [9]) [2, 9]
8-9 <1> (<1> isozyme II, triethanolamine buffer [2]) [2]
Additional information <1> (<1> pI: 5.2 [9]) [9]

Temperature optimum (°C)

30 <4> (<4> assay at [3,7,8]) [3, 7, 8]
37 <1, 3> (<1,3> assay at [6,9]) [6, 9]
38 <1> (<1> assay at [1]) [1]

4 Enzyme Structure

Molecular weight

46000 <1> (<1> gel filtration [9]) [9]
50000 <2> (<2> glycerol gradient centrifugation [4]) [4]
83000 <3> (<3> gel filtration [6]) [6]

Subunits

? <1> (<1> x * 28000, calculated from DNA sequence of trmD gene [10]; <1> x * 31000, runs anomalously slowly, SDS-PAGE, x * 28800, calculated from the amino acid sequence [12]) [10, 12]
monomer <1> (<1> 1 * 32000, enzyme may be a single polypeptide chain of MW 32000, SDS-PAGE [9]; <1> 1 * 28900 [13]) [9, 13]

5 Isolation/Preparation/Mutation/Application

Source/tissue

liver <3> [6]

Localization

soluble <1> [2]

Purification

<1> (partial [1]; 2 isoenzymes: I and II [1,2]; 706fold purification [9]; wild-type and His-tagged enzyme [12]; His-tagged enzyme [13]) [1, 2, 9, 12, 13]
<2> (partial, phosphocellulose chromatography [4]) [4]
<3> (6800fold purification, phosphocellulose chromatography [6]) [6]
<4> (partial, hydroxyapatite chromatography, 2 isoenzymes [7,8]) [3, 7, 8]

Cloning

<1> (multi-copy plasmid pBY03 carrying the structural gene trmD+ expressed in Escherichia coli K12 strain C600 [9]; cloning and sequencing of the trmD gene encoding enzyme [10]; trmD gene is part of an operon of 4 genes: rpsP, rimM, trmD and rplS [10,12]; expression of trmD gene is 40fold less than for ribosomal protein genes of the operon, possibly due to a poor Shine-Dalgarno sequence and secondary/tertiary structure in operon mRNA [12]; trmD gene is sequenced [14]) [9, 10, 12, 14]
<2> (trmD gene encodes tRNA(m1G37)methyltransferase [11,14]; trmD gene is sequenced [14]) [11, 14]
<8> (trmD gene is partly sequenced [14]) [14]
<5-7, 9-15> (trmD gene is sequenced [14]) [14]

Engineering

A145T <2> (<2> trmD11 mutant, no effect on catalytic activity, reduced level of enzyme [14]) [14]
A25T <2> (<2> trmD17 mutant, mutant with altered tRNA substrate specificity, but no effect on catalytic activity [14]) [14]
A8G <2> (<2> trmD23 mutant, mutation in the putative translation initiation region [14]) [14]
E243K <2> (<2> trmD1 mutant, mutant with altered tRNA substrate specificity, but no effect on catalytic activity, reduced level of enzyme in vivo compared to the wild-type enzyme, more unstable than wild-type enzyme [14]) [14]
G140S <2> (<2> trmD10 mutant, with reduced methylation of all potential tRNA substrates, mutation primarily affects catalytic activity [14]) [14]
L94F <2> (<2> trmD2 mutant, mutant with altered tRNA substrate specificity, but no effect on catalytic activity [14]) [14]
P184L <2> (<2> trmD3 mutant with reduced methylation of all potential tRNA substrates, mutation primarily affects catalytic activity, reduced level of enzyme in vivo compared to the wild-type enzyme [14]) [14]

Q246stop <2> (<2> trmD22 mutant, also with substitution V99I, mutant with altered tRNA substrate specificity, but no effect on catalytic activity, reduced level of enzyme in vivo compared to the wild-type enzyme [14]) [14]
V99I <2> (<2> trmD22 mutant, also with substitution Q246stop, mutant with altered tRNA substrate specificity, but no effect on catalytic activity, reduced level of enzyme in vivo compared to the wild-type enzyme [14]) [14]
Additional information <1, 2> (<2> a mutation in the structural gene trmD for tRNA(m1G37)methyltransferase, which results in lack of m1G37, also induces suppression of certain frameshift mutations in the his-operon [11]; <1> deletion of the trmD gene mediates extremely slow growth [11]; <2> isolation and characterization of several trmD mutants, a class of mutant forms of enzyme methylates not the +1 frameshift suppressor sufA6 tRNA, which is methylated by wild-type enzyme, and thereby induce extensive frameshifting resulting in nonviable cell [14]) [11, 14]

6 Stability

Temperature stability

30 <4> (<4> at least 1 min stable [7]) [7]
45 <4> (<4> after 1 min: isoenzyme I: 7% loss of activity, isoenzyme II: inactivation [7]) [7]
60 <4> (<4> inactivation after 1 min [7]) [7]

Organic solvent stability

ethylene glycol <1> (<1> 50% v/v, stable to [9]) [9]

General stability information

<1>, freeze-thawing inactivates, 50% v/v ethylene glycol stabilizes [9]
<3>, glycerol, 40% v/v, stabilizes [6]
<4>, unstable enzyme, freezing without antifreeze agents inactivates, 30% ethylene glycol at -20°C stabilizes [3]

Storage stability

<1>, -20°C, 50% v/v ethylene glycol, at least 2 months, stable [9]
<1>, 0°C, partially purified, 48 h, 15% loss of activity [1]
<1>, 0°C, purified, 6 months, about 70% loss of activity [1]
<3>, -70°C, 0.01 M Tris-HCl buffer, pH 8.15, 5 mM dithiothreitol, 40% v/v glycerol, at least 4 months, stable [6]
<4>, 4°C, isoenzyme I: several days, stable, isoenzyme II: unstable, $t_{1/2}$: about 40 h [7]
<4>, partial purified pooled isoenzyme I, protein concentration 3 mg/ml, 1 year, stable [8]

References

[1] Hurwitz, J.; Gold, M.; Anders, M.: The enzymatic methylation of ribonucleic acid and deoxyribonucleic acid. III. Purification of soluble ribonucleic acid-methylating enzymes. J. Biol. Chem., **239**, 3462-3473 (1964)
[2] Hurwitz, J.; Gold, M.; Anders, M.: The enzymatic methylation of ribonucleic acid and deoxyribonucleic acid. IV. The properties of the soluble ribonucleic acid-methylating enzymes. J. Biol. Chem., **239**, 3474-3482 (1964)
[3] Björk, G.R.; Svensson, I.: Studies on microbial RNA. Fractionation of tRNA methylases from Saccharomyces cerevisiae. Eur. J. Biochem., **9**, 207-215 (1969)
[4] Cimino, F.; Traboni, C.; Colonna, A.; Izzo, P.; Salvatore, F.: Purification and properties of several transfer RNA methyltransferases from S. typhimurium. Mol. Cell. Biochem., **36**, 95-104 (1981)
[5] Holmes, W.M.; Andraos-Selim, C.; Roberts, I.; Wahab, S.Z.: Structural requirements for tRNA methylation. Action of Escherichia coli tRNA(guanosine-1)methyltransferase on $tRNA_1^{Leu}$ structural variants. J. Biol. Chem., **267**, 13440-13445 (1992)
[6] Glick, J.M.; Averyhart, V.M.; Leboy, P.S.: Purification and characterization of two tRNA-(guanine)-methyltransferases from rat liver. Biochim. Biophys. Acta, **518**, 158-171 (1978)
[7] Smolar, N.; Hellman, U.; Svensson, I.: Two transfer RNA (1-methylguanine) methylases from yeast. Nucleic Acids Res., **2**, 993-1004 (1975)
[8] Sindhuphak, T.; Hellman, U.; Svensson, I.: Site specificities of three transfer RNA methyltransferases from yeast. Biochim. Biophys. Acta, **824**, 66-73 (1985)
[9] Hjalmarsson, K.J.; Byström, A.S.; Björk, G.R.: Purification and characterization of transfer RNA (guanine-1)methyltransferase from Escherichia coli. J. Biol. Chem., **258**, 1343-1351 (1983)
[10] Holmes, W.M.; Andraos-Selim, C.; Redlak, M.: tRNA-m1G methyltransferase interactions: Touching bases with structure. Biochimie, **77**, 62-65 (1995)
[11] Qian, Q.; Björk, G.R.: Structural requirements for the formation of 1-methylguanosine in vivo in tRNAGGGPro of Salmonella typhimurium. J. Mol. Biol., **266**, 283-296 (1997)
[12] Redlak, M.; Andraos-Selim, C.; Giege, R.; Florentz, C.; Holmes, W.M.: Interaction of tRNA with tRNA (guanosine-1)methyltransferase: Binding specificity determinants involve the dinucleotide $G^{36}pG^{37}$ and tertiary structure. Biochemistry, **36**, 8699-8709 (1997)
[13] Gabryszuk, J.; Holmes, W.M.: tRNA recognition for modification: Solution probing of tRNA complexed with Escherichia coli tRNA (guanosine-1) methyltransferase. RNA, **3**, 1327-1336 (1997)
[14] Li, J.N.; Björk, G.R.: Structural alterations of the tRNA(m1G37)methyltransferase from Salmonella typhimurium affect tRNA substrate specificity. RNA, **5**, 395-408 (1999)

tRNA (guanine-N^2-)-methyltransferase 2.1.1.32

1 Nomenclature

EC number
2.1.1.32

Systematic name
S-adenosyl-L-methionine:tRNA (guanine-N^2-)-methyltransferase

Recommended name
tRNA (guanine-N^2-)-methyltransferase

Synonyms
guanine-N^2-methylase
methyltransferase, transfer ribonucleate guanine 2-
transfer RNA guanine 2-methyltransferase
transfer ribonucleate guanine 2-methyltransferase
transfer ribonucleate guanine N^2-methyltransferase

CAS registry number
37217-31-5

2 Source Organism

<1> *Oryctolagus cuniculus* [1]
<2> *Mus musculus* (bearing Ehrlich ascites carcinoma [2]; strain BALB/c [14]) [2, 14]
<3> *Gallus gallus* (embryo [3]) [3, 9]
<4> *Rattus norvegicus* (Sprague-Dawley strain [4,16]; leukemic strain Fischer RL [4, 6]; two isozymes [8]) [4, 6-8, 13, 16]
<5> *Saccharomyces cerevisiae* (strain D84 [15]) [15]
<6> *Homo sapiens* (HeLa cells [5,12]) [5, 12]
<7> *Avian myeloblastosis virus* [9]
<8> *Rous sarcoma virus SR4* (clonal line from SR-RSV-D [9]) [9]
<9> *Type C RNA tumor virus* (purified from avian myeloblastosis virions [10]) [10]
<10> *Tetrahymena pyriformis* (strain W [11]) [11]

3 Reaction and Specificity

Catalyzed reaction

S-adenosyl-L-methionine + tRNA = S-adenosyl-L-homocysteine + tRNA containing N^2-methylguanine

Reaction type

methyl group transfer

Natural substrates and products

S S-adenosyl-L-methionine + tRNA <3> (<3> involved in ribonucleic acid modification [3]) (Reversibility: ? <3> [3]) [3]

P ?

Substrates and products

S S-adenosyl-L-methionine + tRNA <1-10> (<4,6> methyl group acceptors are tRNAs from E. coli [5,16]; <1-3,6-8> tRNA from E. coli B [1-3, 9, 12, 14]; <2-4,9> best substrate: $tRNA^{Phe}$ [3, 4, 8, 10, 14, 16]; <4> best substrate, isozyme I: $tRNA^{Val}$ [8]; <3,4,9> $tRNA^{Arg}$ [3, 4, 8, 10, 16]; <3> $tRNA^{Leu}$ [3]; <10> $tRNA_2^{Leu}$ [11]; <3,4,6,10> $tRNA_f^{Met}$ [3, 5, 11, 13]; <10> best substrate: $tRNA^{Ile}$, tRNA3Ser [11]; <3,7,8> tRNA from E. coli K12 [9]; <1,4> isozyme II: $tRNA_f^{Met}$ [1, 8]; <4> methyl-deficient tRNA [6]; <10> best substrates are cation-tRNA complexes [11]) (Reversibility: ? <1-10> [1-16]) [1-16]

P S-adenosyl-L-homocysteine + tRNA containing N^2-methylguanine <5, 6, 10> (<5,10> or N^2,N^2-dimethylguanine [6,11,15]; <10> a low enzyme/tRNA ratio favors N^2-methylguanine formation, a high enzyme/tRNA ratio favors N^2,N^2-dimethylguanine formation [11]) [5, 11, 15]

S Additional information <2-4, 6-10> (<3,4,7,8> poor substrates: yeast tRNA [6, 9]; <9> poor substrates: E. coli $tRNA_4^{Leu}$, $tRNA_1^{Leu}$, $tRNA_f^{Met}$ [10]; <10> no substrates: E. coli $tRNA^{Phe}$, $tRNA^{Val}$ [11]; <4> no substrates: $tRNA_2^{Glu}$, $tRNA_3^{Ser}$, $tRNA_4^{Leu}$, $tRNA_f^{Met}$, [16]; <6> no substrates: $tRNA^{Tyr}$, tRNAs from Bacillus subtilis, E. coli B, HeLa cells [5]; <4,6> Saccharomyces cerevisiae [5,13]; <4,6> rat liver [5,6]; <6> wheat seedlings, calf thymus, chicken embryo fibroblasts, Rous sarcoma virus transformed chicken fibroblast membrane, Virginia creeper, rRNA from rat liver, HeLa cells, [5]; <3,7,8> wheat seedlings, Virginia creeper [9]; <6> no substrate: DNA [5]; <2-4> methylation at position 10 [3,4,14,16]; <2> methylation at position 34 [14]; <3,4,9> methylation of $tRNA^{Phe}$ from E. coli [3, 10, 16]; <4> methylated sequence: A-G(m2)-C [4, 6]; <4> methylated sequence: s4-U-A-G-C [16]; <4> isozyme I acts between acceptor stem and dihydrouridine loop, isozyme II between the anticodon and the dihydrouridine loop [8]; <10> methylates position 26 from 5'-end [11]) [3-6, 8-11, 13, 14, 16]

P ?

Inhibitors

5'-S-(N^6-methyladenosyl)-DL-homocysteine <1> [1]

5'-deoxy-5'-S-isobutyl-adenosine <7, 8> [9]

K^+ <3> (<3> above 100 mM [3]) [3]
Mg^{2+} <4> (<4> in the presence of polyamines [8]) [8]
Na^+ <3> (<3> above 100 mM [3]) [3]
NaCl <4> [6]
S-adenosyl-D-homocysteine <1> (<1> strong, competitive [1]) [1]
S-adenosyl-L-homocysteine <1, 3, 4, 7-9> (<1> product inhibition [1]) [1, 7, 9, 10, 16]
S-adenosyl-L-homocysteine structural analogues <1> [1]
S-adenosylmethionine <7, 8> [9]
homocysteine <9> (<9> weak [10]) [10]
non-dialyzable protein <4> (<4> absent in tumor tissue [4]) [4]
proflavin <9> [10]
selenomethionine <9> (<9> weak [10]) [10]
Additional information <2, 9> (<2,9> not inhibitory: nicotinamide [2, 10]; <9> not inhibitory: several antibiotics, e.g. streptomycin, colcemid, cycloheximide [10]) [2, 10]

Cofactors/prosthetic groups

Additional information <9> (<9> no cofactors are NAD^+, $NADP^+$, NADH, NADPH, nicotinamide, FAD, thiamine, pyridoxine, vitamin B_{12}, purine derivatives, cyclic AMP, emetine, epinephrine, catechol, tropolone, phenol, guaiacol [10]) [10]

Activating compounds

KCl <6> (<6> activation, 150 mM [5]; <4> no stimulation [16]) [5]
NH_4^+ <2, 4, 9> (activation, no requirement [8]) [2, 8, 10, 16]
Nonidet P-40 <3, 7, 8> [9]
adriamycin <9> [10]
daunomycin <9> [10]
polylysine <9> [10]
putrescine <4> [8, 16]
spermidine <4> [8, 16]
spermine <4, 9> [10, 16]
Additional information <2, 9> (<2> no activation by Triton X-100 [14]; <9> no activation by several antibiotics, e.g. streptomycin, colcemid [10]) [10, 14]

Metals, ions

Mg^{2+} <3, 4, 9, 10> (<3,4,9> increase of activity [3,8,10,16]; <3,9> no requirement [3,10]; <10> no activation with tRNA substrate in excess and limiting enzyme concentration [11]; <4> together with polyamines slight inhibition [8]) [3, 8, 10, 11, 16]
Additional information <4, 5> (<5> no divalent metal ions required [15]; <4> no stimulation by K^+, Na^+ [16]) [15, 16]

Specific activity (U/mg)

0.000027 <4> (<4> liver [4]) [4]
0.000073 <4> (<4> leukemic rat spleen [4]) [4]
0.006 <10> [11]

0.021 <4> [16]
Additional information <9> [10]

K_m-Value (mM)

0.0003 <3> (tRNA^{Phe}) [3]
0.00078 <4> (tRNA^{Arg}) [4]
0.001 <8> (S-adenosyl-L-methionine) [9]
0.001 <3> (tRNA, <3> embryo fibroblast membrane [9]) [9]
0.0013 <8> (S-adenosyl-L-methionine) [9]
0.00138 <3> (S-adenosyl-L-methionine) [3]
0.0015 <7> (S-adenosyl-L-methionine) [9]
0.002 <4> (S-adenosyl-L-methionine) [7]
0.002 <7> (S-adenosyl-L-methionine) [9]
0.002 <4> (bulk E. coli tRNA) [16]
0.01 <4> (tRNA, <4> bulk [4]) [4]
0.07 <4> (E. coli tRNA^{Phe}) [16]

K_i-Value (mM)

0.003 <8> (S-adenosyl-L-methionine) [9]
0.004 <7> (S-adenosyl-L-methionine) [9]
0.008 <4> (S-adenosyl-L-homocysteine) [7]
0.04 <1> (S-adenosyl-D-homocysteine) [1]
0.06 <1> (5'-S-(N^6-methyladenosyl)-DL-homocysteine) [1]
4 <7> (5'-deoxy-5'-S-isobutyl-adenosine) [9]
5.2 <8> (5'-deoxy-5'-S-isobutyl-adenosine) [9]

pH-Optimum

7.6-7.9 <9> [10]
7.9 <6> [5]
8 <4> (<4> TEA buffer better than Tris or bicine buffer [16]) [16]
8-8.5 <3> [3]

pH-Range

7.3-8.5 <9> (<9> about half-maximal activity at pH 7.3 and 8.5 [10]) [10]

Temperature optimum (°C)

30 <5> (<5> assay at [15]) [15]
37 <2-4, 7-9> [4, 8-10, 14]

4 Enzyme Structure

Molecular weight

69000 <4> (<4> gel filtration [16]) [16]
77000 <3> (<3> gel filtration [3]) [3]
100000 <4> (<4> gel filtration [4]) [4]
220000 <9> (<9> gel filtration in the presence of Triton X-100 [10]) [10]

Subunits

Additional information <9> (<9> SDS-PAGE fails to give any protein bands [10]) [10]

5 Isolation/Preparation/Mutation/Application

Source/tissue

Ehrlich ascites carcinoma cell <2> [2]
fibroblast <3> [9]
hepatoma cell <4> [13]
liver <1, 4> [1, 4, 6-8, 13, 16]
plasma cell tumor cell <2> (<2> MOPC-104E [14]) [14]
spleen <4> [4]
Additional information <2> (intracisternal A particle) [14]

Localization

cytoplasm <6> [12]
membrane <3> [9]
soluble <2, 4, 9, 10> [2, 6, 10, 11]
viral envelope <7, 8> (<7,8> inside of, not viral core [9]) [9]

Purification

<2> (partial [2]) [2]
<3> (partial, phosphocellulose chromatography, S-adenosyl-L-homocysteine-Sepharose affinity chromatography [3]) [3]
<6> (HeLa cells, partial [5]) [5]
<9> (solubilized with Triton X-100, hydroxyapatite chromatography [10]) [10]
<10> (affinity chromatography [11]) [11]

6 Stability

Temperature stability

37 <4> (<4> stable for 3 h [4]) [4]

General stability information

<3>, dialysis inactivates [3]
<3>, gel filtration inactivates [3]
<6>, freezing inactivates [5]
<6>, glycerol does not stabilize [5]

Storage stability

<2>, -70°C, at least 2 months [2]
<3>, -20°C, at least 6 weeks in 10% v/v glycerol [3]
<4>, liquid N_2: 2 years [4]
<6>, -20°C, $t_{1/2}$: 4 weeks [5]

References

[1] Hildesheim, J.; Hildesheim, R.; Blanchard, P.; Farrugia, G.; Michelot, R.: Studies on synthetic inhibitors of t-RNA methyl transferases: analogs of S-adenosyl homocysteine. Biochimie, **55**, 541-546 (1973)
[2] Chan, T.C.; Fraser, M.J.: Partial purification of a tumour tRNA-guanine methyltransferase. Life Sci., **11**, 793-805 (1972)
[3] Izzo, P.; Gantt, R.: Partial purification and characterization of an N^2-guanine RNA methyltransferase from chicken embryos. Biochemistry, **16**, 3576-3581 (1977)
[4] Kraus, J.; Staehelin, M.: N^2-Guanine specific transfer RNA methyltransferase I from rat liver and leukemic rat spleen. Nucleic Acids Res., **1**, 1455-1478 (1974)
[5] Agris, P.F.; Spremulli, L.L.; Brown, G.M.: tRNA methylases from HeLa cells: purification and properties of an adenine-1-methylase and a guanine-N^2-methylase. Arch. Biochem. Biophys., **162**, 38-47 (1974)
[6] Baguley, B.C.; Staehelin, M.: The specificity of transfer ribonucleic acid methylases from rat liver. Biochemistry, **7**, 45-50 (1968)
[7] Glick, J.M.; Ross, S.; Leboy, P.S.: S-Adenosylhomocysteine inhibition of three purified tRNA methyltransferases from rat liver. Nucleic Acids Res., **2**, 1639-1651 (1975)
[8] Leboy, P.S.; Glick, J.M.: tRNA methyltransferases from rat liver. Differences in response of partially purified enzymes to polyamines and inorganic salts. Biochim. Biophys. Acta, **435**, 30-38 (1976)
[9] Pierre, A.; Berneman, A.; Vedel, M.; Robert-Gero, Vigier, P.: Avian oncornavirus associated N^2-methylguanine transferase, location and origin. Biochem. Biophys. Res. Commun., **81**, 315-321 (1978)
[10] Taylor, M.J.; Gantt, R.: Partial purification and characterization of a ribonucleic acid N^2-guanine methyltransferase associated with avian myeloblastosis virus. Biochemistry, **18**, 5253-5258 (1979)
[11] Reinhart, M.P.; Lewis, J.M.; Leboy, P.S.: A single tRNA (guanine)-methyltransferase from Tetrahymena with both mono- and di-methylating activity. Nucleic Acids Res., **14**, 1131-1148 (1986)
[12] Klagsbrun, M.: Differences in the methylation of transfer ribonucleic acid in vitro by the mitochondrial and cytoplasmic transfer ribonucleic acid methylases of HeLa cells. J. Biol. Chem., **248**, 2606-2611 (1973)
[13] Shershneva, L.P.; Venkstern, T.V.; Bayev, A.A.: A study of tRNA methylase action. FEBS Lett., **29**, 132-134 (1973)
[14] Chepelinsky, A.B.; Gantt, R.; Wivel, N.: Presence of RNA methylases in intracisternal A particles purified from a mouse plasma cell tumor. Eur. J. Biochem., **103**, 339-347 (1980)
[15] Björk, G.R.; Svensson, I.: Studies on microbial RNA. Fractionation of tRNA methylases from Saccharomyces cerevisiae. Eur. J. Biochem., **9**, 207-215 (1969)
[16] Glick, J.M.; Averyhart, V.M.; Leboy, P.S.: Purification and characterization of two tRNA-(guanine)-methyltransferases from rat liver. Biochim. Biophys. Acta, **518**, 158-171 (1978)

tRNA (guanine-N^7-)-methyltransferase 2.1.1.33

1 Nomenclature

EC number
2.1.1.33

Systematic name
S-adenosyl-L-methionine:tRNA (guanine-N^7-)-methyltransferase

Recommended name
tRNA (guanine-N^7-)-methyltransferase

Synonyms
7-methylguanine transfer ribonucleate methylase
N^7-methylguanine methylase
m^7G-methyltransferase
methyltransferase, transfer ribonucleate guanine 7-
tRNA guanine 7-methyltransferase
transfer ribonucleate guanine 7-methyltransferase

CAS registry number
37257-00-4

2 Source Organism

<1> *Salmonella typhimurium* (wild-type [1,9]; wild-type strain TA 265 [7]) [1, 7, 9]
<2> *Parthenocissus tricuspidata* [2]
<3> *Phaseolus vulgaris* [3]
<4> *Escherichia coli* (strain W [4,5]; strain MRE 600, 2 isoenzymes [6]) [4-6]
<5> *Saccharomyces cerevisiae* (strain D84 [8]; strains BY4743, 33523, 33899 and YM317 [10]) [8, 10]
<6> *Homo sapiens* [10]

3 Reaction and Specificity

Catalyzed reaction
S-adenosyl-L-methionine + tRNA = S-adenosyl-L-homocysteine + tRNA containing N^7-methylguanine (<1> mechanism [7])

Reaction type
methyl group transfer

Natural substrates and products

S S-adenosyl-L-methionine + tRNA <1, 5> (<1> tRNA modifying enzyme [1]; <1> specific enzyme involved in the maturation of tRNA [7]; <5> complex of 2 proteins Trm8 and Trm82 is required for efficient m^7G-methyltransferase activity in vitro and in vivo, tRNA site G46 is methylated in vivo [10]) (Reversibility: ? <1, 5> [1, 7, 10]) [1, 7, 10]

P S-adenosyl-L-homocysteine + tRNA containing N^7-methylguanine <1, 5> [1, 7, 10]

Substrates and products

S S-adenosyl-L-methionine + tRNA <1-5> (<4> highly specific for tRNA and S-adenosyl-L-methionine, cannot be replaced by methionine plus ATP, formaldehyde plus tetrahydrofolic acid, methylcobalamin and methyltetrahydrofolic acid [5]; <1> methyl group acceptor: undermethylated tRNA [1]; <2,3> methyl group acceptor: yeast $tRNA_1^{Ala}$, methylates G located in the extra loop at position 46 from the 5'end of $tRNA_1^{Ala}$ [2,3]; <4,5> methyl group acceptor: methyl-deficient tRNA from E. coli [4,5,8]; <4> methyl group acceptors: methyl-deficient bulk tRNA from E. coli K12met-rel-, $tRNA_f^{Met}$ from Bacillus subtilis, formation of 0.8 mol N^7-methylguanine residues per mol Bacillus subtilis tRNA, enzyme recognizes and methylates specifically the extra arm unpaired methyl-deficient guanylate residue [6]; <4> extend of methylation is directly proportional to the amount of tRNA added, 0.19% of nucleotide residues are methylated [5]; <1> methyl group acceptor: undermethylated tRNA from Salmonella typhimurium CS 14, NH_2- and COOH-group of S-adenosyl-L-methionine as well as the length of hydrocarbon chain connecting them to the aromatic ring of adenosine, chirality at the α-C atom and NH_2-group of adenosine are important for substrate binding to enzyme [7]; <5> methyl group acceptors: $tRNA^{Cys}$, $tRNA^{Trp}$, $tRNA^{Pro}$, $tRNA^{Met}$, $tRNA^{Met-i}$ and 2 of each $tRNA^{Phe}$, $tRNA^{Val}$ and $tRNA^{Lys}$, methylation at G46 in the extra loop [10]; <5,6> methyl group acceptor: yeast pre-$tRNA^{Phe}$ [10]) (Reversibility: ir <4> [5]; ? <1-3, 5> [1-4, 6-10]) [1-10]

P S-adenosyl-L-homocysteine + tRNA containing N^7-methylguanine <1-5> [1-8, 10]

S Additional information <4-6> (<4> not as methyl group acceptors: methyl-deficient DNA from E. coli, polyadenylic acid, polycytidylic acid, polyuridylic acid, polyinosinic acid, copolymers of adenylic acid and uridylic acid and of guanylic acid and cytidylic acid, tobacco mosaic viral RNA, turnip yellow mosaic viral RNA, RNA from bacteriophage f2, CMP, AMP, UMP, GMP, rRNA [5]; <4> methyl group transfer at the polynucleotide level of tRNA, two m^7G tRNA-methyltransferases with different specificities for the extra arm unpaired G in different tRNA species [6]; <5> complex of 2 proteins Trm8 and Trm82 is required for efficient m^7G-methyltransferase activity in vitro and in vivo, but Trm82 is not an absolute requirement for activity in vitro, Trm8 may be the catalytically active subunit [10]; <6> proteins METTL1 and WDR4 are required for m^7G-methyltransferase activity [10]) [5, 6, 10]

P Additional information <4> (<4> no products: 2-dimethylguanine, 1-methylguanine, 2-methylguanine, 1-methylhypoxanthine, 7-methylhypoxanthine and 1-methyladenine [4]) [4]

Inhibitors

S-adenosyl-D-homocysteine <1> (<1> competitive inhibitor, L-enantiomer more potent than D-enantiomer by 2 orders of magnitude [7]) [7]
S-adenosyl-DL-homocysteine <1> (<1> competitive inhibitor [7]) [7]
S-adenosyl-L-cysteine <1> (<1> competitive inhibitor [7]) [7]
S-adenosyl-L-homocysteine <1, 4> (<1> potent competitive product inhibition, L-enantiomer more potent than D-enantiomer by 2 orders of magnitude [7]; <4> product inhibition, reversed by S-adenosylmethionine [5]) [5, 7]
S-inosylhomocysteine <1> (<1> weak competitive inhibitor, above 0.5 mM [7]) [7]
methylthioadenosine <1> (<1> competitive inhibitor [7]) [7]
methylthioinosine <1> (<1> weak competitive inhibitor, above 0.5 mM [7]) [7]
p-hydroxymercuribenzoate <4> [5]
spleen phosphodiesterase <4> [5]
Additional information <1, 4> (<4> not inhibited by adenosine, L-homocysteine, snake venom phosphodiesterase [5]; <1> not inhibited by butylthioadenosine, thioethanoladenosine, isobutylthioadenosine [7]) [5, 7]

Cofactors/prosthetic groups

S-adenosyl-L-methionine <1-5> (<4> serves usually as coenzyme [6]) [1-10]

Activating compounds

2-mercaptoethanol <4> (<4> activates [5]) [5]
S-inosylhomocysteine <1> (<1> activates at 0.05-0.2 mM [7]) [7]
methylthioinosine <1> (<1> activates at 0.05-0.2 mM [7]) [7]
spermidine <1> (<1> activates at 5 mM [1]) [1]
spermine <1> (<1> activates at 1 mM [1]) [1]
Additional information <4> (<4> not activated by spermine [5]) [5]

Metals, ions

Ca^{2+} <4> (<4> activates [5]) [5]
Mg^{2+} <1, 4> (<1> activates at 3 mM [1]; <4> in presence of Mg^{2+} 2fold increase of activity [5]) [1, 5]
Mn^{2+} <4> (<4> activates [5]) [5]
cations <1> (<1> mono- and divalent cations activate [1]) [1]

Specific activity (U/mg)

3.95 <4> (<4> $tRNA_f^{Met}$ from Bacillus subtilis [6]) [6]

K_m-Value (mM)

0.0034 <1> (undermethylated tRNA) [1]
0.016 <4> (S-adenosyl-L-methionine) [5]
0.053 <1> (S-adenosyl-L-methionine) [1]
0.21 <4> (methyl deficient tRNA from E. coli) [5]

K_i-Value (mM)

0.008 <1> (S-adenosyl-L-homocysteine) [7]
0.064 <1> (S-adenosyl-DL-homocysteine) [7]
0.46 <1> (methylthioadenosine) [7]
0.55 <1> (S-adenosyl-L-cysteine) [7]
0.84 <1> (S-adenosyl-D-homocysteine) [7]

pH-Optimum

7.5-8 <4> (<4> dimethylglutarate buffer [5]) [5]

Temperature optimum (°C)

30 <5> (<5> assay at [8,10]) [8, 10]
37 <4, 6> (<4,6> assay at [6,10]; <6> 30fold higher activity when assayed at 37°C than at 30°C [10]) [6, 10]
38 <4> (<4> assay at [4]) [4]

4 Enzyme Structure

Molecular weight

25000 <1> (<1> glycerol gradient centrifugation [9]) [9]
100000 <4> (<4> gel filtration, two m^7G tRNA-methyltransferases with different specificities for the extra arm unpaired G in different tRNA species [6]) [6]
300000 <4> (<4> gel filtration, two m^7G tRNA-methyltransferases with different specificities for the extra arm unpaired G in different tRNA species [6]) [6]

Subunits

monomer <1> (<1> 1 * 25000-30000 [1]) [1]

5 Isolation/Preparation/Mutation/Application

Localization

chloroplast <3> [3]
mitochondrion <3> [3]
soluble <4> [4-6]
Additional information <3> (<3> organelle-specific enzyme, not in cytoplasm [3]) [3]

Purification

<1> (partial purification by phosphocellulose column [7]) [1, 7, 9]
<4> (partial [4]; 1000fold purification by affinity chromatography on Escherichia coli tRNA Sepharose [6]) [4-6]
<5> (copurification of 2 proteins, Trm8 and Trm82 [10]) [8, 10]

Cloning

<5> (ORF YDL201w encodes Trm8 and ORF YDR165w encodes Trm82, both proteins form a complex with m^{7}G-methyltransferase activity, Trm8 contains a binding domain for S-adenosyl-L-methionine: GXGXG, coexpression of Trm8/Trm82 in Escherichia coli BL21 Codon Plus (DE3) as His6-fusion proteins under control of the Ptac-promoter [10]) [10]
<6> (coexpression of both METTL1 and WDR4 in Saccharomyces cerevisiae [10]) [10]

Engineering

Additional information <5> (<5> strains lacking either Trm8 or Trm82 yield extracts with no detectable enzyme activity, and have severely reduced m^{7}G-modified tRNA in vivo [10]) [10]

6 Stability

Organic solvent stability

ethylene glycol <5> (<5> 30% stabilizes at -20°C [8]) [8]

General stability information

<1>, freezing inactivates [7]
<5>, freezing without antifreeze agents inactivates, 30% ethylene glycol at -20°C stabilizes [8]

Storage stability

<4>, -70°C, frozen in liquid nitrogen, at least several months, stable [6]
<4>, 0°C, partially purified, 48 h, 15% loss of activity [4]
<4>, 0°C, purified, 6 months, about 70% loss of activity [4]

References

[1] Colonna, A.; Ciliberto, G.; Santamaria, R.; Cimino, F.; Salvatore, F.: Isolation and characterization of a tRNA(guanine-7-)-methyltransferase from Salmonella typhimurium. Mol. Cell. Biochem., **52**, 97-106 (1983)
[2] Dubois, E.G.; Weil, J.H.: A 7-methylguanine tRNA methylase present in crown-gall cells but absent in normal Parthenocissus tricuspidata L. cells. Plant Sci. Lett., **8**, 385-394 (1977)
[3] Dubois, E.G.; Dirheimer, G.; Weil, J.H.: Characterization of an organelle-specific 7-methylguanine tRNA methylase in the chloroplasts and mitochondria of Phaseolus vulgaris. Plant Sci. Lett., **5**, 17-26 (1975)
[4] Hurwitz, J.; Gold, M.; Anders, M.: The enzymatic methylation of ribonucleic acid and deoxyribonucleic acid. III. Purification of soluble ribonucleic acid-methylating enzymes. J. Biol. Chem., **239**, 3462-3473 (1964)
[5] Hurwitz, J.; Gold, M.; Anders, M.: The enzymatic methylation of ribonucleic acid and deoxyribonucleic acid. IV. The properties of the soluble ribonucleic acid-methylating enzymes. J. Biol. Chem., **239**, 3474-3482 (1964)

[6] Aschhoff, H.J.; Elten, H.; Arnold, H.H.; Mahal, G.; Kersten, W.; Kersten, H.: 7-Methylguanine specific tRNA-methyltransferase from Escherichia coli. Nucleic Acids Res., **3**, 3109-3122 (1976)

[7] Paolella, G.; Ciliberto, G.; Traboni, C.; Cimino, F.; Salvatore, F.: Effect of adenosylhomocysteine and other analog thioethers on a prokaryotic tRNA (guanine-7)-methyltransferase. Arch. Biochem. Biophys., **219**, 149-154 (1982)

[8] Björk, G.R.; Svensson, I.: Studies on microbial RNA. Fractionation of tRNA methylases from Saccharomyces cerevisiae. Eur. J. Biochem., **9**, 207-215 (1969)

[9] Cimino, F.; Traboni, C.; Colonna, A.; Izzo, P.; Salvatore, F.: Purification and properties of several transfer RNA methyltransferases from S. typhimurium. Mol. Cell. Biochem., **36**, 95-104 (1981)

[10] Alexandrov, A.; Martzen, M.R.; Phizicky, E.M.: Two proteins that form a complex are required for 7-methylguanosine modification of yeast tRNA. RNA, **8**, 1253-1266 (2002)

tRNA guanosine-2'-O-methyltransferase 2.1.1.34

1 Nomenclature

EC number
2.1.1.34

Systematic name
S-adenosyl-L-methionine:tRNA guanosine-2'-O-methyltransferase

Recommended name
tRNA guanosine-2'-O-methyltransferase

Synonyms
methyltransferase, transfer ribonucleate guanosine 2'-
tRNA (Gm18) 2'-O-methyltransferase <2> [11]
tRNA (Gm18) methyltransferase <4> [10]
tRNA guanosine 2'-methyltransferase
transfer ribonucleate guanosine 2'-methyltransferase

CAS registry number
37257-01-5

2 Source Organism

<1> *yeast* [1]
<2> *Escherichia coli* (diverse strains [11]; strain CA274 [2]) [2, 11]
<3> *Thermus thermophilus* (strain HB27 [3-9]; strain HB8 [5]; high growth temperature enhances activity [3]) [3-9]
<4> *Thermus thermophilus* [10]

3 Reaction and Specificity

Catalyzed reaction
S-adenosyl-L-methionine + tRNA = S-adenosyl-L-homocysteine + tRNA containing 2'-O-methylguanosine (<4> methylation site [10]; <3> G18, G19 and the D-stem of the tRNA are required for Gm-methylase recognition [8,9]; <3> substrate recognition sites and mechanism [4,6-8])

Reaction type
methyl group transfer

Natural substrates and products

S S-adenosyl-L-methionine + tRNA <3> (<3> involved in post-transcriptional tRNA modification [3]) [3]

P S-adenosyl-L-homocysteine + tRNA^{Tyr} containing 2'-O-methylguanosine

Substrates and products

S S-adenosyl-L-methionine + tRNA <2> (<2> in vitro [2]) (Reversibility: ? <2> [2]) [2]

P S-adenosyl-L-homocysteine + 2-methylthio $N^6(\gamma,\gamma$-dimethylallyl)adenosine in tRNA <2> [2]

S S-adenosyl-L-methionine + tRNA <1-4> (<3> substrate tRNAs from E. coli, yeast, Bacillus subtilis [8]; <3,4> methylation at 2'-OH specifically of unmodified G18 of the ribose in the D-loop of tRNA [3-5,8,10]; <2> unmethylated tRNA^{Tyr} [2]; <1> tRNA-Phe with C-34, inosine-34 or U-34 are substrates for the modified enzyme with altered anticodon and 3'-adjacent nucleotide, more efficient than the natural substrate tRNA-Phe with G-34, tRNA-Phe with A-34 is no substrate [1]; <2,3> methyl group acceptors from E. coli: methyl deficient tRNA^{Tyr}, tRNA^{Phe}, tRNA$_f^{Met}$ [2-4,6,7]; <1,3> methyl group acceptor from yeast: tRNA^{Phe} [1,3-9]; <3> methyl group acceptors from Bacillus subtilis: tRNA^{Val}, tRNA^{Lys}, tRNA^{Thr}, tRNA^{Leu} [6]) (Reversibility: ? <1-4> [1-10]) [1-10]

P S-adenosyl-L-homocysteine + tRNA containing 2'-O-methylguanosine <1, 3, 4> [1, 3, 8-10]

S Additional information <1-3> (<1> methylates tRNA at position 34 [1]; <2> transfers a methyl group to the 2'-hydroxy group of a guanosine residue of a GG-sequence in methyl-deficient tRNA^{Tyr} from E. coli 58-161 [2]; <3> specifically methylates guanosine at position 18 in the D-loop of tRNA^{Phe} from E. coli [3-6,8]) [1-6, 8]

P ?

Inhibitors

ΔD-arm <3> (<3> tRNA^{Phe} variant from yeast with a modification in the D-stem, inhibition [8]) [8]

G18A <3> (<3> tRNA^{Phe} variant from yeast with an exchange of G to A at position 18, inhibition [8]) [8]

S-adenosyl-L-homocysteine <3> (<3> strong, product inhibition, competitive [3]) [3]

Cofactors/prosthetic groups

Additional information <1> (<1> inosine formation is not dependent on addition of any cofactor including hypoxanthine [1]) [1]

Activating compounds

spermine <3> (<3> activation, 5 mM [3]) [3]

Metals, ions

Mg^{2+} <3> (<3> activation, 5 mM [3]) [3]

Specific activity (U/mg)

0.0071 <3> [3]

K_m-Value (mM)

0.000005 <3> (Escherichia coli tRNA^{Met}) [8]
0.000005 <3> (yeast tRNA^{Phe}, <3> below [8]) [8]
0.000006 <3> (Escherichia coli tRNA^{Ile}) [8]
0.00001 <3> (Escherichia coli tRNA^{Phe}) [3, 8]
0.00001 <4> (yeast tRNA^{Phe}) [10]
0.00047 <3> (S-adenosyl-L-methionine) [3]
Additional information <3> (<3> overview: K_m of 28 tRNA^{Phe} variants from yeast [8]) [8]

K_i-Value (mM)

0.00079 <3> (G18A) [8]
0.00205 <3> (ΔD-arm) [8]

pH-Optimum

7.2-7.5 <3> [3]
8 <2> (<2> assay at [2]) [2]

Temperature optimum (°C)

37 <2> (<2> assay at [2]) [2]
45 <3> (<3> Bacillus subtilis tRNA^{Leu} [6]) [6]
55 <3> (<3> Bacillus subtilis tRNA^{Thr} [6]) [6]
60 <3> (<3> yeast tRNA^{Phe} [4-6]; <3> Bacillus subtilis tRNA^{Lys} [6]) [4-6]
65 <3> (<3> assay at [3,8]; <3> Escherichia coli tRNA$_f^{Met}$ [4,6]; <3> Bacillus subtilis tRNA^{Val} [6]) [3, 4, 6, 8]
Additional information <3> (<3> optimal temperature for methylation is ca. 20°C lower than the melting temperature of the tRNA, overview [6]) [6]

Temperature range (°C)

40-80 <3> (<3> strain HB 8 shows 80% activity at 40°C, strain HB 27 shows 60% activity [5]) [5]
55-68 <3> (<3> about half-maximal activity at 55°C and 68°C, yeast tRNA^{Phe} [6]) [6]
55-75 <3> (<3> about half-maximal activity at 55°C and 75°C, Escherichia coli tRNA$_f^{Met}$ [6]) [6]

4 Enzyme Structure

Molecular weight

20000 <3> (<3> gel filtration [3]) [3]
21000 <3> (<3> gel filtration [4]) [4]

Subunits

monomer <3> (<3> 1 * 21000, SDS-PAGE [3]) [3]

5 Isolation/Preparation/Mutation/Application

Purification

<3> (affinity chromatography on S-adenosyl-L-homocysteine-Sepharose, elution with 6 M urea [3]) [3, 4]
<4> (native and recombinant from Escherichia coli [10]) [10]

Cloning

<2> (DNA and amino acid sequence analysis of spoU gene, encodes tRNA (Gm18) 2'-O-methyltransferase [11]) [11]
<4> (expression in Escherichia coli with and without His-tag, amino acid sequence analysis and comparison [10]) [10]

Engineering

Additional information <2> (<2> spoU gene mutant contains no residue 18, shows no activity but does not influence growth rate [11]) [11]

6 Stability

Temperature stability

40 <3> (<3> Thermus thermophilus strain HB27: 40% of maximal activity, strain HB8: 80% of maximal activity [5]) [5]
60 <3> (<3> stable below, $tRNA^{Phe}$ [6]) [6]
70 <3> (<3> stable below, $tRNA^{Met}$ [6]) [6]
80 <3> (<3> about 90% of original activity retained after 20 min in the presence of 10% v/v glycerol at pH 7.5 [3]) [3, 4]
90 <3> (<3> $t_{1/2}$: 20 min [3]) [3]

General stability information

<3>, urea, stable to 6 M [3]

References

[1] Droogmans, L.; Grosjean, H.: 2-O-methylation and inosine formation in the wobble position of anticodon-substituted tRNA-Phe in a homologous yeast in vitro system. Biochimie, **73**, 1021-1025 (1991)

[2] Gefter, M.L.: The in vitro synthesis of 2'-omethylguanosine and 2-methylthio ^{6}N (γ,γ, dimethylallyl) adenosine in transfer RNA of Escherichia coli. Biochem. Biophys. Res. Commun., **36**, 435-441 (1969)

[3] Kumagai, I.; Watanabe, K.; Oshima, T.: A thermostable tRNA (guanosine-2)-methyltransferase from Thermus thermophilus HB27 and the effect of ribose methylation on the conformational stability of tRNA. J. Biol. Chem., **257**, 7388-7395 (1982)

[4] Matsumoto, T.; Watanabe, K.; Ohta, T.: Recognition mechanism of tRNA with tRNA(guanosine-2)methyltransferase from Thermus thermophilus HB 27. Nucleic Acids Symp. Ser., **15**, 131-134 (1984)

[5] Hori, H.; Matsumoto, T.; Watanabe, K.; Ohta, T.; Miura, K.: Two types of tRNA(Gm)methylase found in extreme thermophile, Thermus thermophilus strains HB 8 and HB 27. Nucleic Acids Symp. Ser., **16**, 225-228 (1985)
[6] Matsumoto, T.; Ohta, T.; Kumagai, I.; Oshima, T.; Murao, K.; Hasegawa, T.; Ishikura, H.; Watanabe, K.: A thermostable Gm-methylase recognizes the tertiary structure of tRNA. J. Biochem., **101**, 1191-1198 (1987)
[7] Matsumoto, T.; Nishikawa, K.; Hori, H.; Ohta, T.; Miura, K.; Watanabe, K.: Recognition sites of tRNA by a thermostable tRNA(guanosine-2-)-methyltransferase from Thermus thermophilus HB27. J. Biochem., **107**, 331-338 (1990)
[8] Hori, H.; Yamazaki, N.; Matsumoto, T.; Watanabe, Y.-I.; Ueda, T.; Nishikawa, K.; Kumagai, I.; Watanabe, K.: Substrate recognition of tRNA (guanosine-2'-)-methyltransferase from Thermus thermophilus HB27. J. Biol. Chem., **273**, 25721-25727 (1998)
[9] Hori, H.; Yamazaki, N.; Matsumoto, T.; Ueda, T.; Nishikawa, K.; Izumi, K.; Watanabe, K.: Essentially minimal sequence for substrate recognition by tRNA (guanosine-2')-methyltransferase from Thermus thermophilus HB27. Nucleic Acids Symp. Ser., **37**, 189-190 (1997)
[10] Hori, H.; Suzuki, T.; Sugawara, K.; Inoue, Y.; Shibata, T.; Kuramitsu, S.; Yokoyama, S.; Oshima, T.; Watanabe, K.: Identification and characterization of tRNA (Gm18) methyltransferase from Thermus thermophilus HB8: domain structure and conserved amino acid sequence motifs. Genes Cells, **7**, 259-272 (2002)
[11] Persson, B.C.; Jager, G.; Gustafsson, C.: The spoU gene of Escherichia coli, the fourth gene of the spoT operon, is essential for tRNA (Gm18) 2'-O-methyltransferase activity. Nucleic Acids Res., **25**, 4093-4097 (1997)

tRNA (uracil-5-)-methyltransferase 2.1.1.35

1 Nomenclature

EC number
2.1.1.35

Systematic name
S-adenosyl-L-methionine:tRNA (uracil-5-)-methyltransferase

Recommended name
tRNA (uracil-5-)-methyltransferase

Synonyms
RUMT
m5U-methyltransferase
methyltransferase, transfer ribonucleate uracil 5-
rTase
ribothymidyl synthase
tRNA (m^5U54)-methyltransferase
tRNA uracil 5-methyltransferase
transfer RNA uracil 5-methyltransferase
transfer RNA uracil methylase

CAS registry number
37257-02-6

2 Source Organism

<1> *Saccharomyces cerevisiae* (strain D84 [1,15]; several strains [24,25]) [1, 15, 24, 25]
<2> *Escherichia coli* (strain SN01 [3]; MRE 600 [3,4]; strain 1100/K12 [5]; strain K12MO7 [6]; strain GB-1-5-39/pTN102 [7,17]; strain C600/pTN106 [9]; strain HB101/pMTXR1 [10]; strains B and M3S [11]; strain W [12,13]; strain GRB822 [22]; 3 enzyme forms: native protein and 2 different TrmA/RNA complexes [5]; native enzyme, may be in 2 forms of 41 and 42 kDa, and TrmA/RNA complex [9]) [2-13, 16-23]
<3> *Salmonella typhimurium* [14]
<4> *Streptococcus faecalis* [4]
<5> *Bacillus subtilis* [4]
<6> *Bacillus cereus* [4]

3 Reaction and Specificity

Catalyzed reaction

S-adenosyl-L-methionine + tRNA = S-adenosyl-L-homocysteine + tRNA containing thymine (<2> catalytic mechanism [5,7,18,19,23]; <2> kinetic, sequential mechanism [6]; <2> steric course of methyl transfer [10,16,19]; <2> single SN2 displacement mechanism [10,16]; <2> detailed chemical and enzymatic mechanism [16])

Reaction type

methyl group transfer

Natural substrates and products

S S-adenosyl-L-methionine + tRNA <1, 2> (<2> tRNA modifying enzyme [3,7,9,10,21,22]; <2> specific reaction in maturation of tRNA [6]; <2> in vivo present both as 42 kDa polypeptide and as covalent TrmA/RNA complex, to equal amounts, both with methylating activity, in vivo the m^{5}U54 modification confers a small selective growth advantage, two different catalytic activities: synthesis of m^{5}U54 and an unknown essential function, may be associated with binding RNA, either during a maturation process of the rRNA and/or tRNA or as a regulatory device during the assembly and maturation of the translational apparatus such as an RNA chaperone [5]; <2> involved in biosynthesis of all tRNA chains, association of enzyme to tRNA may be of regulatory significance [9]; <2> specific inhibition of enzyme by ethionine in vivo [11]; <2> enzyme-tRNA complexes may exist as dihydrouridine intermediates and may serve as reservoirs of RUMT-tRNA complexes primed for methylation [16]; <2> primary function of nucleoside modification within the tRNA anticodon loop is to negate potential canonical base pairs that would close the loop and, thereby, hinder codon binding [21]; <1> role per se in tRNA maturation, stabilization of tRNA, early event in tRNA maturation: intron-containing pre-tRNAs contain m^{5}U [25]) (Reversibility: ? <1, 2> [3, 5-7, 9-11, 16, 21, 22, 24, 25]) [3, 5-7, 9-11, 16, 21, 22, 24, 25]

P S-adenosyl-L-homocysteine + tRNA containing thymine <1, 2> (<2> ribothymidine is the most common methylated nucleoside found in tRNA, it occurs in all tRNAs of the organism [6]; <1> enzyme is responsible for the formation of 5-methyluridine at position 54 in tRNA [24]) [3, 5-7, 9-11, 16, 21, 22, 24, 25]

Substrates and products

S S-adenosyl-L-methionine + 16S rRNA <2> (<2> methyl group acceptor: E. coli 16S rRNA, site of methylation is U788 which is within a 9-base loop of a predicted stem-loop structure, mechanism is proposed to involve formation of a covalent, reversible Michael adduct with the target U788, poor substrate: methylation of 16S rRNA 40fold slower than of tRNA^{Phe} [20]; <2> unmodified E. coli 16S rRNA can be methylated at U788 [23]) (Reversibility: ? <2> [20, 23]) [20, 23]

P S-adenosyl-L-homocysteine + 16S rRNA containing thymine <2> (<2> in vitro product m^5U is not normally found in wild-type 16S rRNA, U788 is not highly methylated in vivo, 1 mol m^5U per mol of 16S rRNA [20]) [20, 23]

S S-adenosyl-L-methionine + DNA analog of the T-stem/loop domain of yeast $tRNA^{Phe}$ <2> (<2> DNA analog of the T-stem/loop domain of yeast $tRNA^{Phe}$ composed of d(A,U,G,C) is also a substrate, slow methylation of $tDNA^{PheT}$ [21]) (Reversibility: ? <2> [21]) [21]

P S-adenosyl-L-homocysteine + DNA analog of the T-stem/loop domain of yeast $tRNA^{Phe}$ containing thymine <2> [21]

S S-adenosyl-L-methionine + tRNA <1-3> (<2> highly specific for tRNA and S-adenosyl-L-methionine, cannot be replaced by methionine plus ATP, formaldehyde plus tetrahydrofolic acid, methylcobalamin and methyltetrahydrofolic acid [13]; <1> methyl group acceptor: wheat germ tRNA [1]; <2> methyl group acceptors: a variety of tRNA substrates including prokaryotic, eukaryotic, mitochondrial and chloroplastic tRNAs, wheat germ $tRNA_1^{Gly}$, bovine liver $tRNA^{Val}$, Dictyostelium discoideum vegetative stage mitochondrial tRNA and rabbit liver tRNA, methylates E. coli tRNAs at a specific U residue 23 nucleotides from the 3'end [3]; <2> high substrate specificity, methyl group acceptors: wheat germ $tRNA_1^{Gly}$, $tRNA_2^{Gly}$, E. coli $tRNA_1^{Ala}$ [4]; <1,2> tRNAs from wheat germ are the best substrates [1,3]; <2> methyl group acceptor: tRNA from E. coli [5,19,23]; <1,2> methyl group acceptor: methyl-deficient tRNA from E. coli [6,12,13,15,16]; <1> methyl group acceptor: methyl-deficient tRNA from E. coli W6 [1]; <2> methyl group acceptor: thymine-deficient tRNA from E. coli [11]; <1,2> methyl group acceptor: m^5U-deficient tRNA from E. coli GB1-5-39 [7,9,10,24]; <2> methyl group acceptors: E. coli $tRNA^{Phe}$, $tRNA^{Val}$ [8]; <2> methyl group acceptor: thymine-deficient tRNA from E. coli trmA-mutant [11]; <2> methyl group acceptor: tRNA from Micrococcus lysodeikticus [12]; <1> methyl group acceptor: yeast tRNA [24]; <2> methyl group acceptor: unmodified yeast $tRNA^{Phe}$ [8,18,21,23]; <2> methyl group acceptors: unmodified T-stem/loop domain of yeast $tRNA^{Phe}$, methyl-deficient tRNA from E. coli C6 rel met-, cys-, riboT-deficient tRNA from E. coli TrmA- is a specific substrate for U54 methylation, enzyme recognizes almost all tRNAs [21]; <2> methyl group acceptors: E. coli $tRNA_1^{Val}$, 17-oligomer and 11-oligomer of the T-arm of $tRNA_1^{Val}$, essential sequence for tRNA binding: within the T-arm, composed of 7-base T-loop and 5-base pair stem, includes position 52 of T-stem, the T-loop and extends to the 3'-acceptor end, methylation site at U54 [8]; <2> methyl group acceptors: small RNA oligomers corresponding to the T-arm of tRNA, 17-oligomer [19]; <2> methyl group acceptors: E. coli $tRNA_1^{Val}$, 17-mer analog of the T-arm from $tRNA_1^{Val}$ [23]; <2> methyl group acceptors: methyl-deficient and T-deficient E. coli tRNA, enzyme recognizes and methylates TSL: a minimal unmodified yeast $tRNA^{Phe}$ TΨC-stem and loop heptadecamer, composed of a 5 bp stem and seven-membered loop, recognition determinants are contained within the sequence, modification and secondary structure of TSL, enzyme is sensitive to conformational perturbations at or adjacent to the methylation site, positions 54 and 55 [22]; <2> substrate consensus sequence for the T-arm of tRNA consists of a 2-5 base-pair stem and a 7-base loop [20,23]; <2> tRNA recognition reside in the 3-dimensional structure of the 7-member T-loop rather than in its primary structure, base composition of the T-stem is also unimportant, only U54 and C56 are essential for activity, tRNA binding causes a conformational change of tRNA to expose the modification site in the T-arm [23]; <2> extend of methylation is directly proportional to the amount of tRNA added, 0.5% of nucleotide residues are methylated [13]; <2> S-adenosyl-L-methionine/S-adenosyl-L-homocysteine binding site, terminal NH_2-group of the amino acid moiety is nessecary for activity [2]; <2> enzyme transfers methyl group to uridine in position 54 of the TΨ-loop of all E. coli tRNA species, requires the TΨ-arm and, most specifically, the bases U54-U55-C56-Pu57-A58 for tRNA recognition [5]; <2> 5-methylation of U in the TΨC-loop [7]; <2> methylates U54 in the TΨC-loop of tRNA [19, 20, 22]; <1, 2> methylation of U54 [23, 25]; <2> tRNA and S-adenosyl-L-methionine have separate and independent binding sites [6]; <2> enzyme is normally tightly associated with the tRNA [9]; <2> tRNA recognition [16,23]) (Reversibility: ir <2> [13]; ? <1-3> [1-12, 14-25]) [1-25]

P S-adenosyl-L-homocysteine + tRNA containing thymine <1-3> (<1> thymine is 5'-methyluracil, methylation to different extents and at different positions of tRNA by 3 or 4 isoforms [1]; <2> formation of 5-methyluridine [3-5]; <2> formation of 5-methyluridine in position 54 of tRNA, in the TΨC sequence in loop IV [9,19,22]; <2> methylates 5-carbon of U in position 54 of tRNA, product is ribothymidine, which is invariant in the TΨC loop [10,16,17]; <2> one ribothymidine per tRNA molecule, always at the 23rd nucleotide from the 3'end [6]; <1> nearly all tRNAs contain m5U54, initiators excluded [24,25]) [1-25]

S Additional information <1, 2, 4-6> (<1> 3 isoenzymes with different substrate specificities, also some activity with rRNA from E. coli, may be due to a contamination [1]; <2> enzyme is present as 42 kDa native protein or as 2 different covalent TrmA/RNA complexes, with TrmA as a 54 kDa and a 62 kDa polypeptide, respectively, RNA is a subset of undermodified tRNA species as well as the 3'terminus of 16S rRNA, RNA does not affect enzyme activity [5]; <2> enzyme is present as 42 kDa native protein or as 48-57 kDa RNA-protein complex, RNA is not essential, but may increase enzyme activity [9]; <2> up to 50% of enzyme is covalently bound to various tRNA species, including several tRNAs at the 3'end of 16S rRNA [16]; <2> enzyme also catalyzes an S-adenosyl-L-methionine-independent exchange reaction between [5-^{3}H]-Ura-labeled substrate tRNA and protons of water at 1% of the normal methylation reaction rate, mechanism [7,16]; <2> enzyme also catalyzes exchange of tritium of [^{3}H]-Ura-16S rRNA and protons of water [20]; <2> poor acceptors: tRNAs from spinach chloroplasts, Dictyostelium discoideum vegetative stage, Bacillus subtilis and Anacystis nidulans, unfractionated E. coli tRNA [3]; <2> not as methyl group acceptors: methyl-deficient DNA from E. coli, polyadenylic acid, polycytidylic acid, polyuridylic acid, polyinosinic acid, copolymers of adenylic acid and uridylic acid and of guanylic acid and cytidylic acid, tobacco mosaic viral RNA, turnip yellow mosaic viral RNA, RNA from bacteriophage f2, CMP, AMP, UMP, GMP, rRNA [13]; <4-6> tetrahydrofolate derivatives as methyl group donor in the formation of 5-methyluridine [4]; <2> detailed mechanism, Cys-324 is the nucleophilic catalyst of enzyme, involves SN2 attack of C_5 of U54 on the methyl group of AdoMet [16]; <1> absence of TRM2 gene encoding enzyme in strains with mutant forms of $tRNA^{Ser}$CGA decreases the stability of $tRNA^{Ser}$CGA and induces lethality [25]) [1, 3-7, 13, 16, 20, 25]

P Additional information <2> (<2> no product: 3-methyluracil [12]) [12]

Inhibitors

5-fluorouracil substituted tRNA <2> (<2> first-order inactivation in the presence of S-adenosyl-L-methionine, potent mechanism-based irreversible inhibition, formation of a stable covalent complex containing enzyme, FUra-tRNA and the methyl group of AdoMet [7,17]; <2> mechanism of inhibition [7]; <2> potent inhibitor [16,20]; <2> forms inhibitory stable methylated covalent complex with RUMT [19]) [7, 16-20]

DNA analog of the T-stem/loop domain <2> (<2> DNA analog of the T-stem/loop domain of yeast $tRNA^{Phe}$ composed of d(A,U,G,C) inhibits the methylation of unfractionated methyl-deficient E. coli tRNA, significant inhibition when the concentration of $tDNA^{PheT}$ is twice that of the tRNA [21]) [21]

K^+ <2> (<2> inhibitory in presence of Mg^{2+} [9]) [9]

L-ethionine <2> (<2> selective inhibitor of uracil tRNA methylase, non-competitive to S-adenosyl-L-methionine [11]) [11]

L-methionine <2> (<2> 30 mM, weak [11]) [11]

Mg^{2+} <1> (<1> inhibits at low concentrations [15]) [15]

Mn^{2+} <1> (<1> inhibits at low concentrations [15]) [15]

NH_4^+ <1, 2> (<2> inhibitory in presence of Mg^{2+} [9]; <1> at 0.1 M, 50% inhibition [15]) [9, 15]

NO_2-Ura-tRNA <2> (<2> potent mechanism-based inhibitor, competitive with respect to tRNA, can be reversed by dilution into a large excess of tRNA substrate, forms a reversible covalent complex in which Cys-324 of RUMT is linked to the 6-position of NO_2-Ura 54 in NO_2-Ura-tRNA, mechanism [18]) [18]

Na^+ <2> (<2> inhibitory in presence of Mg^{2+} [9]) [9]

Ni^{2+} <2> [13]

S-adenosyl-L-ethionine <2> (<2> 0.6 mM [11]) [11]

S-adenosyl-L-homocysteine <2> (<2> product inhibition, competitive to S-adenosyl-L-methionine [2,6]; <2> non-competitive to tRNA [6]; <2> product inhibition, inhibits both methylation and H-exchange reaction [7]; <2> product inhibition, reversed by S-adenosylmethionine [13]; <2> at 0.002 mM 50% inhibition, E. coli MRE 600 [3,4]; <2> product inhibitor [9]) [2-4, 6, 7, 9, 13]

Zn^{2+} <2> [13]

ethylthioadenosine <2> (<2> 0.2 mM [11]) [11]

mature tRNA <2> (<2> weak, competitive inhibitor to methyl-deficient tRNA, non-competitive to S-adenosyl-L-methionine [6]) [6]
norleucine <2> (<2> 30 mM, weak [11]) [11]
p-chloromercuribenzoate <2> (<2> at 2.5 mM, complete inhibition, completely reversed by addition of 5 mM dithiothreitol [7]; <2> at 2 mM, complete inhibition [9]) [7, 9]
p-hydroxymercuribenzoate <2> [13]
putrescine <2> (<2> inhibition at optimal Mg^{2+}-concentration [9]) [9]
ribonuclease <2> (<2> inactivates [13]) [13]
spermidine <2> (<2> inhibition at optimal Mg^{2+}-concentration [9]) [9]
spleen phosphodiesterase <2> (<2> weak [13]) [13]
structural analogues of S-adenosyl-L-homocysteine <2> (<2> inhibition competitive to S-adenosyl-L-methionine [2]; <2> nitrogen analogue of S-adenosyl-L-methionine AdoMDB inhibits at high concentrations methylation but increases H-exchange reaction [7]) [2, 7]
$tRNA^{Phe}$ <2> (<2> inhibits binding of 16S rRNA to RUMT [20]) [20]
Additional information <2> (<2> not inhibited by alanine, valine, ethionine sulfoxide [11]; <2> not inhibited by spermine, adenosine, L-homocysteine, snake venom phosphodiesterase [13]) [11, 13]

Cofactors/prosthetic groups

S-adenosyl-L-methionine <1-3> (<2> uses S-adenosyl-L-methionine as cofactor [9,16]) [1-25]

Activating compounds

2-mercaptoethanol <1, 2> (<1,2> activates [1,13]) [1, 13]
dithiothreitol <2> (<2> activates [9]) [9]
glutathione <1, 2> (<1> greatly activates at 5 mM [1]; <2> activates [9]) [1, 9]
putrescine <2> (<2> activates, inhibitory at optimal Mg^{2+} concentration [9]) [9]
spermidine <2> (<2> slight activation at 20 mM in presence of 50 mM ammonium acetate [3]; <2> activates, inhibitory at optimal Mg^{2+} concentration [9]) [3, 9]
spermine <2> (<2> slight activation [13]) [13]

Metals, ions

Ca^{2+} <2> (<2> activates [13]) [13]
Mg^{2+} <2> (<2> slight activation [6]; <2> activates at 10 mM [9]; <2> in presence of Mg^{2+} 4fold increase of activity [13]) [6, 9, 13]
Mn^{2+} <2> (<2> activates [13]) [13]
ammonium acetate <2> (<2> 30% activation at 50 mM [3]) [3]
Additional information <1, 2> (<1> no divalent cations required, not activated by Mg^{2+} and Mn^{2+} [15]; <2> not activated by NH_4^+ [6]; <2> no require of cations [9]) [6, 9, 15]

Turnover number (min^{-1})

0.09 <2> (11mer of the T-arm of E. coli $tRNA_1^{Val}$) [8, 23]
1.9 <2> (yeast $tRNA^{Phe}$) [18]

2 <2> (17mer of the T-arm of E. coli tRNA$_1^{Val}$) [8, 23]
2.88 <2> (unmodified E. coli tRNA$_1^{Val}$) [23]
5.5 <2> (yeast tRNA^{Phe}) [8, 23]
19 <2> (tRNA) [9]
Additional information <2> (<2> values for several mutants of yeast tRNA^{Phe} [23]) [23]

Specific activity (U/mg)
0.000167 <2> [6]
0.00134 <2> [4]
0.00165 <2> [17]
0.0034 <2> [7]
0.0051 <2> [12]
0.0064 <2> [13]
0.0104 <2> [3]
0.46 <2> [9]

K_m-Value (mM)
8e-005 <2> (tRNA) [9]
0.00037 <2> (yeast tRNA^{Phe}) [18]
0.0008 <2> (yeast tRNA^{Phe}) [8, 22, 23]
0.0011 <2> (wheat germ tRNA$_1^{Gly}$) [3]
0.002 <2> (S-adenosyl-L-methionine) [6]
0.0028 <2> (unmodified E. coli tRNA$_1^{Val}$) [23]
0.0037 <2> (11mer of the T-arm of E. coli tRNA$_1^{Val}$) [23]
0.004 <2> (11mer of the T-arm of E. coli tRNA$_1^{Val}$) [8]
0.005 <2> (17mer of the T-arm of E. coli tRNA$_1^{Val}$) [8, 23]
0.0057 <2> (unmodified T-stem/loop domain of yeast tRNA^{Phe}) [21]
0.006 <1> (S-adenosyl-L-methionine) [1]
0.0125 <2> (S-adenosyl-L-methionine) [3]
0.0145 <2> (DNA analog of the T-stem/loop domain of yeast tRNA^{Phe}) [21]
0.0156 <2> (unmodified yeast tRNA^{Phe} TΨC-stem and loop heptadecamer) [22]
0.017 <2> (S-adenosyl-L-methionine) [9]
0.018 <2> (S-adenosyl-L-methionine) [13]
0.025 <2> (methyl deficient tRNA) [6]
0.29 <2> (methyl deficient tRNA from E. coli) [13]
Additional information <2> (<2> values for several mutants of yeast tRNA^{Phe} [23]) [23]

K_i-Value (mM)
0.001 <2> (NO_2-Ura-tRNA) [18]
0.0024-0.004 <2> (S-adenosyl-L-homocysteine) [6]
0.02 <2> (S-adenosyl-L-homocysteine) [13]
0.03-0.1 <2> (mature tRNA) [6]

pH-Optimum
7.5-8.5 <1> (<1> Tris-HCl buffer [1]) [1]
8 <2> [9]

8.4 <2> [3, 4]
8.5-9 <2> (<2> triethanolamine buffer [13]) [13]
9 <2> [6]
Additional information <2> (<2> pI: 4.8, E. coli MRE 600 [3]; <2> pI: 4.7 [9]) [3, 9]

pH-Range
7.8-8.8 <2> (<2> 25% inhibition of enzyme activity at pH 7.8 and 8.8 [3]) [3]

Temperature optimum (°C)
15 <2> (<2> assay at [8,18,21-23]) [8, 18, 21-23]
30 <1, 2> (<1,2> assay at [3,4,7,15]) [3, 4, 7, 15]
35 <2> (<2> assay at [11]) [11]
37 <1, 2> (<1,2> assay at [5,9,20,24]) [5, 9, 20, 24]
38 <2> (<2> assay at [12]) [12]
Additional information <2> (<2> assay at room temperature [10,17]) [10, 17]

4 Enzyme Structure

Molecular weight
38000 <2> (<2> gel filtration [3]) [3]
42000 <2> (<2> native enzyme form, sucrose density gradient centrifugation [5]) [5]
54000 <2> (<2> TrmA/RNA complex, sucrose density gradient centrifugation [5]; <2> RNA-enzyme complex: RNA component associated with the 41/42 kDa polypeptide, PAGE [9]) [5, 9]
65000 <3> (<3> glycerol gradient centrifugation [14]) [14]

Subunits
? <2> (<2> x * 54000, first TrmA/RNA-complex, x * 62000, second TrmA/RNA-complex, x * 42000, native enzyme form, SDS-PAGE [5]; <2> x * 41000, SDS-PAGE [7]; <2> x * 42000, SDS-PAGE [17]) [5, 7, 17]
monomer <2> (<2> 1 * 42000, SDS-PAGE [3]; <2> 1 * 42000, native protein, additional bands at 32, 41, 48-57 kDa, polypeptides ranging over 48-57 kDa are dependent on the gel system used and refer to the RNA-enzyme complex, SDS-PAGE [9]) [3, 9]

5 Isolation/Preparation/Mutation/Application

Localization
nucleus <1> (<1> the m^5U modification is found in unspliced tRNAs, suggesting that enzyme action is within the nucleus [24]) [24]
soluble <2> [3, 4, 9, 12, 13]

Purification

<1> (partial purification, hydroxylapatite chromatography [1]; may be 3 isoenzymes [1,15]; purification of glutathione S-transferase tagged recombinant, Trm2p, expressed in Escherichia coli BL21 [24]) [1, 15, 24]
<2> (polyethyleneimine precipitation, phosphocellulose and affinity chromatography [3]; affinity elution and hydroxylapatite chromatography [4]; 3 enzyme forms [5]; partial [6,12]; 5800fold purification [9]; rapid purification scheme [17]) [3, 4-7, 9, 12, 13, 17, 22]
<3> (phosphocellulose chromatography [14]) [14]

Cloning

<1> (cloning and sequencing of the TRM2 gene encoding enzyme, gene is previously identified as RNC1/NUD1 gene, but does not encode the yNucR endo-exonuclease, TRM2 gene is not essential for cell viability and contains alternative translational start sites, glutathione S-transferase tagged recombinant, Trm2p, is expressed in Escherichia coli [24]; enzyme is encoded by the nonessential TRM2 gene [25]) [24, 25]
<2> (trmA gene encodes enzyme, gene is essential for the organism [5,16,20]; hybrid plasmid carrying trmA gene encoding enzyme is used to amplify enzyme production, trmA gene is sequenced [9]; trmA gene encoding for RUMT is cloned into a pET vector and transformed into Escherichia coli BL21 (DE3) cells, extremely high-level expression system [17]; trmA gene is cloned under control of the inducible lac promoter, DH5α [22]) [5, 9, 16, 17, 22]

Engineering

C521A <1> (<1> mutant without methyltransferase activity, putative nucleophilic Cys-521 may be involved in Cys-U54 bond formation [25]) [25]
G428D <1> (<1> mutant without methyltransferase activity, G-428 is located at the presumed S-adenosyl-L-methionine binding site [25]) [25]

Application

molecular biology <2> (<2> enzyme is used for detecting tRNA-like moieties in viral RNA [4]) [4]

6 Stability

Temperature stability

37 <2> (<2> 2 h, stable [9]) [9]

Organic solvent stability

ethylene glycol <1> (<1> 30%, stabilizes at -20°C [15]) [15]

General stability information

<1>, freezing without antifreeze agents inactivates, 30% ethylene glycol at -20°C stabilizes, dialysis inactivates, best pH used for dialysis is 6.5 [15]
<2>, freeze-thawing inactivates [3]

Storage stability

<2>, -70°C, at least 6 months, stable [3, 4]
<2>, -85°C, several months, stable [6]
<2>, 0°C, partially purified, 48 h, about 30% loss of activity [12]
<2>, 0°C, purified, 6 months, stable [12]
<2>, 4°C, crude extract, 10 mM potassium phosphate, pH 7, 1 mM EDTA, 0.1 mM dithiothreitol, 10% v/v ethylene glycol, 0.02% sodium azide, 7 days, 10% loss of activity [9]

References

[1] Svensson, I.; Björk, G.R.; Lundahl, P.: Studies on microbial RNA. Properties of tRNA methylases from Saccharomyces cerevisiae. Eur. J. Biochem., **9**, 216-221 (1969)
[2] Shugart, L.; Chastain, B.: Escherichia coli tRNA (uracil-5-)-methyltransferase: Inhibition by analogues of adenosylhomocysteine. Enzyme, **24**, 353-357 (1979)
[3] Greenberg, R.; Dudock, B.: Isolation and characterization of m^5U-methyltransferase from Escherichia coli. J. Biol. Chem., **255**, 8296-8302 (1980)
[4] Greenberg, R.; Dudock, B.: Bacterial tRNA methyltransferases. Methods Enzymol., **59**, 190-203 (1979)
[5] Gustafsson, C.; Björk, G.R.: The tRNA-(m^5U54)-methyltransferase of Escherichia coli is present in two forms in vivo, one of which is present as bound to tRNA and to a 3-end fragment of 16 S rRNA. J. Biol. Chem., **268**, 1326-1331 (1993)
[6] Shugart, L.: Kinetic studies of Escherichia coli transfer RNA (uracil-5-)-methyltransferase. Biochemistry, **17**, 1068-1072 (1978)
[7] Santi, D.V.; Hardy, L.W.: Catalytic mechanism and inhibition of tRNA (uracil-5-)methyltransferase: Evidence for covalent catalysis. Biochemistry, **26**, 8599-8606 (1987)
[8] Gu, X.; Santi, D.V.: The T-arm of tRNA is a substrate for tRNA (m^5U54)-methyltransferase. Biochemistry, **30**, 2999-3002 (1991)
[9] Ny, T.; Lindström, P.H.R.; Hagervall, T.G.; Björk, G.R.: Purification of transfer RNA (m^5U54)-methyltransferase from Escherichia coli. Association with RNA. Eur. J. Biochem., **177**, 467-475 (1988)
[10] Kealey, J.T.; Lee, S.; Floss, H.G.; Santi, D.V.: Stereochemistry of methyl transfer catalyzed by tRNA (m^5U54)-methyltransferase–evidence for a single displacement mechanism. Nucleic Acids Res., **19**, 6465-6468 (1991)
[11] Tscherne, J.S.; Wainfan, E.: Selective inhibition of uracil tRNA methylases of E. coli by ethionine. Nucleic Acids Res., **5**, 451-461 (1978)
[12] Hurwitz, J.; Gold, M.; Anders, M.: The enzymatic methylation of ribonucleic acid and deoxyribonucleic acid. III. Purification of soluble ribonucleic acid-methylating enzymes. J. Biol. Chem., **239**, 3462-3473 (1964)
[13] Hurwitz, J.; Gold, M.; Anders, M.: The enzymatic methylation of ribonucleic acid and deoxyribonucleic acid. IV. The properties of the soluble ribonucleic acid-methylating enzymes. J. Biol. Chem., **239**, 3474-3482 (1964)

[14] Cimino, F.; Traboni, C.; Colonna, A.; Izzo, P.; Salvatore, F.: Purification and properties of several transfer RNA methyltransferases from S. typhimurium. Mol. Cell. Biochem., **36**, 95-104 (1981)
[15] Björk, G.R.; Svensson, I.: Studies on microbial RNA. Fractionation of tRNA methylases from Saccharomyces cerevisiae. Eur. J. Biochem., **9**, 207-215 (1969)
[16] Kealey, J.T.; Gu, X.; Santi, D.V.: Enzymatic mechanism of tRNA (m^5U54)methyltransferase. Biochimie, **76**, 1133-1142 (1994)
[17] Kealey, J.T.; Santi, D.V.: High-level expression and rapid purification of tRNA (m^5U54)-methyltransferase. Protein Expr. Purif., **5**, 149-152 (1994)
[18] Gu, X.; Matsuda, A.; Ivanetich, K.M.; Santi, D.V.: Interaction of tRNA (uracil-5-)-methyltransferase with NO_2-Ura-tRNA. Nucleic Acids Res., **24**, 1059-1064 (1996)
[19] Kealey, J.T.; Santi, D.V.: Stereochemistry of tRNA(m^5U54)-methyltransferase catalysis: ^{19}F NMR spectroscopy of an enzyme-FUraRNA covalent complex. Biochemistry, **34**, 2441-2446 (1995)
[20] Gu, X.; Ofengand, J.; Santi, D.V.: In vitro methylation of Escherichia coli 16S rRNA by tRNA (m^5U54)-methyltransferase. Biochemistry, **33**, 2255-2261 (1994)
[21] Guenther, R.H.; Bakal, R.S.; Forrest, B.; Chen, Y.; Sengupta, R.; Nawrot, B.; Sochacka, E.; Jankowska, J.; Kraszewski, A.; Malkiewicz, A.; Agris, P.F.: Aminoacyl-tRNA synthetase and U54 methyltransferase recognize conformations of the yeast $tRNA^{Phe}$ anticodon and T stem/loop domain. Biochimie, **76**, 1143-1151 (1994)
[22] Sengupta, R.; Vainauskas, S.; Yarian, C.; Sochacka, E.; Malkiewicz, A.; Guenther, R.H.; Koshlap, K.M.; Agris, P.F.: Modified constructs of the tRNA TΨC domain to probe substrate conformational requirements of m^1A58 and m^5U54 tRNA methyltransferases. Nucleic Acids Res., **28**, 1374-1380 (2000)
[23] Gu, X.; Ivanetich, K.M.; Santi, D.V.: Recognition of the T-arm of tRNA by tRNA (m^5U54)-methyltransferase is not sequence specific. Biochemistry, **35**, 11652-11659 (1996)
[24] Nordlund, M.E.; Johansson, J.O.M.; von Pawel-Rammingen, U.; Byström, A.S.: Identification of the TRM2 gene encoding the tRNA(m^5U54)methyltransferase of Saccharomyces cerevisiae. RNA, **6**, 844-860 (2000)
[25] Johansson, M.J.O.; Byström, A.S.: Dual function of the tRNA(m^5U54)-methyltransferase in tRNA maturation. RNA, **8**, 324-335 (2002)

tRNA (adenine-N^1-)-methyltransferase 2.1.1.36

1 Nomenclature

EC number
2.1.1.36

Systematic name
S-adenosyl-L-methionine:tRNA (adenine-N^1-)-methyltransferase

Recommended name
tRNA (adenine-N^1-)-methyltransferase

Synonyms
1-methyladenine transfer RNA methyltransferase
RAMT <12> [26]
adenine-1-methylase
methyltransferase, transfer ribonucleate adenine 1-
tRNA:m^1A58 methyltransferase <3, 12> [23, 26]
transfer RNA (adenine-1) methyltransferase
transfer ribonucleate adenine 1-methyltransferase

CAS registry number
37257-03-7

2 Source Organism

<1> *Rattus norvegicus* (Sprague-Dawley [3,6,17]; leukemic strain Fisher RL [17]; Fisher [18]) [1, 3, 4, 6, 12, 15, 17-19, 28, 29]
<2> *Mus musculus* (strain BALB/c [16]) [8, 16]
<3> *Saccharomyces cerevisiae* (enzyme is built of a Gcd10p/Gcd14p complex [27]; strain D84 [22]) [22, 23, 27]
<4> *Phaseolus vulgaris* [2]
<5> *Bacillus subtilis* [5]
<6> *Mycobacterium smegmatis* (strain SN_2 [7]) [7]
<7> *Thermus flavus* (strain 71 [9]) [9]
<8> *Thermus thermophilus* (HB27 [21,24]) [21, 24]
<9> *Dictyostelium discoideum* (strain Ax2, slime mould [10]) [10]
<10> *Homo sapiens* (HeLa cells [11]) [11]
<11> *Escherichia coli* (strain W [13,14]) [13, 14]
<12> *Tetrahymena pyriformis* (strain W [20]; strain GL, micronuclear [26]) [20, 26]
<13> *Sulfolobus solfataricus* [23]
<14> *Haloferax volcanii* (archaebacterium, strain DS2 [25]) [25]

3 Reaction and Specificity

Catalyzed reaction

S-adenosyl-L-methionine + tRNA = S-adenosyl-L-homocysteine + tRNA containing N^1-methyladenine (<3> subunit interaction of Gcd10p/Gcd14p complex, mechanism [27]; <3, 13> amino acid sequence determination and comparison [23]; <3> tRNA:m^{1}A58 methyltransferase: evolutionary conclusions based on analysis of the substrate binding structure and amino acid sequence [23]; <1, 3, 5, 7, 8, 12> methylation site of tRNA [4, 5, 9, 15, 21, 23, 24, 26]; <5> mechanism [5])

Reaction type

methyl group transfer

Natural substrates and products

S S-adenosyl-L-methionine + tRNA <1, 3, 8> (<8> activity is not dependent on growth temperatur [21]; <1> reaction in tRNA maturation [4]) (Reversibility: ? <1, 3, 8> [4, 21, 27]) [4, 21, 27]

P S-adenosyl-L-homocysteine + tRNA containing N^1-methyladenine

Substrates and products

S S-adenosyl-L-methionine + tRNA <1-14> (<12> recognizes and methylizes A58 in an unmodified heptamer of E. coli RNA [26]; <1> best substrate is E. coli tRNA^{Glu} [12, 19]; <2, 10> specific, methyl group acceptors are tRNAs from E. coli: methyl deficient tRNA, E. coli B tRNA [11, 16]; <1, 7, 10> tRNA$_f^{Met}$ [3, 9, 11, 12, 15, 29]; <1> tRNA$_2^{Glu}$, tRNA^{Phe}, tRNA^{Val} less effective [3]; <1, 7> increase of activity from tRNA^{Phe} to tRNA^{Val} to bulk tRNA to tRNA$_2^{Glu}$ [1, 4, 9]; <1, 5> tRNA$_1^{Ser}$, tRNA$_3^{Ser}$ [4, 5]; <8> tRNA$_2^{Glu}$ [24]; <1, 5, 10> tRNA^{Tyr} [4, 5, 11]; <1> tRNA^{Arg} [4]; <5> tRNA$_1^{Leu}$, tRNA$_2^{Leu}$, tRNA$_{1,2}^{Gln}$ [5]; <1> tRNA^{Asp} [6]; <1, 2, 7> tRNA^{Phe} [9, 12, 16]; <1, 7> tRNA^{Val} [9, 12]; <1, 4> tRNA^{Asp} from yeast [2, 6]; <5> tRNA^{Tyr}, tRNA$_{1a,b}^{Thr}$, tRNA$_1^{Leu}$, tRNA$_{1,2}^{Ser}$ [5]; <6> bulk, isozyme II [7]; <7> tRNA^{Val}, tRNA$_{1,2}^{Ser}$, tRNA^{Asp} [9]; <9> tRNA^{Phe} [10]; <5> tRNA$_1^{Ser}$ [5]; <6> poly-A [7]; <1> E. coli tRNA [1]) (Reversibility: ? <1-14> [1-29]) [1-29]

P S-adenosyl-L-homocysteine + tRNA containing N^1-methyladenine <1, 5, 8, 10, 11, 14> (<1> methylation at the invariant adenine residue in the TGψC loop of E. coli tRNA$_2^{Glu}$ [29]; <5> sole products [5]; <11> assay products: 6-dimethylaminopurine, 6-methyladenosine, 2-methyladenosine [13]; <11> no products are 7-methyl-, 9-methyl-, and 3-methyladenine [13]) [1, 5, 11, 13, 24-26, 29]

S Additional information <1, 2, 4, 5, 7-10, 14> (<14> 1-methyladenosine-57 is an intermediate in the formation of 1-methylinosine-57 [25]; <14> methylation specifically at position 57 of tRNA^{Ile} from Haloferax volcanii, differing from eukaryotes and prokaryotes [25]; <8> the main recognition site is within the 3'-half of E. coli tRNA$_2^{Glu}$ [24]; <4> methylation of the T-ψ-C-A-A-U sequence of the Tψ-loop [2]; <4> an organelle-specific methylase acts on position 7 from 5'-end of the tRNA^{Asp} [2]; <1> methylates invariant A of GTψC-loop [3]; <5> methylates 3'-terminal adenosine residue in the dihydrouridine loop of yeast tRNA^{Thr} or tRNA^{Tyr} [5]; <5> -A-A+-G-G-C-m^{2}G-sequence specific [5]; <1> methylation at position 14 of D-loop [6]; <3,7,12> methylates adenosine at position 58 of the Tψ-loop [9,23,26]; <1> methylation in vitro and in vivo at the same position [15]; <2> methylates position 73 of tRNA^{Phe} [16]; <1> methylation of non-terminal oligonucleotide G-MA-A-U [18]; <1> methylates adenosine of T-ψ-C-loop [19]; <1> poor substrates are E. coli tRNA^{Phe}, tRNA^{Val}, tRNA$_1^{Leu}$, tRNA$_2^{Leu}$ or tRNA$_3^{Leu}$, tRNA$_f^{Met}$ [4]; <1> yeast bulk tRNA, tRNA Asp, rat liver [4]; <9> calf liver tRNA [10]; <5> no substrates: tRNA from Torulopsis utilis [5]; <5,10> various Saccharomyces cerevisiae tRNA [5,11]; <10> rat liver tRNA [11]; <5> E. coli tRNA^{Tyr}-dimer, heat-denatured Saccharomyces tRNA$_3^{Leu}$ [5]; <7> yeast tRNA^{Trp} is no substrate [9]; <10> rat or HeLa rRNA, HeLa or E. coli DNA and polyA [11]) [2-5, 9-11, 15, 16, 19, 23-26]

P ?

Inhibitors

5'-deoxy-5'-S-isobutylthioadenosine <6> (<6> isozyme II [7]) [7]
AMP-Sepharose <6> (<6> isozyme II [7]) [7]
Ca^{2+} <8, 9> [10, 24]
KCl <2> (<2> inactivation in the presence of putrescine or Mg^{2+} [8]) [8]
Mg^{2+} <1, 2, 9> (<1> at concentrations above 2 mM [19]; <1> in the presence of polyamines [3,19,29]; <2> inactivation in the presence of KCl [8]) [3, 8, 10, 19, 29]
Mn^{2+} <8> [24]
NH_4^+ <7> (<7> 100 mM and above [9]) [9]
S-adenosyl-D-homocysteine <1> [1]
S-adenosyl-L-cysteine <1> [1]
S-adenosyl-L-homocysteine <1, 6, 9, 11> (<1> product inhibition, competitive [12]) [1, 3, 7, 10, 12, 14]
ethionine <6> [7]
monovalent cations <9> (<9> above 0.3 M [10]) [10]
putrescine <2> (<2> inactivation in the presence of KCl [8]) [8]
spermidine <9> [10]
tRNA with T54 in Tψ-loop <12> (<12> strongly reduced activity [26]) [26]
Additional information <1> (<1> no inhibition by nicotinamide [12]) [12]

Activating compounds

EDTA <9> (<9> activation [10]) [10]
EGTA <9> (<9> activation [10]) [10]
putrescine <1> (<1> activation, 20-40 mM, only in free form, not as tRNA cation complex [3]; <2> inactivation in presence of KCl [8]) [3, 19, 29]
spermidine <1, 8> (<8> activates [24]; <1> activation, only in free form, not as tRNA-cation complex, requirement [6]; <1> not as effective as putrescine [19]; <9> inhibitory [10]) [3, 6, 19, 24]
spermine <8> (<8> activates [24]) [24]
Additional information <1, 2> (<1> dependent on added cation [3]; <2> no activation by Triton X-100 [16]) [3, 16]

Metals, ions

K^+ <1, 2, 10> (<10> optimal concentration is 0.15 M [11]; <1,2,10> activation [3,8,11]) [3, 8, 11]
Mg^{2+} <1, 3, 6-8, 12> (<8,12> requires Mg^{2+} at concentration above 5 mM [24,26]; <2> inhibits [8]; <1,7> requirement [6,9]; <7> activation, together with NH_4^+ and Na^+ [9]) [6, 7, 9, 24, 26, 27]
NH_4^+ <1, 7> (<7> activation, together with Mg^{2+} and Na^+ [9]) [3, 9]
Na^+ <1, 7> (<7> activation, together with Mg^{2+} and NH_4^+ [9]) [3, 9]
Additional information <3> (<3> no divalent metal ions required [22]) [22]

Specific activity (U/mg)

0.0000117 <1> [4]
0.000046 <9> (<9> above, increasing activity during cell differentiation from vegetative to aggregative state [10]) [10]
0.00019 <8> (<8> partially purified enzyme [21]) [21]
0.00022 <1> [12]
0.00041 <6> (<6> isozyme I [7]) [7]
0.00044 <6> (<6> isozyme II [7]) [7]
0.0141 <8> (<8> purified enzyme [24]) [24]
0.04 <2> [8]
0.125 <7> [9]
3.5 <11> (<11> purified enzyme [13]) [13]

K_m-Value (mM)

0.0000025 <3> ($tRNA_i^{Met}$) [27]
0.000012 <1> ($tRNA_f^{Met}$) [3, 29]
0.000033 <1> ($tRNA_2^{Glu}$) [3, 29]
0.0001 <8> ($tRNA_2^{Glu}$) [24]
0.00027 <9> (S-adenosyl-L-methionine) [10]
0.0003 <1> (S-adenosyl-L-methionine) [3, 29]
0.00031 <1> ($tRNA_2^{Glu}$, <1> from E. coli [3]) [3]
0.00034 <1> ($tRNA_f^{Met}$, <1> from E. coli [3]) [3]
0.0004-0.0005 <7> ($tRNA_2^{Gln}$, <7> from E. coli [9]) [9]
0.0015 <1> (S-adenosyl-L-methionine) [12]
0.0017 <12> ($tRNA^{Phe}$, <12> from E. coli [26]) [26]
0.005 <3> (S-adenosyl-L-methionine) [27]
0.006 <1, 7> (S-adenosyl-L-methionine) [1, 9]
0.0078 <8> (S-adenosyl-L-methionine) [24]
Additional information <2, 12> (<12> modified tRNAs [26]; <2> K_m of tRNA substrates and S-adenosyl-L-methionine at different KCl concentrations [8]) [8, 26]

K_i-Value (mM)

0.00085 <1> (S-adenosyl-L-homocysteine) [3]
0.0024 <1> (S-adenosyl-L-homocysteine) [12]
0.0034 <1> (S-adenosyl-L-homocysteine) [1]

pH-Optimum

6.5-7 <8> (<8> at 65°C [24]) [24]
6.9 <7> (<7> around [9]) [9]
7.3 <9> [10]
7.4 <6> (<6> around, isozyme II [7]) [7]
7.5 <8, 12> (<8,12> assay at [21,26]) [21, 26]
7.7 <10> [11]
7.8 <2> (<2> assay at [8]) [8]
8 <1, 6, 14> (<14> assay at [25]; <1> triethanolamine-HCl buffer is better than Tris-HCl or bicine buffer [3]; <6> around, isozyme I [7]) [3, 7, 25]
8.5 <1, 4> (<1,4> assay at [2,12]) [2, 12]

pH-Range

5.8-7.7 <7> (<7> about half-maximal activity at pH 5.8 and 7.7 [9]) [9]
6.4-8.2 <9> (<9> about half-maximal activity at pH 6.4 and 8.2 [10]) [10]

Temperature optimum (°C)

30 <4, 12> (<4,12> assay at [2,26]) [2, 26]
37 <1, 2, 6, 10, 14> (<1,2,6,10,14> assay at [3,7,8,11,19,25]) [3, 7, 8, 11, 19, 25]
65 <8> (<8> assay at [21,24]) [21, 24]
75 <6> [7]
75-80 <8> [24]

Temperature range (°C)

68-80 <7> (<7> about 60% of maximal activity at 68°C and 80°C [9]) [9]

4 Enzyme Structure

Molecular weight

78000 <7> (<7> gel filtration [9]) [9]
95000 <1> (<1> gel filtration, glycerol gradient centifugation [3,29]) [3, 29]
200000 <3> (<3> Gcd10p and Gcd14p coelutes as a complex, gel filtration [27]) [27]
200000-230000 <2> (<2> gel filtration, density gradient centrifugation [8]) [8]

Subunits

? <3, 8> (<3> x * 56000, Gcd10p-Flag, + x * 44000, Gcd14p-Flag, SDS-PAGE [27]; <8> x * 60000, SDS-PAGE [24]) [24, 27]
Additional information <1> (<1> SDS-PAGE unsuccessful [3]) [3]

5 Isolation/Preparation/Mutation/Application

Source/tissue

amoeba <9> [10]
brain <1> (<1> cortex [1,6]) [1, 6]
cell culture <10> (<10> HeLa cells [11]) [11]

hepatoma cell <1> [15]
hypocotyl <4> (<4> dark-grown [2]) [2]
leaf <4> (<4> chloroplasts [2]) [2]
liver <1> [3, 4, 12, 15, 17-19, 28, 29]
neuron <1> [1]
plasmacytoma cell <2> (<2> MOPC 173 [8]) [8]
plasmacytoma cell <2> (<2> MOPC-104E [16]) [16]
spleen <1> (<1> leukemic rat [17]) [17, 28]
vegetative <9> [10]

Localization

chloroplast <4> [2]
cytosol <1, 4, 6, 7, 9, 11> [2, 7, 9, 10, 12, 13, 18]
mitochondrion <4> [2]

Purification

<1> (hydroxylapatite and phosphocellulose chromatography [6]; partial [1,12]) [1, 3, 6, 12, 28, 29]
<2> (partial, RNA-Sepharose-affinity chromatography [8]) [8]
<3> (Gcd10p and Gcd14p as complex, Flag-tagged, from Saccharomyces cerevisiae [27]) [27]
<6> (affinity chromatography on AMP-Sepharose, two isozymes with different substrate specificity, isozyme II is inactivated by affinity chromatography [7]) [7]
<7> (affinity chromatography [9]) [9]
<8> (partial [21]) [21, 24]
<9> (affinity chromatography on S-adenosine-L-homocysteine-Sepharose [10]) [10]
<10> (partial [11]) [11]
<12> [26]
<14> (partial [25]) [25]

Cloning

<3> (expression of isolated Gcdp10/Gcdp14 complex and Gcdp14/Gcdp10 complex, Flag-tagged, in Saccharomyces cerevisiae [27]) [27]

Engineering

Additional information <3, 8> (<3> Δgcdp10 and Δgcdp14 mutants are inactive and lethal [27]; <8> nucleoside exchange mutantS: A58G, U54A and U8G of $tRNA^{Phe}$ from yeast: A58G at the methylation site leads to decreased methyl-accepting activity, U8G and U54A lead to altered interaction with the enzyme [24]) [24, 27]

6 Stability

Temperature stability

37 <2> (<2> 3 h full activity [8]) [8]
47 <2> (<2> $t_{1/2}$: 2-4 min [8]) [8]

60 <7> (<7> at least 1 h stable [9]) [9]
70 <7> (<7> 1 h stable [9]) [9]
80 <7> (<7> inactivation after 40 min [9]) [9]
90 <7> (<7> inactivation after 5 min [9]) [9]

General stability information

<1>, DTT stabilizes [12]
<2>, dilution inactivates [8]
<7>, albumin stabilizes [9]
<9>, EDTA stabilizes [10]
<10>, glycerol does not stabilize [11]
<11>, purification, unstable upon [13]
<1, 11>, freeze-thawing inactivates [3, 13]
<1, 2, 9>, glycerol, 20-50% v/v, stabilizes [3, 8, 10, 12]
<3, 10>, freezing inactivates [11, 22]

Storage stability

<1>, -15°C, 1 week with 20% v/v glycerol and 1 mM DTT [12]
<1>, -20°C, unstable [3]
<1>, -70°C, at least 1 year in the presence of 40% glycerol [3, 29]
<1>, 0°C, ammonium sulfate saturated, stable [28]
<1>, 0°C, in solution, pH 8.0, 20% loss of activity per day [28]
<1>, 4°C, 2 weeks in the presence of glycerol [3]
<2>, -20°C, 50% glycerol, stable for more than 2 months [8]
<2>, 4°C, $t_{1/2}$: 2 weeks [8]
<7>, 0°C, crude, at least 1 month [9]
<8>, 4°C, stable for at least 5 months [24]
<9>, -20°C, crude or partially purified, $t_{1/2}$: overnight [10]
<9>, -20°C, purified, at least 3 months with 50% v/v glycerol and 1 mM EDTA [10]
<9>, 4°C, crude or partially purified, $t_{1/2}$: overnight [10]
<10>, -20°C, $t_{1/2}$: 4 weeks [11]
<11>, 0°C, crude extract, 2 months [13]
<14>, 4°C, stable for 1 month [25]

References

[1] Salas, C.E.; Sellinger, O.Z.: Methylation of E. coli transfer ribonucleic acids by a tRNA adenine-1-methyltransferase from rat brain cortex and bulk-isolated neurons. J. Neurochem., **31**, 85-91 (1978)

[2] Dubois, E.G.; Dirheimer, G.; Weil, J.H.: Methylation of yeast $tRNA^{Asp}$ by enzymes from cytoplasm, chloroplasts and mitochondria of phaseolus vulgaris. Biochim. Biophys. Acta, **374**, 332-341 (1974)

[3] Glick, J.M.; Leboy, P.S.: Purification and properties of tRNA(adenine-1)-methyltransferase from rat liver. J. Biol. Chem., **252**, 4790-4795 (1977)

[4] Kraus, J.: Recognition of individual Escherichia coli transfer ribonucleic acids by 1-adenine-specific methyltransferase from rat liver. Biochem. J., **169**, 247-249 (1978)
[5] Kersten, H.; Raettig, R.; Weissenbach, J.; Dirheimer, G.: Recognition of individual procaryotic and eucaryotic transfer-ribonucleic acids by B subtilis adenine-1-methyltransferase specific for the dihydrouridine loop. Nucleic Acids Res., **5**, 3033-3042 (1978)
[6] Salas, C.E.; Dirheimer, G.: In vitro methylation of yeast $tRNA^{Asp}$ by rat brain cortical tRNA-(adenine-1) methyltransferase. Nucleic Acids Res., **6**, 1123-1133 (1979)
[7] Brahmachari, V.; Ramakrishnan, T.: Studies on 1-methyl adenine transfer RNA methyltransferase of Mycobacterium smegmatis. Arch. Microbiol., **140**, 91-95 (1984)
[8] Nau, F.; Pham-Coeur-Joly, G.; Dubert, J.M.: A study of some molecular and kinetic properties of two tRNA methyltransferases from mouse plasmocytoma. Eur. J. Biochem., **130**, 261-268 (1983)
[9] Morozov, I.A.; Gambaryan, A.S.; Lvova, T.N.; Nedospasov, A.A.; Venkstern, T.V.: Purification and characterization of tRNA (adenine-1-)-methyltransferase from Thermus flavus strain 71. Eur. J. Biochem., **129**, 429-436 (1982)
[10] Mutzel, R.; Malchow, D.; Meyer, D.; Kersten,H.: tRNA (adenine-N^1)-methyltransferase from Dictyostelium discoideum. Purification, characterization and developmental changes in activity. Eur. J. Biochem., **160**, 101-108 (1986)
[11] Agris, P.F.; Spremulli, L.L.; Brown, G.M.: tRNA methylases from HeLa cells: purification and properties of an adenine-1-methylase and a guanine-N^2-methylase. Arch. Biochem. Biophys., **162**, 38-47 (1974)
[12] Glick, J.M.; Ross, S.; Leboy, P.S.: S-Adenosylhomocysteine inhibition of three purified tRNA methyltransferases from rat liver. Nucleic Acids Res., **2**, 1639-1651 (1975)
[13] Hurwitz, J.; Gold, M.; Anders, M.: The enzymatic methylation of ribonucleic acid and deoxyribonucleic acid. III. Purification of soluble ribonucleic acid-methylating enzymes. J. Biol. Chem., **239**, 3462-3473 (1964)
[14] Hurwitz, J.; Gold, M.; Anders, M.: The enzymatic methylation of ribonucleic acid and deoxyribonucleic acid. IV. The properties of the soluble ribonucleic acid-methylating enzymes. J. Biol. Chem., **239**, 3474-3482 (1964)
[15] Shershneva, L.P.; Venkstern, T.V.; Bayev, A.A.: A study of tRNA methylase action. FEBS Lett., **29**, 132-134 (1973)
[16] Chepelinsky, A.B.; Gantt, R.; Wivel, N.: Presence of RNA methylases in intracisternal A particles purified from a mouse plasma cell tumor. Eur. J. Biochem., **103**, 339-347 (1980)
[17] Kraus, J.; Staehelin, M.: N^2-guanine specific transfer RNA methyltransferase I from rat liver and leukemic rat spleen. Nucleic Acids Res., **1**, 1455-1478 (1974)
[18] Baguley, B.C.; Staehelin, M.: The specificity of transfer ribonucleic acid methylases from rat liver. Biochemistry, **7**, 45-50 (1968)

[19] Leboy, P.S.; Glick, J.M.: tRNA methyltransferases from rat liver. Differences in response of partially purified enzymes to polyamines and inorganic salts. Biochim. Biophys. Acta, **435**, 30-38 (1976)
[20] Reinhart, M.P.; Lewis, J.M.; Leboy, P.S.: A single tRNA (guanine)-methyltransferase from Tetrahymena with both mono- and di-methylating activity. Nucleic Acids Res., **14**, 1131-1148 (1986)
[21] Kumagai, I.; Watanabe, K.; Oshima, T.: Thermally induced biosynthesis of 2-O-methylguanosine in tRNA from an extreme thermophile, Thermus thermophilus HB27. Proc. Natl. Acad. Sci. USA, **77**, 1922-1926 (1980)
[22] Björk, G.R.; Svensson, I.: Studies on microbial RNA. Fractionation of tRNA methylases from Saccharomyces cerevisiae. Eur. J. Biochem., **9**, 207-215 (1969)
[23] Bujnicki, J.M.: In silico analysis of the tRNA:m^{1}A58 methyltransferase family: homology-based fold prediction and identification of new members from Eubacteria and Archaea. FEBS Lett., **507**, 123-127 (2001)
[24] Yamazaki, N.; Hori, H.; Ozawa, K.; Nakanishi, S.; Ueda, T.; Kumagai, I.; Watanabe, K.; Nishikawa, K.: Substrate specificity of tRNA (adenine-1-)-methyltransferase from Thermus thermophilus HB27. Biosci. Biotechnol. Biochem., **58**, 1128-1133 (1994)
[25] Grosjean, H.; Constantinesco, F.; Foiret, D.; Benachenhou, N.: A novel enzymic pathway leading to 1-methylinosine modification in Haloferax volcanii tRNA. Nucleic Acids Res., **23**, 4312-4319 (1995)
[26] Sengupta, R.; Vainauskas, S.; Yarian, C.; Sochacka, E.; Malkiewicz, A.; Guenther, R.H.; Koshlap, K.M.; Agris, P.F.: Modified constructs of the tRNA TPsiC domain to probe substrate conformational requirements of m^{1}A58 and m^{5}U54 tRNA methyltransferases. Nucleic Acids Res., **28**, 1374-1380 (2000)
[27] Anderson, J.; Phan, L.; Hinnebusch, A.G.: The Gcd10p/Gcd14p complex is the essential two-subunit tRNA(1-methyladenosine) methyltransferase of Saccharomyces cerevisiae. Proc. Natl. Acad. Sci. USA, **97**, 5173-5178 (2000)
[28] Kehr, S.J.: tRNA methyltransferases. Methods Enzymol., **29**, 716-726 (1974)
[29] Söll, D.; Kline, L.K.: RNA methylation. The Enzymes, 3rd Ed. (Boyer, P.D., ed.), **15B**, 557-566 (1982)

DNA (cytosine-5-)-methyltransferase 2.1.1.37

1 Nomenclature

EC number
2.1.1.37

Systematic name
S-adenosyl-L-methionine:DNA (cytosine-5-)-methyltransferase

Recommended name
DNA (cytosine-5-)-methyltransferase

Synonyms
C5-MTase
DCMT
DNA 5-cytosine methylase
DNA cytosine c^5 methylase
DNA cytosine methylase
DNA methylase
DNA methyltransferase
DNA transmethylase
DNA-cytosine 5-methylase
DNA-cytosine methyltransferase
Dnmt1 <11> [19]
EcoRI methylase
EcoRII DNA-cytosine methylase <4> [15]
M.BssHII <10> [18]
M.BsuRIa <5> [10]
M.BsuRIb <5> [10]
M.MspI <9> [17]
MspI DNA methyltransferase <9> [17]
cytosine 5-methyltransferase
cytosine DNA methylase
cytosine DNA methyltransferase
cytosine-specific DNA methyltransferase
deoxyribonucleate methylase
deoxyribonucleate methyltransferase
deoxyribonucleic (cytosine-5-)-methyltransferase
deoxyribonucleic acid (cytosine-5-)-methyltransferase
deoxyribonucleic acid methylase
deoxyribonucleic acid methyltransferase
deoxyribonucleic acid modification methylase
deoxyribonucleic methylase

methylphosphotriester-DNA methyltransferase
methyltransferase, deoxyribonucleate
type II DNA methylase
Additional information <17> (<17>, this is a large group of enzymes most of which, with enzymes of similar site specificity listed as EC 3.1.21.3, 4 or 5, form so-called "restriction-modification systems". A complete listing of all these enzymes has been produced by R.J. Roberts, this list is updated annually [35]) [35]

CAS registry number
9037-42-7

2 Source Organism

<1> *Homo sapiens* [1, 9, 12, 14, 23, 29, 30]
<2> *Rattus norvegicus* [2, 4, 6, 20, 24]
<3> *Mus musculus* [3, 7, 9, 21, 26, 27, 28, 32, 33]
<4> *Escherichia coli* (strain W [5]; strain K12 [15]) [5, 8, 15]
<5> *Bacillus subtilis* [10]
<6> *Oryza sativa* (L. cv Roncarolo [11]) [11]
<7> *Bos taurus* [13]
<8> *Triticum aestivum* [16]
<9> *Moraxella sp.* [17]
<10> *Geobacillus stearothermophilus* (H3 [18]; enzyme M.BssHII expressed as two isoforms: isoform I and isoform II [37]) [18, 37]
<11> *Paracentrotus lividus* [19]
<12> *Bacillus subtilis phage PHI3T* [22]
<13> *Bacillus subtilis phage sigma11S* [22]
<14> *Chaetopterus variopedatus* [25]
<15> *Pisum sativum* [31]
<16> *Haemophilus parahaemolyticus* [34]
<17> *cellular organism* [35]
<18> *Bacillus sphaericus* (enzyme M.BspRI [36]) [36]
<19> *Haemophilus aegypticus* (enzyme M.HaeIII [36]) [36]
<20> *Salmonella infantis* (enzyme M.SInI [36]) [36]
<21> *Staphylococcus aureus* (enzyme M.Sau961 [36]) [36]
<22> *Haemophilus parainfluenzae* (enzyme M.HpaII [36]) [36]
<23> *Haemophilus haemolyticus* (enzyme M.HhaI [36,38]) [36, 38]

3 Reaction and Specificity

Catalyzed reaction
S-adenosyl-L-methionine + DNA = S-adenosyl-L-homocysteine + DNA containing 5-methylcytosine (<7>, random sequential mechanism [13]; <3>, sequential ordered bi-bi kinetic mechanism in which DNA binds to the enzyme

first, followed by S-adenosyl-L-homocysteine and methylated DNA are released, formation of an enzyme-DNA-DNA ternary complex [28]; <1>, sequential mechanism [30])

Reaction type

methyl group transfer

Natural substrates and products

S Additional information <1, 2, 3, 16, 18, 19, 20, 21, 22, 23> (<2>, the enzyme catalyzes the methylation of DNA during replication, in NGF-induced PC12 cell. The enzyme activity is sharply reduced 4 days after induction of differentiation [20]; <3>, enzymatic DNA methylation of carbon 5 of cytosines is an epigenetic modification that plays a role in regulating gene expression, differentiation, and tumorigenesis. DNA (cytosine-C^5)-methyltransferase-1 is the enzyme responsible for maintaining established methylation patterns during replication in mammalian cells [21]; <2>, methylation of genomic DNA is involved in the basic mechanism of gene inactivation, chromatin organization, X-chromosome inactivation, and genomic imprinting [24]; <3>, the enzyme is essential for viable mammalian development and has a central function in the determination and maintenance of epigenetic methalation pattern [28]; <1>, role of the enzyme in maintaining the methylation patterns throughout development, the enzyme may be involved in the etiology of fragile X, a syndrome characterized by de novo methylation of a greatly expanded CGG-CCG triplet repeat sequence [30]; <3>, intrinsic sequence-specificity of Dnmt1 on linear duplex DNA is unlikely to be important in the establishment of genomic methylation patterns [32]; <16>, the enzyme is capable of protecting plasmid DNA in vivo against action of the cognate restriction endonuclease [34]; <18,19,20,21,22,23>, enzyme induces DNA bending, a mechanism to establish specific interface between proteins and DNA. Enzymes recognizing a cytosine 3' to the target cytosine tend to induce greater bends than enzymes with guanine in this position [36]) [20, 21, 24, 28, 30, 32, 34, 36]

P ?

Substrates and products

S S-adenosyl-L-methionine + DNA <1-23> (<1>, the enzyme is able to methylate homologues HeLa DNA, altough to a lesser extent than heterologous DNA. The enzyme can methylate single-stranded DNA three times higher than that of the corresponding double-stranded DNA, the 5' neighbor can be either G or C while the 3' neighbor is always G, enzyme can methylate poly(dG,dC) [1]; <2>, native and heat-denatured DNA from ethionine-treated synchronized Novikoff cells. The native DNA sample is about 3.5 fold better as a methyl acceptor than the heat-denatured sample [2]; <3>, both native and denatured DNA are methylated, with calf thymus DNA the double stranded form is the better substrate but the enzyme preferentially methylates single stranded E. coli DNA even in the native preparation. Homologous ascites DNA can be methylated, but only to about 5% of the level of the best substrate, undermethylated mouse L929 cell DNA. DNA isolated from growing cells or tissues is a better substrate than DNA from normal liver or pancreas or from stationary cells [3]; <2>, the enzyme is able to methylate single stranded and double stranded DNA. Previously unmethylated double-stranded DNA from prokaryotes, Micrococcus luteus, as well as from eukaryotes, Ascaris suis, can serve as substrates. The synthetic copolymers (dG-dC)n*(dC-dG)n and (dG,dC)n are also methylated. SV40 DNA is almost not methylated. PM2 DNA is a good substrate even in the supercoiled form. The enzyme methylates 1 in 17 bases in heterologous Micrococcus luteus DNA, but only 1 in 590 in homologous rat liver DNA [4]; <4>, transfer of methyl groups to heterologous DNA [5]) (Reversibility: ir <4> [5]; ? <1-23> [1-4, 6-38]) [1-38]

P S-adenosyl-L-homocysteine + DNA containing 5-methylcytosine <1-23> [1-38]

S S-adenosyl-L-methionine + DNA <1-23> (<1>, the enzyme methylates DNA from various sources in native and heat-denatured forms. The synthetic copolymer poly(dG-dC)*poly(dG-dC) is methylated in B-conformation and in Z-conformation to about the same extent, determination of maintenance DNA methylase activity with P815 DNA and determination of de novo DNA methylase activity with 5-methylcytosine-free procaryotic DNA [12]; <7>, about 60% of the total methylation occurs in the 5'd(C-G)3' doublet. Single-stranded and hemi-methylated DNAs are methylated at an elevated rate [13]; <4>, site-specificity [15]) (Reversibility: ir <4> [5]; ? <1-23> [1-4, 6-38]) [1-38]

P S-adenosyl-L-homocysteine + DNA containing 5-methylcytosine <1-23> [1-38]

S S-adenosyl-L-methionine + DNA <1-23> (<1,3>, methylation of double-stranded Micrococcus luteus DNA. The enzyme transfers methyl groups to both double-stranded and single-stranded natural DNAs. The methyl-accepting activity of the DNA is correlated to their GC content, homologous DNAs are poor substrates. Very low rate of methylation with poly[(-dA-dC)*(dG-dT)], highest methyl-accepting ability with poly[(dG-dC)*(dG-dC)] [9]; <6>, native plant DNA or hemimethylated poly(-dI-MedC)*poly(dI-dC) [11]) (Reversibility: ir <4> [5]; ? <1-23> [1-4, 6-38]) [1-38]

P S-adenosyl-L-homocysteine + DNA containing 5-methylcytosine <1-23> [1-38]

S S-adenosyl-L-methionine + DNA <1-23> (<3>, higher methylating activity with chemically synthesized hemimethylated oligodeoxynucleotide than with poly(dI*dC)*poly(dI*dC) [27]; <3>, poly(dI*dC-dI*dC) of chain length 100, 500, 2000 or 5000 base pairs [28]; <1>, poly(dI-dC)*poly(dI-dC) and unmethylated and hemimethylated 36-mer and 75-mer oligonucleotides. The preference of the enzyme for hemimethylated, over unmethylated DNA is 7-21fold. Poly(dI-dC)'poly(dI-dC) [29]; <1>, DNA topology strongly influences the reaction rate, which increases with increasing negative superhelical tension [30]; <3>, specificity is confined to the sequence 5'-CpG-3'. Any sequence-specific de novo methylation mediated by Dnmt1 is either under the control of regulatory factors that interact [32]; <3>, isoenzymes Dnmt3a, Dnmt3b1 and Dnmt3b2 show similar activity towards poly(dG-dC)-poly(dG-dC) [33]) (Reversibility: ir <4> [5]; ? <1-23> [1-4, 6-38]) [1-38]

P S-adenosyl-L-homocysteine + DNA containing 5-methylcytosine <1-23> [1-38]

S S-adenosyl-L-methionine + DNA <1-23> (<3>, poly(dIdC:dIdC) is efficiently methylated [21]; <12,13>, enzyme methylates the cytosine of the DNA target sequence TCGA [22]; <1>, methylation of a series of snapback oligodeoxynucleotides of differing stem length, each containing a centrally located CG dinucleotide recognition site. The enzyme requires a minimum of 22 base pairs in the stem for maximum activity. Extrahelical cytosines in slipped duplexes that are 30 base pairs in length act as effective methyl acceptors and are more rapidly methylated than cytosines that are Watson-Crick paired. Duplexes containing hairpins of CCG repeats in cruciform structures in which the enzyme recognition sequence is disrupted by a C-C mispair are also more rapidly methylated than the control Watson-Crick-paired duplexes [23]; <14>, the enzyme is unable to make de novo methylation in double stranded DNA, methylation occurs at maximal rate on hemimethylated ds DNA [25]) (Reversibility: ir <4> [5]; ? <1-23> [1-4, 6-38]) [1-38]

P S-adenosyl-L-homocysteine + DNA containing 5-methylcytosine <1-23> [1-38]

S S-adenosyl-L-methionine + DNA <1-23> (<8>, methylation of cytosine residues in double-stranded DNAs isolated from wheat, maize, calf thymus or bacteria. Single-stranded DNAs from all sources are methylated to only about one-fourth to one-third the degree of double-stranded substrate DNA. Synthetic polydeoxyribonucleotides of alternating base sequence, poly[d(G-C)]*poly[d(G-C)], poly[d(I-C)]*poly[d(I-C)], poly[d(A-C)]*poly[d(G-T)], are methylated to low but clearly measurable extent. Preference for endogenous, double-stranded DNA [16]; <9>, the enzyme recognizes the sequence 5'-CCGG-3' and methylates the outer cytosine residue [17]; <10>, methylation targets of M.BssHII: ACGCGT/CCGCGG MluI/SacII, PuGCGCPy HaeII, PuCCGGPy Cfr10I and GCGCGC BssHII [18]; <3>, the purified enzyme has at least a 30fold higher catalytic efficiency with hemimethylated double-stranded oligodeoxyribonucleotide substrates than unmethylated substrates and is most active with small oligodeoxyribonucleotide substrates [21]) (Reversibility: ir <4> [5]; ? <1-23> [1-4, 6-38]) [1-38]

P S-adenosyl-L-homocysteine + DNA containing 5-methylcytosine <1-23> [1-38]

Inhibitors

5-azacytosine <4> (<4>, mechanism-based inhibitor [8]) [8]
5-fluorocytosine <4> (<4>, mechanism-based inhibitor [8]) [8]
Ca^{2+} <8> (<8>, above 1 mM [16]) [16]
EDTA <1> (<1>, above 10 mM [12]) [12]

KCl <3, 8> (<8>, above 0.1 M, inhibition may be simply due to dissociation of the DNA-enzyme complex [16]; <3>, above 100 mM [33]) [16, 33]
NaCl <3> (<3>, incubation of native calf thymus DNA in presence of 40 mM NaCl results in 50% inhibition, more than 90% inhibition at 200 mM. With denatured calf thymus DNA, low concentrations of NaCl, up to 90 mM stimulate, 50% inhibition at 175 mM [3]; <3>, above 100 mM [33]; <8>, above 0.1 M, inhibition may be simply due to dissociation of the DNA-enzyme complex [16]) [3, 16, 33]
S-adenosyl-L-ethionine <7, 8> (<7>, non-competitive inhibitor against DNA [13]; <8>, IC50: 0.05 mM [16]) [13, 16]
S-adenosyl-L-homocysteine <2, 4, 8> (<2>, potent competitive inhibitor [4]; <8>, half-inhibition at 0.02 mM [16]) [4, 5, 16]
iodoacetamide <1> [1]
native DNA <7> (<7>, non-competitive inhibitor against S-adenosyl-L-methionine [13]) [13]
poly[d(G-5-azacytidine)] <1> [14]
putrescine <2> (<2>, 1 mM, 42% inhibition [6]) [6]
spermidine <2, 8> (<2>, 1 mM, 59% inhibition [6]; <8>, at millimolar concentrations [16]) [6, 16]
spermine <2, 8> (<2>, 1 mM, 97% inhibition [6]; <8>, at millimolar concentrations [16]) [6, 16]
Additional information <3> (<3>, stimulating proteins from murine P815 mastocytoma cells stimulate both de novo and maintenance activity of DNA methyltransferase about 3fold. They enhance the methylation of any natural DNA and of poly[(dI-dC)*(dI-dC)] but inhibit methylation of poly[(dG-dC)*(dG-dC)] [7]) [7]

Activating compounds

stimulating proteins <3> (<3>, the proteins from murine P815 mastocytoma cells stimulate both de novo and maintenance activity of DNA methyltransferase about 3fold. They enhance the methylation of any natural DNA and of poly[(dI$^-$dC)*(dI$^-$dC)] but inhibit methylation of poly[(dG-dC)*(dG-dC)] [7]) [7]

Metals, ions

NaCl <3> (<3>, incubation of native calf thymus DNA in presence of 40 mM NaCl results in 50% inhibition, more than 90% inhibition at 200 mM. With denatured calf thymus DNA, low concentrations of NaCl, up to 90 mM stimulate, 50% inhibition at 175 mM [3]) [3]

Turnover number (min^{-1})

0.033 <1> ((CGG*CCG)73) [30]
0.033 <1> (pRW3602, relaxed circular) [30]
0.1 <1> (pRW3602, linear) [30]
0.167 <1> (pRW3602, supercoiled) [30]
0.183 <1> ((CGG*CGG)12) [30]
0.358 <3> (poly(dI*dC-dI*dC) of chain length 100) [28]
0.366 <3> (poly(dIdC:dIdC)) [21]

0.38 <3> (poly(dI*dC-dI*dC) of chain length 500) [28]
0.415 <3> (poly(dI*dC-dI*dC) of chain length 2000) [28]
0.52 <3> (poly(dI*dC-dI*dC) of chain length 5000) [28]
3.067 <1> (poly(dI-dC)'poly(dI-dC)) [29]
8.71 <1> (poly(dI-dC)*poly(dI-dC)) [30]
Additional information <1, 3> (<3>, turnover-numbers for hybrid enzymes [26]) [21, 26, 29, 30]

Specific activity (U/mg)
0.000013 <1> [1]
0.0000166 <1> [12]
0.000107 <3> [3]
0.0011 <3> [9]
0.00115 <1> [9]
Additional information <3, 6, 7, 8> [11, 13, 16, 21]

K_m-Value (mM)
0.0001 <1> ((CGG*CCG)73) [30]
0.000125 <3> (poly(dI*dC-dI*dC) of chain length 2000) [28]
0.00014 <3> (poly(dI*dC-dI*dC) of chain length 5000) [28]
0.00025 <2> (S-adenosyl-L-methionine, <2>, in presence of Micrococcus luteus DNA [4]) [4]
0.0003 <3> (S-adenosyl-L-methionine, <3>, isoenzyme Dnmt3a, reaction with poly(dG-dC)-poly(dG-dC) [33]) [33]
0.0003 <3> (poly(dI*dC-dI*dC) of chain length 500) [28]
0.0003 <3> (poly(dI-dC)-poly(dI-dC), <3> isoenzyme Dnmt3a [33]) [33]
0.0004 <3> (S-adenosyl-L-methionine, <3>, isoenzyme Dnmt3a, reaction with poly(dI-dC)-poly(dI-dC) [33]) [33]
0.0005 <1> (poly(dI-dC)*poly(dI-dC)) [29]
0.0007 <1> (poly(dI-dC)*poly(dI-dC)) [30]
0.0008 <1> ((CGG*CCG)12) [30]
0.0008 <3> (S-adenosyl-L-methionine, <3>, isoenzyme Dnmt3b2, reaction with poly(dG-dC)-poly(dG-dC) [33]) [33]
0.0009 <3> (S-adenosyl-L-methionine, <3>, isoenzyme Dnmt3b2, reaction with poly(dI-dC)-poly(dI-dC) [33]) [33]
0.0009 <3> (poly(dG-dC)-poly(dG-dC), <3>, isoenzyme Dnmt3b2 [33]) [33]
0.000973 <3> (poly(dIdC:dIdC)) [21]
0.001 <3> (poly(dG-dC)-poly(dG-dC), <3>, isoenzyme Dnmt3b1 [33]) [33]
0.0012 <3> (S-adenosyl-L-methionine, <3>, isoenzyme Dnmt3b1, reaction with poly(dI-dC)-poly(dI-dC) por poly(dG-dC)*poly(dG-dC) [33]) [33]
0.0013 <3> (poly(dI-dC)-poly(dI-dC), <3> isoenzyme Dnmt3b1 [33]) [33]
0.0015 <3> (poly(dI-dC)-poly(dI-dC), <3>, isoenzyme Dnmt3b2 [33]) [33]
0.00189 <3> (poly(dI*dC-dI*dC) of chain length 100) [28]
0.0026 <6> (S-adenosyl-L-methionine) [11]
0.0027 <3> (poly(dG-dC)-poly(dG-dC), <3>, isoenzyme Dnmt3a [33]) [33]
0.003 <1> (pRW3602, relaxed circular) [30]
0.00325 <1> (S-adenosyl-L-methionine) [1]
0.0035 <3> (poly(dG-dC)-poly(dG-dC), <3>, isoenzyme Dnmt3a [33]) [33]

0.004 <1> (S-adenosyl-L-methionine, <1>, reaction with (CGG‘CCG)12 [30]) [12, 30]
0.005-0.006 <8> (S-adenosyl-L-methionine) [16]
0.006 <1> (S-adenosyl-L-methionine, <1>, reaction with relaxed circular pRW3602 [30]) [30]
0.007 <1> (S-adenosyl-L-methionine, <1>, reaction with linear pRW3602 [30]) [30]
0.0072 <1> (S-adenosyl-L-methionine) [29]
0.0077 <1> (native Micrococcus lysodeikticus DNA) [12]
0.009 <1> (S-adenosyl-L-methionine, <1>, reaction with poly(dI-dC)*poly (dI-dC) [30]) [30]
0.011 <1> (S-adenosyl-L-methionine, <1>, reaction with pRW3602, supercoiled [30]) [30]
0.013 <3> (S-adenosyl-L-methionine) [3]
0.014 <1> (S-adenosyl-L-methionine, <1>, reaction with (CGG*CCG)73 [30]) [30]
0.015 <1> (P815 DNA) [12]
0.02 <1> (pRW3602, linear) [30]
0.021 <14> (S-adenosyl-L-methionine) [25]
0.023 <1> (pRW3602, supercoiled) [30]
0.086 <1> (Micrococcus luteus DNA) [1]
Additional information <1, 3, 6> (<6>, K_m-value for poly (dI-MedC)*poly(dI-dC): 0.017 mg/ml [11]; <3>, K_m-values for hybrid enzymes [26]) [11, 21, 26, 29, 30]

K_i-Value (mM)

0.00033 <2> (S-adenosyl-L-homocysteine) [4]

pH-Optimum

6.5 <1> [1]
7-7.5 <1, 2> [2, 12]
7.5-8 <8> [16]
8 <6> [11]

pH-Range

6-9 <8> (<8>, about 50% of maximal activity at pH 6 and pH 9 [16]) [16]
7.6-8.7 <6> (<6>, 50% of maximal activity at pH 7.6 and at pH 8.7 [11]) [11]

4 Enzyme Structure

Molecular weight

48000-54000 <9> (<9>, gel filtration [17]) [17]
50000-55000 <8> (<8>, gel filtration, sucrose density gradient centrifugation [16]) [16]
54000 <6> (<6>, sucrose density gradient centrifugation [11]) [11]
115000 <2> (<2>, glycerol gradient centrifugation [4]) [4]
120000 <1> (<1>, glycerol gradient sedimentation [1]) [1]

130000 <7> (<7>, glycerol density gradient centrifugation [13]) [13]
135000 <1> (<1>, glycerol gradient centrifugation [12]) [12]
150000-200000 <14> (<14>, gel filtration [25]) [25]
170000 <3> (<3>, hybrid enzymes Dnmt1-HhaI and Dnmt1-HpaII [26]) [26]
180000-205000 <3> (<3>, gel filtration [27]) [27]
240000 <3> (<3>, gel filtration [3]) [3]

Subunits

? <3, 10> (<10>, x * 59000, isoform II, SDS-PAGE [37]; <10>, x * 66000, isoform I, SDS-PAGE [37]; <3>, x * 80000, SDS-PAGE [3]; <3>, x * 190000, SDS-PAGE [21,27]) [3, 21, 27, 37]
monomer <9, 14> (<9>, 1 * 49000, at low concentrations, less than 0.4 mg/ml, the enzyme is a monomer in solution but at higher concentrations, greater than 3.0 mg/ml it exists predominantly as a dimer, SDS-PAGE [17]; <14>, 1 * 200000, SDS-PAGE [25]) [17, 25]
Additional information <3> (<3>, the apparent molecular weights of the GST-fused Dnmt3a, Dnmt3b1 and Dnmt3b2 are 130000 Da, 120000 Da and 115000 Da, as determined by SDS-PAGE [33]) [33]

Posttranslational modification

proteolytic modification <11> (<11>, at blastula stage a proteolytic mechanism hydrolyses the enzyme present in all embryonic cells by removing a peptide of about 45000 Da from the amino terminal region of the 190000 Da enzyme initially synthesized on maternal transcripts. The resulting 145000 Da enzyme shows modified catalytic properties, different antibody reactivity and is rapidly destroyed in the few hours before gastrulation [19]) [19]

5 Isolation/Preparation/Mutation/Application

Source/tissue

HeLa cell <1> [1]
Krebs II ascites cell <3> [3]
Novikoff hepatoma cell <2> [2]
P-815 cell <3> [7]
Rcho-1 cell <2> (<2>, the enzyme is regulated at the posttranscriptional level in terminally differentiating Rcho-1 cell [24]) [24]
embryo <8, 11, 14> (<8>, germinating embryo [16];<11>, Dnmt1 is removed from embryonic cells before gastrulation to be synthesized again at different levels in different cell types [19]) [16, 19, 25]
erythroleukemia cell <3> [28, 32]
liver <2> [4, 6]
placenta <1, 2> (<1>, soft tissue of [12]; <2>, the enzyme is highly expressed in placenta during mid-to late-pregnancy [24]) [9, 12, 14, 23, 24]
shoot tip <15> [31]
suspension culture <6> [11]
thymus <7> [13]
trophoblast <2> (giant cell) [24]

- Additional information <11, 15> (<11>, Dnmt1 is removed from embryonic cells before gastrulation to be synthesized again at different levels in different cell types [19]; <15>, expression is confined mostly to the rapidly dividing tissues of the plant [31]) [19, 31]

Localization

nucleus <1, 2, 8, 15> [1, 4, 16, 31]

Purification

<1> [1, 9, 12]
<2> (partial [2]) [2, 4]
<3> (Dnmt1-HhaI and Dnmt1-HpaII hybrid enzymes [26]; recombinant enzyme [27]; recombinant Dnmt3a, Dnmt3b1 and Dnmt3b2, GST-fused Dnmt3 [33]) [3, 7, 9, 21, 26, 27, 33]
<4> [5]
<5> (M.BsuRIa and M.BsuRIb [10]) [10]
<7> (partial [13]) [13]
<8> [16]
<10> (expressed as N-terminal fusion at ATC149-151 or ATG344-345 to histidine taggs [37]) [37]
<15> [31]
<23> [38]

Crystallization

<23> (crystal structures of HhaI methyltransferase complexed with cognate unmethylated or methylated DNA together with S-adenosyl-L-homocysteine [38]) [38]

Cloning

<1> (expression of the enzyme by using a protein splicing fusion partner in a baculovirus expression vector [29]) [29]
<2> [24]
<3> (DNA (cytosine C5) methyltransferase-1, fused to an amino-terminal hexahistidine tag, is expressed by infecting Spodoptera frugiperda cells for 46 h with a recombinant baculovirus carrying the DNA (cytosine-C5)methyltransferase-1 cDNA [21]; construction of three hybrid methyltransferases, containing the intact N-terminus of the murine Dnmt1 and most of the coding sequences from M.HhaI (GCGC), M.HpaII (CCGG) or M.SssI (CG). The hybrids are biologically active when expressed in a baculovirus system and show the specificity of the parental C-terminal domain [26]; <3>, expression in Escherichia coli [27]) [21, 26, 27]
<10> (expression in Escherichia coli and Bacillus stearothermophilus, expressed as N-terminal fusion at ATC149-151 or ATG344-345 to histidine taggs [37]) [18, 37]
<12> [22]
<15> [31]
<16> (expression in Escherichia coli [34]) [34]

6 Stability

Temperature stability

4 <6> (<6>, stable for 5 days [11]) [11]
30 <6> (<6>, 15 min, stable [11]) [11]
37 <1, 3> (<1,3>, rather unstable [9]) [9]
42 <6> (<6>, half-life: 48 s [11]) [11]

General stability information

<1>, glycerol stabilizes [1]

Storage stability

<1>, -70°C, stable for at least several weeks [1]
<1>, -80°C, 30% glycerol, activity decays at a rate of about 15% in 2 weeks [12]
<1>, -80°C, little loss of activity after 4 weeks, significant loss of activity after 6 weeks [29]
<2>, -70°C, the purified enzyme preparation is stable for at least 6 months [4]
<2>, 4°C, the purified enzyme preparation is rather unstable [4]
<3>, -20°C, hexahistidine-tagged enzyme is stable for several months [21]
<3>, -20°C, hybrid enzymes Dnmt1-HhaI and Dnmt1-HpaII are stable for several weeks [26]
<5>, -20°C, 100 mM Tris-HCl, pH 8.4, 1 mM dithioerythritol, 0.005% Brij 58, 50% glycerol, M.BsuRIa loses 10% of ist activity and M.BsuRIb loses about 30% of ist activity after 3 months [10]
<6>, -20°C, stable for more than 2 months [11]
<6>, 4°C, stable for 5 days [11]
<7>, -20°C, enzyme loses activity within several weeks [13]
<1, 3>, -80°C, stable for at least 3 weeks [9]

References

[1] Roy, P.H.; Weissbach, A.: DNA methylase from HeLa cell nuclei. Nucleic Acids Res., **2**, 1669-1684 (1975)
[2] Sneider, T.W.; Teague, W.M.; Rogachevsky, L.M.: S-Adenosylmethionine: DNA-cytosine 5-methyltransferase from a Novikoff rat hepatoma cell line. Nucleic Acids Res., **2**, 1685-1700 (1975)
[3] Turnbull, J.F.; Adams, R.L. P.: DNA methylase: purification from ascites cells and the effect of various DNA substrates on its activity. Nucleic Acids Res., **3**, 677-695 (1976)
[4] Simon, D.; Grunert, F.; v. Acken, U.; Döring, H.P.; Kröger, H.: DNA-methylase from regenerating rat liver: purification and characterisation. Nucleic Acids Res., **5**, 2153-2167 (1978)
[5] Gold, M.; Hurwitz, J.: The enzymatic methylation of ribonucleic acid and deoxyribonucleic acid. J. Biol. Chem., **239**, 3858-3865 (1964)

[6] Cox, R.: Polyamines inhibit DNA methylation in vitro. Biochem. Biophys. Res. Commun., **86**, 594-598 (1979)
[7] Tomassetti, A.; Driever, P.H.; Pfeifer, G.P.; Drahovsky, D.: Isolation and characterization of proteins that stimulate the activity of mammalian DNA methyltransferase. Biochim. Biophys. Acta, **951**, 201-212 (1988)
[8] Santi, D.V.; Garrett, C.E.; Barr, P.J.: On the mechanism of inhibition of DNA-cytosine methyltransferases by cytosine analogs. Cell, **33**, 9-10 (1983)
[9] Pfeifer, G.P.; Grunwald, S.; Palitti, F.; Kaul, S.; Boehm, T.L.J.; Hirth, H.P.; Drahovsky, D.: Purification and characterization of mammalian DNA methyltransferases by use of monoclonal antibodies. J. Biol. Chem., **260**, 13787-13793 (1985)
[10] Gunthert, U.; Freund, M.; Trautner, T.A.: Restriction and modification in Bacillus subtilis: two DNA methyltransferases with BsuRI specificity. I. Purification and physical properties. J. Biol. Chem., **256**, 9340-9345 (1981)
[11] Giordano, M.; Mattachini, M.E.; Cella, R.; Pedrali-Noy, G.: Purification and properties of a novel DNA methyltransferase from cultured rice cells. Biochem. Biophys. Res. Commun., **177**, 711-719 (1991)
[12] Pfeifer, G.P.; Grunwald, S.; Boehm, T.L.J.; Drahovsky, D.: Isolation and characterization of DNA cytosine 5-methyltransferase from human placenta. Biochim. Biophys. Acta, **740**, 323-330 (1983)
[13] Sano, H.; Noguchi, H.; Sager, R.: Characterization of DNA methyltransferase from bovine thymus cells. Eur. J. Biochem., **135**, 181-185 (1983)
[14] McIntosh, L.P.; Zielinski, W.S.; Kalisch, B.W.; Pfeifer, G.P.; Sprinzl, M.; Drahovsky, D.; van de Sande, J.H.; Jovin, T.M.: Synthesis and characterization of poly[d(G-z5C)]. B-Z transition and inhibition of DNA methylase. Biochemistry, **24**, 4806-4814 (1985)
[15] Buryanov, Y.I.; Bogdarina, I.G.; Bayev, A.A.: Site specificity and chromatographic properties of E. coli K12 and EcoRIIDNA-cytosine methylases. FEBS Lett., **88**, 251-254 (1978)
[16] Theiss, G.; Schleicher, R.; Schimpff-Weiland, G.; Follmann, H.: DNA methylation in wheat. Purification and properties of DNA methyltransferase. Eur. J. Biochem., **167**, 89-96 (1987)
[17] Dubey, A.K.; Mollet, B.; Roberts, R.J.: Purification and characterization of the MspI DNA methyltransferase cloned and overexpressed in E. coli. Nucleic Acids Res., **20**, 1579-1585 (1992)
[18] Schumann, J.; Walter, J.; Willert, J.; Wild, C.; Koch, D.; Trautner, T.A.: M.BssHII, a multispecific cytosine-C5-DNA-methyltransferase with unusual target recognizing properties. J. Mol. Biol., **257**, 949-959 (1996)
[19] di Giaimo, R.; Locascio, A.; Aniello, F.; Branno, M.; del Gaudio, R.; Potenza, N.; Geraci, G.: DNA (cytosine-5) methyltransferase turnover and cellular localization in developing Paracentrotus lividus sea urchin embryo. Gene, **272**, 199-208 (2001)
[20] Deng, J.; Szyf, M.: Downregulation of DNA (cytosine-5-)methyltransferase is a late event in NGF-induced PC12 cell differentiation. Mol. Brain Res., **71**, 23-31 (1999)
[21] Brank, A.S.; Van Bemmel, D.M.; Christman, J.K.: Optimization of baculovirus-mediated expression and purification of hexahistidine-tagged murine

DNA (cytosine-C5)-methyltransferase-1 in Spodoptera frugiperda 9 cells. Protein Expr. Purif., **25**, 31-40 (2002)
[22] Noyer-Weidner, M.; Walter, J.; Terschueren, P.A.; Chai, S.; Trautner, T.A.: M.f3TII: a new monospecific DNA (cytosine-C5) methyltransferase with pronounced amino acid sequence similarity to a family of adenine-N^6-DNA-methyltransferases. Nucleic Acids Res., **22**, 5517-5523 (1994)
[23] Laayoun, A.; Smith, S.S.: Methylation of slipped duplexes, snapbacks and cruciforms by human DNA(cytosine-5)methyltransferase. Nucleic Acids Res., **23**, 1584-1589 (1995)
[24] Kimura, H.; Takeda, T.; Tanaka, S.; Ogawa, T.; Shiota, K.: Expression of rat DNA (cytosine-5) methyltransferase (DNA MTase) in rodent trophoblast giant cells: molecular cloning and characterization of rat DNA MTase. Biochem. Biophys. Res. Commun., **253**, 495-501 (1998)
[25] del Gaudio, R.; Di Giaimo, R.; Potenza, N.; Branno, M.; Aniello, F.; Geraci, G.: Characterization of a new variant DNA (cytosine-5)-methyltransferase unable to methylate double stranded DNA isolated from the marine annelid worm Chaetopterus variopedatus. FEBS Lett., **460**, 380-384 (1999)
[26] Pradhan, S.; Roberts, R.J.: Hybrid mouse-prokaryotic DNA (cytosine-5) methyltransferases retain the specificity of the parental C-terminal domain. EMBO J., **19**, 2103-2114 (2000)
[27] Tollefsbol, T.O.; Hutchison, C.A.: Mammalian DNA (cytosine-5-)-methyltransferase expressed in Escherichia coli, purified and characterized. J. Biol. Chem., **270**, 18543-18550 (1995)
[28] Flynn, J.; Reich, N.: Murine DNA (cytosine-5-)methyltransferase: steady-state and substrate trapping analyses of the kinetic mechanism. Biochemistry, **37**, 15162-15169 (1989)
[29] Pradhan, S.; Bacolla, A.; Wells, R.D.; Roberts, R.J.: Recombinant human DNA (cytosine-5) methyltransferase. I. Expression, purification, and comparison of de novo and maintenance methylation. J. Biol. Chem., **274**, 33002-33010 (1999)
[30] Bacolla, A.; Pradhan, S.; Larson, J.E.; Roberts, R.J.; Wells, R.D.: Recombinant human DNA (cytosine-5) methyltransferase: III. Allosteric control, reaction order, and influence of plasmid topology and triplet repeat length on methylation of the fragile X CGG*CCG sequence. J. Biol. Chem., **276**, 18605-18613 (2001)
[31] Pradhan, S.; Cummings, M.; Roberts, R.J.; Adams, R.L.: Isolation, characterization and baculovirus-mediated expression of the cDNA encoding cytosine DNA methyltransferase from Pisum sativum. Nucleic Acids Res., **26**, 1214-1222 (1998)
[32] Yoder, J.A.; Soman, N.S.; Verdine, G.L.; Bestor, T.H.: DNA (cytosine-5)-methyltransferases in mouse cells and tissues. Studies with a mechanism-based probe. J. Mol. Biol., **270**, 385-395 (1997)
[33] Aoki, A.; Suetake, I.; Miyagawa, J.; Fujio, T.; Chijiwa, T.; Sasaki, H.; Tajima, S.: Enzymatic properties of de novo-type mouse DNA (cytosine-5) methyltransferases. Nucleic Acids Res., **29**, 3506-3512 (2001)
[34] Lubys, A.; Lubienne, J.; Kulakauskas, S.; Stankevivius, K.; Timinskas, A.; Janulaitis, A.: Cloning and analysis of the genes encoding the type IIS re-

striction-modification system HpHI from Haemophilus parahaemolyticus. Nucleic Acids Res., **24**, 2760-2766 (1996)
[35] Roberts, R.J.; Macelis, D.: Restriction enzymes and their isoschizomers. Nucleic Acids Res., **20**, 2167-2180 (1992)
[36] Rasko, T.; Finta, C.; Kiss, A.: DNA bending induced by DNA (cytosine-5) methyltransferases. Nucleic Acids Res., **28**, 3083-3091 (2000)
[37] Sethmann, S.; Ceglowski, P.; Willert, J.; Iwanicka-Nowicka, R.; Trautner, T.A.; Walter, J.: M.(f)BssHII, a novel cytosine-C5-DNA-methyltransferase with target-recognizing domains at separated locations of the enzyme. EMBO J., **18**, 3502-3508 (1999)
[38] O'Gara, M.; Klimasauskas, S.; Roberts, R.J.; Cheng, X.: Enzymic C5-cytosine methylation of DNA: mechanistic implications of new crystal structures for HhaI methyltransferase-DNA-AdoHcy complexes. J. Mol. Biol., **261**, 634-645 (1996)

O-Demethylpuromycin O-methyltransferase 2.1.1.38

1 Nomenclature

EC number
2.1.1.38

Systematic name
S-adenosyl-L-methionine:O-demethylpuromycin O-methyltransferase

Recommended name
O-demethylpuromycin O-methyltransferase

Synonyms
O-demethylpuromycin methyltransferase

CAS registry number
37257-04-8

2 Source Organism

<1> *Streptomyces alboniger* (wild-type organism ATCC 12461 [2, 4]; mutant strain ATCC 12462 [2]) [1-5]

3 Reaction and Specificity

Catalyzed reaction
S-adenosyl-L-methionine + O-demethylpuromycin = S-adenosyl-L-homocysteine + puromycin (puromycin is the antibiotic derived from N^6-dimethyladenosine by replacing the 3'-hydroxyl group with an amino group and acylating this with 4-O-methyltyrosine)

Reaction type
methyl group transfer

Natural substrates and products
S S-adenosyl-L-methionine + O-demethylpuromycin <1> (<1> last step in the biosynthesis of puromycin [4]) (Reversibility: ? <1> [1-5]) [1-5]
P S-adenosyl-L-homocysteine + puromycin

Substrates and products
S S-adenosyl-L-methionine + N-acetyl-O-demethylpuromycin <1> (Reversibility: ? <1> [4]) [4]
P S-adenosyl-L-homocysteine + N-acetylpuromycin

S S-adenosyl-L-methionine + O-demethylpuromycin <1> (Reversibility: ? <1> [1-5]) [1-5]
P S-adenosyl-L-homocysteine + puromycin
S Additional information <1> (<1> L-tyrosine, L-tyrosine ethyl ester, L-tyrosine amide, catechol and epinephrine are not substrates [1]) [1]
P ?

Inhibitors
D-glucose <1> (<1> reduced enzyme form, represses puromycin synthesis [3]) [3]
puromycin <1> [3]

Specific activity (U/mg)
0.0165 <1> [2]

K_m-Value (mM)
0.01 <1> (S-adenosyl-L-methionine) [1]
0.21 <1> (O-demethylpuromycin) [1]

pH-Optimum
7-9 <1> [1, 2]
8 <1> (<1> broad pH optimum over the range 7.0-9.0 with a maximum at pH 8.0 [1]) [1]

pH-Range
6-9 <1> [1]

4 Enzyme Structure

Molecular weight
68000 <1> (<1> sucrose density gradient centrifugation [1]) [1]

5 Isolation/Preparation/Mutation/Application

Purification
<1> (partial [4]) [2-4]

Cloning
<1> (molecular analysis of the dmpM gene encoding an O-demethyl puromycin O-methyltransferase, cloning of the complete gene cluster for puromycin and expression in heterologous hosts [5]) [5]

Application
medicine <1> (<1> product puromycin is an antibiotic and antitumor agent [1,2]) [1, 2]

6 Stability

pH-Stability

7.5 <1> (<1> very unstable at pH 7.5 and 38°C [2]) [2]

Temperature stability

37 <1> (<1> rapidly inactivated at 37°C, very unstable even at low temperatures, enzyme activity is destroyed upon boiling [3]) [3]
38 <1> (<1> very unstable at 38°C and pH 7.5 [2]) [2]

General stability information

<1>, stabilized by addition of S-adenosyl-L-methionine [3]

Storage stability

<1>, -65°C, can be stored for several months without loss of activity [1]

References

[1] Rao, M. M.; Rebello, P.F.; Pogell, B.M.: Biosynthesis of puromycin in Streptomyces alboniger. Enzymatic methylation of O-demethylpuromycin. J. Biol. Chem., **244**, 112-118 (1969)

[2] Pogell, B.M.: S-Adenosylmethionine: O-demethylpuromycin O-methyltransferase. Methods Enzymol., **43**, 508-515 (1975)

[3] Sankaran, L.; Pogell, B.M.: Biosynthesis of puromycin in Streptomyces alboniger: regulation and properties of O-demethylpuromycin O-methyltransferase. Antimicrob. Agents Chemother., **8**, 721-732 (1975)

[4] Vara, J.; Perez-Gonzalez, J.A.; Jimenez, A.: Biosynthesis of puromycin by Streptomyces alboniger: Characterization of puromycin N-acetyltransferase. Biochemistry, **24**, 8074-8081 (1985)

[5] Lacalle, R.A.; Ruiz, D.; Jimenez, A: Molecular analysis of the dmpM gene encoding an O-demethyl puromycin O-methyltransferase from Streptomyces alboniger. Gene, **109**, 55-61 (1991)

Inositol 3-O-methyltransferase 2.1.1.39

1 Nomenclature

EC number
2.1.1.39

Systematic name
S-adenosyl-L-methionine:1D-myo-inositol 3-O-methyltransferase

Recommended name
inositol 3-O-methyltransferase

Synonyms
S-adenosyl-L-methionine:myo-inositol 1-O-methyltransferase
S-adenosylmethionine:myo-inositol 1-methyltransferase
inositol L-1-methyltransferase
methyltransferase, inositol L-1-
myo-inositol 1-methyltransferase

CAS registry number
37257-05-9

2 Source Organism

<1> *Vinca rosea* [1]
<2> *Vinca minor* [2]

3 Reaction and Specificity

Catalyzed reaction
S-adenosyl-L-methionine + myo-inositol = S-adenosyl-L-homocysteine + 1D-3-O-methyl-myo-inositol

Reaction type
methyl group transfer

Natural substrates and products
S S-adenosyl-L-methionine + myo-inositol <1, 2> (Reversibility: ? [1, 2]) [1, 2]
P 1D-3-O-methyl-myo-inositol <1, 2> [1, 2]

Substrates and products

S S-adenosyl-L-methionine + myo-inositol <1, 2> (Reversibility: ? [1, 2]) [1, 2]

P 1D-3-O-methyl-myo-inositol <1, 2> (i.e. pinitol) [1, 2]

Inhibitors

N-ethylmaleimide <1> [1]
iodoacetate <1> [1]
p-chloromercuribenzoate <1> [1]

Specific activity (U/mg)

0.0078 <2> [2]
Additional information <1> [1]

K_m-Value (mM)

0.8 <2> (S-adenosyl-L-methionine) [2]
2.5 <2> (myo-inositol) [2]

pH-Optimum

7 <1, 2> [1, 2]

pH-Range

6.5-8 <1> (<1> pH 6.5: about 20% of activity maximum, pH 8.0: about 15% of activity maximum [1]) [1]

4 Enzyme Structure

Molecular weight

27000 <2> (<2> gel filtration [2]) [2]
30000 <1> (<1> gel filtration [1]) [1]

Subunits

monomer <2> (<2> 1 * 27000, SDS-PAGE [2]) [2]

5 Isolation/Preparation/Mutation/Application

Localization

soluble <1> [1]

Purification

<1> (partial [1]) [1]
<2> [2]

References

[1] Hofmann, H.; Hoffmann-Ostenhof, O.: Untersuchungen über die Biosynthese der Cyclite, XXIV. Über ein lösliches Enzym aus Vinca rosea, das myo-Inosit zu L-Bornesit methyliert. Hoppe-Seyler's Z. Physiol. Chem., **350**, 1465-1468 (1969)

[2] Koller, F.; Hoffmann-Ostenhof, O.: Studies on the biosynthesis of cyclitols, XXXIV. Purification of myo-inositol 3-methyltransferase from Pisum sativum and of myo-inositol 1-methyltransferase from Vinca minor to homogeneity by affinity chromatography. Hoppe-Seyler's Z. Physiol. Chem., **357**, 1465-1468 (1976)

Inositol 1-methyltransferase 2.1.1.40

1 Nomenclature

EC number
2.1.1.40

Systematic name
S-adenosyl-L-methionine:1D-myo-inositol 1-O-methyltransferase

Recommended name
inositol 1-methyltransferase

Synonyms
S-adenosyl-L-methionine:1D-myo-inositol 3-O-methyltransferase
S-adenosylmethionine:myo-inositol 3-O-methyltransferase
S-adenosylmethionine:myo-inositol 3-methyltransferase
inositol 3-O-methyltransferase
inositol D-1-methyltransferase
methyltransferase, inositol, D-1-
myo-inositol 3-O-methyltransferase

CAS registry number
37257-06-0

2 Source Organism

<1> *Pisum sativum* (pea) [1, 2]
<2> *Mesembryanthemum crystallinum* [3]

3 Reaction and Specificity

Catalyzed reaction
S-adenosyl-L-methionine + myo-inositol = S-adenosyl-L-homocysteine + 1D-1-O-methyl-myo-inositol

Reaction type
methyl group transfer

Natural substrates and products
S S-adenosyl-L-methionine + myo-inositol <1> (<1> biosynthesis of cyclitols [1]) (Reversibility: ? <1> [1]) [1]
P S-adenosyl-L-homocysteine + D-bornesitol

Substrates and products

S D-bornesitol + S-adenosyl-L-methionine <1> (<1> low reaction rate [1]) (Reversibility: ? <1> [1]) [1]
P 1,3-di-O-methyl-myo-inositol + S-adenosyl-L-homocysteine
S D-chiro-inositol + S-adenosyl-L-methionine <1> (Reversibility: ? <1> [1]) [1]
P quebrachitol + S-adenosyl-L-homocysteine <1> [1]
S L-bornesitol + S-adenosyl-L-methionine <1> (<1> low reaction rate [1]) (Reversibility: ? <1> [1]) [1]
P 1,3-di-O-methyl-myo-inositol + S-adenosyl-L-homocysteine
S L-chiro-inositol + S-adenosyl-L-methionine <1> (Reversibility: ? <1> [1]) [1]
P quebrachitol + pinitol + S-adenosyl-L-homocysteine <1> [1]
S S-adenosyl-L-methionine + myo-inositol <1> (Reversibility: ? <1> [1]) [1]
P S-adenosyl-L-homocysteine + D-bornesitol <1> [1]
S scyllo-inositol + S-adenosyl-L-methionine <1> (<1> low reaction rate [1]) (Reversibility: ? <1> [1]) [1]
P ?

Inhibitors

N-ethylmaleimide <1> [1]
iodoacetamide <1> [1]
p-chloromercuribenzoate <1> [1]

Specific activity (U/mg)

0.0078 <1> [2]

K_m-Value (mM)

2.3 <1> (myo-inositol) [1]
2.5 <1> (S-adenosyl-L-methionine) [2]
2.9 <1> (myo-inositol) [2]
3.2 <1> (S-adenosyl-L-methionine) [1]

pH-Optimum

7 <1> [1, 2]

pH-Range

6.7-7.6 <1> (<1> about 50% of activity maximum at pH 6.7 and 7.6 [1]) [1]

4 Enzyme Structure

Molecular weight

30000 <1> (<1> gel filtration [1]) [1]
42000 <1> (<1> gel filtration, PAGE) [2]

Subunits

monomer <1> (<1> 1 * 42000, SDS-PAGE [2]) [2]

5 Isolation/Preparation/Mutation/Application

Source/tissue
leaf <2> [3]
leaf axil <2> [3]
seedling <1> [1]

Purification
<1> (partial [1]) [1, 2]

6 Stability

Temperature stability
3 <1> (<1> 24 h, 40% loss of activity [1]) [1]

General stability information
<1>, glutathione stabilizes slightly [1]

References

[1] Wagner, I.; Hofmann, H.; Hoffmann-Ostenhof, O.: Biosynthesis of cyclitols. 23. A soluble enzyme from pea seedlings methylating myo-inositol to D-bornesitol. Hoppe-Seyler's Z. Physiol. Chem., **350**, 1460-1464 (1969)
[2] Koller, F.; Hoffmann-Ostenhof, O.: Studies on the biosynthesis of cyclitols, XXXIV[l] Purification of myo-inositol 3-methyltransferase from Pisum sativum and of myo-inositol 1-methyltransferase from Vinca minor to homogeneity by affinity chromatography. Hoppe-Seyler's Z. Physiol. Chem., **357**, 1465-1468 (1976)
[3] Vernon, D.M.; Bohnert, H.J.: Increased expression of a myo-inositol methyl transferase in Mesembryanthemum crystallinum is part of a stress response distinct from Crassulacean acid metabolism induction. Plant Physiol., **99**, 1695-1698 (1992)

Sterol 24-C-methyltransferase 2.1.1.41

1 Nomenclature

EC number
2.1.1.41

Systematic name
S-adenosyl-L-methionine:zymosterol 24-C-methyltransferase

Recommended name
sterol 24-C-methyltransferase

Synonyms
(S)-adenosyl-L-methionine:$\Delta^{24(25)}$-sterol methyl transferase <1> [11]
24-sterol C-methyltransferase
Δ^{24}-methyltransferase
Δ^{24}-sterol methyltransferase
S-adenosyl-4-methionine:sterol Δ^{24}-methyltransferase
S-adenosyl-L-methionine:$\Delta^{24(23)}$-sterol methyltransferase
SMT1
methyltransferase, Δ^{24}-sterol
phytosterol methyltransferase
zymosterol-24-methyltransferase

CAS registry number
37257-07-1

2 Source Organism

<1> *Saccharomyces cerevisiae* (strain A184D erg+ and sterol mutant strain +erg2 [16]; mutant strain GL7 [11]) [1, 4, 8, 10-13, 16, 20]
<2> *Helianthus annuus* (sunflower [2]) [2, 10]
<3> *Candida albicans* [3, 10]
<4> *Zea mays* (corn [5]) [5]
<5> *Uromyces phaseoli* (bean rust [6,7]) [6, 7]
<6> *Candida tropicalis* [3]
<7> *Kluyveromyces lactis* (yeast [9]) [9]
<8> *Prototheca wickerhamii* (algae [17]) [10, 17]
<9> *Glycine max* (gene SMT1 [19]) [10, 19]
<10> *Arabidopsis thaliana* (3 isozymes encoded on SMT1, SMT2, SMT3 [18, 21]) [10, 18, 21]
<11> *Zea mays* [14]

<12> *Pneumocystis carinii* [15]
<13> *Glycine max* [19]

3 Reaction and Specificity

Catalyzed reaction

S-adenosyl-L-methionine + 5α-cholesta-8,24-dien-3β-ol = S-adenosyl-L-homocysteine + 24-methylene-5α-cholest-8-en-3β-ol (<1> substrate analogue inhibitor binding site structure [20]; <8> steric substrate preferences: 3β-hydroxy group, planar nucleus, side chain orientated into right-handed structure [17]; <1, 8> kinetic and inhibition mechanism [11, 12, 16, 17]; <1-3, 8-10, 12> substrate binding, amino acid sequence [10,13,15]; <1-3, 8-10> active site model, termed steric-electric plug model, consists of a non-covalent mechanism involving the intermediacy of a 24β-methyl (or ethyl) sterol bound to the ternary complex [10]; <1-3, 8-10> mechanism [2, 3, 10-13, 16, 17]; <8> structure-function relationship [17])

Reaction type

methyl group transfer

Natural substrates and products

S cycloartenol + S-adenosyl-L-methionine <13> (<9> rate-limiting initial step in the conversion of phytosterol [19]) (Reversibility: ? <13> [19]) [19]
P 24-methylenecycloartenol + S-adenosyl-L-homocysteine <13> [19]
S zymosterol + S-adenosyl-L-methionine <1-13> (<1-3,8-10> role in biosynthesis of plant and fungi sterols [10]) (Reversibility: ? <1-13> [1-21]) [1-21]
P fecosterol + S-adenosyl-L-homocysteine

Substrates and products

S 14α-methylzymosterol + S-adenosyl-L-methionine <2> (<2> 14α-methylzymosterol is 14α-methylcholesta-8,24-dien-3β-ol [2]) (Reversibility: ? <2> [2]) [2]
P 14α-methyl-24-methylenezymosterol + S-adenosyl-L-homocysteine
S 24,25-dehydropollinastanol + S-adenosyl-L-methionine <2> (<2> 24,25-dehydropollinastanol is 14α-methyl-9β,19-cyclo-5α-cholest-24-en-3β-ol [2]) (Reversibility: ? <2> [2]) [2]
P ?
S 24-methylenecholesterol + S-adenosyl-L-methionine <5> (<5> 5% the reaction rate of zymosterol [7]) (Reversibility: ? <5> [7]) [7]
P 24-ethylidenecholesterol + S-adenosyl-L-homocysteine <5> [7]
S 24-methylenelanosterol + S-adenosyl-L-methionine <12> (<12> preferred substrate [15]) (Reversibility: ? <12> [15]) [15]
P pneumocysterol + S-adenosyl-L-homocysteine <12> [15]
S 26,27-dehydrozymosterol + S-adenosyl-L-methionine <1> (<1> mutants [20]) (Reversibility: ? <1> [20]) [20]

P 26-homo-cholesta-8(9),23(24)E,26(26')-trienol + 26-homo-cholesta-8(9),26(26')-3β,24β-dienol + S-adenosyl-L-homocysteine <1> [20]

S 31-norcycloartenol + S-adenosyl-L-methionine <2> (<2> 31-norcycloartenol is 4,14α-dimethyl-9β,19-cyclo-5α-cholest-24-en-3β-ol [2]) (Reversibility: ? <2> [2]) [2]

P 24-methylene-31-norcycloartenol + S-adenosyl-L-homocysteine

S agnosterol + S-adenosyl-L-methionine <2> (<2> agnosterol is 4,4,14α-trimethyl-5α-cholesta-7(8),9(11),24-trien-3β-ol [2]) (Reversibility: ? <2> [2]) [2]

P 24-methyleneagnosterol + S-adenosyl-L-homocysteine

S cholesta-5,7,24-trienol + S-adenosyl-L-methionine <2> (Reversibility: ? <2> [2]) [2]

P ?

S cycloartenol + S-adenosyl-L-methionine <1-4, 8-10, 13> (<12> no substrate [15]; <2,4> cycloartenol is 4,4,14α-trimethyl-9β,19-cyclo-5α-cholest-24-en-3β-ol [2,5]) (Reversibility: ? <1-4, 8-10, 13> [2, 5, 10, 17, 19]) [2, 5, 10, 17, 19]

P 24-methylenecycloartenol + S-adenosyl-L-homocysteine <1-3, 8-10, 13> [2, 10, 19]

S cycloartenol 3-ketone + S-adenosyl-L-methionine <2> (<2> cycloartenol 3-ketone is 4,4,14α-trimethyl-9β,19-cyclo-5α-cholest-24-en-3-one [2]) (Reversibility: ? <2> [2]) [2]

P ?

S desmosterol + S-adenosyl-L-methionine <2, 5> (<2,5> desmosterol is 5α-cholesta-5,24-dien-3β-ol [2,6,7]) (Reversibility: ? <2, 5> [2, 6, 7]) [2, 6, 7]

P 24-methylenedesmosterol + S-adenosyl-L-homocysteine <5> [7]

S lanost-24-enol + S-adenosyl-L-methionine <2> (<2> lanost-24-enol is 4,4,14α-trimethyl-5α-cholesta-9(11),24-dien-3β-ol [2]) (Reversibility: ? <2> [2]) [2]

P 24-methylene-lanost-24-erol + S-adenosyl-L-homocysteine

S lanosta-7,24-dienol + S-adenosyl-L-methionine <2> (<2> lanosta-7,24-dienol is 4,4,14α-trimethyl-5α-cholesta-7(8),24-dien-3β-ol [2]) (Reversibility: ? <2> [2]) [2]

P 24-methylene-lanost-7,24-dienol + S-adenosyl-L-homocysteine

S lanosterol + S-adenosyl-L-methionine <1-3, 5, 8-10, 12> (<8> regiospecific [17]; <12> preferred substrate [15]; <3> no reaction with [3]; <2,5> lanosterol is 4,4,14α-trimethyl-5α-cholesta-8,24-dien-3-ol [2,6,7]) (Reversibility: ? <1-3, 5, 8-10, 12> [2, 6, 7, 10, 15, 17]) [2, 6, 7, 10, 15, 17]

P 24-methylene-24,25-dihydrolanosterol + S-adenosyl-L-homocysteine <1-3, 8-10, 12> [2, 10, 15]

S zymosterol + S-adenosyl-L-methionine <1> (<1> mutants D79L and E82L [20]) (Reversibility: ? <1> [20]) [20]

P 24-methylzymosta-8,25(27)-dienol + S-adenosyl-L-homocysteine <1> [20]

S zymosterol + S-adenosyl-L-methionine <1-13> (<1> wild-type and mutants [20]; <1> regio- and stereospecific for substrates [11,13]; <1-3,5>

zymosterol is 5α-cholesta-8,24-dien-3β-ol [1-4,6-8]) (Reversibility: ? <1-13> [1-21]) [1-21]

P fecosterol + S-adenosyl-L-homocysteine <1> (<1> fecosterol is 24-exo-methylene-5α-cholesta-8,24-dien-3β-ol [1,12]) [1, 13]

S Additional information <1> (<1> overview about a large number of sterol substrates of various origin, substrate specificity analysis, structures: overview [11]) [11]

P ?

Inhibitors

(24R)-methyl-25-thiacholesteryl iodide <1> [16]
(24R,S),25-epiminolanosterol <1, 2> [10, 11]
(24R,S),25-epiminozymosterol <1> [11]
(24R,S)-24-ethyl-25-azacycloartenol <4> [5]
(24R,S)-24-methyl-25,26,27-trisnor-24-dimethylamino-N-oxide cycloartenol <4> [5]
(24R,S)-24-methyl-25,26,27-trisnor-24-dimethylsulfonium cycloartenol iodide <4> [5]
(24R,S)-24-methyl-25,26,27-trisnor-24-trimethylarsonium cycloartenol iodide <4> [5]
(24R,S)-24-methyl-25-azacycloartenol <4> [5]
(24S)-methyl-25-thiacholesteryl iodide <1> [16]
(3β,24,R,S)-24-methyl-24-thioniacholest-5-enol <3> [10]
22,25-diazacholesterol <1> (<1> 20R and 20S isomer [4]) [4]
23-azacholesterol <1, 4> [4, 5]
24(28)-methylenecycloartenol <8> [17]
24(S)-24-methyl-25-thiacholest-5-enol iodide <1> [10]
24-azacholesterol <1, 4> [4, 5]
24-azacycloartanyl acetate <4> [5]
24-methyl-25,26,27-trisnor-24-trimethylammonium cycloartenol iodide <4> (<4> 24R and 24S isomer [5]) [5]
24-methyl-25-azacycloartenol <4> (<4> 24R and 24S isomer [5]) [5]
24-oxo-25-azacycloartenol <4> [5]
25,26,27-trisnor-24-trimethylammonium cycloartenol iodide <4> [5]
25-aza-24,25-dihydrolanosterol <1> [4]
25-aza-24,25-dihydrozymosterol <1> [4, 10]
25-azacholestane <1> [11]
25-azacholestanol <1> [4]
25-azacholesterol <1, 4> [4, 5, 11]
25-azacholesterol hydrochloride <1> [4, 11]
25-azacycloartenol <4, 8-10> [5, 10, 17]
25-azalanosterol <1> [11]
25-thiacholesterol iodide <1> [16]
26,27-dehydrozymosterol <1> (<1> mechanism-based irreversible inhibition [20]) [20]
Cs^{+} <1> [8]
K^{+} <1> [8]

Li^+ <1> [8]
NAD^+ <5> [7]
$NADP^+$ <5> [7]
NH_4^+ <1> [8]
Na^+ <1> [8]
S-adenosyl-L-homocysteine <3> [3]
amphotericine B <7> (<7> inhibits the wild-type at concentrations above 60 nM, inhibition can be hindered by zymosterol [9]) [9]
cyclolaudanol <8> [17]
desmosterol <3, 6> (<3,6> competitive versus sterol and S-adenosyl-L-methionine [3]) [3]
ergosterol <8> [17]
filipin <7> (<7> slightly inhibiting at low concentration [9]) [9]
nystatin <7> (<7> slightly inhibiting at low concentration [9]) [9]
solasodine <8> [10]
substrate analogues <3> (<3> dead-end inhibitors [3]) [3]
triparanol <5> [7]
zymosterol <3, 6> (<3,6> competitive versus sterol and S-adenosyl-L-methionine [3]) [3]
Additional information <1, 8> (<1,8> overview, diverse substrate and transition state analogues [12,17]) [12, 17]

Activating compounds
amphotericine B <7> (<7> activates at concentrations below 5.4 nM, optimal at 2.1 nM, wild-type and mutant [9]) [9]
candicidin <7> (<7> stimulates, best at low concentration [9]) [9]
filipin <7> (<7> slightly activating at high concentration [9]) [9]
glutathione <1> (<1> required [1]) [1]
nystatin <7> (<7> slightly activating at high concentration [9]) [9]
pimaricin <7> (<7> strong activation, activation level decreases with increasing concentration of pimaricin [9]) [9]
Additional information <2> (<2> no cofactor requirement [2]) [2]

Metals, ions
Mg^{2+} <1> (<1> 5 mM, required [1]) [1]
Additional information <2> (<2> no metal ion required [2]) [2]

Turnover number (min^{-1})
0.048 <1> (26,27-dehydrozymosterol) [20]
0.06 <1> (zymosterol, <1> mutant E98L [20]) [20]
0.3 <1> (zymosterol, <1> mutants E82L and E64L [20]) [20]
0.42 <1> (zymosterol, <1> mutant D65L [20]) [20]
0.6 <1> (zymosterol, <1> wild-type and mutant D79L [20]) [13, 20]

Specific activity (U/mg)
0.0008 <1> (<1> purified enzyme [13]) [13]
0.00124 <5> (<5> purified enzyme [6]) [6]
0.53 <1> (<1> purified enzyme [1]) [1]

Additional information <1, 5> (<1> comparison recombinant and wild-type [13]; <1> activity at various concentrations of glutathione [1]) [1, 6, 7, 13]

K_m-Value (mM)

0.0106 <12> (lanosterol, <12> recombinant enzyme [15]) [15]
0.013 <1> (26,27-dehydrozymosterol, <1> mutant E64L [20]) [20]
0.015 <1> (26,27-dehydrozymosterol, <1> mutant D65L [20]) [20]
0.015-0.017 <1> (26,27-dehydrozymosterol, <1> wild-type [20]) [20]
0.017 <1> (26,27-dehydrozymosterol, <1> mutant D79L [20]) [20]
0.019 <12> (24-methylenelanosterol, <12> recombinant enzyme [15]) [15]
0.022 <1> (26,27-dehydrozymosterol, <1> mutant E82L [20]) [20]
0.028 <8> (cycloartenol) [17]
0.028 <2> (lanosterol) [2]
0.032 <2> (cycloartenol) [2]
0.033 <2> (14α-methylzymosterol) [2]
0.0334 <5> (S-adenosyl-L-methionine) [7]
0.036 <2> (31-norcycloartenol) [2]
0.043 <2> (24,25-dehydropollinastanol) [2]
0.052 <6> (zymosterol) [3]
0.055-0.075 <5> (zymosterol, <5> value depends on concentration of detergent [7]) [7]
0.063 <1, 2> (zymosterol) [1, 2]
0.067 <2> (desmosterol, <2> lanosta-7,24-dienol [2]) [2]
0.071 <3> (zymosterol) [3]
0.074 <1> (26,27-dehydrozymosterol, <1> mutant E98L [20]) [20]
0.125 <2> (cholesta-5,7,24-trienol) [2]
0.13 <2> (agnosterol) [2]
0.236 <6> (desmosterol) [3]
0.281 <3> (desmosterol) [3]
Additional information <1, 8> (<1,8> overview [11,17]) [8, 11, 17]

K_i-Value (mM)

0.00000000075 <8> (solasodine) [10]
0.0000011 <1> (25-thiacholesterol iodide, <1> versus S-adenosyl-L-methionine [16]) [16]
0.00000125 <1> (25-aza-24,25-dihydrozymosterol) [4]
0.0000017 <1> ((24R)-methyl-25-thiacholesteryl iodide, <1> versus S-adenosyl-L-methionine [16]) [16]
0.000002 <8> (25-azacycloartenol) [17]
0.0000023 <1> ((24S)-methyl-25-thiacholesteryl iodide, <1> versus S-adenosyl-L-methionine [16]) [16]
0.0000024 <1> (25-thiacholesterol iodide, <1> versus zymosterol [16]) [16]
0.000003 <1> ((24R)-methyl-25-thiacholesteryl iodide, <1> versus zymosterol [16]) [16]
0.000003 <2> (24(R,S),25-epiminolanosterol) [10]
0.000003 <1> (25-aza-24,25-dihydrozymosterol) [10]
0.000003 <8> (25-azacycloartenol) [10]

0.0000046 <1> ((24S)-methyl-25-thiacholesteryl iodide, <1> versus zymosterol [16]) [16]
0.00001 <1> ((24,S),25-epiminozymosterol) [11]
0.00001 <1> ((24R,S),25-epiminolanosterol) [11]
0.000015 <4> ((24R)-24-methyl-25,26,27-trisnor-24-dimethylamino-N-oxide cycloartenol) [5]
0.000015 <1> (25-azacholesterol hydrochloride) [11]
0.000015 <9, 10> (25-azacycloartenol) [10]
0.000016 <3> ((3β,24,R,S)-24-methyl-24-thioniacholest-5-enol) [10]
0.00002 <4> ((24R)-24-methyl-25-azacycloartenol) [5]
0.00002 <4> ((24R,S)-24-methyl-25-azacycloartenol) [5]
0.000025 <4> ((24R,S)-24-methyl-25,26,27-trisnor-24-trimethylarsonium cycloartenol iodide) [5]
0.00003 <4> ((24S)-24-methyl-25-azacycloartenol) [5]
0.00003 <4> (24-azacycloartanyl acetate) [5]
0.00003 <4> (25-azacycloartenol) [5]
0.000035 <4> (24-methyl-25,26,27-trisnor-24-trimethylammonium cycloartenol iodide) [5]
0.000035 <4> (25,26,27-trisnor-24-trimethylammonium cycloartenol iodide) [5]
0.000045 <4> ((24S)-24-methyl-25,26,27-trisnor-24-dimethylamino-N-oxide cycloartenol) [5]
0.000045 <4> (25-azacholesterol) [5]
0.000045 <1> (25-azalanosterol) [11]
0.000048 <1> (25-azacholestane) [11]
0.00005 <4> ((24R,S)-24-methyl-25,26,27-trisnor-24-dimethylsulfonium cycloartenol iodide) [5]
0.000085 <4> ((24R,S)-24-ethyl-25-azacycloartenol) [5]
0.00009 <4> (23-azacholesterol) [5]
0.00017 <4> (24-oxo-25-azacycloartenol) [5]
0.00018 <4> (24-azacholesterol) [5]
0.0011 <1> (26,27-dehydrozymosterol) [20]
0.0045 <6> (zymosterol, <6> versus S-adenosyl-L-methionine [3]) [3]
0.009 <6> (desmosterol, <6> versus S-adenosyl-L-methionine [3]) [3]
0.0193 <3> (desmosterol, <3> versus S-adenosyl-L-methionine [3]) [3]
0.022 <6> (zymosterol, <6> versus sterol [3]) [3]
0.023 <8> (cyclolaudanol) [17]
0.08 <8> (ergosterol) [17]
0.085 <8> (24(28)-methylenecycloartenol) [17]
0.449 <6> (desmosterol, <6> versus sterol [3]) [3]
0.489 <3> (desmosterol, <3> versus sterol [3]) [3]
Additional information <1, 8> (<1,8> overview, diverse substrate and transition state analogues [12,17]) [12, 17]

pH-Optimum

7-7.7 <1, 5> [7, 8]

7.5 <1-3, 6, 12> (<12> recombinant enzyme [15]; <1,3,6> assay at [3,13]) [1-3, 13, 15]
7.5-8 <8> (<8> assay at [17]) [17]
7.6 <7> (<7> assay at [9]) [9]

pH-Range
5.5-8.5 <5> [7]
5.5-9 <12> (<12> recombinant enzyme [15]) [15]
5.5-10 <2> (<2> slightly active at pH 5.5, no activity at pH 10 [2]) [2]
6.8-7.7 <1> (<1> less than half-maximal activity above and below [1]) [1]

Temperature optimum (°C)
23 <5> [7]
30 <3, 6, 13> (<3,6,13> assay at [3,19]) [3, 19]
32 <1> (<1> assay at [13]) [13]
35 <7, 8> (<7,8> assay at [9,17]) [9, 17]
37 <12> (<12> recombinant enzyme [15]) [15]

Temperature range (°C)
15-30 <5> (<5> less than half-maximal activity above and below [7]) [7]
22-45 <12> (<12> recombinant enzyme [15]) [15]

4 Enzyme Structure

Molecular weight
150000 <3> (<3> gel filtration [3]) [3]
172000 <1> (<1> gel filtration [13]) [13]

Subunits
tetramer <1> [13]
Additional information <1> (<1> cooperativity among the subunits [13]) [13]

5 Isolation/Preparation/Mutation/Application

Source/tissue
embryo <2> [2]
endosperm <11> [14]
flower <13> [19]
hypocotyl <13> (<13> etiolated [19]) [19]
seedling <4> [5]

Localization
cytosol <1> [1]
membrane <8> [17]
microsome <1-4, 6> [2, 3, 5, 10]
mitochondrial inner membrane <1, 5> (<1,5> some activity [7,8]) [7, 8]
mitochondrion <1, 5> (<5> major activity [7,8]) [7, 8]

Purification

<1> (recombinant mutant from E. coli [20]; recombinant wild-type from E. coli [13,20]; partial [10]) [1, 10, 13, 20]
<3> [3]
<5> (partial [7]) [6, 7]
<6> [3]
<9> (partial, recombinant from E. coli [10]) [10]
<10> (partial, recombinant from E. coli [10]) [10]

Cloning

<1> (expression of exchange mutants in Escherichia coli [20]; overexpression in Escherichia coli [10,13,20]) [10, 13, 20]
<9> (expression in Escherichia coli [10]) [10]
<10> (expression in Saccharomyces cerevisiae mutant line erg6, which is deleted in the SMT1 gene, complementation of the yeast [18]; expression in Escherichia coli [10]) [10, 18]
<11> (expression in Saccharomyces cerevisiae mutant line erg6, which is deleted in the C-24-sterol methyltransferase, via Escherichia coli, complementation of the yeast, amino acid sequence comparison [14]) [14]
<12> (expression in Escherichia coli, His-tagged protein [15]) [15]
<13> (expression as FLAG-fusion protein in Escherichia coli, amino acid sequence analysis [19]) [19]

Engineering

D65L <1> (<1> potentially responsible for substrate binding to leucine by site directed mutagenesis, unchanged activity [20]) [20]
D79L <1> (<1> potentially responsible for substrate binding to leucine by site directed mutagenesis, not decreased activity, 1 new activity to form 24-methylzymosterol from zymosterol [20]) [20]
E64L <1> (<1> potentially responsible for substrate binding to leucine by site directed mutagenesis, unchanged activity [20]) [20]
E68L <1> (<1> potentially responsible for substrate binding to leucine by site directed mutagenesis, no activity [20]) [20]
E82L <1> (<1> potentially responsible for substrate binding to leucine by site directed mutagenesis, not decreased activity, 1 new activity to form 24-methylzymosterol from zymosterol [20]) [20]
E98L <1> (<1> potentially responsible for substrate binding to leucine by site directed mutagenesis, unchanged activity [20]) [20]
Additional information <7, 10> (<10> overexpression of SMT2 gene in a mutant strain leads to an accumulation of sitosterol and to an altered phenotype, some of the changes can be restored by brassinosteroid treatment, some cannot, sterol profile of wild-type and mutant plants [21]; <10> construction of SMT1-deficient mutants with altered phenotype, reduced fertility and root growth defects, transgenic plant seedlings are sensitive against brassinosteroids and transfection with a genomic clone or a SMT1 cDNA can complement the mutants [18]; <7> natural amphotericine B-resistant mutant strain [9]) [9, 18, 21]

Application

agriculture <1-3, 8-10> (<1-3,8-10> construction of transgenic plants with modified sterol profile via recombinant sterol 24-C-methyltransferase shall protect crops against damage by insect infestation [10]) [10]
medicine <12> (<12> target for drug development against pneumonitis caused by this pathogen based on its unique substrate preference for lanosterol and 24-methylenelanosterol [15]) [15]
synthesis <8> (<8> rational drug target design based on structure-function-relation [17]) [17]

6 Stability

General stability information

<3>, no stabilization by $MgCl_2$, $CaCl_2$, DTT, reduced glutathione, phosphatidylcholine [3]

Storage stability

<1>, 4°C, 0.1 M Tris-HCl buffer, 1 mM Mg^{2+}, 90% loss of activity overnight [1]
<3>, -70°C, 1 month, less than 10% loss of activity [3]

References

[1] Moore, J.T.; Gaylor, J.L.: Isolation and purification of an S-adenosylmethionine: Δ^{24}-sterol methyltransferase from yeast. J. Biol. Chem., **244**, 6334-6340 (1969)
[2] Nes, W.D.; Janssen, G.G.; Bergenstrahle, A.: Structural requirements for transformation of substrates by the (S)-adenosyl-L-methionine:$\Delta^{24(25)}$-sterol methyl transferase. J. Biol. Chem., **266**, 15202-15212 (1991)
[3] Ator, M.A.; Schmidt, S.J.; Adams, J.L.; Dolle, R.E.: Mechanism and inhibition of Δ^{24}-sterol methyltransferase from Candida albicans and Candida tropicalis. Biochemistry, **28**, 9633-9640 (1989)
[4] Oelschlager, A.C.; Angus, R.H.; Pierce, A.M.; Pierce, H.D.; Srinivasan, R.: Azasterol inhibition of Δ^{24}-sterol methyltransferase in Saccharomyces cerevisiae. Biochemistry, **23**, 3582-3589 (1984)
[5] Rahier, A.; Genot, J.C.; Schuber, F.; Benveniste, P.; Narula, A.S.: Inhibition of S-adenosyl-L-methionine sterol-C-24-methyltransferase by analogues of a carbocationic ion high-energy intermediate. Structure activity relationships for C-25 heteroatoms (N, As, S) substituted triterpenoid derivatives. J. Biol. Chem., **259**, 15215-15223 (1984)
[6] Bansal, S.K.; Knoche, H.W.: Sterol methyltransferase from Uromyces phaseoli: an investigation of the first and the second transmethylation reactions. Phytochemistry, **20**, 1269-1277 (1981)
[7] Lin, H.K.; Knoche, H.W.: A sterol methyltransferase from bean rust uredospores Uromyces phaseoli. Phytochemistry, **15**, 683-687 (1976)

[8] Bailey, R.B.; Thompson, E.D.; Parks, L.W.: Kinetic properties of S-adenosyl-L-methionine:Δ^{24}-sterol methyltransferase enzyme(s) in mitochondrial structures of Saccharomyces cerevisiae. Biochim. Biophys. Acta, **334**, 127-136 (1974)
[9] Mukhtar, H.; Hakkou, A.; Bonaly, R.: Studies on the activity of Kluyveromyces lactis S-adenosylmethionine:Δ^{24}-sterol methyltransferase in presence of polyenic antifungal agents. Mycopathologia, **126**, 75-83 (1994)
[10] Nes, W.D.: Sterol methyl transferase: enzymology and inhibition. Biochim. Biophys. Acta, **1529**, 63-88 (2000)
[11] Venkatramesh, M.; Guo, D.A..; Jia, Z.; Nes, W.D.: Mechanism and structural requirements for transformation of substrates by the (S)-adenosyl-L-methionine:$\Delta^{24(25)}$-sterol methyl transferase from Saccharomyces cerevisiae. Biochim. Biophys. Acta, **1299**, 313-324 (1996)
[12] Nes, W.D.; Guo, D.a.; Zhou, W.: Substrate-based inhibitors of the (S)-adenosyl-L-methionine: $\Delta^{24(25)}$- to $\Delta^{24(28)}$-sterol methyltransferase from Saccharomyces cerevisiae. Arch. Biochem. Biophys., **342**, 68-81 (1997)
[13] Nes, W.D.; McCourt, B.S.; Zhou, W.X.; Ma, J.; Marshall, J.A.; Peek, L.A.; Brennan, M.: Overexpression, purification, and stereochemical studies of the recombinant (S)-adenosyl-L-methionine: $\Delta^{24(25)}$-to $\Delta^{24(28)}$-sterol methyl transferase enzyme from Saccharomyces cerevisiae. Arch. Biochem. Biophys., **353**, 297-311 (1998)
[14] Grebenok, R.J.; Galbraith, D.W.; Penna, D.D.: Characterization of Zea mays endosperm C-24 sterol methyltransferase: one of two types of sterol methyltransferase in higher plants. Plant Mol. Biol., **34**, 891-896 (1997)
[15] Kaneshiro, E.S.; Rosenfeld, J.A.; Basselin-Eiweida, M.; Stringer, J.R.; Keely, S.P.; Smulian, A.G.; Giner, J.L.: The Pneumocystis carinii drug target s-adenosyl-L-methionine:Sterol C-24 methyl transferase has a unique substrate preference. Mol. Microbiol., **44**, 989-999 (2002)
[16] Acuna-Johnson, A.P.; Oehlschlager, A.C.; Pierce, A.M.; Pierce, H.D., Jr.; Czyzewska, E.K.: Stereochemistry of yeast Δ^{24}-sterol methyl transferase. Bioorg. Med. Chem., **5**, 821-832 (1997)
[17] Mangla, A.T.; Nes, W.D.: Sterol C-methyl transferase from Prototheca wickerhamii mechanism, sterol specificity and inhibition. Bioorg. Med. Chem., **8**, 925-936 (2000)
[18] Diener, A.C.; Li, H.; Zhou, W.X.; Whoriskey, W.J.; Nes, W.D.; Fink, G.R.: Sterol methyltransferase 1 controls the level of cholesterol in plants. Plant Cell, **12**, 853-870 (2000)
[19] Shi, J.; Gonzales, R.A.; Bhattacharyya, M.K.: Identification and characterization of an S-adenosyl-L-methionine:Δ^{24}-sterol-C-methyltransferase cDNA from soybean. J. Biol. Chem., **271**, 9384-9389 (1996)
[20] Nes, W.D.; Marshall, J.A.; Jia, Z.; Jaradat, T.T.; Song, Z.; Jayasimha, P.: Active site mapping and substrate channeling in the sterol methyltransferase pathway. J. Biol. Chem., **277**, 42549-42556 (2002)
[21] Schaeffer, A.; Bronner, R.; Benveniste, P.; Schaller, H.: The ratio of campesterol to sitosterol that modulates growth in Arabidopsis is controlled by sterol methyltransferase 2;1. Plant J., **25**, 605-615 (2001)

Luteolin O-methyltransferase 2.1.1.42

1 Nomenclature

EC number
2.1.1.42

Systematic name
S-adenosyl-L-methionine:5,7,3',4'-tetrahydroxyflavone 3'-O-methyltransferase

Recommended name
luteolin O-methyltransferase

Synonyms
S-adenosylmethionine:flavone/flavonol 3'-O-methyltransferase
luteolin 3'-O-methyltransferase
luteolin methyltransferase
o-dihydric phenol methyltransferase
o-dihydric phenol *meta*-O-methyltransferase
o-diphenol *m*-O-methyltransferase

CAS registry number
37205-55-3

2 Source Organism

<1> *Glycine max* (soybean, L. var. Mandarin [3]) [3]
<2> *Nicotiana tabacum* (tobacco, Samsun N N. [4]) [4]
<3> *Petroselinum hortense* (parsley [1,3]; Hoffm. [2]) [1-3]

3 Reaction and Specificity

Catalyzed reaction
S-adenosyl-L-methionine + 5,7,3',4'-tetrahydroxyflavone = S-adenosyl-L-homocysteine + 5,7,4'-trihydroxy-3'-methoxyflavone (also acts on luteolin 7-O-β-D-glucoside)

Reaction type
methyl group transfer

Natural substrates and products

S S-adenosyl-L-methionine + 5,7,3',4'-tetrahydroxyflavone <3> (<3> luteolin, enzyme directly related to flavone glycoside biosynthesis, biosynthesis of 3'-methoxyflavonoids [1]) (Reversibility: ? <3> [1]) [1]

P S-adenosyl-L-homocysteine + 5,7,4'-trihydroxy-3'-methoxyflavone <3> (<3> chrysoeriol [1]) [1]

Substrates and products

S S-adenosyl-L-methionine + 3,3',4',5,7-pentahydroxy flavone <1, 2> (<1,2> quercetin [2-4]) (Reversibility: ? <1, 2> [2-4]) [2-4]

P S-adenosyl-L-homocysteine + isorhamnetin

S S-adenosyl-L-methionine + 5,7,3',4'-tetrahydroxyflavanone <1, 3> (<1,3> eriodictyol [1,3]) (Reversibility: ? <1, 3> [1, 3]) [1, 3]

P S-adenosyl-L-homocysteine + 5,7,4'-trihydroxy-3-methoxyflavanone

S S-adenosyl-L-methionine + 5,7,3',4'-tetrahydroxyflavone <1, 3> (<1,3> luteolin [1,3]) (Reversibility: ? <1, 3> [1, 3]) [1, 3]

P S-adenosyl-L-homocysteine + 5,7,4'-trihydroxy-3'-methoxyflavone <3> (<3> chrysoeriol [1]) [1]

S S-adenosyl-L-methionine + 5,7,3'-trihydroxyflavone 7-O-glucoside <1, 3> (<3> luteolin 7-O-glucoside [1]) (Reversibility: ? <1, 3> [1, 3]) [1, 3]

P S-adenosyl-L-homocysteine + 5,7,4'-trihydroxy-3'-methoxyflavone 7-O-glucoside <3> (<1,3> chrysoeriol-7-O-glucoside [1,3]) [1]

S S-adenosyl-L-methionine + 5-hydroxy-3-methoxy-4-hydroxycinnamic acid <1, 2> (<1,2> 5-hydroxy ferulic acid [3,4]) (Reversibility: ? <1, 2> [3, 4]) [3, 4]

P S-adenosyl-L-homocysteine + ?

S S-adenosyl-L-methionine + 6,7-dihydroxy-3'-methoxyisoflavone <1, 3> (<1,3> texasin [3]) (Reversibility: ? <1, 3> [3]) [3]

P S-adenosyl-L-homocysteine + ?

S S-adenosyl-L-methionine + caffeic acid <1-3> (Reversibility: ? <1-3> [1, 3, 4]) [1, 3, 4]

P S-adenosyl-L-homocysteine + 3-methoxy-4-hydroxycinnamic acid <3> (<3> ferulic acid [1]) [1]

S S-adenosyl-L-methionine + dihydro-3,3',4',5,7-pentahydroxy flavone <1> (<1,2> dihydroquercetin [2-4]) (Reversibility: ? <1> [3]) [3]

P S-adenosyl-L-homocysteine + dihydroisorhamnetin

S S-adenosyl-L-methionine + quercetin 3-O-glucoside <1> (<1> poor substrate [3]) (Reversibility: ? <1> [3]) [3]

P S-adenosyl-L-homocysteine + ?

S S-adenosyl-L-methionine + rutin <1> (<1> poor substrate [3]) (Reversibility: ? <1> [3]) [3]

P S-adenosyl-L-homocysteine + ?

S Additional information <1-3> (<3> only *ortho*-dihydric phenols can function as substrates and only the *meta*-hydroxyl group is methylated, no reaction takes place if one of the two phenolic hydroxyls is methylated or if only one hydroxyl group is present [1]; <3> isomeric substances isoferulic acid and diosmetin, chrysoeriol, apigenin *p*-cumaric acid and ferulic acid are not substrates [1]; <1> methylation proceedes only in the *meta*-position, enzyme is unable to catalyze the methylation of *p*-coumaric acid, *m*-coumaric acid, ferulic acid, isoferulic acid, sinapic acid, apigenin and naringenin [3]; <1> isoflavones biochanin A and daidzein do not serve as substrates [3]; <2> rutin is no substrate [4]) [1, 3, 4]

P ?

Inhibitors

EDTA <1, 3> [1, 3]
Mn^{2+} <3> [1]
S-adenosyl-L-homocysteine <1> (<1> inhibits the methylation of luteolin by S-adenosylmethionine [3]) [3]

Activating compounds

Additional information <3> (<3> extractable activity drastically increases by illuminating cells with white light [1]) [1]

Metals, ions

Mg^{2+} <1, 3> (<1,3> required for activity [1,3]) [1, 3]

Specific activity (U/mg)

0.000027 <3> [1]
0.0029 <1> [3]

K_m-Value (mM)

0.011 <1> (S-adenosyl-L-methionine, <1> with luteolin as substrate [3]) [3]
0.016 <1> (5,7,3',4'-tetrahydroxyflavone) [3]
0.028 <1> (5,7,3'-trihydroxyflavone 7-O-glucoside) [3]
0.031 <3> (5,7,3'-trihydroxyflavone 7-O-glucoside) [1]
0.035 <1> (6,7-dihydroxy-3'-methoxyisoflavone) [3]
0.035 <3> (quercetin) [1]
0.046 <3> (5,7,3'-trihydroxyflavone) [1]
0.075 <1> (5,7,3',4-tetrahydroxyflavanone) [3]
0.15 <3> (S-adenosyl-L-methionine, <3> with luteolin as substrate [1]) [1]
0.227 <1> (5-hydroxy-3-methoxy-4-hydroxycinnamic acid) [3]
0.435 <1> (dihydroquercetin) [3]
0.77 <1> (caffeic acid) [3]
1.2 <3> (5,7,3',4-tetrahydroxyflavanone) [1]
1.6 <3> (caffeic acid) [1]

pH-Optimum

8.6-8.9 <1> (<1> methylation of caffeic acid and quercetin [3]) [3]
9.6-9.8 <3> (<3> glycine-NaOH buffer with caffeic acid and luteolin as substrates [1]) [1]
9.7 <3> [1]

pH-Range

6-10 <1> [3]

4 Enzyme Structure

Molecular weight

48000 <3> (<3> gel filtration [1]) [1]

5 Isolation/Preparation/Mutation/Application

Source/tissue

cell suspension culture <1> [2, 3]
leaf <2> [4]

Purification

<1> [3]
<3> (partial [1]) [1]

6 Stability

pH-Stability

9.8 <3> (<3> at higher pH values activity rapidly declines [1]) [1]

Storage stability

<1>, -20°C, stored in 10% glycerol for 2 months, no significant loss in enzyme activity [3]
<1>, 4°C, highly unstable, stored with Triton X-100, bovine serum albumin, ethylene glycol, dimethyl sulfoxide, 2-mercaptoethanol, dithiothreitol or ethylene glycol monoethylether, loses over 90% of its activity within 94 h [3]
<3>, -20°C, 1-3 mg protein/ml in 0.02 M Tris-HCl buffer loses about 30% of its activity after 1 month [1]
<3>, 4°C, acetone powder of fresh cells as source of the enzyme can be kept without significant loss of enzyme activity [1]

References

[1] Ebel, J.; Hahlbrock, K.; Grisebach, H.: Purification and properties of an *o*-dihydricphenol *meta*-O-methyltransferase from cell suspension cultures of parsley and its relation to flavonoid biosynthesis. Biochim. Biophys. Acta, **269**, 313-326 (1972)
[2] Ebel, J.; Hahlbrock, K.: Enzymes of flavone and flavonol-glycoside biosynthesis. Coordinated and selective induction in cell-suspension cultures of Petroselinum hortense. Eur. J. Biochem., **75**, 201-209 (1977)
[3] Poulton, J.E.; Hahlbrock, K.; Grisebach, H.: O-Methylation of flavonoid substrates by a partially purified enzyme from soybean cell suspension cultures. Arch. Biochem. Biophys., **180**, 543-549 (1977)
[4] Legrand, M.; Fritig, B.; Hirth, L.: O-Diphenol O-methyltransferases of healthy and tobacco-mosaic-virus-infected hypersensitive tobacco. Planta, **144**, 101-108 (1978)

Histone-lysine N-methyltransferase 2.1.1.43

1 Nomenclature

EC number

2.1.1.43

Systematic name

S-adenosyl-L-methionine:histone-L-lysine N^6-methyltransferase

Recommended name

histone-lysine N-methyltransferase

Synonyms

histone H1-specific S-adenosylmethionine:protein-lysine N-methyltransferase
histone lysine (K) methyltransferase <1, 2, 7-11> [6]
methyltransferase, protein (lysine)
protein (lysine) methyltransferase
protein methylase 3
protein methylase III
protein methyltransferase II
protein-lysine N-methyltransferase

CAS registry number

9055-08-7

2 Source Organism

<1> *Saccharomyces cerevisiae* (similar enzyme with low activity towards calf thymus histone but high activity towards ribosomal protein [3]) [3, 6]
<2> *Homo sapiens* (SET9, also called SET7 [9,10]) [4, 6, 9, 10, 11]
<3> *Bos taurus* (calf [1]) [1]
<4> *Euglena gracilis* (2 enzyme forms: VA and VB [2]) [2]
<5> *Physarum polycephalum* [5]
<6> *Rattus norvegicus* [1]
<7> *Schizosaccharomyces pombe* [6]
<8> *Neurospora crassa* (DIM-5 protein [10]) [6, 10]
<9> *Arabidopsis thaliana* [6]
<10> *Drosophila melanogaster* (SET8 sequence [7]) [6, 7, 11]
<11> *Caenorhabditis elegans* [6]
<12> *Mus musculus* (G9a histone methyltransferase [8]) [6, 8, 11]
<13> *Homo sapiens* (HeLa cells [7]; SET8 sequence [7]) [7]
<14> *Drosophila elegans* [7]
<15> *Pisum sativum* [9, 10]

3 Reaction and Specificity

Catalyzed reaction

S-adenosyl-L-methionine + histone L-lysine = S-adenosyl-L-homocysteine + histone N^6-methyl-L-lysine (<2, 8, 15> SET structure [10]; <2, 15> catalytic mechanism [9]; <15> substrate and inhibitor binding site [9]; <8, 12> H3-Lys9 methylation is catalyzed by the Suv39h family proteins [8, 10, 11]; <2, 13-15> amino acid sequence comparison [7,9]; <1, 2, 7, 8, 10, 13-15> enzyme is encoded on genes containing the SET-domain sequence, substrate specificity is connected to SET sequence [6, 7, 9, 10]; <1> formation of ε-N-dimethyllysine in a 2-step process via ε-N-monomethyl-lysine [3])

Reaction type

methyl group transfer

Natural substrates and products

S S-adenosyl-L-methionine + histone L-lysine <1-15> (<12> Lys9 of histone 3 [8]; <13,14> Lys20 of histone 4, SET8 [7]) (Reversibility: ? <1-15> [1-11]) [1-11]

P S-adenosyl-L-homocysteine + histone N^6-methyl-L-lysine <1-15> [1-11]

S Additional information <2, 10, 12-15> (<2,10,12> deregulation of SET-domain function has an important role in carcinogenesis, overview, effect in knock-out mice [11]; <10,12> methylation of H3 Lys9 correlates with transcription repression, H3 Lys4 correlates with activation [11]; <12-15> histone methylation has significant effects on heterochromatin formation and transcriptional regulation [7-9]; <10> H3-Lys9-methylation is an initiating and a maintenance modification for the X-chromosome silencing, in opposite to H3-Lys4-methylation, by a putative RNA/Xi-specific enzyme [6]) [6-9, 11]

P ?

Substrates and products

S S-adenosyl-L-methionine + histone L-lysine <1-15> (<1,2,7-13> high substrate and methylation site specifity [6,7]; <10,13> Lys20 of histone 4 [7]; <2> Lys20 of histone 4 [6]; <7> Lys36 of histone 3 [6]; <2> Lys27 of histone 3 [6]; <1,2,7,11> Lys4 of histone 3 [6]; <2,7-10,12> Lys9 of histone H3 [6,8]; <3,6> various types of histones [1]; <5> lysine-rich histones [5]; <3,5,6> arginine-rich histones [1,5]; <3,6> heat-denatured histone [1]; <4> highly specific towards histone H1 [2]) (Reversibility: ? <1-15> [1-11]) [1-11]

P S-adenosyl-L-homocysteine + histone N^6-methyl-L-lysine <1-15> (<4,5,10,12> a mixture of ε-N-mono-, di- and trimethyllysine ratio 4:1:1 [2,5,11]; <4> the 2 enzyme forms VA and VB methylate different sites in the histone H1 molecule with high specificity, VB is more specific than VA [2]) [1-11]

S polylysine + S-adenosylmethionine <3> (<5> not [5]) (Reversibility: ? <3> [1]) [1]

P S-adenosyl-L-homocysteine + N^6-methylpolylysine

S Additional information <1, 4, 5> (<4> low activity with myosin [2]; <5> not: lysine, ribonuclease A, cytochrome c, bovine serum albumin [5]; <1> a similar enzyme from Saccharomyces cerevisiae has low activity towards calf thymus histone but high activity towards ribosomal protein [3]) [2, 3, 5]

P ?

Inhibitors

Cu^{2+} <3> [1]
DNA <3> [1]
EDTA <5> [5]
Hg^{2+} <3> [1]
RNA <3> [1]
S-adenosyl-L-cysteine <4> [2]
S-adenosyl-L-ethionine <4> [2]
S-adenosyl-L-homocysteine <1, 4, 5, 15> [2, 3, 5, 9]
Zn^{2+} <3> [1]
sinefungin <4> [2]

Activating compounds

sulfhydryl reducing agent <5> (<5> required for optimal activity [5]) [5]

Specific activity (U/mg)

0.000285 <4> (<4> isozyme VA [2]) [2]
0.00202 <4> (<4> isozyme VB [2]) [2]
0.0027 <3> [1]
Additional information <1, 4> [2, 3]

K_m-Value (mM)

0.00031 <4> (histone H1, <4> isozyme VA [2]) [2]
0.00044 <4> (histone H1, <4> isozyme VB [2]) [2]
0.0017 <1> (S-adenosyl-L-methionine) [3]
0.003 <3> (S-adenosyl-L-methionine) [1]
0.027 <4> (S-adenosyl-L-methionine, <4> isozyme VA [2]) [2]
0.0345 <4> (S-adenosyl-L-methionine, <4> isozyme VB [2]) [2]
Additional information <5> [5]

K_i-Value (mM)

0.00148 <4> (S-adenosyl-L-homocysteine, <4> isozyme VA [2]) [2]
0.0015 <4> (sinefungin, <4> isozyme VB [2]) [2]
0.0016 <4> (S-adenosyl-L-homocysteine, <4> isozyme VB [2]) [2]
0.0019 <4> (sinefungin, <4> isozyme VA [2]) [2]
0.04 <4> (S-adenosyl-L-ethionine, <4> isozyme VA [2]) [2]
0.044 <4> (S-adenosyl-L-ethionine, <4> isozyme VB [2]) [2]
0.071 <1> (S-adenosyl-L-homocysteine) [3]

pH-Optimum

7.5 <1> [3]
8 <5> [5]
8.5-9 <3> [1]
9 <3, 4> (<3> assay at [1]) [1, 2]

pH-Range

5.5-9.5 <1> (<1> at pH 5.5 and pH 9.5 about 50% of activity maximum [3]) [3]
7.5-9.9 <3> (<3> about 50% of activity maximum at pH 7.5 and 9.9 [1]) [1]

Temperature optimum (°C)

30 <15> (<15> assay at [9]) [9]
37 <2, 3, 6> (<2,3,6> assay at [1,9]) [1, 9]

4 Enzyme Structure

Molecular weight

34000 <4> (<4> isozyme VB, gel filtration [2]) [2]
45000 <13, 14> (<13,14> SDS-PAGE [7]) [7]
55000 <4> (<4> isozyme VA, gel filtration [2]) [2]
82000 <1> (<1> gel filtration [3]) [3]

Subunits

monomer <13, 14> (<13,14> 1 * 45000, SDS-PAGE [7]) [7]

5 Isolation/Preparation/Mutation/Application

Source/tissue

brain <6> [1]
cerebrospinal fluid <2> [4]
embryo <12> [8]
heart <6> [1]
kidney <6> [1]
liver <6> [1]
lung <6> [1]
plasmodium <5> [5]
spleen <6> [1]
testis <6> [1]
thymus <3, 12> [1, 8]
Additional information <3, 6> (<3,6> found in all rat organs tested [1]) [1]

Localization

cytosol <5> (<5> subcellular distribution is not definitely determined [5]) [5]
microsome <5> (<5> subcellular distribution is not definitely determined [5]) [5]
mitochondrion <5> (<5> subcellular distribution is not definitely determined [5]) [5]
nucleus <3, 5, 12, 13> (<5> subcellular distribution is not definitely determined [5]) [1, 5, 7, 8]

Purification

<1> (partial, similar enzyme with low activity towards calf thymus histone but high activity towards ribosomal protein [3]) [3]
<3> (partial [1]) [1]
<4> (2 isozymes [2]) [2]
<5> (partial [5]) [5]
<13> (partial [7]) [7]
<15> (recombinant from E. coli [9]) [9]

Crystallization

<15> (complexed with S-adenosyl-L-homocysteine, hanging drop method, 25°C, reservoir solution contains HEPES 100 mM, pH 6.8, sodium acetate 1.2-1.35 M, structure analysis [9]) [9]

Cloning

<2> (expression of wild-type SET9 and mutants in Escherichia coli, His-tagged [9]) [9]
<13> (glutathione-S-transferase-fusion protein, SET gene [7]) [7]
<14> (glutathione-S-transferase-fusion protein, SET gene [7]) [7]
<15> (overexpression in Escherichia coli BL21, His-tagged protein [9]) [9]

Engineering

H299A <14> (<14> overlap PCR, destroyed SET gene, no activity remaining [7]) [7]
Additional information <12> (<12> construction of several SET domain family deficient or knock-out mutants, analysis of effect on carcinogenesis [11]; <12> G9a deficient embryos, drastically decreased activity in vivo [8]) [8, 11]

6 Stability

pH-Stability

5 <1> (<1> enzyme becomes irreversibly inactivated at pH lower than 5.0 [3]) [3]

Temperature stability

45 <5> (<5> 5 min, 50% loss of activity without addition of histone, no loss of activity with addition of histone, 20 min, 72% loss of activity without addition of histone, 65% loss of activity with addition of histone [5]) [5]
60 <5> (<5> 3 min, 97% loss of activity [5]) [5]

General stability information

<5>, histones protect against heat inactivation at 45°C, no effect at 60°C [5]

References

[1] Paik, W.K.; Kim, S.: Solubilization and partial purification of protein methylase 3 from calf thymus nuclei. J. Biol. Chem., **245**, 6010-6015 (1970)

[2] Tuck, M.T.; Farooqui, J.Z.; Paik, W.K.: Two histone H1-specific protein-lysine N-methyltransferases from Euglena gracilis. Purification and characterization. J. Biol. Chem., **260**, 7114-7121 (1985)
[3] Lobet, Y.; Lhoest, J.; Colson, C.: Partial purification and characterization of the specific protein-lysine N-methyltransferase of YL32, a yeast ribosomal protein. Biochim. Biophys. Acta, **997**, 224-231 (1989)
[4] Park, J.; Greenstein, J.I.; Paik, W.K.; Kim, S.: Studies on protein methyltransferase in human cerebrospinal fluid. J. Mol. Neurosci., **1**, 151-157 (1989)
[5] Venkatesan, M.; McManus, I.R.: Partial purification and characterization of a protein lysine methyltransferase from plasmodia of Physarum polycephalum. Biochemistry, **18**, 5365-5371 (1979)
[6] Lachner, M.; Jenuwein, T.: The many faces of histone lysine methylation. Curr. Opin. Cell Biol., **14**, 286-298 (2002)
[7] Fang, J.; Feng, Q.; Ketel, C.S.; Wang, H.; Cao, R.; Xia, L.; Erdjument-Bromage, H.; Tempst, P.; Simon, J.A.; Zhang, Y.: Purification and functional characterization of SET8, a nucleosomal histone H4-lysine 20-specific methyltransferase. Curr. Biol., **12**, 1086-1099 (2002)
[8] Tachibana, M.; Sugimoto, K.; Nozaki, M.; Ueda, J.; Ohta, T.; Ohki, M.; Fukuda, M.; Takeda, N.; Niida, H.; Kato, H.; Shinkai, Y.: G9a histone methyltransferase plays a dominant role in euchromatic histone H3 lysine 9 methylation and is essential for early embryogenesis. Genes Dev., **16**, 1779-1791 (2002)
[9] Trievel, R.C.; Beach, B.M.; Dirk, L.M.A.; Houtz, R.L.; Hurley, J.H.: Structure and catalytic mechanism of a SET domain protein methyltransferase. Cell, **111**, 91-103 (2002)
[10] Yeates, T.O.: Structures of SET domain proteins: protein lysine methyltransferases make their mark. Cell, **111**, 5-7 (2002)
[11] Schneider, R.; Bannister, A.J.; Kouzarides, T.: Unsafe SETs: histone lysine methyltransferases and cancer. Trends Biochem. Sci., **27**, 396-402 (2002)

Dimethylhistidine N-methyltransferase 2.1.1.44

1 Nomenclature

EC number
2.1.1.44

Systematic name
S-adenosyl-L-methionine:N^{α},N^{α}-dimethyl-L-histidine N^{α}-methyltransferase

Recommended name
dimethylhistidine N-methyltransferase

Synonyms
dimethylhistidine methyltransferase
histidine-α-N-methyltransferase
methyltransferase, histidine N^{α}-

CAS registry number
62213-53-0

2 Source Organism

<1> *Neurospora crassa* [1]

3 Reaction and Specificity

Catalyzed reaction
S-adenosyl-L-methionine + N^{α},N^{α}-dimethyl-L-histidine = S-adenosyl-L-homocysteine + N^{α},N^{α},N^{α}-trimethyl-L-histidine

Reaction type
methyl group transfer

Natural substrates and products
S S-adenosyl-L-methionine + N^{α},N^{α}-dimethyl-L-histidine <1> (Reversibility: ? <1> [1]) [1]
P S-adenosyl-L-homocysteine + N^{α},N^{α},N^{α}-trimethyl-L-histidine <1> [1]
S S-adenosyl-L-methionine + histidine <1> (Reversibility: ? <1> [1]) [1]
P S-adenosyl-L-homocysteine + N^{α}-methylhistidine
S S-adenosyl-L-methionine + methylhistidine <1> (Reversibility: ? <1> [1]) [1]
P S-adenosyl-L-homocysteine + N^{α},N^{α}-dimethyl-L-histidine

Substrates and products

- **S** S-adenosyl-L-methionine + N^{α},N^{α}-dimethyl-L-histidine <1> (Reversibility: ? <1> [1]) [1]
- **P** S-adenosyl-L-homocysteine + N^{α},N^{α},N^{α}-trimethyl-L-histidine <1> [1]
- **S** S-adenosyl-L-methionine + N^{α}-methylhistidine <1> (Reversibility: ? <1> [1]) [1]
- **P** S-adenosyl-L-homocysteine + N^{α},N^{α}-dimethyl-L-histidine
- **S** S-adenosyl-L-methionine + histidine <1> (Reversibility: ? <1> [1]) [1]
- **P** S-adenosyl-L-homocysteine + N^{α}-methylhistidine
- **S** Additional information <1> (<1> conversion of histidine to hercynine is mediated either by one enzyme or by two or three enzymes with chemical and physical properties so similar that common methods of protein fractionation fail to effect any separation [1]) [1]
- **P** ?

Inhibitors

Cd^{2+} <1> [1]
Co^{2+} <1> [1]
Cu^{2+} <1> [1]
Fe^{2+} <1> [1]
Fe^{3+} <1> [1]
Hg^{2+} <1> [1]
Ni^{2+} <1> [1]
S-adenosyl-L-homocysteine <1> [1]
Sn^{2+} <1> [1]
Zn^{2+} <1> [1]
hercynine <1> [1]
histidine <1> (<1> inhibits reaction with dimethylhistidine [1]) [1]
imidazolepropionate <1> [1]
methylhistidine <1> (<1> inhibits reaction with dimethylhistidine [1]) [1]

Specific activity (U/mg)

0.0434 <1> [1]

pH-Optimum

8-8.5 <1> (<1> histidine [1]) [1]
8.5-9 <1> (<1> dimethylhistidine [1]) [1]
9-9.5 <1> (<1> methylhistidine [1]) [1]

4 Enzyme Structure

Molecular weight

270000-300000 <1> (<1> gel filtration [1]) [1]

5 Isolation/Preparation/Mutation/Application

Purification

<1> [1]

References

[1] Ishikawa, Y.; Melville, D.B.: The enzymatic α-N-methylation of histidine. J. Biol. Chem., **245**, 5967-5973 (1970)

Thymidylate synthase 2.1.1.45

1 Nomenclature

EC number
2.1.1.45

Systematic name
5,10-methylenetetrahydrofolate:dUMP C-methyltransferase

Recommended name
thymidylate synthase

Synonyms
TMP synthetase
TS <19> [41]
dTMP synthase
methylenetetrahydrofolate:dUMP C-methyltransferase
thymidylate synthetase

CAS registry number
9031-61-2

2 Source Organism

<1> *Abies sp.* (pine [36]) [36]
<2> *Aedes aegypti* [48]
<3> *Angiostrongylus cantonensis* (metastrongyloid nematode, synonym Parastrongylus [48]) [48]
<4> *Bacillus subtilis* [32, 58, 62, 65, 69]
<5> *bacteriophage T2* [7, 13, 24, 25, 32]
<6> *bacteriophage T4* [10, 13, 32, 47, 52, 53, 55, 59]
<7> *bacteriophage T5* [10]
<8> *bacteriophage T6* [7, 13]
<9> *Bos taurus* (calf [3,7,13,17,25,27,36,46,48]) [3, 7, 13, 17, 25, 27, 36, 46, 48]
<10> *Brugia pahangi* [48]
<11> *Candida albicans* [47, 48, 59]
<12> *Chlamydomonas reinhardtii* [36]
<13> *Crithidia fasciculata* (host Anopheles strain ATCC 11,745 [18]) [18, 33, 46, 48, 57]
<14> *Crithidia oncopelti* [18]
<15> *Cryptococcus neoformans* (encapsulated yeast-like fungus, GenBank accession No. [47]) [47, 63]

<16> *Daucus carota* (carrot, wild and culture form [36,38]) [36, 38]
<17> *Diplococcus pneumoniae* [16, 19, 24, 33]
<18> *Drosophila melanogaster* [11]
<19> *Escherichia coli* (strain B [1, 9, 15-17, 19, 20, 24, 27, 32]; strain K12 [24, 32, 33, 46]; strain EM 20031 K12 [24]; strain R2 [20]) [1-3, 7, 9, 13, 15-20, 24-27, 32, 33, 41, 46, 49, 50, 52, 53, 57, 59-63, 65-67, 69, 70, 73-76]
<20> *Gallus gallus* (chicken [3, 7, 28, 31, 48]) [3, 7, 16, 28, 31, 48]
<21> *Haemophilus influenzae* [58]
<22> *Homo sapiens* (human [2, 19, 20, 25, 33, 39, 44, 47, 48, 66, 70, 74, 77]; neoplastic leukemic CCRF-CEM cells [25, 37]; MCF-7 breast cancer cells [39, 42, 52]; AML cells [39]; HeLa cells [26, 42]; human colorectal tumor cell line C [73]; human colon tumor cell line HCT116/200-10 [67]; human sarcome HT1080 cells [63]) [2, 3, 8, 19, 20, 25, 26, 28, 31, 33, 37, 39, 42, 44, 47, 48, 51, 52, 57, 59, 63, 66, 67, 70, 72-74, 77]
<23> *Lactobacillus casei* (var. rhamnosus, ATCC 7469 [21]; strain TS N229D, TS N229C [54]; thymidylate synthase chemically synthesized based on the amino acid sequence of TS from Lactobacillus casei [53]) [3-5, 7-10, 12-14, 16, 19-29, 31-33, 37, 40, 41, 43-45, 47, 49, 50, 53-56, 59, 60, 62, 63, 67, 71, 73-75]
<24> *Lactococcus lactis* [62]
<25> *Leishmania major* (formerly called Leishmania tropica [57,61]) [33, 36, 48, 57, 61]
<26> *Mus musculus* (mouse, sarcoma 180 [17, 27]; Gardner lymphosarcoma [17, 27]; Ehrlich ascites carcinoma cells [3-5, 7, 9, 12, 13, 16, 19, 26, 28, 34, 37, 42, 46, 48]; L1210 cells [26, 30, 31, 35, 44, 74]; albino Swiss mice [3]; L5178 murine leukemia cell line [71]) [3-5, 7, 9, 12, 13, 17, 19, 26-28, 30, 31, 34-37, 42, 44, 46-48, 57, 59, 70, 71, 74]
<27> *Neisseria gonorrhoeae* (strain MS11 [58]) [58]
<28> *Nicotiana tabacum* (tobacco [36]) [36]
<29> *phage Phi3T* [62]
<30> *Plasmodium berghei* [33, 57]
<31> *Plasmodium falciparum* [48, 53, 57]
<32> *Plasmodium yoelii* [57]
<33> *Pneumocystis carinii* [47, 64, 68, 74, 75]
<34> *Rattus norvegicus* (rat [42]; H35 hepatoma cells [32,42]; strain Wistar [37]) [13, 18, 32, 37, 42, 48, 71, 74]
<35> *Saccharomyces cerevisiae* (bakers' yeast [28, 36, 37, 46]; haploid strain [28]) [28, 36, 37, 46, 48]
<36> *Scenedesmus obliquus* (unicellular green algae [36]) [36]
<37> *Streptococcus faecalis* (synonym Enterococcus faecalis, strain R [2]) [1-3, 7, 19, 28]
<38> *Streptococcus faecium* (synonym Enterococcus faecium [52]; var. durans [16]) [16, 52]
<39> *Streptomyces aureofaciens* (industrial strain BMK [46]) [46]
<40> *Sus scrofa* (pig [6, 39, 48]) [6, 39, 48]
<41> *Toxoplasma gondii* [48, 61]
<42> *Trichinella pseudospiralis* [74]

<43> *Trichinella spiralis* [71, 74]
<44> *Trypanosoma brucei* [18]
<45> *Trypanosoma congolense* (causes nagana, the equivalent disease of cattle [18]) [18]
<46> *Trypanosoma cruzi* (causes Chagas' disease in man [18]; sonya strain, maintained in ASH/XP mice [18]) [18]
<47> *Trypanosoma gambiense* [18]
<48> *Trypanosoma lewisi* (natural pathogen of rats [18]) [18]
<49> *Trypanosoma rhodesiense* (causes human sleeping sickness [18]) [18]

3 Reaction and Specificity

Catalyzed reaction
5,10-methylenetetrahydrofolate + dUMP = dihydrofolate + dTMP

Reaction type
dehalogenation
methylene group transfer
reductive methylation

Natural substrates and products
S 5,10-methylenetetrahydrofolate + deoxyuridine phosphate <1-49> (<3, 15, 19, 22, 27, 33, 37> biosynthesis of thymidylic acid, only de novo source of thymidine, crucial for DNA replication in every organism [2, 3, 8, 11, 19, 20, 24, 25, 28, 30, 35, 39, 47, 48, 52, 58, 59, 64, 67, 68, 74, 75, 77]; <3, 19, 22, 23, 26, 30, 37> key function in DNA synthesis, repair and cell division [2, 5, 17, 19, 25, 29, 33, 44, 48, 49, 52, 63, 66]) (Reversibility: ? <1-49> [1-77]) [1-77]
P dihydrofolate + thymidine phosphate

Substrates and products
S (-)-tetrahydropteroylglutamate + 2'-deoxyuridine 5'-monophosphate <19> (Reversibility: ? <19> [19]) [19]
P 7,8-dihydropteroylglutamate + thymidine phosphate
S (6R)-N^5,N^{10}-methylene-5,6,7,8-tetrahydrofolate + 2'-deoxyuridylate <22, 23> (Reversibility: r <22, 23> [27, 77]) [27, 77]
P 7,8-dihydrofolate + 5-methyl-2'-deoxyuridylate
S (6S)-methylenetetrahydropteroylhexaglutamate + 2'-deoxyuridylate <23> (Reversibility: r <23> [14]) [14]
P 7,8-dihydropteroylhexaglutamate + thymidine phosphate
S (6S)-tetrahydropteroylglutamate + 2-deoxyuridylate <23> (Reversibility: r <23> [14]) [14]
P 7,8-dihydropteroylglutamate + thymidine phosphate
S (6S)-tetrahydropteroyltriglutamate + 2'-deoxyuridylate <23> (Reversibility: r <23> [14]) [14]
P 7,8-dihydropteroyltriglutamate + thymidine phosphate

S 5,10-methylene-5,6,7,8-tetrahydrofolate + dUMP <19, 22, 37> (Reversibility: ? <19, 22, 37> [2]) [2]
P 7,8-dihydrofolate + dTMP
S 5,10-methylenetetrahydrofolate + 2'-deoxycytidine 5'-monophosphate <23> (<23> only mutant TS N229D [54]) (Reversibility: ? <23> [54]) [54]
P dihydrofolate + cytidine phosphate
S 5,10-methylenetetrahydrofolate + 2'-fluoro-2'-deoxyuridine 5'-phosphate <19> (Reversibility: ? <19> [24]) [24]
P 7,8-dihydrofolate + ?
S 5,10-methylenetetrahydrofolate + 5'-fluoro-2'-deoxyuridine 5'-phosphate <23> (<23> some mutants are capable of forming covalent thymidylate synthase-5-fluoro-dUMP-methylenetetrahydrofolate complex [60]) (Reversibility: ? <23> [60]) [60]
P 7,8-dihydrofolate + ?
S 5,10-methylenetetrahydrofolate + dUMP <1-49> (<19,22,37> reverse reaction is at least 630times slower than the forward reaction [2]) (Reversibility: r <1-49> [1-77]) [1-77]
P dihydrofolate + dTMP
S 5,10-methylenetetrahydrofolate + uridine 5'-phosphate <23, 40> (Reversibility: r <23, 40> [6, 21]) [6, 21]
P dihydrofolate + TMP
S 5,10-methylenetetrahydrofolylheptaglutamate + 2'-deoxyuridylate <34> (Reversibility: r <34> [42]) [42]
P 7,8-dihydrofolylheptaglutamate + thymidine phosphate
S 5,10-methylenetetrahydrofolylpentaglutamate + 2'-deoxyuridylate <34> (Reversibility: r <34> [42]) [42]
P 7,8-dihydrofolylpentaglutamate + thymidine phosphate
S 5,10-methylenetetrahydropteroylheptaglutamate + 2'-deoxyuridylate <22, 34> (Reversibility: r <22, 34> [42]) [42]
P 7,8-dihydropteroylheptaglutamate + thymidine phosphate
S 5,10-methylenetetrahydropteroylpentaglutamate + 2'-deoxyuridylate <22, 34> (Reversibility: r <22, 34> [20, 42]) [20, 42]
P 7,8-dihydropteroylpentaglutamate + thymidine phosphate
S 5-bromo-2'-deoxyuridine 5'-monophosphate + ? <19, 23, 26, 34> (<19,23,26> dehalogenation [59,60,71]; <19> dehalogenation does not require methylenetetrahydrofolate as cosubstrate [73]; <34> dehalogenated by recombinant rat hepatoma enzyme [71]) (Reversibility: ? <19, 23, 26, 34> [59, 60, 71, 73]) [59, 60, 71]
P ?
S 5-iodo-2'-deoxyuridine 5'-monophosphate + ? <19, 26, 34> (<26> dehalogenation [71]; <34> dehalogenated by recombinant rat hepatoma enzyme [71]) (Reversibility: ? <19, 26, 34> [59, 71]) [59, 71]
P ?
S 7,8-dihydrofolate + NADPH + H^+ <19, 22, 36, 37> (Reversibility: ? <19, 22, 36, 37> [2, 36]) [2, 36]
P 5,6,7,8-tetrahydrofolate + $NADP^+$

S $N^{5,10}$-methylenetetrahydrofolate + 2'-deoxyuridine 5'-dithiophosphate <26, 34> (Reversibility: ? <26, 34> [74]) [74]
P 7,8-dihydrofolate + ?
S $N^{5,10}$-methylenetetrahydrofolate + 2'-deoxyuridine 5'-thiophosphate <26, 34> (Reversibility: ? <26, 34> [74]) [74]
P 7,8-dihydrofolate + ?
S $N^{5,10}$-methylenetetrahydrofolate + 5-hydroxy-2'-deoxyuridine 5'-monophosphate <22> (Reversibility: ? <22> [72]) [72]
P 7,8-dihydrofolate + ?
S [6R,S]-5,10-methylenetetrahydropteroylglutamate + 2'-deoxyuridylate <26> (Reversibility: r <26> [42]) [42]
P ?
S [6R]-methylenetetrahydropteroylglutamate + 2'-deoxyuridylate <22> (Reversibility: r <22> [39]) [39]
P ?
S Additional information <17, 19, 22, 23, 36, 37, 39, 41> (<19> (-)-tetrahydropteroylglutamates are substrates, (+) forms are inhibitors [19]; <19> enzyme catalyzes the exchange of a hydrogen atom between water and position 5 of dUMP, occurrence of reactions other than a methyl-group transfer during the interaction of the enzyme with its 2 substrates [24]; <19> nucleotides with large substituents on the 2' position are no substrates [24]; <37> UMP, dCMP and deoxyuridine are no substrates [2]; <22> inhibitors targeted at human TS designed from structures of bacterial TS may not be optimal [67]; <23> dCMP is no substrate, exception mutant TS N229D [54]; <17> uridylate, cytidylate and deoxycytidylate are not substrates [16]; <36> bifunctional enzyme thymidilate synthase-dihydrofolate reductase, not possible to separate the 2 activities [36]; <39> monofunctional enzyme [46]; <25,31,41> bifunctional enzyme dihydrofolate reductase-thymidylate synthase [48,57,61]) [2, 19, 24, 36, 46, 48, 54, 57, 61, 67]
P ?

Inhibitors

(+)tetrahydropteroylglutamate <19> [19]
(6R)-tetrahydropteroylglutamate <23> [14]
(S)-2-(5-(((1,2-dihydro3-methyl-1-oxobenzo[f]quinazolin-9-yl)methyl)amino)-1-oxo-2-isoindolinyl)glutaric acid <22> (<22> BW1843U89 [63]) [63]
1-(2'-deoxyribosyl)8-azapurin-2-one 5'-monophosphate <23> [29]
1-(β-D-2'-deoxyribofuranosyl)1,6-dihydro-8-azapurin-2(3H,6H)-one 5'-monophosphate <23> [29]
1-(β-D-2'-deoxyribofuranosyl)8-azapurin-2-one 5'-monophosphate <23> [29]
1-(β-D-2-deoxyribofuranosyl)-4-hydroxyamino-5-bromopyrimidin-2(1H)-one 5'-monophosphate <26> [71]
1-(β-D-2-deoxyribofuranosyl)-4-hydroxyamino-5-chloropyrimidin-2(1H)-one 5'-monophosphate <26> [71]
1-(β-D-2-deoxyribofuranosyl)-4-hydroxyamino-5-fluoropyrimidin-2(1H)-one 5'-monophosphate <26> [71]

1-(β-D-2-deoxyribofuranosyl)-4-hydroxyamino-5-hydroxymethylpyrimidin-2(1H)-one 5'-monophosphate <26> [71]
1-(β-D-2-deoxyribofuranosyl)-4-hydroxyamino-5-iodopyrimidin-2(1H)-one 5'-monophosphate <26> [71]
1-(β-D-2-deoxyribofuranosyl)-4-hydroxyaminopyrimidin-2(1H)-one 5'-monophosphate <26> [71]
1-ethyl-3-dimethylaminopropyl carbodiimide <23> [45]
1-phenyl-3-(3-dimethylaminopropyl) carbodiimide <23> [45]
1-phenyl-3-(3-trimethylaminopropyl) carbodiimide <23> [45]
10-formyl folic acid <26> [31]
10-formyl-5,8-dideazafolate <26> [26]
10-formyldihydropteroylheptaglutamate <34> [42]
10-formyldihydropteroylpentaglutamate <22> [42]
10-formyltetrahydrofolate <34> [42]
10-formyltetrahydropteroylheptaglutamate <34> [42]
10-formyltetrahydropteroylpolyglutamate <34> [42]
10-methenyldihydropteroylpentaglutamate <22> [42]
10-methenyltetrahydropteroylheptaglutamate <22> [42]
10-propargyl-5,8-dideazafolate <19> [59]
2'-deoxyuridine <34> [37]
2,4-diamino pteridine <17> [16]
2,4-diamino quinazoline antifolate <17> [16]
2-amino-4-hydroxyquinazoline <17> [16]
2-amino-4-hydroxyquinazoline folic acid <17, 23, 26, 37> [19]
2-hydroxyethyl disulfide <23> [21]
2-mercaptoethanol <13, 14, 44-46, 48> [18]
3-β-D-ribofuranosylmaleimide <23> (<23> showdomycin [19]) [19]
3-β-D-ribofuranosylmaleimide 5'-phosphate <23> (<23> showdomycin 5'-phosphate [19]) [19]
5,10-methylenetetrahydrofolate methenyltetrahydropteroylglutamate <19> [41]
5,5'-dithiobis(2-nitrobenzoic acid) <23> [21]
5-bromo-2'-deoxyuridylic acid <35> [28]
5-bromo-N^4-hydroxy-2'-deoxycytidine <26> [71]
5-bromoacetamido-5'-deoxythymidine <26> [35]
5-chloro-N^4-hydroxy-2'-deoxycytidine <26> [71]
5-chloroacetamido-5'-deoxythymidine <26> [35]
5-fluoro-2'-deoxyuridine 5'-(*p*-aminophenyl phosphate) <23, 37> [2, 23]
5-fluoro-2'-deoxyuridine 5'-monophosphate <23> [22, 49]
5-fluoro-2'-deoxyuridylic acid <3, 6, 16, 19, 22, 23, 25, 26, 31, 34-37, 39> (<37> very strong inhibition, pseudo-noncompetitive [2]) [2, 3, 10, 13, 19, 21-23, 27-29, 31, 36-38, 41, 44, 46, 48, 57, 61, 63, 67, 74]
5-fluorodeoxypyrimidine <13, 46> [18]
5-fluoroorotate <31, 32> [57]
5-fluorouracil <13, 18, 19, 22, 31, 46> [11, 18, 39, 41, 57, 63, 67]
5-fluorouridine <22> [19]
5-formyl-2'-deoxyuridylic acid <19, 22, 23> [9, 19]

5-haloacetamido-5'-deoxythymidine <26> [35]
5-hydroseleno-2'-deoxyuridylic acid <22> [47]
5-hydroxy-methyl-2'-deoxyuridine 5'-monophosphate <22> [72]
5-iodo-2'-deoxyuridylic acid <23> [22]
5-iodoacetamido-5'-deoxythymidine <26> [35]
5-iodoacetamidomethyl-2'-deoxyuridine 5'-phosphate <26> [19]
5-mercapto-2'-deoxyuridylic acid <19, 22, 26> (<26> possesses antitumor activity against Ehrlich ascites cells [19]) [19, 72]
5-nitro-2'-deoxyuridine 5'-monophosphate <23> [22]
5-nitro-2'-deoxyuridylic acid <23> [22, 29]
5-trifluoromethyl-2'-deoxyuridylic acid <22> [19]
ATP <9> [7]
CB3717 <33> [64]
Cibacron blue <35> [28]
D1694 <26, 31> [57]
D1694 pentaglutamate <26, 31> [57]
FdUSSO3 <26> [74]
H_2O_2 <23> [21]
KCl <30> [33]
Mg^{2+} <13, 14, 44-46, 48> [18]
N-ethylmaleimide <19, 23> [1, 21]
N^{10}-formyltetrahydrofolylmonoglutamate <34> [15, 42]
N^4-hydroxy-2'-deoxycytidine <26> [71]
N^4-hydroxy-5-methyl-2'-deoxycytidine <26> [71]
N^5-formiminotetrahydropteroyltriglutamate <19> [15]
N^5-formyltetrahydropteroyloligoglutamate <19> [15, 19]
NaCl <30> [33]
Procion blue <34> [37]
SK 28,757 <17> [16]
SK 28,758 <17> [16]
SK 28,815 <17> [16]
Thymitaq <22> [63]
TomudexR <22, 34> [63, 74]
UDP-glucuronate <34> [37]
amethopterin <17, 19, 22, 34> [9, 16, 20, 37]
aminofolate <3> [48]
aminopterin <13, 17, 35, 46> [16, 18, 28]
antrycide <13, 14, 44-46, 48> [18]
arabinouridylate <36> [36]
bromodeoxyuridine <22> [19]
chloroacetamide <19> [1]
cyanide <23> [21]
cytosine arabinoside <22> [19]
deaza-aminopterin <17> [16]
diglutamate <19> [19]
dihydrofolate <3, 19, 37> [2, 48, 59]
dihydropteroylglutamate <23> [20]

dihydropteroylhexaglutamate <23> [14]
dihydropteroylpentaglutamate <22> [42]
dihydropteroyltriglutamate <23> [14]
dimidium <13, 14, 44-46, 48> [18]
ethidium <13, 14, 44-46, 48> [18]
ethyl methanesulfonate <22> [63]
ethylene glycol <19> [75]
fluorodeoxyuridine monophosphate <18, 26> [11, 74]
fluoroorotic acid <22> [19]
folic acid oligoglutamate <19> [15]
glutamate <19> [19]
glyceraldehyde 3-phosphate <23> [43]
glycerol <19> [75]
hydroxyurea <36> (<36> supresses thymidilate synthase in growing algae, but no effect on purified enzyme [36]) [36]
iodoacetamide <23> [19, 21]
iodoacetate <19> [1]
iododeoxyuridine <22> [19]
methasquin <17> [16]
methotrexate <3, 5, 6, 18, 22, 30, 35, 36> [11, 20, 28, 32, 33, 36, 48]
p-chloromercuribenzoate <23> [3, 21]
pentamidine <13, 14, 44-46, 48> [18]
phenylglyoxal <23> [53]
prothidium <13, 14, 44-46, 48> [18]
pteroyldiglutamate <22> [39]
pteroylheptaglutamate <22> [39]
pteroylhexaglutamate <19, 22, 23> [14, 19, 39]
pteroylmonoglutamate <22> [20, 39]
pteroylpentaglutamate <22> [39]
pteroylpolyglutamate <22, 23> [14, 39]
pteroyltetraglutamate <22> [39]
pteroyltriglutamate <22> [39]
pyridoxal 5-phosphate <23> [40]
quinaspar <17> [16]
quinazoline <5, 6, 23> [30, 32]
suramin <13, 14, 44-46, 48> [18]
tetrahydrohomofolate <19> [19]
tetrahydromethotrexate <9, 19, 23, 26> [17, 23, 27]
thiophosphate <26> [74]
thymidine phosphate <3, 9, 19, 26, 37> (<37> product inhibition [2]) [2, 7, 48, 59, 71]
trifluoromethyldeoxyuridine 5'-monophosphate <6, 18, 26> [3, 10, 11]
trimethoprim <3, 13, 46> [18, 48]
urea <30> [33]
vanoxonin-vanadium <26> [34]

Cofactors/prosthetic groups

(+)-methylenetetrahydrofolate <19, 23> [21, 29, 41, 49]
(6R)-5,10-methylene-5,6,7,8-tetrahydrofolate <19> [76]
N^5,N^{10}-methylenetetrahydrofolate <22, 26> [66, 71]
NADPH <39> (<39> absolute requirement [46]) [46]
methenyltetrahydropteroylglutamate <19> [41]
methylenetetrahydrofolate <22> [63]

Activating compounds

$(NH_4)_2SO_4$ <9> (<9> 10 mM, relative activity 108% [7]) [7]
2,3-dimercaptopropanol <19, 22, 37> (<19,22,37> absolute requirement for a thiol, synthetase shows greatly dimished activity in absence [2]) [2]
ATP <9> (<9> 0.1 mM, relative activity 102% [7]) [7]
EDTA <9> (<9> 5.0 mM, relative activity 103% [7]) [7]
KCN <19, 22, 37> [2]
β-mercaptoethanol <26, 39> (<39> essential for maximum activity [46]) [3, 46]
dithioerythritol <39> [46]
dithiothreitol <4, 19, 22, 26, 39, 40> (<22> enzyme activity is stimulated 2fold by 8-20 mM [25]; <26> thiols enhances enzyme activity [3]; <40> maximal catalytic activity only in presence of sulfhydryl compounds [6]) [3, 6, 25, 46, 69]

Metals, ions

Mg^{2+} <17, 19, 23> (<19> high requirement [75]; <23> in Tris-acetate buffers 10 mM enhances enzyme activity by some 30% [21]; <17> requirement depending upon the buffer [16]) [1, 16, 21, 24, 75]
Na^+ <39> (<39> stimulates activity [46]) [46]

Turnover number (min^{-1})

35 <34> (2'-deoxyuridine) [37]
100 <25> (2'-deoxyuridine) [61]
250 <22> (deoxyuridine phosphate) [25]
960 <5> (2'-deoxyuridine) [13]

Specific activity (U/mg)

0.00012 <44> [18]
0.00014 <45> [18]
0.0003367 <3> [48]
0.00035 <9> [7]
0.00035 <46> (<46> intracellular form [18]) [18]
0.00038 <48> [18]
0.00054 <14> [18]
0.00119 <46> (<46> blood form [18]) [18]
0.00156 <46> (<46> culture form [18]) [18]
0.002 <17> [16]
0.002135 <26> (<26> Gardner lymphosarcoma [27]) [27]
0.00218 <13> [18]
0.00218 <13> [18]

0.00235 <30> [33]
0.003519 <26> (<26> sarcoma 180 [27]) [27]
0.0063 <22> [39]
0.00652 <39> [46]
0.01 <26> [30]
0.0156 <19> [17]
0.016 <19> [1]
0.0255 <31> [57]
0.0257 <36> [36]
0.1024 <9> [27]
0.17 <37> [19]
0.27 <37> [19]
0.28 <19> [26]
0.6 <34> [71]
0.6778 <19> [27]
1.2 <35> [28]
1.3 <26> (<26> affinity chromatography [44]) [44]
1.33 <26> [44]
1.47 <19> [24]
1.52 <19> [24]
2.5 <23> [21]
2.5-3 <23> [49]
3 <15> [47]
3.3 <23> [23]
3.8 <5, 22> (<5> 50% pure preparation [13]) [13, 25]
14.2 <7> [10]
14.45 <6> [10]
15 <5> [13]
16 <5> [13]
54 <37> [7]
150 <23> [7]

K_m-Value (mM)

0.00068 <34> (5,10-formyltetrahydropteroylheptaglutamate) [42]
0.0007 <22> (2'-deoxyuridylate, <22> mutant 362(T51S,K82Q,N171S) [77]) [77]
0.00074 <34> (5,10-methenyltetrahydropteroylpentaglutamate) [42]
0.001 <22> (2'-deoxyuridylate, <22> mutant Y258F [77]) [77]
0.001 <19> (methylenetetrahydropteroylgluamate, <19> mutant P254S [70]) [70]
0.0011 <19> (2'-deoxyuridylate, <19> wild-type [66,73]) [66, 73]
0.0011 <22> (2'-deoxyuridylate, <22> mutant 302(T53S,Y258F) [77]) [77]
0.0011 <19> (methylenetetrahydropteroylgluamate, <19> wild-type [70]) [70]
0.0012 <19> (2'-deoxyuridylate, <19> mutant Y4F [66]; <19> wild-type [70]) [2, 66, 70]

0.0012 <19> (methylenetetrahydropteroylgluamate, <19> mutant P254G [70]) [70]
0.0013 <31> (2'-deoxyuridylate, <31> native enzyme) [57]
0.0014 <19> (2'-deoxyuridylate, <19> mutant Y4I, Y4V [66]) [66]
0.0015 <23> (2'-deoxyuridine 5'-monophosphate) [22]
0.0015 <19> (2'-deoxyuridylate, <19> mutant P254G [70]) [70]
0.0018 <22> (2'-deoxyuridylate) [57]
0.0018 <19> (2'-deoxyuridylate, <19> mutant P254F [70]) [70]
0.0018 <19> (methylenetetrahydropteroylgluamate, <19> mutant P254F [70]) [70]
0.002 <22> (2'-deoxyuridylate, <22> wild type [77]) [77]
0.002 <31> (2'-deoxyuridylate, <31> recombinant enzyme) [57]
0.0022 <22> (5,10-methylenetetrahydropteroylpentaglutamate) [20]
0.0026 <15> (2'-deoxyuridylate) [47]
0.0026 <22> (2'-deoxyuridylate, <22> mutant T51S [77]) [77]
0.0027 <19> (2'-deoxyuridylate, <19> mutant P254S [70]) [70]
0.003 <19> (2'-deoxyuridylate, <19> mutant Y4H [66]) [66]
0.00315 <26> (2'-deoxyuridylate) [74]
0.0033 <13> (2'-deoxyuridylate) [46]
0.0034 <19> (2'-deoxyuridylate, <19> mutant P254D [70]) [57, 70]
0.0035 <22> ([6R]methylenetetrahydropteroyltriglutamate) [39]
0.0036 <22> (2'-deoxyuridylate) [39]
0.00365 <34> (2'-deoxyuridylate) [74]
0.0037 <22> (2'-deoxyuridylate) [20]
0.00378 <39> (2'-deoxyuridylate) [46]
0.004 <19> (2'-deoxyuridylate) [1, 7]
0.0043 <22> ([6R]methylenetetrahydropteroylglutamate) [39]
0.0043 <22> ([6R]methylenetetrahydropteroylheptaglutamate) [39]
0.0045 <19> (L-tetrahydrofolate) [1]
0.005 <35> (2'-deoxyuridylate) [28]
0.005 <23> (2'-deoxyuridylate, <23> wild-type [54,60]) [54, 60]
0.0051 <23> (2'-deoxyuridylate, <23> in presence of 10 mM Mg^{2+} [7,21]) [7, 21]
0.0053 <19> (2'-deoxyuridylate, <19> mutant S167A [73]) [73]
0.0057 <37> (2'-deoxyuridylate) [2, 7]
0.0057 <23> (5-bromo-2'-deoxyuridylate, <19> wild-type [60]) [60]
0.0063 <26> (2'-deoxyuridylate) [3]
0.0063 <23> (2'-deoxyuridylate, <19> mutant D221S [60]) [60]
0.00632 <26> (2'-deoxyuridylate 5'-thiophosphate) [74]
0.0065 <22> ([6R]methylenetetrahydropteroyldiglutamate) [39]
0.0068 <23, 34> (2'-deoxyuridylate) [21, 37]
0.0075 <20> (2'-deoxyuridylate) [7]
0.008 <22> (methylenetetrahydrofolate) [57]
0.008 <19> (methylenetetrahydrofolate, <19> mutant Y4F [66]) [66]
0.0081 <19> (methylenetetrahydrofolate, <19> wild-type [73]) [73]
0.0083 <22> ([6R]methylenetetrahydropteroylpentaglutamate) [39]
0.0085 <19> (methylenetetrahydrofolate, <19> wild-type [73]) [73]

0.0087 <22> ([6R]methylenetetrahydropteroylhexaglutamate) [39]
0.009 <9> (2'-deoxyuridylate) [7]
0.0092 <26> (2'-deoxyuridylate) [35]
0.0097 <19> (5-bromo-2'deoxyuridine, <19> wild-type [73]) [73]
0.0097 <19> (methylenetetrahydrofolate, <19> mutant S167A [73]) [73]
0.01 <3, 19> (2'-deoxyuridylate) [24, 48]
0.01 <19> (methylenetetrahydrofolate, <19> wild-type [66]) [66]
0.01 <23> (methylenetetrahydrofolate, <23> wild-type [54]) [54]
0.0104 <19> (2'-deoxyuridylate) [24]
0.0109 <22> ([6R]methylenetetrahydropteroyltetraglutamate) [39]
0.011 <23> (2'-deoxyuridylate, <19> mutant D221E [60]) [60]
0.011 <19> (methylenetetrahydrofolate) [59]
0.0118 <19> (methylenetetrahydropteroylgluamate, <19> mutant P254D [70]) [70]
0.012 <23> ((+-)methylenetetrahydrofolate, <23> in presence of 10 mM Mg^{2+} [21]) [21]
0.012 <19> (methylenetetrahydrofolate, <19> mutant S167A [73]) [73]
0.013 <19> (methylenetetrahydrofolate, <19> mutant Y4H [66]) [66]
0.0137 <26> (2'-deoxyuridylate 5'-dithiophosphate) [74]
0.014 <19> ((+-)5,10-methylene-5,6,7,8-tetrahydrofolate) [24]
0.014 <26> (2'-deoxyuridylate, <26> Ehrlich ascites carcinoma cells [7]) [7]
0.015 <23> (2'-deoxyuridylate, <19> mutant D221C [60]) [60]
0.015 <19> (methylenetetrahydrofolate, <19> mutant Y4V [66]) [66]
0.0155 <34> (2'-deoxyuridylate 5'-dithiophosphate) [74]
0.016 <37> (methylenetetrahydrofolate) [2]
0.017 <23> (2'-deoxyuridylate, <23> mutant TS N229C [54]) [54]
0.018 <5> (2'-deoxyuridylate, <5> phage infected E. coli [7]) [7]
0.018 <37> (methylenetetrahydrofolate) [2]
0.018 <19> (methylenetetrahydrofolate, <19> mutant Y4I [66]) [66]
0.02 <9> (2'-deoxyuridylate, <9> less purified enzyme [7]) [7]
0.02 <23> (5-bromo-2'deoxyuridylate, <19> mutant D221A [60]) [60]
0.0211 <39> (N^5,N^{10}-methylenetetrahydrofolate) [46]
0.025 <23> ((l)-tetrahydropteroyltriglutamate) [14]
0.025 <23> (5-bromo-2'deoxyuridylate, <19> mutant D221C [60]) [60]
0.0253 <26> (methylenetetrahydrofolate) [46]
0.027 <23> (5-bromo-2'deoxyuridylate, <19> mutant D221S [60]) [60]
0.027 <26> (methylenetetrahydrofolate) [20, 57]
0.03 <8> (2'-deoxyuridylate, <8> phage infected E. coli [7]) [7]
0.03 <31> (methylenetetrahydrofolate, <31> native enzyme [57]) [57]
0.0308 <17> (2'-deoxyuridylate) [16]
0.031 <22> (5,10-methylenetetrahydropteroylglutamate) [20]
0.031 <22> (methylenetetrahydrofolate) [57]
0.032 <23> ((+-)methylenetetrahydrofolate) [21]
0.0339 <19> (2'-deoxyuridylate, <19> mutant S167G [73]) [73]
0.035 <37> (methylenetetrahydrofolate) [2]
0.037 <26> (2'-deoxyuridylate, <26> Ehrlich ascites carcinoma cells [7]) [7]
0.037 <34> (2'-deoxyuridylate 5'-thiophosphate) [74]

0.037 <37> (methylenetetrahydrofolate) [2]
0.0382 <15> (methylenetetrahydrofolate) [47]
0.0389 <39> (NADPH) [46]
0.039 <37> (L-methylenetetrahydrofolate) [2]
0.039 <31> (methylenetetrahydrofolate, <31> recombinant enzyme [57]) [57]
0.0417 <19> (2'-deoxyuridylate, <19> mutant S167T [73]) [73]
0.0426 <19> (methylenetetrahydrofolate, <19> mutant S167G [73]) [73]
0.043 <37> (L-methylenetetrahydrofolate) [3]
0.045 <23> (5-bromo-2'-deoxyuridylate, <19> mutant D221N [60]) [60]
0.046 <23> (5-bromo-2'-deoxyuridylate, <19> mutant D221E [60]) [60]
0.05 <23> (2'-deoxyuridylate, <23> mutant TS N229D [54]) [54]
0.0521 <19> (methylenetetrahydrofolate, <19> mutant S167T [73]) [73]
0.061 <23> ((l)-tetrahydropteroylglutamate) [14]
0.0627 <19> (methylenetetrahydrofolate, <19> mutant S167T [73]) [73]
0.065 <34> (L-5,10-methylenetetrahydrofolate) [37]
0.07 <35> ((-)-5,10-methylene-tetrahydropteroylglutamate) [28]
0.0714 <30> (2'-deoxyuridylate) [33]
0.0723 <19> (5-bromo-2'-deoxyuridine, <19> mutant S167A [73]) [73]
0.077 <23> (tetrahydropteroylglutamate) [14]
0.092 <23> (methylenetetrahydrofolate, <23> mutant TS N229C [54]) [54]
0.092 <23> (methylenetetrahydrofolate, <23> mutant TS N229D [54]) [54]
0.0981 <19> (2'-deoxyuridylate, <19> mutant S167C [73]) [73]
0.11 <19> (2'-fluoro-2'-deoxyuridine 5'-phosphate) [24]
0.113 <19> (2'-fluoro-2'-deoxyuridine 5'-phosphate) [24]
0.13 <19> (methylenetetrahydrofolate, <19> mutant S167C [73]) [73]
0.15 <23> (2'-deoxycytidine 5'-monophosphate, <23> mutant TS N229D [54]) [54]
0.17 <23> (methylenetetrahydrofolate, <23> mutant TS N229D, dCMP as substrate [54]) [54]
0.185 <3> (5,10-methylenetetrahydrofolate) [48]
0.266 <17> (L-5,10-methylenetetrahydrofolate) [16]
0.3125 <30> (methylenetetrahydrofolate) [33]
0.514 <19> (5-bromo-2'-deoxyuridine, <19> mutant S167G [73]) [73]

K_i-Value (mM)

0.00000008 <26> (D1694) [57]
0.000001 <4> (5-fluoro-2'-deoxyuridylic acid) [58]
0.000001 <26> (D1694 pentaglutamate) [57]
0.0000015 <31> (D1694 pentaglutamate) [57]
0.0000017 <22> (5-fluoro-2'-deoxyuridylic acid) [57]
0.000002 <31> (5-fluoro-2'-deoxyuridylic acid) [57]
0.0000028 <31> (D1694) [57]
0.000004 <37> (5-fluoro-2'-deoxyuridylic acid) [2]
0.000004 <35> (Cibacron blue) [28]
0.000005 <35> (5-fluoro-2'-deoxyuridylic acid) [28]
0.000005 <34> (5-fluoro-2'-deoxyuridylic acid) [37]
0.000005 <23> (5-fluoro-2'-deoxyuridylic acid) [55]

0.00001 <22> (5-formyl-2'-deoxyuridylic acid) [19]
0.00001 <22> (5-mercapto-2'-deoxyuridylic acid) [59]
0.00001 <22> (5-trifluoromethyl-2'-deoxyuridylic acid) [19]
0.000012 <23> (5-formyl-2'-deoxyuridylic acid) [9]
0.000013 <19> (5-formyl-2'-deoxyuridylic acid) [9]
0.000014 <23> (5-fluoro-2'-deoxyuridine 5'-monophosphate) [22]
0.00002 <26> (5-fluoro-2'-deoxyuridylic acid, <26> L1210 parental enzyme [71]) [71]
0.000021 <3> (5-fluoro-2'-deoxyuridylic acid, <3> 50 nM [48]) [48]
0.0000235 <3> (5-fluoro-2'-deoxyuridylic acid) [48]
0.000026 <3> (5-fluoro-2'-deoxyuridylic acid, <3> 200 nM [48]) [48]
0.000029 <23> (5-nitro-2'-deoxyuridine 5'-monophosphate) [22]
0.000039 <23> (5-trifluoromethyl-2'-deoxyuridine 5'-monophosphate) [22]
0.00004 <22> (5-mercapto-2'-deoxyuridylic acid) [19]
0.00005 <30> (5-fluoro-2'-deoxyuridylic acid) [33]
0.00009 <33> (CB3717) [64]
0.0001 <34> (10-formyltetrahydrofolylheptaglutamate) [42]
0.0001 <34> (Procion blue, <34> variable substrate 2'-deoxyuridine [37]) [37]
0.00012 <19> (tetrahydromethotrexate) [17]
0.00022 <34> (Procion blue, <34> variable substrate 5,10-methylenetetrahydropteroylglutamate [37]) [37]
0.0003 <26> (chloro-2'-deoxyuridine, <26> L1210 parental enzyme [71]) [71]
0.0006 <22> (pteroyltetraglutamate) [39]
0.0007 <34> (10-formyldihydropteroylheptaglutamate) [42]
0.0008 <26> (1-(β-D-2-deoxyribofuranosyl)-4-hydroxyamino-5-fluoropyrimidin-2(1H)-one 5'-monophosphate, <26> L1210 parental enzyme [71]) [71]
0.000833 <3> (trimethoprim, <3> 1.0 mM [48]) [48]
0.00087 <3> (trimethoprim, <3> 2.0 mM [48]) [48]
0.001 <22> (pteroylpentaglutamate) [39]
0.001 <3> (thymidine phosphate, <3> 0.25 mM, noncompetitive inhibition, 5,10-mehylenetetrahydrofolate as variable substrate [48]) [48]
0.00124 <3> (thymidine phosphate, <3> 1.0 mM, noncompetitive inhibition, 5,10-methylenetetrahydrofolate as variable substrate [48]) [48]
0.0013 <26> (1-(β-D-2-deoxyribofuranosyl)-4-hydroxyamino-5-hydroxymethylpyrimidin-2(1H)-one 5'-monophosphate, <26> L1210 recombinant enzyme [71]) [71]
0.0013 <17> (2,4-diamino quinazoline) [16]
0.0013 <17> (SK 28,757) [16]
0.0014 <26> (1-(β-D-2-deoxyribofuranosyl)-4-hydroxyamino-5-fluoropyrimidin-2(1H)-one 5'-monophosphate, <26> L1210 recombinant enzyme [71]) [71]
0.0014 <3> (dihydrofolate, <3> 1.0 mM, noncompetitive inhibition, 5,10-methylenetetrahydrofolate as variable substrate [48]) [48]
0.0015 <17> (SK 28,758) [16]

0.0015 <26> (bromo-2'-deoxyuridine, <26> L1210 parental enzyme [71]) [71]

0.0016 <22> (10-formyldihydropteroylpentaglutamate) [42]

0.0017 <26> (1-(β-D-2-deoxyribofuranosyl)-4-hydroxyaminopyrimidin-2(1H)-one 5'-monophosphate, <26> L1210 parental enzyme [71]) [71]

0.0017 <26> (thiophosphate) [74]

0.00178 <3> (dihydrofolate, <3> 5.0 mM, noncompetitive inhibition, 5,10-methylenetetrahydrofolate as variable substrate [48]) [48]

0.00191 <26> (1-(β-D-2-deoxyribofuranosyl)-4-hydroxyamino-5-hydroxymethylpyrimidin-2(1H)-one 5'-monophosphate, <26> L1210 parental enzyme [71]) [71]

0.002 <31> (5-fluoro-2'-deoxyuridylic acid) [57]

0.002 <22> (pteroylhexaglutamate) [39]

0.0024 <34> (10-formyltetrahydrofolylmonoglutamate) [42]

0.0029 <17> (SK 28,815) [16]

0.0029 <17> (methasquin) [16]

0.00294 <3> (dihydrofolate, <3> 0.5 mM, noncompetitive inhibition, dUMP as variable substrate [48]) [48]

0.003 <17> (2,4-diamino quinazoline antifolate) [16]

0.00303 <3> (dihydrofolate, <3> 1.0 mM, noncompetitive inhibition, dUMP as variable substrate [48]) [48]

0.0034 <26> (1-(β-D-2-deoxyribofuranosyl)-4-hydroxyamino-5-iodopyrimidin-2(1H)-one 5'-monophosphate, <26> L1210 recombinant enzyme [71]) [71]

0.0034 <22> (pteroylheptaglutamate) [39]

0.0034 <22> (pteroyltriglutamate) [39]

0.0037 <26> (hydroxymethyl-2'-deoxyuridine, <26> L1210 parental enzyme [71]) [71]

0.0044 <17> (quinaspar) [16]

0.0049 <26> (1-(β-D-2-deoxyribofranosyl)-4-hydroxyamino-5-chloropyrimidin-2(1H)-one 5'-monophosphate, <26> L1210 recombinant enzyme [71]) [71]

0.0054 <26> (5-bromoacetamido-5'-deoxythymidine, <26> competitive inhibition [35]) [35]

0.0066 <26> (5-bromoacetamido-5'-deoxythymidine) [35]

0.0073 <26> (thymidine phosphate, <26> L1210 parental enzyme [71]) [71]

0.00746 <26> (1-(β-D-2-deoxyribofuranosyl)-4-hydroxyamino-5-iodopyrimidin-2(1H)-one 5'-monophosphate, <26> L1210 parental enzyme [71]) [71]

0.0075 <34> (10-formyldihydropteroylglutamate) [42]

0.0075 <17> (deaza-aminopterin) [16]

0.0076 <22> (pteroylmonoglutamate) [39]

0.00768 <26> (iodo-2'-deoxyuridine, <26> L1210 parental enzyme [71]) [71]

0.0084 <26> (dithiophosphate) [74]

0.0085 <26> (1-(β-D-2-deoxyribofuranosyl)-4-hydroxyamino-5-chloropyrimidin-2(1H)-one 5'-monophosphate, <26> L1210 parental enzyme [71]) [71]

0.0093 <26> (1-(β-D-2-deoxyribofuranosyl)-4-hydroxyaminopyrimidin-2(1H)-one 5'-monophosphate, <26> L1210 recombinant enzyme [71]) [71]

0.01 <35> (5-bromo-2-deoxyuridine 5'-monophosphate) [28]
0.0132 <22> (pteroyldiglutamate) [39]
0.015 <35> (aminopterin) [28]
0.016 <23> (1-(2'-deoxyribosyl)-8-azapurin-2-one 5'-monophosphate) [29]
0.02 <26> (10-formyltetrahydropteroylmonoglutamate) [42]
0.02 <35> (aminopterin) [28]
0.02 <35> (methotrexate) [28]
0.022 <26> (1-(β-D-2-deoxyribofuranosyl)-4-hydroxyamino-5-bromopyrimidin-2(1H)-one 5'-monophosphate, <26> L1210 recombinant enzyme [71]) [71]
0.023 <34> (amethopterin) [37]
0.023 <22> (methotrexate, <22> uncompetitive with methylene tetrahydrofolate [2]) [2]
0.024 <37> (dTMP) [2, 7]
0.0284 <17> (amethopterin) [16]
0.03 <17> (2,4-diamino pteridine) [16]
0.0321 <26> (1-(β-D-2-deoxyribofuranosyl)-4-hydroxyamino-5-bromopyrimidin-2(1H)-one 5'-monophosphate, <26> L1210 parental enzyme [71]) [71]
0.0347 <17> (aminopterin) [16]
0.059 <9> (dTMP) [7]
0.07 <34> (dTMP, <34> variable substrate 2'-deoxyuridine [37]) [37]
0.073 <34> (amethopterin, <34> variable substrate 2'-deoxyuridine [37]) [37]
0.08 <35> (thymidine phosphate) [28]
0.1 <35> (dTMP) [28]
0.103 <30> (methotrexate, <30> uncompetitive with dUMP [33]) [33]
0.12 <23> (1-(β-D-2'-deoxyribofuranosyl)-1,6-dihydro-8-azapurin-2(3H,6H)-one 5'-monophosphate) [29]
0.12 <35> (thymidine phosphate) [28]
0.14 <20> (dTMP) [7]
0.238 <3> (methotrexate, <3> 0.05 mM [48]) [48]
0.24 <34> (dTMP, <34> variable substrate 5,10-methylenetetrahydropteroylglutamate [37]) [37]
0.267 <3> (methotrexate, <3> 0.02 mM [48]) [48]
0.3125 <3> (thymidine phosphate, <3> 0.25 mM, competitive inhibition, dUMP as variable substrate [48]) [48]
0.333 <3> (thymidine phosphate) [48]
0.406 <3> (methotrexate, <3> 0.5 mM [48]) [48]
0.413 <3> (methotrexate, <3> 1.0 mM [48]) [48]
0.6-0.9 <23> (pyridoxal 5'-phosphate) [40]
0.71 <26> (S-thiosulfate) [27]
0.852 <3> (trimethoprim) [48]
5 <3> (aminofolate, <3> 1.0 mM [48]) [48]
5.2 <3> (aminofolate, <3> 5.0 mM [48]) [48]
12 <23> (glyceraldehyde 3-phosphate) [43]

pH-Optimum

5.9 <23> (<23> UMP as substrate [21]) [21]
6.5-6.7 <47, 48> [18]
6.5-6.8 <9, 23> (<9> phosphate buffer [7]) [7, 21]
6.5-7 <34> [74]
6.5-8 <30> [33]
6.6 <9> [7]
6.6-6.8 <40> [6]
6.7 <26> [3]
6.8-7.2 <35> [28]
6.8-8.7 <5> [13]
7 <9, 15, 19, 26, 35> [44, 46, 47]
7-7.7 <17> [16]
7-8 <19> (<19> activity is higher in Tris buffer than in phosphate buffer [1]) [1, 24]
7-8.1 <22> [25]
7.2-7.4 <40> (<40> double pH-optimum in 2-N-morpholinoethane sulfonic acid buffer [6]) [6]
7.3-8 <34> (<34> half-maximal activity at pH 6.3 [37]) [37]
7.5 <20, 36> [7, 36]
7.5-8 <18, 19> [7, 11]
7.8 <13> [18]

pH-Range

4-8 <34> [74]
4-9 <26> [44]
5-8 <17> [16]
5.5-7.8 <19> (<19> phosphate buffer [1]) [1]
5.5-8 <9> [7]
6-8 <26> [3]
6-9 <36> (<36> activity is negligible below pH 6.0 and above pH 9.0 [36]) [36]
6.2-9.5 <19> (<19> Tris buffer [1]) [1]

Temperature optimum (°C)

25-30 <36> [36]
30 <13, 39> [18, 46]
55 <23> (<23> mutant enzyme, 2 disulfide bridges across the dimer interface of the enzyme engineered [55]) [55]

Temperature range (°C)

4-40 <19> [75]
15-60 <39> [46]
16-35 <23> [49]
30-70 <23> (<23> mutant enzyme, 2 disulfide bridges across the dimer interface of the enzyme engineered [55]) [55]

4 Enzyme Structure

Molecular weight

34000-38000 <17> (<17> gel filtration, sucrose density-gradient centrifugation [16]) [16]

35000 <15> (<15> SDS-PAGE [47]) [47]

35840 <15> (<15> cDNA sequence [47]) [47]

36620 <23> (<23> calculated molecular mass based on the sequence [55]) [55]

55000 <7> (<7> gel filtration [10]) [10]

58000 <6, 20> (<6> sucrose gradient analysis [10]) [3, 10, 16, 31]

60000 <19> [59]

63000-64000 <19> (<19> gel filtration [24]) [24]

64000 <19> [24]

64400 <5> (<5> sedimentation equilibrium measurement [13]) [13]

65000 <35> (<35> gel filtration, velocity sedimentation electrophoresis [28]) [28]

66000 <3, 22> (<3,22> gel filtration [25,48]) [25, 48]

67000 <23, 26> (<26> Ehrlich ascites carcinoma cells, gel filtration [3,26]) [3, 26]

68000 <34, 35> (<34,35> gel filtration [28,37]) [28, 37]

69000 <26> (<26> Ehrlich ascites carcinoma cells [28,37]) [28, 37]

70000 <9, 17, 19, 23, 26> (<9,19,23> gel filtration [21,27]) [3, 16, 21, 27, 52]

71000 <23, 26, 31> (<23> analytical ultracentrifugation [55]; <26> Ehrlich ascites carcinoma cells, SDS-PAGE [3,26]) [3, 26, 55, 57]

72000 <22> [77]

75000 <26> (<26> L1210 cells, gel filtration [26]) [26]

76000 <22, 26> (<26> L1210 cells, purified enzyme, gel filtration [31]) [31]

77000 <39> (<39> gel filtration [46]) [46]

78000 <26> (<26> L1210 cells, gel filtration [31]) [31]

100000 <36> (<36> sedimentation in sucrose gradients [36]) [36]

110000 <25> [36]

132000 <30> (<30> gel filtration [33]) [33]

150000 <13, 25, 30> (<13,25,30> gel filtration [33]) [33]

175000 <13> (<13> gel filtration [18]) [18]

185000 <16> (<16> gel filtration [38]) [38]

200000 <46> (<46> gel filtration [18]) [18]

Subunits

dimer <3-6, 9, 17, 19, 22, 23, 25, 26, 30, 31, 33-36, 39> (<3> 2 * 31000, SDS-PAGE [48]; <5> 2 * 31500, SDS-PAGE [13]; <5> 2 * 32000, SDS-PAGE [32]; <6> 2 * 29000, SDS-PAGE [10]; <6> 2 * 32000, SDS-PAGE [32]; <9> 2 * 35000, SDS-PAGE [27]; <17> 2 * 34000-38000 [16]; <19> 2 * 28000, crystallographic studies [41]; <19> 2 * 31000, SDS-PAGE [57]; <19> 2 * 32000 [24]; <19> 2 * 32500, SDS-PAGE [24]; <22> 2 * 33000, SDS-PAGE [25]; <22> 2 * 35000, SDS-PAGE [73]; <22> homodimer [77]; <23> 2 * 35000, SDS-PAGE [16,21,30,56]; <23> 2 * 36000 [49]; <25> 2 * 56000 [36]; <26> 2 * 35000,

SDS-PAGE [27,30]; <26> 2 * 38000, SDS-PAGE [31]; <26> 2 + 38500, SDS-PAGE [26,31]; <30> 2 * 65000, SDS-PAGE [33]; <31> 2 * 31000, SDS-PAGE [57]; <34> 2 * 35000, SDS-PAGE [37]; <35> 2 * 30000, SDS-PAGE [28]; <36> 2 * 50000, SDS-PAGE [36]; <39> 2 * 28000, SDS-PAGE [46]) [8, 10, 13, 16, 21, 24-28, 30, 32, 33, 36, 37, 41, 46, 48-50, 56, 57, 59, 64, 65, 73, 75, 77]
monomer <26> (<26> 1 * 71000, Ehrlich ascites carcinoma cells, SDS-PAGE [3]; <26> 1 * 69000, Ehrlich ascites carcinoma cells [28,37]) [3, 28, 37]

5 Isolation/Preparation/Mutation/Application

Source/tissue

blast cell <22> (<22> from patients with acute myelocytic leukemia [20, 31, 33]) [20, 25, 28, 31, 33, 48]
blood <13, 14, 44-46, 48> (<13, 14, 44-46, 48> intracellular parasites, host blood [18]; <44, 45, 48> obtained from infected rat blood [18]) [18]
bone marrow <22, 26> [70, 77]
colon <22> (<22> adenocarcinoma [39]) [39]
embryo <20> [3, 7, 16, 28, 31]
fetus <34, 40> [18, 39, 48]
fibroblast <26> [48]
heart <3> (<3> parasitic helminth, obtained from the pulmonary hearts and vessels of infected Wistar albino rats [48]) [48]
liver <22, 26, 34, 40> (<26> neonatal mouse liver [30]) [13, 18, 19, 30, 37, 39, 42, 48, 71, 74]
neuroblastoma cell <26> [19]
peritoneum <26> (<26> L1210 leukemia cells grown in the peritoneal cavity of CDF1 mice [31]) [31]
thymus <9, 22, 26, 40> (<9> thymus gland [7]) [2, 3, 6, 7, 13, 17, 25, 27, 31, 36, 48]

Purification

<3> [48]
<4> [69]
<5> (purified from phage infected E. coli [13,32]) [13, 32]
<6> (purified from phage infected E. coli [10,32]) [10, 32]
<7> (purified from phage infected E. coli [10]) [10]
<9> [7, 17, 25, 31, 48]
<11> [48]
<15> (recombinant enzyme, expressed in E. coli [47]) [47]
<16> (partially [38]) [38]
<19> (recombinant enzyme [52]) [1, 9, 17, 24, 39, 52, 59, 63, 69, 70, 75, 76]
<20> [7, 31, 48]
<22> (partially [2]; recombinant enzyme [52,77]; Pro-254 mutant enzymes [70]) [2, 20, 25, 26, 33, 39, 42, 48, 52, 63, 67, 70, 77]
<23> (partially [2,12,25]; recombinant enzyme [49]; mutant enzymes D221A, N, R, F, C, E, and S [60]) [2, 4, 5, 8, 12, 19, 21-23, 25, 27, 29-31, 49, 60, 67]

<26> (Ehrlich ascites carcinoma cells [3, 7, 12, 13]; recombinant enzyme [44, 48]; L1210 cells [26, 31]) [3, 4, 7, 12, 13, 17, 26, 30, 31, 35, 44, 48, 74]
<27> [58]
<30> (partially, copurified with dihydrofolate reductase [33]) [33]
<31> (recombinant enzyme, expressed in E. coli [57]) [48, 57]
<33> [64]
<34> (recombinant enzyme, expressed in E. coli [74]) [37, 74]
<35> [28, 37]
<36> [36]
<37> (partially [2]) [2]
<39> [46]

Crystallization
<4> [62, 65]
<19> (crystals of TS apoprotein grown using hanging drop method of vapor diffusion, cubic Laue group m3 with a = 133 A [41]; crystallized in the I2(1)3 space group [73]; crystals grown by the hanging drop vapor diffusion method [76]) [41, 49, 50, 52, 63, 66, 67, 70, 73-76]
<22> (mutant P254S [70]; crystals grown by vapor diffusion method in hanging drop setup [73]) [51, 66, 70, 72, 73, 77]
<23> [7, 49, 50, 53-56, 60, 67, 74, 75]
<25> [61]
<33> (crystals belong to space group P2(1)2(1)2(1), a = 54.05 A, b = 66.16 A and c = 178.76 A [64]) [64, 68, 75]
<34> [74]

Cloning
<4> (contains 2 genes for thymidilate synthase, ThyA and ThyB [65]; 2 different thymidilate synthases, bsTS-A and bsTS-B [62]) [62, 65]
<6> (T4 thymidilate synthase gene in amplification plasmid pKC30 [32]) [32, 59]
<11> [48, 59]
<15> (cDNA of TS gene subcloned into T7 polymerase vector pET15-b, production of active TS by complementation of Escherichia coli [47]) [47]
<19> (Escherichia coli transformed with a high amplifying expression plasmid containing the Escherichia coli thymidylate synthase gene [41]; overexpression of the encoded gene by transforming Escherichia coli Chi2913 [75]) [41, 52, 58, 59, 66, 75]
<22> (mutant P254D cDNA cloned into SacII/XhoI sites of the HaMSV retroviral vector and transfected into the TS negative cell line RJK88.13 [70]; cDNAs from Thymitaq resistant clones, gene amplification with plasmid pET-17xb, expressed in E. coli BL21(DE3) [63]; bacterial expression vector for overproduction in Escherichia coli [73]; Escherichia coli NM522 cells used for cloning and library construction, E. coli chi2913recA lacking TS used for expression by complementation [73,77]) [44, 59, 63, 66, 67, 70, 73, 77]
<23> (Escherichia coli chi2913 transformed with pKPTSd containing the complete coding sequence for Lactobacillus casei TS [49]; plasmid pSCTS9, synthetic TS gene in pUC18 [55]; cloned and overexpressed in Escherichia

coli [56]; synthetic DNA for TS pSCTS9 transfromed in Escherichia coli x2913recA [53]) [49, 53, 55, 56, 58, 59]
<25> (clone harbors the pO2CLSA-4 plasmid expressed in an Escherichia coli Rue 10 expression vector [61]) [61]
<26> (expression vector pETSM, cloned and expressed in Escherichia coli BL21(DE3)pLysS [44]) [44, 59]
<27> (thyA gene cloned, sequenced and expressed in Escherichia coli KLC1597, transformation with pBr322 [58]) [58]
<31> (Escherichia coli cells carrying an expression plasmid with the malarial dihydrofolate reductase-thymidylate synthase [57]) [53, 57]
<33> [47, 59]
<34> [71, 74]

Engineering

T155C/E188C/C244T <23> (<23> formation of 2 disulfide crosslinks across the subunit interface results in dramatic stabilization of the overall structure of the protein, retention of secondary structure at temperatures as high as 90°C, engineering of an intersubunit crosslink yields a fully active enzyme [55]) [55]

Application

medicine <3, 5, 6, 13-15, 22, 26, 30, 31, 33, 44-46, 48> (<5, 6, 22, 23> enzyme is believed to be the major site of the cytotoxic action of the cancer chemotherapeutic drug 5-fluorouracil, development of inhibitors as clinical useful drugs [5, 8, 19, 20, 25, 30-32, 35, 39, 49, 51, 52, 59, 63, 66, 72-74, 77]; <22> human TS mutants have potential applications in gene therapy by protecting hematopoietic progenitors from TS inhibitor toxicity, drug-resistant variants can be used to decrease the myelosuppressive side effects of TS-directed anticancer agents or to select genetically modified cells in vivo, clincal trials introduce chemoprotecting genes into hematopoietic cells [63, 70, 73]; <30> target enzyme for antitumor and antibacterial agents [33]; <30> target enzyme for malarial chemotherapy [33]; <13, 14, 44-46, 48> possible target for chemotherapeutic attack [18]; <33> important drug target, fungus causes opportunistic pneumonia infections in immune-compromised patients and is among the leading causes of death of AIDS patients, inhibitors selective for fungal TS over human TS will be greatly beneficial in combating this disease [68]; <3> development of effective chemotherapy of the eosinophilic meningoencephalitis angiostrongylosis [48]; <15> drug development against cryptococcosis, a pulmonary subclinical infection, in immunocompromised individuals having diseases such as leukemia, lymphoma or AIDS the infection seminates to the central nervous system, the fourth most common opportunistic infection complicating AIDS [47]; <31> development of antimalarial drugs, selective chemoptherapy directed at malarial TS, inhibitors of pyrimidine biosynthesis shall be considered for malaria chemotherapy [57]) [5, 8, 18-20, 25, 30-33, 35, 39, 44, 47-49, 51, 52, 57, 59, 63, 68, 70, 72-74, 77]

6 Stability

pH-Stability

5-10 <30> (<30> stable in this range but loses activity completely below pH 3.5 or above pH 11.5 [33]) [33]
5.8-8.5 <35> (<35> half-maximal activity at either pH 5.8 or pH 8.5 [28]) [28]
6.4-8.6 <15> (<15> 50% activity at pH 6.4 and 8.6 [47]) [47]
7-9 <34> (<34> 70% maximal activity at either pH 7.0 or 9.0, half-maximal activity at pH 6.3 [37]) [37]

Temperature stability

25 <19, 23> (<19> rather unstable [9]; <23> considerably stable, no loss of activity as long as 50 min [9]) [9]
37 <22> (<22> half-life of wild-type enzyme is 90 min, half-life of mutants P254F, P254S and P254D are approximately 50 min [70]; <22> no decrease in activity of recombinant enzymes up to [67]) [67, 70]
38 <9> (<9> unstable above [7]) [7]
40 <19> (<19> TS activity at 40°C falls rapidly with time [75]) [75]
43 <19> (<19> fairly stable, incubation for 10 min results in loss of 50% enzyme activity [69]) [69]
50 <4> (<4> fairly stable, incubation for 10 min results in loss of 50% enzyme activity [69]) [69]
52 <23> (<23> wild-type enzyme, precipitates at 52°C [55]) [55]
90 <23> (<23> mutant enzyme remains soluble, enhancement of activity is observed up to 60°C, possesses appreciable enzymatic activity at 65°C and retains secondary structure even at 90°C [55]) [55]

General stability information

<6>, quite unstable [10]
<7>, quite unstable [10]
<9>, enzyme can be kept active for about 2-3 weeks in buffer F in presence of 0.001 mM methotrexate, dialysis against dilute buffer leads to rapid irreversible inactivation as does freeze-drying or simple freezing [27]
<15>, stable upon quick-freezing in liquid nitrogen in buffer containing 150 mM KCl [47]
<19>, enzyme ist significantly stabilized by dUMP [9]
<19>, freezing and freeze-drying or dialysis causes rapid inactivation [27]
<22>, enzyme is stabilized by addition of 20-120 mM dithiothreitol + bovine serum albumin [25]
<22>, stable to protein denaturants and limited proteolysis [25]
<23>, thawing of frozen enzyme leads to irreversible precipitation of the protein and total loss of activity, no successfully lyophilization possible [21]
<26>, comparatively unstable, more stable in 20% sucrose solution [7]
<26>, enzyme is not inactivated by freezing and thawing [3]
<26>, glycerol and Triton X-100 stabilize the enzyme [44]
<36>, -80°C, storage in 30% glycerol stabilizes enzyme [36]
<39>, glycerol and Triton X-100 stabilize the enzyme [46]

Storage stability

<3>, -70°C preparation with 0.1% bovine serum albumin is stable up to 8 weeks of storage [48]

<5>, 0°C, enzyme can be stored in a concentrated solution for 2 weeks without a loss in activity [13]

<9>, -50°C-to -20°C, activity of purified enzyme remains unchanged for 2-3 months [27]

<9>, 0-7°C, concentrated enzyme keeps its activity for 14 days [27]

<15>, -20°C, 40% glycerol, stable over 2 months, 50% loss of activity by 24 h storage without glycerol [47]

<15>, -80°C, 40% glycerol, stable over 2 months [47]

<15>, -80°C, stable upon storage in Tris-HCl pH 7.4 and thiol [47]

<19>, -20°C, enzyme can be stored frozen for several months without significant loss of activity [24]

<19>, 0-10°C enzyme is stable in buffer D for several weeks [27]

<19>, 0°C, enzyme can be maintained in its active form in presence of saccharose or glycerol for several days [17]

<19>, 4°C enzyme can be maintained for at least 4 months with little or no loss in activity [69]

<22>, -25°C enzyme retains activity for 3 months when stored in presence of thiols, without thiols enzyme loses all activity in 2 days [25]

<22>, 4°C enzyme retains activity for 3 months when stored in presence of thiols, without thiols enzyme loses all activity in 2 days [25]

<23>, 0-5°C, enzyme is stable over a period of months in presence of 10 mM 2-mercaptoethanol in capped vessels layered with argon [21]

<26>, -20°C, activity remains unchanged after 1 month [44]

<26>, -20°C, after 5 months less than 25% activity is lost [26]

<26>, -20°C, can be stored in 10% glycerol, 0.05 M Tris-HCl, pH7.4 and 10 mM 2-mercaptoethanol for 30 days with a loss of approximately 50% activity [30]

<26>, -20°C, in presence of 0.1% Triton X-100 and 10% glycerol, no loss of activity when stored in lysis buffer for several weeks [44]

<26>, -20°C, occasionally loses activity [27]

<26>, -20°C, sulfate-precipitated enzyme in 0.1 M potassium phosphate, 0.01 M dithiothreitol, pH 7.1can be kept at least 4 months with only 25% loss of activity [3]

<26>, 0°C, enzyme rapidly loses its activity, more than 90% in 2 h [27]

<26>, 4°C, in presence of 0.1% Triton X-100 and 10% glycerol, no loss of activity when stored in lysis buffer for at least 2 weeks [44]

<30>, 4°C, enzyme loses half of its activity upon storage for a few days [33]

<34>, 4°C, enzyme retains activity for about 3 weeks when stored with 20 mM dithiothreitol and 0.1 mM dUMP, without dithiothreitol purified enzyme is inactivated totally in 2 days, without dUMP enzyme keeps 50% activity after 1 week [37]

<36>, -80°C, storage in 30% glycerol stabilizes enzyme [36]

<36>, 5°C half-life in buffered solution only 2 days [36]

<39>, -20°C, TS activity of crude enzyme remains unchanged for 1 month of storage, purified enzyme loses 50% of its activity within 3-4 weeks [46]

References

[1] Wahba, A.J.; Friedkin, M.: The enzymatic synthesis of thymidylate. I. Early steps in the purification of thymidylate synthetase of Escherichia coli. J. Biol. Chem., **237**, 3794-3801 (1962)
[2] Blakeley, R.L.: The biosythesis of thymidylic acid. IV. Further studies on thymidylate synthetase. J. Biol. Chem., **238**, 2113-2118 (1963)
[3] Fridland, A.; Langenbach, R.J.; Heidelberger, C.: Purification of thymidylate synthetase from Ehrlich ascites carcinoma cells. J. Biol. Chem., **246**, 7110-7114 (1971)
[4] Kalman, T.I.: On the mechanism of the reaction catalyzed by thymidylate synthetase. Ann. N.Y. Acad. Sci., **186**, 166-167 (1971)
[5] Danenberg, P.V.; Langenbach, R.J.; Heidelberger, C.: Purification of thymidylate synthetase from L. casei by affinity chromatography. Biochem. Biophys. Res. Commun., **49**, 1029-1033 (1972)
[6] Gupta, V.S.; Meldrum, J.B.: Purification and properties of thymidylate synthetase from pig thymus. Can. J. Biochem., **50**, 353-362 (1972)
[7] Horinishi, H.; Greenberg, D.M.: Purification and properties of thymidylate synthase from calf thymus. Biochim. Biophys. Acta, **258**, 741-752 (1972)
[8] Loeble, R.B.; Dunlap, R.B.: Characterization of the subunits of thymidylate synthetase. Biochem. Biophys. Res. Commun., **49**, 1671-1677 (1972)
[9] Santi, D.V.; Sakai, T.T.: Irreversible inhibition of thymidylate synthetase by 5-formyl-2-deoxyuridylic acid. Biochem. Biophys. Res. Commun., **46**, 1320-1325 (1972)
[10] Capco, G.R.; Krupp, J.R.; Mathews, C.K.: Bacteriophage-coded thymidylate synthetase: characteristics of the T4 and T5 enzymes. Arch. Biochem. Biophys., **158**, 726-735 (1973)
[11] Carpenter, N.J.: Properties and inhibition of thymidylate synthetase in Drosophila melanogaster. J. Insect Physiol., **20**, 1389-1401 (1974)
[12] Danenberg, P.V.; Heidelberger, C.: Purification of thymidylate synthetase with 2'-deoxyuridylate-agarose. Methods Enzymol., **34 B**, 520-523 (1974)
[13] Galivan, J.; Maley, G.F.; Maley, F.: Purification and properties of T2 bacteriophage-induced thymidylate synthetase. Biochemistry, **13**, 2282-2289 (1974)
[14] Kisliuk, R.L.; Gaumont, Y.: Polyglutamyl derivatives of folate as substrates and inhibitors of thymidylate synthetase. J. Biol. Chem., **249**, 4100-4103 (1974)
[15] Friedkin, M.; Plante, L.T.; Crawford, E.J.; Crumm, M.: Inhibition of thymidylate synthetase and dihydrofolate reductase by naturally occurring oligoglutamate derivatives of folic acid. J. Biol. Chem., **250**, 5614-5621 (1975)
[16] McCuen, R.W.; Sirotnak, F.M.: Thymidylate synthetase from Diplococcus pneumoniae, properties and inhibition by folate analogs. Biochim. Biophys. Acta, **384**, 369-380 (1975)

[17] Slavik, K.; Rode, W.; Slavikova, V.: Purification of thymidylate synthetase from enzyme-poor sources by affinity chromatography. Biochemistry, **15**, 4222-4227 (1976)
[18] Al Chalabi, K.; Gutteridge, W.E.: Presence and properties of thymidylate synthase in trypanosomatids. Biochim. Biophys. Acta, **481**, 71-79 (1977)
[19] Danenberg, P.V.: Thymidylate synthetase - a target enzyme in cancer chemotherapy. Biochim. Biophys. Acta, **473**, 73-92 (1977)
[20] Dolnick, B.J.; Cheng, Y.C.: Human thymidylate synthetase. II. Derivatives of pteroylmono- and -polyglutamates as substrates and inhibitors. J. Biol. Chem., **253**, 3563-3567 (1978)
[21] Dunlap, R.B.: TMP synthetase from Lactobacillus casei. Methods Enzymol., **51**, 90-97 (1978)
[22] Mertes, M.P.; Chang, C.T.C.; de Clercq, E.; Huang, G.F.; Torrence, P.F.: 5-Nitro-2-deoxyuridine 5-monophosphate is a potent irreversible inhibitor of Lactobacillus caesi thymidylate synthetase. Biochem. Biophys. Res. Commun., **84**, 1054-1059 (1978)
[23] Whiteley, J.M.: 5-Fluoro-2-deoxyuridylate-agarose in the affinity-chromatographic purification of thymidylate synthetase. Methods Enzymol., **51**, 98-104 (1978)
[24] Haertle, T.; Wohlrab, F.; Guschlbauer, W.: Thymidylate synthetase from Escherichia coli K12. Purification, and dependence of kinetic properties on sugar conformation and size of the 2 substituent. Eur. J. Biochem., **102**, 223-230 (1979)
[25] Lockshin, A.; Moran, R.G.; Danenberg, P.V.: Thymidylate synthetase purified to homogeneity from human leukemic cells. Proc. Natl. Acad. Sci. USA, **76**, 750-754 (1979)
[26] Rode, W.; Scanlon, K.J.; Hynes, J.; Bertino, J.R.: Purification of mammalian tumor (L1210) thymidylate synthetase by affinity chromatography on stable biospecific adsorbent. Stabilization of the enzyme with neutral detergents. J. Biol. Chem., **254**, 11538-11543 (1979)
[27] Slavik, K.; Slavikova, V.: Purification of thymidylate synthetase from enzyme-poor sources by affinity chromatography. Methods Enzymol., **66**, 709-723 (1980)
[28] Bisson, L.F.; Thorner, J.: Thymidylate synthetase from Saccharomyces cerevisiae. Purification and enzymic properties. J. Biol. Chem., **256**, 12456-12462 (1981)
[29] Kalman, T.I.; Goldman, D.: Inactivation of thymidylate synthetase by a novel mechanism-based enzyme inhibitor: 1-(β-D-2-deoxyribofuranosyl) 8-azapurin-2-one 5-monophosphate. Biochem. Biophys. Res. Commun., **102**, 682-689 (1981)
[30] Priest, D.G.; Doig, M.T.; Hynes, J.B.: Purification of mouse liver thymidylate synthetase by affinity chromatography using 10-methyl-5,8-dideazafolate as the affinant. Experientia, **37**, 119-120 (1981)
[31] Banerjee, C.K.; Bennett, L.L.; Brockman, R.W.; Sani, B.P.; Temple, C.: A convenient procedure for purification of thymidylate synthase from L1210 cells. Anal. Biochem., **121**, 275-280 (1982)

[32] Belfort, M.; Moelleken, A.; Maley, G.F.; Maley, F.: Purification and properties of T4 phage thymidylate synthetase produced by the cloned gene in an amplification vector. J. Biol. Chem., **258**, 2045-2051 (1983)
[33] Pattanakitsakul, S.N.; Ruenwongsa, P.: Characterization of thymidylate synthetase and dihydrofolate reductase from Plasmodium berghei. Int. J. Parasitol., **14**, 513-520 (1984)
[34] Kanai, F.; Kaneko, T.; Morishima, H.; Isshiki, K.; Takita, T.; Takeuchi, T.; Umezawa, H.: Vanoxonin, a new inhibitor of thymidylate synthetase. III. Inhibition of thymidylate synthetase by vanoxonin-vanadium complex. J. Antibiot., **38**, 39-50 (1985)
[35] Sani, B.P.; Vaid, A.; Cory, J.G.; Brockman, R.W.; Elliott, R.D.; Montgomery, J.A.: 5-Haloacetamido-5-deoxythymidines: novel inhibitors of thymidylate synthase. Biochim. Biophys. Acta, **881**, 175-184 (1986)
[36] Bachmann, B.; Follmann, H.: Deoxyribonucleotide biosynthesis in green algae: characterization of thymidylate synthase-dihydrofolate reductase in Scenedesmus obliquus. Arch. Biochem. Biophys., **256**, 244-252 (1987)
[37] Nakata, R.; Tsukamoto, I.; Miyoshi, M.; Kojo, S.: Purification and characterization of thymidylate synthetase from rat regenerating liver. Biochim. Biophys. Acta, **924**, 297-302 (1987)
[38] Nielsen, E.; Cella, R.: Thymidylate synthase in plant cells: Kinetic and molecular properties of the enzyme from Daucus carota L. cell cultures. Plant Cell Physiol., **29**, 503-508 (1988)
[39] Radparvar, S.; Houghton, P.J.; Houghton, J.A.: Characteristics of thymidylate synthase purified from a human colon adenocarcinoma. Arch. Biochem. Biophys., **260**, 342-350 (1988)
[40] Chen, S.C.; Daron, H.H.; Aull, J.L.: Inhibition of thymidylate synthase by pyridoxal phosphate. Int. J. Biochem., **21**, 1217-1221 (1989)
[41] Matthews, D.A.; Villafranca, J.E.; Janson, C.A.; Smith, W.W.; Welsh, K.; Freer, S.: Stereochemical mechanism of action for thymidylate synthase based on the X-ray structure of the covalent inhibitory ternary complex with 5-fluoro-2-deoxyuridylate and 5,10-methylenetetrahydrofolate. J. Mol. Biol., **214**, 937-948 (1990)
[42] Balinska, M.; Rhee, M.; Whiteley, J.M.; Priest, D.G.; Galivan, J.: Inhibition of mammalian thymidylate synthase by 10-formyltetrahydropteroylpolyglutamate. Arch. Biochem. Biophys., **284**, 219-222 (1991)
[43] Bures, A.K.; Daron, H.H.; Aull, J.L.: Inhibition of thymidylate synthase by glyceraldehyde 3-phosphate. Int. J. Biochem., **23**, 733-736 (1991)
[44] Zhang, H.; Cisneros, R.J.; Deng, W.; Zapf, J.W.; Johnson, L.F.; Dunlap, R.B.: Purification and characterization of recombinant mouse thymidylate synthase. Biochim. Biophys. Acta, **1077**, 35-46 (1991)
[45] Chen, D.H.; Daron, H.H.; Aull, J.L.: 1-Phenyl-3-trimethylaminopropyl carbodiimide: a new inhibitor of thymidylate synthase. J. Enzyme Inhib., **5**, 259-268 (1992)
[46] Galova, M.; Koptidesova, D.; Rusznakova, D.; Racay, P.; Kollarova, M.: Characteristics of NADPH-dependent thymidylate synthetase purified from Streptomyces aureofaciens. Arch. Biochem. Biophys., **296**, 81-87 (1992)

[47] Livi, L.L.; Edman, U.; Schneider, G.P.; Greene, P.J.; Santi, D.V.: Cloning, expression and characterization of thymidylate synthase from Cryptococcus neoformans. Gene, **150**, 221-226 (1994)
[48] So, N.N.C.; Wong, P.C.L.; Ko, R.C.: Angiostrongylus cantonensis: Characterization of thymidylate synthetase. Exp. Parasitol., **79**, 526-535 (1994)
[49] Garcia-Fuentes, L.; Reche, P.; Lopez-Mayorga, O.; Santi, D.V.; Gonzalez-Pacanowska, D.; Baron, C.: Thermodynamic analysis of the binding of 5-fluoro-2'-deoxyuridine 5'-monophosphate to thymidylate synthase over a range of temperatures. Eur. J. Biochem., **232**, 641-645 (1995)
[50] Maley, F.; Pedersen-Lane, J.; Changchien, L.: Complete restoration of activity to inactive mutants of Escherichia coli thymidylate synthase: Evidence that E. coli thymidylate synthase is a half-the-sites activity enzyme. Biochemistry, **34**, 1469-1474 (1995)
[51] Schiffer, C.A.; Clifton, I.J.; Davisson, V.J.; Santi, D.V.; Stroud, R.M.: Crystal structure of human thymidylate synthase: a structural mechanism for guiding substrates into the active site. Biochemistry, **34**, 16279-16287 (1995)
[52] Voeller, D.M.; Changchien, L.M.; Maley, G.F.; Maley, F.; Takechi, T.; Turner, R.E.; Montfort, W.R.; Allegra, C.J.; Chu, E.: Characterization of a specific interaction between Escherichia coli thymidylate synthase and Escherichia coli thymidylate synthase mRNA. Nucleic Acids Res., **23**, 869-875 (1995)
[53] Cho, S.W.; Lee, J.; Cho, E.H.: Effects of mutation at arginine-218 residue on the reaction of synthetic thymidylate synthase and 5-fluorouracil. Exp. Mol. Med., **28**, 77-83 (1996)
[54] Finer-Moore, J.S.; Liu, L.; Schafmeister, C.E.; Birdsall, D.L.; Mau, T.; Santi, D.V.; Stroud, R.M.: Partitioning roles of side chains in affinity, orientation, and catalysis with structures for mutant complexes: Asparagine-229 in thymidylate synthase. Biochemistry, **35**, 5125-5136 (1996)
[55] Gokhale, R.S.; Agarwalla, S.; Santi, D.V.; Balaram, P.: Covalent reinforcement of a fragile region in the dimeric enzyme thymidylate synthase stabilizes the protein against chaotrope-induced unfolding. Biochemistry, **35**, 7150-7158 (1996)
[56] Gokhale, R.S.; Agarwalla, S.; Santi, D.V.; Balaram, P.: Covalent reinforcement of a fragile region in the dimeric enzyme thymidylate synthase stabilizes the protein against chaotrope-induced unfolding. Biochemistry, **35**, 7150-7158 (1996)
[57] Hekmat-Nejad, M.; Rathod, P.K.: Kinetics of Plasmodium flaciparum thymidylate synthase: interactions with high-affinity metabolites of 5-fluoroorotate and D1694. Antimicrob. Agents Chemother., **40**, 1628-1632 (1996)
[58] Carlson, J.H.; Hill, S.A.: Identification and characterization of thymidylate synthase from Neisseria gonorrhoeae. FEMS Microbiol. Lett., **151**, 225-230 (1997)
[59] Spencer, H.T.; Villafranca, J.E.; Appleman, J.R.: Kinetic scheme for thymidylate synthase from Escherichia coli: determination from measurements of ligand binding, primary and secondary isotope effects, and pre-steady-state catalysis. Biochemistry, **36**, 4212-4222 (1997)

[60] Chiericatti, G.; Santi, D.V.: Aspartate 221 of thymidylate synthase is involved in folate cofactor binding and in catalysis. Biochemistry, **37**, 9038-9042 (1998)
[61] Liang, P.H.; Anderson, K.S.: Substrate channeling and domain-domain interactions in bifunctional thymidylate synthase-dihydrofolate reductase. Biochemistry, **37**, 12195-12205 (1998)
[62] Stout, T.J.; Schellenberger, U.; Santi, D.V.; Stroud, R.M.: Crystal structures of a unique thermal-stable thymidylate synthase from Bacillus subtilis. Biochemistry, **37**, 14736-14747 (1998)
[63] Tong, Y.; Liu-Chen, X.; Ercikan-Abali, E.A.; Capiaux, G.M.; Zhao, S.C.; Banerjee, D.; Bertino, J.R.: Isolation and characterization of Thymitaq (AG337) and 5-fluoro-2-deoxyuridylate-resistant mutants of human thymidylate synthase from ethyl methanesulfonate-exposed human sarcoma HT1080 cells. J. Biol. Chem., **273**, 11611-11618 (1998)
[64] Anderson, A.C.; O'Neil, R.H.; DeLano, W.L.; Stroud, R.M.: The structural mechanism for half-the-sites reactivity in an enzyme, thymidylate synthase, involves a relay of changes between subunits. Biochemistry, **38**, 13829-13836 (1999)
[65] Fox, K.M.; Maley, F.; Garibian, A.; Changchien, L.M.; Van Roey, P.: Crystal structure of thymidylate synthase A from Bacillus subtilis. Protein Sci., **8**, 538-544 (1999)
[66] Mahdavian, E.; Spencer, H.T.; Dunlap, R.B.: Kinetic studies on drug-resistant variants of Escherichia coli thymidylate sythase: Functional effects of amino acid sunstitutions at residue 4. Arch. Biochem. Biophys., **368**, 257-264 (1999)
[67] Steadman, D.J.; Spencer, H.T.; Dunlap, R.B.; Berger, S.H.: Substitution at residue 214 of human thymidylate synthase alters nucleotide binding and isomerization of ligand-protein complexes. Biochemistry, **38**, 5582-5587 (1999)
[68] Anderson, A.C.; Perry, K.M.; Freymann, D.M.; Stroud, R.M.: The crystal structure of thymidylate synthase from Pneumocystis carinii reveals a fungal insert important for drug design. J. Mol. Biol., **297**, 645-657 (2000)
[69] Changchien, L.M.; Garibian, A.; Frasca, V.; Lobo, A.; Maley, G.F.; Maley, F.: High-Level expression of Escherichia coli and Bacillus subtilis thymidylate synthases. Protein Expr. Purif., **19**, 265-270 (2000)
[70] Fantz, C.; Shaw, D.; Jennings, W.; Forsthoefel, A.; Kitchens, M.; Phan, J.; Minor, W.; Lebioda, L.; Berger, F.G.; Spencer, H.T.: Drug-resistant variants of Escherichia coli thymidylate synthase: effects of substitutions at Pro-254. Mol. Pharmacol., **57**, 359-366 (2000)
[71] Felczak, K.; Miazga, A.; Poznanski, J.; Bretner, M.; Kulikowski, T.; Dzik, J.M.; Golos, B.; Zielinski, Z.; Ciesla, J.; Rode, W.: 5-Substituted N^4-hydroxy-2'-deoxycytidines and their 5'-monophosphates: Synthesis, conformation, interaction with tumor thymidylate synthase, and in vitro antitumor activity. J. Med. Chem., **43**, 4647-4656 (2000)
[72] Jarmula, A.; Les, A.; Rode, W.: Different activities of 5-hydroxy-dUMP and 5-hydroxymethyl-dUMP in thymidylate synthase-catalyzed reaction in view

of molecular modeling and structural studies. Bioorg. Chem., **28**, 156-162 (2000)

[73] Phan, J.; Mahdavian, E.; Nivens, M.C.; Minor, W.; Berger, S.; Spencer, H.T.; Dunlap, R.B.; Lebioda, L.: Catalytic cysteine of thymidylate synthase is activated upon substrate binding. Biochemistry, **39**, 6969-6978 (2000)

[74] Golos, B.; Dzik, J.M.; Kazimierczuk, Z.; Ciesla, J.; Zielinski, Z.; Jankowska, J.; Kraszewski, A.; Stawinski, J.; Rode, W.; Shugar, D.: Interaction of thymidylate synthase with the 5'-thiophosphates, 5'-dithiophosphates, 5'-H-phosphonates and 5'-S-thiosulfates of 2'-deoxyuridine, thymidine and 5-fluoro-2'-deoxyuridine. Biol. Chem., **382**, 1439-1445 (2001)

[75] Saxl, R.L.; Changchien, L.M.; Hardy, L.W.; Maley, F.: Parameters affecting the restoration of activity to inactive mutants of thymidylate synthase via subunit exchange: Further evidence that thymidylate synthase is a half-of-the-sites activity enzyme. Biochemistry, **40**, 5275-5282 (2001)

[76] Fritz, T.A.; Liu, L.; Finer-Moore, J.S.; Stroud, R.M.: Tryptophan 80 and leucine 143 are critical for the hydride transfer step of thymidylate synthase by controlling active site access. Biochemistry, **41**, 7021-7029 (2002)

[77] Kawate, H.; Landis, D.M.; Loeb, L.A.: Distribution of mutations in human thymidylate synthase yielding resistance to 5-fluorodeoxyuridine. J. Biol. Chem., **277**, 36304-36311 (2002)

Isoflavone 4'-O-methyltransferase 2.1.1.46

1 Nomenclature

EC number
2.1.1.46

Systematic name
S-adenosyl-L-methionine:isoflavone 4'-O-methyltransferase

Recommended name
isoflavone 4'-O-methyltransferase

Synonyms
2-hydroxyisoflavone 4'-O-methyltransferase
HI4'OMT
IOMT
isoflavone O-methyltransferase
isoflavone methyltransferase
methyltransferase, 4'-hydroxyisoflavone

CAS registry number
55071-80-2

2 Source Organism

<1> *Cicer arietinum* (chick pea [1]) [1]
<2> *Medicago sativa* (alfalfa [2]) [2-4, 6]
<3> *Medicago sativa* (IOMT8 [5]) [5]
<4> *Glycyrrhiza echinata* (Fabaceae [7]) [7]

3 Reaction and Specificity

Catalyzed reaction
S-adenosyl-L-methionine + an isoflavone = S-adenosyl-L-homocysteine + a 4'-O-methylisoflavone (<1> ordered bi-bi-mechanism [1]; <1> last step in the biosynthesis of 4'-methoxyisoflavones [1])

Reaction type
methyl group transfer

Natural substrates and products

- **S** S-adenosyl-L-methionine + 2,4',7-trihydroxyisoflavanone <2> (Reversibility: ? <2> [6]) [6]
- **P** S-adenosyl-L-homocysteine + 2,7-dihydroxy-4'-methoxyisoflavanone <2> (<2> main reaction in vivo [6]) [6]
- **S** S-adenosyl-L-methionine + 2,4',7-trihydroxyisoflavanone <2, 4> (<4> main pathway of formononetin biosynthesis [7]) (Reversibility: ? <2, 4> [6, 7]) [6, 7]
- **P** S-adenosyl-L-homocysteine + 2,7-dihydroxy-4'-methoxyisoflavanone <2, 4> [6, 7]
- **S** S-adenosyl-L-methionine + isoflavone <1> (<1> last step in the biosynthesis of 4'-methoxyisoflavones [1]) (Reversibility: ? <1> [1]) [1]
- **P** ?
- **S** Additional information <2, 3> (<2> IOMT shows specificity for 7-hydroxy-position in vitro, but in contrast, modulation of enzyme in transgenic alfalfa has corresponding effects on the amounts of 4'-O-methylated isoflavonoids present [3]; <2,3> in vivo reaction is 4'-O-methylation, in vitro reaction is 7-O-methylation [4,5]; <2> colocalization with isoflavone synthase cytochrome P_{450} ensures rapid B-ring methylation of unstable product of isoflavone synthase [6]) [3-6]
- **P** ?

Substrates and products

- **S** S-adenosyl-L-methionine + (+)6a-hydroxymaackiain <3> (Reversibility: ? <3> [5]) [5]
- **P** S-adenosyl-L-homocysteine + pisatin <3> [5]
- **S** S-adenosyl-L-methionine + 2,4',7-trihydroxyisoflavanone <2, 4> (Reversibility: ? <2, 4> [6, 7]) [6, 7]
- **P** S-adenosyl-L-homocysteine + 2,7-dihydroxy-4'-methoxyisoflavanone <2, 4> [6, 7]
- **S** S-adenosyl-L-methionine + 6,7,4'-trihydroxyisoflavone <3> (Reversibility: ? <3> [5]) [5]
- **P** S-adenosyl-L-homocysteine + 7-methoxy-6,4'-dihydroxyisoflavone <3> [5]
- **S** S-adenosyl-L-methionine + daidzein <1, 2> (<1> i.e. 7,4'-dihydroxyisoflavone [1]) (Reversibility: ? <1, 2> [1-4]) [1-4]
- **P** S-adenosyl-L-homocysteine + formononetin <1, 2> (<1> i.e. 7-hydroxy-4'-methoxyisoflavone [1]) [1-4]
- **S** S-adenosyl-L-methionine + daidzein <2, 3> (<2> i.e. 7,4'-dihydroxyisoflavone [3]) (Reversibility: ? <2, 3> [3, 4, 5]) [3, 4, 5]
- **P** S-adenosyl-L-homocysteine + isoformononetin <2, 3> (<2> i.e. 7-methoxy-4'-hydroxyisoflavone, reaction product of ectopically expressed enzyme and if substrate is exogenously supplied to unchallenged tissue [3]) [3, 4, 5]
- **S** S-adenosyl-L-methionine + genistein <1, 2> (Reversibility: ? <1, 2> [1, 3]) [1, 3]
- **P** S-adenosyl-L-homocysteine + biochanin A <1, 2> [1, 3]

S S-adenosyl-L-methionine + genistein <2, 3> (Reversibility: ? <2, 3> [3, 5]) [3, 5]

P S-adenosyl-L-homocysteine + prunetin <2, 3> (<2> i.e. 7-methoxy-5,4'-dihydroxy-isoflavone, reaction product of ectopically expressed enzyme and if substrate is exogenously supplied to unchallenged tissue [3]) [3, 5]

S Additional information <1-3> (<1> no substrates: 4-hydroxycinnamic acid, 4-hydroxybenzoic acid, 3,4-dihydroxycinnamic acid, naringenin, apigenin, luteolin, kaempferol, wedelolactone [1]; <2,3> in vivo reaction is 4'-O-methylation, in vitro reaction is 7-O-methylation [4,5]) [1, 4, 5]

P ?

Inhibitors

Cd^{2+} <1> (<1> strong [1]) [1]
Cu^{2+} <1> (<1> strong [1]) [1]
Fe^{2+} <1> [1]
Mn^{2+} <1> [1]
S-adenosyl-L-homocysteine <1> (<1> product inhibition, competitive [1]) [1]
Zn^{2+} <1> (<1> strong [1]) [1]
p-chloromercuribenzoate <1> [1]

Activating compounds

SH-group-protecting agents <1> (<1> activation, dithioerythritol more effective than 2-mercaptoethanol [1]) [1]

Metals, ions

Additional information <1> (<1> no activation by Mg^{2+} or other divalent cations [1]) [1]

Specific activity (U/mg)

27.9 <1> [1]

K_m-Value (mM)

0.08 <1> (daidzein) [1]
0.16 <1> (S-adenosyl-L-methionine) [1]

pH-Optimum

8.5 <2> (<2> 7-IOMT activity, no clear pH-optimum for 4'-IOMT activity [4]) [4]
8.6 <1> (<1> Tris-HCl buffer [1]) [1]
9 <1> (<1> glycine-NaOH buffer [1]) [1]

Temperature optimum (°C)

30 <1> (<1> assay at [1]) [1]

4 Enzyme Structure

Molecular weight

110000 <1> (<1> gel filtration) [1]

Subunits

? <2, 3> (<2> x * 41000, SDS-PAGE [4]; <3> x * 41000, SDS-PAGE, recombinant enzyme [5]) [4, 5]

5 Isolation/Preparation/Mutation/Application

Source/tissue

cell suspension culture <1, 2> [1, 2, 6]
seedling <1> [1]

Localization

endoplasmic reticulum <2> (<2> colocalizes with isoflavone synthase cytochrome P_{450} [6]) [6]
soluble <1> [1]

Purification

<1> [1]
<2> [4]

Cloning

<3> [5]
<4> [7]

6 Stability

Organic solvent stability

dimethyl sulfoxide <1> (<1> 10% v/v, stable to [1]) [1]

Storage stability

<1>, 4°C, concentrated ammonium sulfate in the presence of dithioerythritol, several weeks [1]

References

[1] Wengenmayer, H.; Ebel, J.; Grisebach, H.: Purification and properties of a S-adenosylmethionine:isoflavone 4-O-methyltransferase from cell suspension cultures of Cicer arietinum L. Eur. J. Biochem., **50**, 135-143 (1974)
[2] Edwards, R.; Dixon, R.A.: Isoflavone O-methyltransferase activities in elicitor-treated cell suspension cultures of Medicago sativa. Phytochemistry, **30**, 2597-2606 (1991)
[3] He, X.Z.; Dixon, R.A.: Genetic manipulation of isoflavone 7-O-methyltransferase enhances biosynthesis of 4'-O-methylated isoflavonoid phytoalexins and disease resistance in alfalfa. Plant Cell, **12**, 1689-1702 (2000)
[4] He, X.Z.; Dixon, R.A.: Affinity chromatography, substrate/product specificity, and amino acid sequence analysis of an isoflavone O-methyltransferase from alfalfa (Medicago sativa L.). Arch. Biochem. Biophys., **336**, 121-129 (1996)

[5] He, X.Z.; Reddy, J.T.; Dixon, R.A.: Stress responses in alfalfa (Medicago sativa L). XXII. cDNA cloning and characterization of an elicitor-inducible isoflavone 7-O-methyltransferase. Plant Mol. Biol., **36**, 43-54 (1998)
[6] Liu, C.J.; Dixon, R.A.: Elicitor-induced association of isoflavone O-methyltransferase with endomembranes prevents the formation and 7-O-methylation of daidzein during isoflavonoid phytoalexin biosynthesis. Plant Cell, **13**, 2643-2658 (2001)
[7] Akashi, T.; Sawada, Y.; Shimada, N.; Sakurai, N.; Aoki, T.; Ayabe, S.: cDNA cloning and biochemical characterization of S-adenosyl-L-methionine:2,7,4'-trihydroxyisoflavanone 4'-O-methyltransferase, a critical enzyme of the legume isoflavonoid phytoalexin pathway. Plant Cell Physiol., **44**, 103-112 (2003)

Indolepyruvate C-methyltransferase 2.1.1.47

1 Nomenclature

EC number
2.1.1.47

Systematic name
S-adenosyl-L-methionine:indolepyruvate C-methyltransferase

Recommended name
indolepyruvate C-methyltransferase

Synonyms
indolepyruvate 3-methyltransferase
indolepyruvate methyltransferase
indolepyruvic acid methyltransferase
methyltransferase, indolepyruvate

CAS registry number
54576-88-4

2 Source Organism

<1> *Streptomyces griseus* [1, 2]

3 Reaction and Specificity

Catalyzed reaction
S-adenosyl-L-methionine + indolepyruvate = S-adenosyl-L-homocysteine + (S)-β-methylindolepyruvate

Reaction type
methyl group transfer

Natural substrates and products
S S-adenosyl-L-methionine + indolepyruvate <1> (<1> reaction in indolmycin biosynthesis [2]) (Reversibility: ? <1> [2]) [2]
P ?

Substrates and products
S S-adenosyl-L-methionine + indolepyruvate <1> (Reversibility: ? <1> [1, 2]) [1, 2]
P S-adenosyl-L-homocysteine + (S)-β-methylindolepyruvate <1> [1, 2]

S Additional information <1> (<1> no substrates are phenylpyruvate, *p*-hydroxyphenylpyruvate [1, 2]; <1> no substrate: L-tryptophan [2]) [1, 2]
P ?

Inhibitors

1,10-phenanthroline <1> [1, 2]
2,2'-bipyridyl <1> [1, 2]
N-ethylmaleimide <1> [1, 2]
iodoacetamide <1> (<1> weak [2]) [2]
iodoacetic acid <1> (<1> weak [2]) [2]
p-chloromercuribenzoate <1> (<1> strong, reversible by cysteine [2]) [1, 2]
Additional information <1> (<1> no inhibition by EDTA [1, 2]; <1> not inhibitory: indolmycin, L-tryptophan, diethyldithiocarbamate, L-tryptophanylglycine, L-tryptophanamide [2]) [1, 2]

Metals, ions

Additional information <1> (<1> no requirement for Mg^{2+}, Mn^{2+}, Ca^{2+} [2]) [2]

Specific activity (U/mg)

0.0083 <1> [1]

K_m-Value (mM)

0.0048 <1> (indolepyruvate) [1, 2]
0.013 <1> (S-adenosyl-L-methionine) [1, 2]

pH-Optimum

7.5 <1> (<1> phosphate buffer [1,2]) [1, 2]
7.5-8.5 <1> [1, 2]

4 Enzyme Structure

Molecular weight

55000 <1> (<1> gel filtration) [1, 2]
58500 <1> (<1> sucrose density gradient centrifugation) [1]

5 Isolation/Preparation/Mutation/Application

Localization

soluble <1> [1]

Purification

<1> [1, 2]

6 Stability

pH-Stability
5.5 <1> (<1> and below, irreversible inactivation [1,2]) [1, 2]

Temperature stability
Additional information <1> (<1> heating inactivates [2]) [2]

General stability information
<1>, freezing inactivates [2]
<1>, heating inactivates [2]

Storage stability
<1>, 2°C, up to 1 month [2]

References

[1] Speedie, M.K.; Hornemann, U.; Floss, H.G.: S-Adenosylmethionine:indolepyruvate 3-methyltransferase. Methods Enzymol., **43**, 498-502 (1975)
[2] Speedie, M.K.; Hornemann, U.; Floss, H.G.: Isolation and characterization of tryptophan transaminase and indolepyruvate C-methyltransferase. Enzymes involved in indolmycin biosynthesis in Streptomyces griseus. J. Biol. Chem., **250**, 7819-7825 (1975)

rRNA (adenine-N^6-)-methyltransferase 2.1.1.48

1 Nomenclature

EC number
2.1.1.48

Systematic name
S-adenosyl-L-methionine:rRNA (adenine-N^6-)-methyltransferase

Recommended name
rRNA (adenine-N^6-)-methyltransferase

Synonyms
Erm methyltransferase
ErmC 23S rRNA methyltransferase
S-adenosylmethionine:ribosomal ribonucleic acid-adenine (N^6-)methyltransferase
ermC 23 S rRNA methyltransferase
erythromycin-resistance methylase
gene ksgA methyltransferase
ksgA methyltransferase
methyltransferase, ribosomal ribonucleate adenine 6-
rRNA-adenine(N^6-)methylase
ribonucleic acid-adenine (N^6) methylase
ribosomal RNA adenine N^6-methyltransferase

CAS registry number
9076-81-7

2 Source Organism

<1> *Escherichia coli* (strain B [1,3]) [1, 3, 9]
<2> *Streptomyces erythraeus* [2]
<3> *Bacillus subtilis* (BD170 [4]) [4, 5, 8]
<4> *Saccharomyces erythraea* [6]
<5> *Streptomyces fradiae* (NRRL 2702 [7]) [7]

3 Reaction and Specificity

Catalyzed reaction
S-adenosyl-L-methionine + rRNA = S-adenosyl-L-homocysteine + rRNA containing N^6-methyladenine

Reaction type

methyl group transfer

Natural substrates and products

S S-adenosyl-L-methionine + rRNA <2, 3, 4> (<2>, the enzyme confers resistance to erythromycin in the producing organism Streptomyces erythraeus [2]; <4>, the enzyme confers resistance to macrolide, lincosamide and streptogramin B antibiotics by methylation of adenosine2058 in 23S rRNA [6]; <3>, the enzyme mediates bacterial resistance to erythromycin and other antibiotics of the macrolide-lincosamide-streptogramin B group by methylation of the 23S rRNA component of the 50S subunit in bacterial ribosomes [8]) (Reversibility: ? <2, 3, 4> [2, 6, 8]) [2, 6, 8]

P S-adenosyl-L-homocysteine + rRNA containing N^6-methyladenine

Substrates and products

S S-adenosyl-L-methionine + 9-β-ribosyl-2,6-diaminopurine <1> [1]

P S-adenosyl-L-homocysteine +9-β-ribosyl-2-methylamino-6-aminopurine

S S-adenosyl-L-methionine + rRNA <1, 2, 3, 4, 5> (<1>, the enzyme methylates a limited number of adenine moieties in heterologous rRNA from Micrococcus lysodeikticus and Bacillus subtilis and methyl-deficient homologous rRNA [1]; <2>, methylates E. coli 23S rRNA residue adenosine2058 [2]; <1>, E. coli strain B possesses two species of rRNA-adenine (N^6-)methylase, each of which recognizes a specific adenine moiety in a unique pentapurine nucleotide sequence in single-stranded region of rRNA [3]; <3>, enzyme incorporates two methyl groups per Bacillus subtilis 23S rRNA molecule [4]; <3>, the enzyme produces both N^6-mono and N^6,N^6-dimethylated adenine residues in Bacillus subtilis 23S rRNA. An almost total conversion of monomethylated intermediates into dimethylated products is observed upon completion of the reaction. The addition of the two methyl groups to each 23S rRNA molecule takes place through a monomethylated intermediate and suggests that the enzyme dissociates from its RNA substrate between the two consecutive methylation reactions. The reaction involves a sequential mechanism occuring by two consecutive random bi bi reactions [5]; <4>, the enzyme methylates adenosine2058 in 23S rRNA. Methylation is reduced by substitution in the 2051-2060 region of the 23S rRNA and is abolished by changes at position 2058 [6]; <1>, methylation of the conserved adenosine1518-adenosine1519 in Escherichia coli 16S ribosomal RNA by the ksgA methyltransferase is influenced by methylations around the similarly conserved uracil1512-guanosine1523 base pair in the 3' terminal hairpin [9]) (Reversibility: ? <1-5> [1-9]) [1, 2, 3, 4, 5, 6, 7, 8, 9]

P S-adenosyl-L-homocysteine + rRNA containing N^6-methyladenine <3> (<3>,the enzyme produces both N^6-mono and N^6,N^6-dimethylated adenine residues in Bacillus subtilis 23S rRNA [5]) [5]

Inhibitors

N^6,N^6-dimethyladenine-23S rRNA <3> [5]
S-adenosyl-L-homocysteine <3> [5]

SO_4^{2-} <1> [3]
nonmethylated rRNA <1> (<1>, competitive [3]) [3]
polyvinylsulfate <1> (<1>, 54% inhibition by 0.11 mg/ml [3]) [3]

Metals, ions

K^+ <1> (<1>, weak increase of reaction rate, inhibition at elevated concentrations [1]; <1>, optimal stimulation at 45 mM KCl. Dimethylsulfoxide plus KCl synergistically increase both the rate and the extent of methylation to a greater degree than the combination of Mg^{2+} and K^+ [3]) [3]
Mg^{2+} <1> (<1>, increases reaction rate, inhibition at elevated concentrations [1]; <1>, optimal stimulation at 5 mM $MgCl_2$ [3]) [1, 3]
Na^+ <1> (<1>, increases reaction rate, inhibition at elevated concentrations [1]) [1]

K_m-Value (mM)

0.00012 <1> (Micrococcus lysodeikticus rRNA) [1]
0.000354 <3> (Bacillus subtilis 23S rRNA) [5]
0.000375 <3> (Bacillus subtilis 23S rRNA) [4]
0.012 <3> (S-adenosyl-L-methionine) [4, 5]
0.036 <1> (9-β-ribosyl-2,6-diaminopurine) [1]

K_i-Value (mM)

0.000128 <3> (N^6,N^6-dimethyladenine-23S rRNA, <3>, with 23S rRNA as fixed substrate and S-adenosyl-L-methionine as variable substrate [5]) [5]
0.00036 <3> (S-adenosyl-L-homocysteine, <3>, with 23S rRNA as variable substrate and S-adenosyl-L-methionine as fixed substrate [5]) [5]
0.069 <3> (S-adenosyl-L-homocysteine, <3>, with 23S rRNA as variable substrate and S-adenosyl-L-methionine as fixed substrate [5]) [5]
0.086 <3> (S-adenosyl-L-homocysteine, <3>, with 23S rRNA as fixed substrate and S-adenosyl-L-methionine as variable substrate [5]) [5]

pH-Optimum

7.5-8.2 <1> [1]

5 Isolation/Preparation/Mutation/Application

Purification

<5> (recombinant enzyme [7]) [7]

Crystallization

<3> (hanging-drop vapor diffusion method, recombinant enzyme [8]) [8]

Cloning

<3> [8]
<5> (overexpression in Escherichia coli BL21(DE3) as both a soluble protein and insoluble aggregate - inclusion body - using the T7 promoter driver expression vector, the ermSF gene confers macrolide-lincosamide-streptogramin B resistance to Escherichia coli [7]) [7]

6 Stability

Temperature stability

50 <1> (<1>, 40% increase in activity during the first 6 min, 32-38% decay of activity after 18 min [1]) [1]

Storage stability

<1>, -20°C, stable for at least 9 months [1]

References

[1] Sipe, J.E.; Anderson, W.M.; Remy, C.N.; Love, S.H.: Characterization of S-adenosylmethionine: ribosomal ribonucleic acid-adenine N^6-methyltransferase of Escherichia coli strain B. J. Bacteriol., **110**, 81-91 (1972)
[2] Skinner, R.; Cundliffe, E.; Schmidt, F.J.: Site of action of a ribosomal RNA methylase responsible for resistance to erythromycin and other antibiotics. J. Biol. Chem., **258**, 12702-12706 (1983)
[3] Anderson, W.M.; Remy, C.N.; Sipe, J.E.: Ribosomal ribonucleic acid-adenine N^6-methylase of Escherichia coli strain B: ionic and substrate site requirements. J. Bacteriol., **114**, 988-998 (1973)
[4] Denoya, C.; Dubnau, D.: Site and substrate specificity of the ermC 23S rRNA methyltransferase. J. Bacteriol., **169**, 3857-3860 (1987)
[5] Denoya, C.; Dubnau, D.: Mono- and dimethylating activities and kinetic studies of the ermC 23S rRNA methyltransferase. J. Biol. Chem., **264**, 2615-2624 (1989)
[6] Villsen, I.D.; Vester, B.; Douthwaite, S.: ErmE methyltransferase recognizes features of the primary and secondary structure in a motif within domain V of 23 S rRNA. J. Mol. Biol., **286**, 365-374 (1999)
[7] Jin, H.J.: ErmSF, a ribosomal RNA adenine N^6-methyltransferase gene from Streptomyces fradiae, confers MLS (macrolide-lincosamide-streptogramin B) resistance to E. coli when it is expressed. Mol. Cell., **9**, 252-257 (1999)
[8] Bussiere, D.E.; Muchmore, S.W.; Dealwis, C.G.; Schluckebier, G.; Nienaber, V.L.; Edalji, R.P.; Walter, K.A.; Ladror, U.S.; Holzman, T.F.; Abad-Zapatero, C.: Crystal structure of ErmC', an rRNA methyltransferase which mediates antibiotic resistance in bacteria. Biochemistry, **37**, 7103-7112 (1998)
[9] Formenoy, L.J.; Cunningham, P.R.; Nurse, K.; Pleij, C.W.A.; Ofengand, J.: Methylation of the conserved A1518-A1519 in Escherichia coli 16S ribosomal RNA by the ksgA methyltransferase is influenced by methylations around the similarly conserved U1512-G1523 base pair in the 3' terminal hairpin. Biochimie, **76**, 1123-1128 (1994)

Amine N-methyltransferase 2.1.1.49

1 Nomenclature

EC number
2.1.1.49

Systematic name
S-adenosyl-L-methionine:amine N-methyltransferase

Recommended name
amine N-methyltransferase

Synonyms
aromatic alkylamine-N-methyltransferase <2> [2]
arylamine N-methyltransferase
methyltransferase, tryptamine
nicotine N-methyltransferase
tryptamine N-methyltransferase
tryptamine methyltransferase

CAS registry number
51377-47-0

2 Source Organism

<1> *Oryctolagus cuniculus* (strain New Zealand, 10-16 week old [1,3]; amine N-methyltransferase A and B [3]) [1, 3]
<2> *Rattus norvegicus* (Sprague-Dawley strain male rats [2,5]; rats with common bile duct ligation and obstructive jaundice respectively [5]) [2, 5]
<3> *Bos taurus* (amine N-methyltransferase A [4]) [4]

3 Reaction and Specificity

Catalyzed reaction
S-adenosyl-L-methionine + an amine = S-adenosyl-L-homocysteine + a methylated amine

Reaction type
methyl group transfer

Natural substrates and products

S (R)-isosalsoline + methyl group donor <3> (Reversibility: ? <3> [4]) [4]
P N-methyl-(R)-isosalsoline + demethylated methyl group donor <3> (<3> N-methyl-(R)-isosalsoline is enzymatically formed in liver and can be transported into the brain from blood, it may act as neurotransmitter [4]) [4]
S S-adenosyl-L-methionine + an amine <1, 2> (<1> enzymes participate in detoxication, recycling of drugs after initial demethylation [3]; <2> xenobiotic biotransforming enzyme, enzyme is leaked easily into the blood stream due to an increased membrane permeability of hepatocytes caused by cholestasis [5]) (Reversibility: ? <1, 2> [3, 5]) [3, 5]
P S-adenosyl-L-homocysteine + an N-methylamine

Substrates and products

S S-adenosyl-L-methionine + 1,2,3,4-tetrahydroisoquinoline <3> (<3> amine N-methyltransferase A [4]) (Reversibility: ? <3> [4]) [4]
P S-adenosyl-L-homocysteine + N-methyl-1,2,3,4-tetrahydroisoquinoline <3> [4]
S S-adenosyl-L-methionine + 5-methyloxytryptamine <1> (<1> higher activity with transferase A [3]) (Reversibility: ? <1> [3]) [3]
P ?
S S-adenosyl-L-methionine + 7,8-dichloro-1,2,3,4-tetrahydroisoquinoline <1> (<1> substrate with highest apparent k_{cat} [3]) (Reversibility: ? <1> [3]) [3]
P ?
S S-adenosyl-L-methionine + N-methylbutylamine <1> (<1> only transferase A [3]) (Reversibility: ? <1> [3]) [3]
P S-adenosyl-L-homocysteine + N,N-dimethylbutylamine
S S-adenosyl-L-methionine + N-methylcyclohexylamine <1> (<1> only transferase A [3]) (Reversibility: ? <1> [3]) [3]
P S-adenosyl-L-homocysteine + N,N-dimethylcyclohexylamine
S S-adenosyl-L-methionine + N-methyltryptamine <1> (Reversibility: ? <1> [3]) [3]
P S-adenosyl-L-homocysteine + N,N-dimethyltryptamine
S S-adenosyl-L-methionine + α-N-methyltryptamine <1, 2> (Reversibility: ? <1, 2> [1-3]) [1-3]
P S-adenosyl-L-homocysteine + α,α-N,N-dimethyltryptamine <1, 2> (<1> no methylation of indolic nitrogen [1]; <2> N,N-dimethyltryptamine as product [2]) [1, 2]
S S-adenosyl-L-methionine + aniline <1> (<1> poor substrate [3]) (Reversibility: ? <1> [3]) [3]
P ?
S S-adenosyl-L-methionine + anisidine <1> (<1> higher activity with transferase B [3]) (Reversibility: ? <1> [3]) [3]
P ?

S S-adenosyl-L-methionine + β-phenylethylamine <2> (<2> 5-methyltetrahydrofolate can replace S-adenosyl-L-methionine as methyl donor [2]) (Reversibility: ? <2> [2]) [2]
P ?
S S-adenosyl-L-methionine + desmethylimipramine <1> (<1> higher activity with transferase A [3]) (Reversibility: ? <1> [3]) [3]
P ?
S S-adenosyl-L-methionine + imidazole <1> (<1> higher activity with transferase A [3]) (Reversibility: ? <1> [3]) [3]
P ?
S S-adenosyl-L-methionine + isosalsoline <3> (<3> isosalsoline is identical with 7-hydroxy-6-methoxy-1-methyl-1,2,3,4-tetrahydroisoquinoline, stereospecific N-methylation by amine N-methyltransferase A, (+)-(R)-enantiomer is preferentially methylated [4]) (Reversibility: ? <3> [4]) [4]
P S-adenosyl-L-homocysteine + N-methylisosalsoline <3> [4]
S S-adenosyl-L-methionine + phenyltetrahydropyridine <1> (<1> higher activity with transferase A [3]) (Reversibility: ? <1> [3]) [3]
P ?
S S-adenosyl-L-methionine + pyrazole <1> (<1> higher activity with transferase A [3]) (Reversibility: ? <1> [3]) [3]
P ?
S S-adenosyl-L-methionine + salsolidine <3> (<3> salsolidine is identical with 6,7-dimethoxy-1-methyl-1,2,3,4-tetrahydroisoquinoline, stereospecific N-methylation by amine N-methyltransferase A, (-)-(S)-enantiomer is preferentially methylated [4]) (Reversibility: ? <3> [4]) [4]
P S-adenosyl-L-homocysteine + N-methylsalsolidine <3> [4]
S S-adenosyl-L-methionine + tryptamine <1, 2> (<1,2> i.e. 3-[2-aminoethyl] indole, best substrate, two N-methyltransferases: A and B, non-specific [1-3]; <2> two N-methyltransferases, non-specific or specific for tryptamine, 5-methyltetrahydrofolate can replace S-adenosyl-L-methionine as methyl donor [2]; <2> S-adenosyl-L-methionine-dependent enzyme [5]) (Reversibility: ? <1, 2> [1-5]) [1-5]
P S-adenosyl-L-homocysteine + α-N-methyltryptamine <1, 2> [2, 3, 5]
S Additional information <1> (<1> two N-methyltransferases A and B with very broad but overlapping substrate specificity for primary and secondary amines, lipophilic substrates, overview poor substrates, no substrates: N-acetylserotonine, 4-aminoacetophenone, 4-aminophenol, benzamide, 4-chloroaniline, deoxyepinephrine, 1,4-diaminobutane, dopamine, 1-ethanol-2-amine, ethylenediamine, histamine, histidine methylester, 3-hydroxytryptamine, indole, pyrrole, methylamine, N-methyldopamine, nicotinamide, octopamine, phenylalanine, phenylethanolamine, theophylline, tryptophan, tyrosine, N,N-dimethyltryptamine [3]) [3]
P ?

Inhibitors

N,N-dimethyltryptamine <1> (<1> competitive inhibitor [3]) [3]
Additional information <1> (<1> substrate inhibition, but not by tryptamine and 7,8-dichloro-1,2,3,4-tetrahydroisoquinoline [3]) [3]

Cofactors/prosthetic groups

S-adenosyl-L-methionine <1, 2> [1-5]

Activating compounds

Additional information <2> (<2> biosynthesis of enzyme is induced in response to obstructive jaundice [5]) [5]

Turnover number (min^{-1})

2 <1> (aniline, <1> transferase A, pH 7.6 [3]) [3]
2 <1> (desmethylimipramine, <1> transferase B, pH 8.5 [3]) [3]
5 <1> (imidazole, <1> transferase A, pH 8.5 [3]) [3]
6 <1> (pyrazole, <1> transferase B, pH 7.6 [3]) [3]
7 <1> (S-adenosyl-L-methionine, <1> transferase B, pH 8.5 [3]) [3]
10 <1> (imidazole, <1> transferase B, pH 8.5 [3]) [3]
12 <1> (S-adenosyl-L-methionine, <1> transferase A, pH 8.5 [3]) [3]
14 <1> (desmethylimipramine, <1> transferase A, pH 8.5 [3]) [3]
15 <1> (pyrazole, <1> transferase A, pH 7.6 [3]) [3]
22 <1> (aniline, <1> transferase B, pH 7.6 [3]) [3]
24 <1> (tryptamine, <1> transferase A, pH 8.5 [3]) [3]
27 <1> (tryptamine, <1> transferase B, pH 8.5 [3]) [3]
40 <1> (7,8-dichloro-1,2,3,4-tetrahydroisoquinoline, <1> transferase A, pH 7.6 [3]) [3]
220 <1> (7,8-dichloro-1,2,3,4-tetrahydroisoquinoline, <1> transferase B, pH 7.6 [3]) [3]

Specific activity (U/mg)

0.00000038 <2> (<2> β-phenylethylamine + S-adenosyl-L-methionine [2]) [2]
0.00000118 <2> (<2> tryptamine + S-adenosyl-L-methionine [2]) [2]
0.0000012 <3> (<3> N-methyltransferase A, (-)-(S)-isosalsoline [4]) [4]
0.0000021 <3> (<3> N-methyltransferase A, (-)-(S)-salsolidine [4]) [4]
0.0000033 <3> (<3> N-methyltransferase A, (+)-(R)-isosalsoline [4]) [4]
0.000065 <2> (<2> β-phenylethylamine + 5-methyltetrahydrofolate [2]) [2]
0.000068 <1> (<1> N-methyltransferase A, aniline [3]) [3]
0.000198 <2> (<2> tryptamine + 5-methyltetrahydrofolate [2]) [2]
0.00022 <1> (<1> N-methyltransferase B, desmethylimipramine [3]) [3]
0.00026 <1> (<1> N-methyltransferase A, desmethylimipramine [3]) [3]
0.00044 <1> (<1> N-methyltransferase A, tryptamine [3]) [3]
0.00127 <1> (<1> N-methyltransferase B, aniline [3]) [3]
0.0031 <1> (<1> N-methyltransferase B, tryptamine [3]) [3]
Additional information <2> (<2> specific activities of serum, control and cholestatic liver in cytosolic, mitochondrial and microsomal preparations for 0.5, 1, 2, 3, 7, 14, 28 and 42 days after common bile duct ligation, activities of mitochondrial and microsomal enzyme in cholestatic liver and of serum enzyme increase after common bile duct ligation, V_{max}-values [5]) [5]

K_m-Value (mM)

0.01 <1> (7,8-dichloro-1,2,3,4-tetrahydroisoquinoline, <1> transferase A, pH 7.6 [3]) [3]
0.015 <1> (S-adenosyl-L-methionine, <1> 1 mM tryptamine as methyl acceptor, pH 8.5 [3]) [3]
0.12 <1> (tryptamine, <1> transferase B, pH 8.5 [3]) [3]
0.2 <1> (imidazole, <1> transferase A, pH 8.5 [3]) [3]
0.24 <1> (desmethylimipramine, <1> transferase B, pH 8.5 [3]) [3]
0.3 <1> (desmethylimipramine, <1> transferase A, pH 8.5 [3]) [3]
0.3 <1> (tryptamine, <1> transferase A, pH 8.5 [3]) [3]
1.5 <1> (pyrazole, <1> transferase A, pH 7.6 [3]) [3]
2 <1> (7,8-dichloro-1,2,3,4-tetrahydroisoquinoline, <1> transferase B, pH 7.6 [3]) [3]
3.2 <1> (pyrazole, <1> transferase B, pH 7.6 [3]) [3]
5.1 <1> (imidazole, <1> transferase B, pH 8.5 [3]) [3]
6.25 <2> (β-phenylethylamine, <2> 5-methyltetrahydrofolate as methyl donor [2]) [2]
7.6 <1> (aniline, <1> transferase A, pH 7.6 [3]) [3]
7.8 <1> (aniline, <1> transferase B, pH 7.6 [3]) [3]
8.3 <2> (tryptamine, <2> 5-methyltetrahydrofolate as methyl donor [2]) [2]
28.3 <2> (tryptamine, <2> mitochondrial enzyme, cholestatic liver, S-adenosyl-L-methionine [5]) [5]
32.5 <2> (tryptamine, <2> mitochondrial enzyme, control liver, S-adenosyl-L-methionine [5]) [5]
34.1 <2> (tryptamine, <2> microsomal enzyme, cholestatic liver, S-adenosyl-L-methionine [5]) [5]
35.6 <2> (tryptamine, <2> microsomal enzyme, control liver, S-adenosyl-L-methionine [5]) [5]

K_i-Value (mM)

0.006 <1> (N,N-dimethyltryptamine, <1> N-methyltransferase B [3]) [3]
0.01 <1> (N,N-dimethyltryptamine, <1> N-methyltransferase A [3]) [3]

pH-Optimum

6.5-7 <2> (<2> 5-methyltetrahydrofolate as donor [2]) [2]
7.9 <2> (<2> S-adenosyl-L-methionine as donor [2]) [2]
8.5 <1> [3]
Additional information <1> (<1> transferase A: pI 4.9, transferase B: pI 5.1 [3]) [3]

pH-Range

7.2-8.9 <1> (<1> tryptamine, about half-maximal activity at pH 7.2 and 8.9, transferase A [3]) [3]
7.5-8.8 <1> (<1> tryptamine, about half-maximal activity at pH 7.5 and 90% at pH 8.8, transferase B [3]) [3]

Temperature optimum (°C)

37 <1-3> (<1-3> assay at [2-5]) [2-5]

4 Enzyme Structure

Molecular weight

30000 <1> (<1> gel filtration [3]) [3]
Additional information <1> (<1> amino acid composition of transferase A [3]) [3]

Subunits

monomer <1> (<1> 1 * 30000, SDS-PAGE [3]) [3]
Additional information <1> (<1> aggregation upon denaturation leads to 60 kDa and 90 kDa fractions [3]) [3]

5 Isolation/Preparation/Mutation/Application

Source/tissue

brain <2> [2, 5]
heart <2> [5]
liver <1-3> (<1> two amine N-methyltransferases, A and B [1,3]; <2> investigation of enzyme in cholestatic liver, induced by common bile duct ligation for a period of 42 days [5]) [1, 3-5]
lung <1, 2> [1, 5]
salivary gland <2> [5]
serum <2> (<2> investigation of enzyme in serum, after common bile duct ligation for a period of 42 days [5]) [5]
small intestine <2> [5]
stomach <2> [5]

Localization

cytosol <2> (<2> liver [5]) [5]
membrane <2> (<2> hepatic membrane-bound enzyme [5]) [5]
microsome <2> (<2> liver [5]) [5]
mitochondrion <2> (<2> liver [5]) [5]

Purification

<1> (two N-methyltransferases A and B separable by ammonium sulfate fractionation, HPLC-ion exchange chromatography [3]) [3]
<2> (partial [2]) [2]
<3> (amine N-methyltransferase A [4]) [4]

6 Stability

pH-Stability

6.2 <1> (<1> optimal for transferase A [3]) [3]
7.8 <1> (<1> optimal for transferase B [3]) [3]

Storage stability

<1>, -80°C, partially purified, per month, less than 10% loss of activity [3]
<1>, -80°C, purified, per month, about 25% loss of activity [3]
<1>, 2°C, per week, 25% loss of activity [3]

References

[1] Crooks, P.A.; Godin, C.S.; Nwosu, C.G.; Ansher, S.S.; Jakoby, W.B.: Reevaluation of the products of tryptamine catalyzed by rabbit liver N-methyltransferases. Biochem. Pharmacol., **35**, 1600-1603 (1986)
[2] Hsu, L.L.; Mandell, A.J.: Multiple N-methyltransferases for aromatic alkylamines in brain. Life Sci., **13**, 847-858 (1973)
[3] Ansher, S.S.; Jakoby, W.B.: Amine N-methyltransferases from rabbit liver. J. Biol. Chem., **261**, 3996-4001 (1986)
[4] Bahnmaier, A.H.; Woesle, B.; Thomas, H.: Stereospecific N-methylation of the tetrahydroisoquinoline alkaloids isosalsoline and salsolidine by amine N-methyltransferase A from bovine liver. Chirality, **11**, 160-165 (1999)
[5] Kim, Y.H.; Joo, I.: Arylamine N-methyltransferase and thiol methyltransferase activities in cholestatic rat liver induced by common bile duct ligation. Exp. Mol. Med., **33**, 23-28 (2001)

Loganate O-methyltransferase 2.1.1.50

1 Nomenclature

EC number
2.1.1.50

Systematic name
S-adenosyl-L-methionine:loganate 11-O-methyltransferase

Recommended name
loganate O-methyltransferase

Synonyms
S-adenosyl-L-methionine:loganic acid methyltransferase
loganate methyltransferase
methyltransferase, loganate

CAS registry number
39391-10-1

2 Source Organism

<1> *Vinca rosea* [1]

3 Reaction and Specificity

Catalyzed reaction
S-adenosyl-L-methionine + loganate = S-adenosyl-L-homocysteine + loganin

Reaction type
methyl group transfer

Natural substrates and products
S S-adenosyl-L-methionine + loganate <1> (<1> biosynthesis of indole alkaloids [1]) (Reversibility: ? <1> [1]) [1]
P S-adenosyl-L-homocysteine + loganin
S S-adenosyl-L-methionine + secologanate <1> (Reversibility: ? <1> [1]) [1]
P S-adenosyl-L-homocysteine + secologanin

Substrates and products
S S-adenosyl-L-methionine + loganate <1> (Reversibility: ? <1> [1]) [1]
P S-adenosyl-L-homocysteine + loganin

S S-adenosyl-L-methionine + secologanate <1> (Reversibility: ? <1> [1]) [1]
P S-adenosyl-L-homocysteine + secologanin
S Additional information <1> (<1> no significant methylation of: 7-deoxyloganic acid, geniposidic acid, 7-epiloganic acid [1]) [1]
P ?

Inhibitors
KCN <1> [1]
N-methylmaleimide <1> [1]
iodoacetamide <1> [1]
p-chloromercuribenzoate <1> [1]

Metals, ions
Additional information <1> (<1> dependency on divalent metal cations not detected [1]) [1]

K_m-Value (mM)
12.5 <1> (loganic acid) [1]

pH-Optimum
7.2 <1> (<1> assay at [1]) [1]

Temperature optimum (°C)
32-33 <1> (<1> assay at [1]) [1]

5 Isolation/Preparation/Mutation/Application

Purification
<1> [1]

References

[1] Madyastha, K.M.; Guarnaccia, R.; Baxter, C.; Coscia, C.L.: S-Adenosyl-L-methionine: loganic acid methyltransferase. A carboxyl-alkylating enzyme from Vinca rosea. J. Biol. Chem., **248**, 2497-2501 (1973)

rRNA (guanine-N^1-)-methyltransferase 2.1.1.51

1 Nomenclature

EC number
2.1.1.51

Systematic name
S-adenosyl-L-methionine:rRNA (guanine-N^1-)-methyltransferase

Recommended name
rRNA (guanine-N^1-)-methyltransferase

Synonyms
RrmA
methyltransferase, ribosomal ribonucleate guanine 1-
ribosomal ribonucleate guanine 1-methyltransferase

CAS registry number
50812-25-4

2 Source Organism

<1> *Escherichia coli* (K12, strain LMUR (wild-type), mutant IB7 of strain CP79 [1]) [1-4]
<2> *Streptomyces fradiae* [3]
<3> *Streptomyces sp.* [4]

3 Reaction and Specificity

Catalyzed reaction
S-adenosyl-L-methionine + rRNA = S-adenosyl-L-homocysteine + rRNA containing N^1-methylguanine

Reaction type
N-methylation
methyl group transfer

Natural substrates and products
S S-adenosyl-L-methionine + rRNA <1> (Reversibility: ? <1> [1]) [1]
P S-adenosyl-L-homocysteine + rRNA containing N^1-methylguanine

S Additional information <1, 2> (<1,2> methylation of G745 or G748 matches with Gram-negative or Gram-positive bacterial strains, resp. [3]) [3]
P ?

Substrates and products
S S-adenosyl-L-methionine + rRNA <1-3> (Reversibility: ? <1-3> [1-4]) [1-4]
P S-adenosyl-L-homocysteine + rRNA containing N^1-methylguanine <1> [1]
S Additional information <1> (<1> structures in stem loops 33, 34, 35 of substrate are required, methylation only occurs in free RNA [2]) [2]
P ?

Inhibitors
Mg^{2+} <1> (<1> weak [1]) [1]
unidentified protein factor A1 <1> (<1> trypsin-sensitive [1]) [1]

Cofactors/prosthetic groups
unidentified protein factor B <1> (<1> activation, papain-sensitive [1]) [1]

Activating compounds
unidentified protein factor B <1> (<1> activation, papain-sensitive [1]) [1]

Metals, ions
Zn^{2+} <1, 3> (<1,3> putative Zn-finger domain [4]) [4]

Specific activity (U/mg)
Additional information <1> [1]

pH-Optimum
8 <1> (<1> assay at [1]) [1]

Temperature optimum (°C)
30 <1> (<1> assay at [1]) [1]

4 Enzyme Structure

Subunits
Additional information <1, 3> (<1,3> sequence-to-structure threading, structural organization [4]) [4]

5 Isolation/Preparation/Mutation/Application

Purification
<1> (partial [1]; recombinant protein [2]) [1, 2]

References

[1] Isaksson, L.A.: Partial purification of ribosomal RNA(m^1G)- and rRNA(m^2G)-methylases from Escherichia coli and demonstration of some proteins affecting their apparent activity. Biochim. Biophys. Acta, **312**, 122-133 (1973)

[2] Hansen, L.H.; Kirpekar, F.; Douthwaite, S.: Recognition of nucleotide G745 in 23S ribosomal RNA by the RrmA methyltransferase. J. Mol. Biol., **310**, 1001-1010 (2001)

[3] Liu, M.; Douthwaite, S.: Methylation at nucleotide G745 or G748 in 23S rRNA distinguishes gram-negative from gram-positive bacteria. Mol. Microbiol., **44**, 195-204 (2002)

[4] Bujnicki, J.M.; Blumenthal, R.M.; Rychlewski, L.: Sequence analysis and structure prediction of 23S rRNA:m1G methyltransferases reveals a conserved core augmented with a putative Zn-binding domain in the N-terminus and family-specific elaborations in the C-terminus. J. Mol. Microbiol. Biotechnol., **4**, 93-99 (2002)

rRNA (guanine-N^2-)-methyltransferase 2.1.1.52

1 Nomenclature

EC number
2.1.1.52

Systematic name
S-adenosyl-L-methionine:rRNA (guanine-N^2-)-methyltransferase

Recommended name
rRNA (guanine-N^2-)-methyltransferase

Synonyms
16S rRNA:(guanine-N^2) methyltransferase
methyltransferase, ribosomal ribonucleate guanine 2-
ribosomal ribonucleate guanine-2-methyltransferase

CAS registry number
50812-26-5

2 Source Organism

<1> *Escherichia coli* (K12, strain LMUR (wild-type), mutant IB10 of strain CP79 [1]) [1, 2]
<2> *Escherichia coli* (16S RNA m^2G1207 methyltransferase [3]) [3]

3 Reaction and Specificity

Catalyzed reaction
S-adenosyl-L-methionine + rRNA = S-adenosyl-L-homocysteine + rRNA containing N^2-methylguanine

Reaction type
methyl group transfer

Natural substrates and products
S S-adenosyl-L-methionine + rRNA <1> (Reversibility: ? <1> [1]) [1]
P S-adenosyl-L-homocysteine + rRNA containing N^2-methylguanine <1> [1]

Substrates and products

- **S** S-adenosyl-L-methionine + 16S RNA <2> (<2> reacts with 30S subunits reconstituted from 16S transkripts, not with free RNA [3]) (Reversibility: ? <2> [3]) [3]
- **P** S-adenosyl-L-homocysteine + m^2G1207 16S RNA <2> [3]
- **S** S-adenosyl-L-methionine + rRNA <1> (Reversibility: ? <1> [1]) [1]
- **P** S-adenosyl-L-homocysteine + rRNA containing N^2-methylguanine <1> [1]

Inhibitors

EDTA <2> (<2> at 2 mM, 80% inhibition [3]) [3]
unidentified protein factor A2 <1> [1]
Additional information <1> (<1> no inhibition by Mg^{2+} [1]) [1]

Metals, ions

Mg^{2+} <2> [3]
Additional information <1> (<1> no Mg^{2+} requirement [1]) [1]

Specific activity (U/mg)

Additional information <1> (<1> two different forms of the same protein [1]) [1]

pH-Optimum

8 <1> (<1> assay at [1]) [1]

Temperature optimum (°C)

30 <1> (<1> assay at [1]) [1]

5 Isolation/Preparation/Mutation/Application

Purification

<1> (partial [1]) [1]

Cloning

<2> [3]

Engineering

Additional information <1> (<1> phylogenomic analysis of related enzymes [2]) [2]

References

[1] Isaksson, L.A.: Partial purification of ribosomal RNA(m^1G)- and rRNA(m^2G)-methylases from Escherichia coli and demonstration of some proteins affecting their apparent activity. Biochim. Biophys. Acta, **312**, 122-133 (1973)
[2] Bujnicki, J.M.: Phylogenomic analysis of 16S rRNA:(guanine-N2) methyltransferases suggests new family members and reveals highly conserved mo-

tifs and a domain structure similar to other nucleic acid amino-methyltransferases. FASEB J., **14**, 2365-2368 (2000)
[3] Tscherne, J.S.; Nurse, K.; Popienick, P.; Ofengand, J.: Purification, cloning, and characterization of the 16S RNA m^2G1207 methyltransferase from Escherichia coli. J. Biol. Chem., **274**, 924-929 (1999)

Putrescine N-methyltransferase 2.1.1.53

1 Nomenclature

EC number
2.1.1.53

Systematic name
S-adenosyl-L-methionine:putrescine N-methyltransferase

Recommended name
putrescine N-methyltransferase

Synonyms
methyltransferase, putrescine
putrescine methyltransferase

CAS registry number
9075-39-2

2 Source Organism

<1> *Nicotiana tabacum* (var. Samsun [1]; Bright Yellow [2]; cv. Burley 21 seeds [4]; strain cv. SR1 [4, 6]) [1, 2, 4, 6]
<2> *Datura stramonium* [1, 3]
<3> *Atropa belladonna* [4]
<4> *Atropa belladonna* [7]
<5> *Atropa belladonna* [7]
<6> *Atropa belladonna* [7]
<7> *Hyoscyamus niger* [7]
<8> *Hyoscyamus niger* [4]
<9> *Nicotiana tomentosiformis* [4]
<10> *Nicotiana otophora* [4]
<11> *Nicotiana sylvestris* [4, 5]
<12> *Nicotiana sylvestris* [4, 5]
<13> *Nicotiana sylvestris* [4, 5]
<14> *Nicotiana sylvestris* [4]

3 Reaction and Specificity

Catalyzed reaction
S-adenosyl-L-methionine + putrescine = S-adenosyl-L-homocysteine + N-methylputrescine

Reaction type

methyl group transfer

Natural substrates and products

- **S** S-adenosyl-L-methionine + putrescine <1-14> (<1,3-14> enzyme is probably involved in nicotine biosynthesis [2,4,5,7]; <1-14> enzyme catalyzes the initial reaction in the biosynthesis of tropane and pyrrolidine alkaloids from putrescine [3-7]; <1> enzyme possibly may play an additional, subtle role in alkaloid metabolism in leaf tissue [6]) (Reversibility: ? <1-14> [2-7]) [2-7]
- **P** S-adenosyl-L-homocysteine + N-methylputrescine <1-3, 8-14> [2-4]

Substrates and products

- **S** S-adenosyl-L-methionine + 1,4-diamino-2,3-dibromo-trans-but-2-ene <2> (Reversibility: ? <2> [3]) [3]
- **P** S-adenosyl-L-homocysteine + ?
- **S** S-adenosyl-L-methionine + 1,4-diamino-2-hydroxybutane <2> (Reversibility: ? <2> [3]) [3]
- **P** S-adenosyl-L-homocysteine + ?
- **S** S-adenosyl-L-methionine + 1,4-diamino-2-oxobutane <2> (Reversibility: ? <2> [3]) [3]
- **P** S-adenosyl-L-homocysteine + ?
- **S** S-adenosyl-L-methionine + 1,4-diamino-trans-but-2-ene <2> (Reversibility: ? <2> [3]) [3]
- **P** S-adenosyl-L-homocysteine + ?
- **S** S-adenosyl-L-methionine + 4-(aminomethyl)piperidine <2> (Reversibility: ? <2> [3]) [3]
- **P** S-adenosyl-L-homocysteine + ?
- **S** S-adenosyl-L-methionine + N-methylputrescine <1> (<1> about 8% of activity compared to putrescine [2]) (Reversibility: ? <1> [2]) [2]
- **P** S-adenosyl-L-homocysteine + N,N-dimethylputrescine
- **S** S-adenosyl-L-methionine + putrescine <1-14> (<2> best substrate [3]) (Reversibility: ? <1-14> [1-7]) [1-7]
- **P** S-adenosyl-L-homocysteine + N-methylputrescine <1-3, 8-14> [1-5]
- **S** S-adenosyl-L-methionine + trans-1,4-diamonocyclohexane <2> (Reversibility: ? <2> [3]) [3]
- **P** S-adenosyl-L-homocysteine + ?
- **S** Additional information <2> (<2> substrate with little activity: 1,4 -diamino-cis-but-2-ene, not: aliphatic monoamines, 1,3-diaminopropane, 1,5-diaminopentane, N,N'-dimethyl-1,4-diamino-trans-but-2-ene, 1,3-diamino-2-hydroxypropane, 1,3-diaminoacetone, 1,4-diaminobut-2-yne, N-carbamoylputrescine [3]; <2> requirement for substrate activity: two amino groups in a trans conformation, separated by four carbon atoms [3]) [3]
- **P** ?

Inhibitors

2-(aminomethyl)pyridine <2> [3]

3-methylcadaverine <2> [3]

4-(aminomethyl)pyridine <2> [3]
Ag^+ <1> (<1> 0.1 mM, strong [2]) [2]
N-methylputrescine <2> [3]
Na_2SO_4 <2> [3]
S-adenosyl-L-homocysteine <2> (<2> competitive inhibitor [3]) [3]
cadaverine <2> [3]
p-chloromercuribenzoate <1> (<1> strong [2]) [2]
Additional information <2> (<2> monoamines p.e. cyclohexylamine, exo-2-aminonorborane, n-butylamine, 4-methylcyclohexylamine, cyclopentylamine [3]; <2> not: 1,3-diaminopropane, 2-(2-aminoethyl)pyridine, 2-aminopyridine, 3-aminopyridine, 4-aminopyridine [3]) [3]

Cofactors/prosthetic groups

Additional information <1> (<1> no requirement of cofactors [2]) [2]

Activating compounds

cysteine <1> (<1> slight activatory effect [2]) [2]

Metals, ions

Additional information <1> (<1> Mg^{2+}, Mn^{2+} and EDTA at 0.001 M have no influence [2]) [2]

Specific activity (U/mg)

0.00016 <1> (<1> leaves [6]) [6]
0.0018 <1> (<1> roots [6]) [6]
0.051 <1> (<1> putrescine [2]) [2]
Additional information <2> [3]

K_m-Value (mM)

0.1 <2> (S-adenosyl-L-methionine) [3]
0.11 <1> (S-adenosyl-L-methionine) [2]
0.15 <2> (S-adenosyl-L-methionine) [1]
0.31 <2> (putrescine) [3]
0.4 <1> (putrescine) [2]
0.88 <2> (putrescine) [1]

K_i-Value (mM)

0.01 <2> (S-adenosyl-L-homocysteine) [3]
0.04 <2, 2> (cadaverine) [3]
0.15 <2> (N-methylputrescine) [3]

pH-Optimum

8-9 <1> [2]
9 <2> [3]
Additional information <2> (<2> pI: 5.3 [3]) [3]

Temperature optimum (°C)

26 <1> (<1> assay at [2]) [2]
30 <1, 2> (<1,2> assay at [1]) [1]

4 Enzyme Structure

Molecular weight

40000 <2> (<2> gel filtration [3]) [3]
60000 <1> (<1> gel filtration [2]) [2]
360000 <2> (<2> SDS-PAGE [3]) [3]

5 Isolation/Preparation/Mutation/Application

Source/tissue

bark <11-13> [5]
endodermis <11-13> [5]
leaf <1, 3, 8-14> (<1> low levels of activity [6]) [4, 6]
root <1-14> (<1> roots of decapitated plants show highest activity, no activity in leaves [2]) [1-7]
xylem <11-13> [5]

Purification

<1> [1, 2]
<2> [1]
<2> [3]

Cloning

<1> (expressed in roots and leaves [6]) [6]
<11-13> (expressed in Escherichia coli, three genes isolated from Nicotiana sylvestris: NsPMT1, NsPMT2, NsPMT3 [4]; expressed in xylem-tissues of roots, three genes isolated from Nicotiana sylvestris: NsPMT1, NsPMT2, NsPMT3, a highly conserved NsPMT 5'-flanking region is sufficient for tissue-specific and jasmonate-responsive expression of all three PMT genes [5]) [4, 5]
<4-7> (AbPMT1 is expressed more strongly than ABPMT2 in the root [7]) [7]

6 Stability

pH-Stability

9.5 <2> (<2> 20% loss of activity [3]) [3]
9.8 <2> (<2> 35% loss of activity [3]) [3]

Temperature stability

4 <1, 2> (3 h: 75% loss of activity, presence of 12.5% w/v glucose: 35% loss of activity [1]) [1]

General stability information

<1>, 2-mercaptoethanol stabilizes [2]

Storage stability

<1>, -20°C, 0.05 M phosphate buffer, pH 7.5, 0.01 M 2-mercaptoethanol, 12.5% glucose, stable for several weeks [2]

<1>, 0°C, without addition of a sulfhydryl compound, most activity is lost within a few days [2]

<2>, -20°C, 12.5% glucose for at least a week [3]

References

[1] Feth, F.; Arfmann, H.A.; Wray, V.; Wagner, K. G.: Determination of putrescine N-methyltransferase by high performance liquid chromatography. Phytochemistry, **24**, 921-923 (1985)

[2] Mizusaki, S.; Tanabe, Y.; Noguchi, M.; Tamaki, E.: Phytochemical studies on tobacco alkaloids XIV, the occurrence. Plant Cell Physiol., **12**, 633-640 (1971)

[3] Walton, N.J.; Peerless, A.C.J.; Robins, R.J.; Rhodes, M.J.C.; Boswell, H.D.; Robins, D.J.: Purification and properties of putrescine N-methyltransferase from transformed roots of Datura stramonium L. Planta, **193**, 9-15 (1994)

[4] Hashimoto, T.; Shoji, T.; Mihara, T.; Oguri, H.; Tamaki, K.; Suzuki, K.I.; Yamada, Y.: Intraspecific variability of the tandem repeats in Nicotiana putrescine N-methyltransferases. Plant Mol. Biol., **37**, 25-37 (1998)

[5] Shoji, T.; Yamada, Y.; Hashimoto, T.: Jasmonate induction of putrescine N-methyltransferase genes in the root of Nicotiana sylvestris. Plant Cell Physiol., **41**, 831-839 (2000)

[6] Sachan, N.; Falcone, D.L.: Wound-induced gene expression of putrescine N-methyltransferase in leaves of Nicotiana tabacum. Phytochemistry, **61**, 797-805 (2002)

[7] Suzuki, K.I.; Yamada, Y.; Hashimoto, T.: Expression of Atropa belladonna putrescine N-methyltransferase gene in root pericycle. Plant Cell Physiol., **40**, 289-297 (1999)

Deoxycytidylate C-methyltransferase 2.1.1.54

1 Nomenclature

EC number
2.1.1.54

Systematic name
5,10-methylenetetrahydrofolate:dCMP C-methyltransferase

Recommended name
deoxycytidylate C-methyltransferase

Synonyms
dCMP methyltransferase
deoxycytidylate methyltransferase
methyltransferase, deoxycytidylate

CAS registry number
61970-01-2

2 Source Organism

<1> *Xanthomonas oryzae* (phage Xp12-infected [1-3]) [1-3]

3 Reaction and Specificity

Catalyzed reaction
5,10-methylenetetrahydrofolate + dCMP = dihydrofolate + deoxy-5-methylcytidylate

Reaction type
methyl group transfer

Natural substrates and products
S 5,10-methylenetetrahydrofolate + dCMP <1> [1]
P dihydrofolate + deoxy-5-methylcytidylate

Substrates and products
S 5,10-methylenetetrahydrofolate + CMP <1> (Reversibility: ? <1> [3]) [3]
P dihydrofolate + 5-methylcytidylate
S 5,10-methylenetetrahydrofolate + CTP <1> (<1> low activity [3]) [3]
P dihydrofolate + 5-methylcytidine 5'-triphosphate

S 5,10-methylenetetrahydrofolate + dCMP <1> (Reversibility: ? <1> [1, 2]) [1, 2]
P dihydrofolate + deoxy-5-methylcytidylate <1> [1]
S 5,10-methylenetetrahydrofolate + dCTP <1> (<1> low activity [3]) (Reversibility: ? <1> [3]) [3]
P dihydrofolate + deoxy-5-methylcytidine 5'-triphosphate
S Additional information <1> (<1> dCMP is methylated by formaldehyde in the presence of tetrahydrofolate [1,2]) [1, 2]
P ?

Inhibitors
Cu^{2+} <1> (<1> above 1 mM [3]) [3]
Mg^{2+} <1> (<1> above 50 mM [3]) [3]
Zn^{2+} <1> (<1> above 1 mM [3]) [3]

Metals, ions
Ca^{2+} <1> (<1> 25 mM, slight stimulation [3]) [3]
Cu^{2+} <1> (<1> below 1 mM, stimulation [3]) [3]
K^{+} <1> (<1> 150 mM, stimulation [3]) [3]
Mg^{2+} <1> (<1> required [1]; <1> stimulates below 25 mM [3]) [1, 3]
Mn^{2+} <1> (<1> has a similar effect in stimulation but not as strong as Mg^{2+} [3]) [3]
Zn^{2+} <1> (<1> below 1 mM, stimulation [3]) [3]
Additional information <1> (<1> among the buffers tested, 6-morpholinoethane sulfonate/Mg^{2+} is the best [3]) [3]

Specific activity (U/mg)
0.11 <1> [3]

K_m-Value (mM)
0.0313 <1> (dCMP) [3]
0.0714 <1> (tetrahydrofolic acid) [3]

pH-Optimum
6 <1> [3]

pH-Range
Additional information <1> [3]

Temperature optimum (°C)
30 <1> (<1> assay at [3]) [3]

5 Isolation/Preparation/Mutation/Application

Purification
<1> (phage Xp12-infected, partial [3]) [3]

6 Stability

Temperature stability

50 <1> (<1> 3 min, 90% loss of activity [3]) [3]
70 <1> (<1> 3 min, complete loss of activity [3]) [3]

Storage stability

<1>, 4°C, stable for at least 2 weeks [1]

References

[1] Kuo, T.T.; Tu, J.: Enzymatic synthesis of deoxy-5-methyl-cytidylic acid replacing deoxycytidylic acid in Xanthomonas oryzae phage Xp12 DNA. Nature, **263**, 615 (1976)
[2] Wang, R. Y.H.; Huang, L.H.; Ehrlich, M.: Enzymatic conversion of deoxycytidine 5-monophosphate to 5-methyldeoxycytidine 5-triphosphate. Anal. Biochem., **122**, 89-93 (1982)
[3] Feng, T.Y.; Tu, J.; Kuo, T.T.: Characterization of deoxycytidylate methyltransferase in Xanthomonas oryzae infected with bacteriophage Xp12. Eur. J. Biochem., **87**, 29-36 (1978)

tRNA (adenine-N^6-)-methyltransferase 2.1.1.55

1 Nomenclature

EC number
2.1.1.55

Systematic name
S-adenosyl-L-methionine:tRNA (adenine-N^6-)-methyltransferase

Recommended name
tRNA (adenine-N^6-)-methyltransferase

CAS registry number
9014-53-3

2 Source Organism

<1> *Rattus norvegicus* [1]

3 Reaction and Specificity

Catalyzed reaction
S-adenosyl-L-methionine + tRNA = S-adenosyl-L-homocysteine + tRNA containing N^6-methyladenine

Reaction type
methyl group transfer

Natural substrates and products
- **S** S-adenosyl-L-methionine + tRNA <1> (Reversibility: ? <1> [1]) [1]
- **P** S-adenosyl-L-homocysteine + tRNA containing N^6-methyladenine

Substrates and products
- **S** S-adenosyl-L-methionine + tRNA <1> (<1> methyl group acceptor is yeast tRNA, poor acceptors are E. coli or mycoplasma tRNAs [1]) (Reversibility: ? <1> [1]) [1]
- **P** S-adenosyl-L-homocysteine + tRNA containing N^6-methyladenine <1> [1]

pH-Optimum
8.2 <1> (<1> assay at [1]) [1]

Temperature optimum (°C)

37 <1> (<1> assay at [1]) [1]

5 Isolation/Preparation/Mutation/Application

Source/tissue

Novikoff hepatoma cell <1> [1]
liver <1> [1]

References

[1] Sharma, O.K.: Differences in the transfer RNA methyltransferases from normal rat liver and Novikoff hepatoma. Biochim. Biophys. Acta, **299**, 415-427 (1973)

mRNA (guanine-N^7-)-methyltransferase 2.1.1.56

1 Nomenclature

EC number
2.1.1.56

Systematic name
S-adenosyl-L-methionine:mRNA (guanine-N^7-)-methyltransferase

Recommended name
mRNA (guanine-N^7-)-methyltransferase

Synonyms
guanine-7-methyltransferase
messenger RNA guanine 7-methyltransferase
messenger ribonucleate guanine 7-methyltransferase
methyltransferase, messenger ribonucleate guanine 7-

CAS registry number
56941-25-4

2 Source Organism

<1> *Vaccinia virus* (strain WR, from infected HeLa cells [3,9,15]) [1, 3-7, 9, 12, 14, 15, 18, 19, 23]
<2> *Newcastle disease virus* [10]
<3> *Neurospora crassa* (mutant met- [11]) [11]
<4> *Saccharomyces cerevisiae* (strain VY1160, osmotic sensitive [13]) [13]
<5> *Homo sapiens* (HeLa cells [8]) [8, 22]
<6> *Mus musculus* (L929-cells [12]) [12]
<7> *Semliki forest virus* (enzyme activity belongs to NSP1 protein [17]; in infected BHK cells [2]) [2, 17]
<8> *bluetongue virus* (enzyme activity belongs to VP4 protein [16]) [16]
<9> *Homo sapiens* [20]
<10> *Saccharomyces cerevisiae* [20]
<11> *Xenopus laevis* (<11> mRNAcapping enzyme [21]) [21]
<12> *Xenopus laevis* (<12> mRNA (guanine-7-)methyltransferase component of the capping enzyme [21]) [21, 21]
<13> *Xenopus laevis* (<13> fragment of 402 amino acids [21]) [21, 21]
<14> *bamboo mosaic virus* (activity belongs to the N-terminal part of the replicase [24]) [24]

3 Reaction and Specificity

Catalyzed reaction

S-adenosyl-L-methionine + G(5')pppR-RNA = S-adenosyl-L-homocysteine + m^7G(5')pppR-RNA (mRNA containing an N^7-methylguanine cap) (R may be guanosine or adenosine; <8> lipid-binding site, membrane binding activity of wild-type and diverse mutants [16]; <1, 8, 11-13> RNA triphosphatase, RNA guanylyltransferase, and RNA (guanine-N^7-)methyltransferase form the mRNA capping enzyme complex consiting of a small and a large subunit [6, 7, 14, 16, 18, 19, 21, 23]; <1> subunit interaction in vitro and in vivo is a necessary condition for full enzyme activity [19]; <1> active site is located at His682 and Tyr683 [19]; <1> substrate binding sites [23])

Reaction type

methyl group transfer

Natural substrates and products

S S-adenosyl-L-methionine + G(5')pppR-RNA <1, 3, 7> (<7> in infected BHK cells [2]; <1> R may be guanosine or adenosine, involved in synthesis of methylated capped structure at 5'-terminal of viral mRNA [1]; <3> associated with mRNA-guanylyltransferase [11]; <1> one of a series of three reactions in capping of vaccinia virus mRNA catalyzed by one enzyme, which is a heterodimer: no methylase activity with one of the subunits alone [6,7,14]; <1> the presence of the small one is required for methylase activity, the large subunit bears functional domains for binding of the small subunit and for methylase activity [6,14]; <1> guanylyltransferase and RNA-triphosphatase are located on the large subunit [7,14]) (Reversibility: ? <1, 3, 7> [1-3, 6, 7, 11, 14]) [1-3, 6, 7, 11, 14]

P S-adenosyl-L-homocysteine + m^7G(5')pppR-RNA <7> [2]

S Additional information <11-13> (<11-13> methylation of the cap structure is also implicated in polyadenylation-mediated translational activation during Xenopus oocyte maturation [21]) [21]

P ?

Substrates and products

S S-adenosyl-L-methionine + G(5')pppN <1, 5, 8> (<8> unmethylated cap analogue, serves as in vitro substrate [16]; <1,3> N: A [9,11]; <3,5,8> N: G [8,11,16]) (Reversibility: ? <1, 5, 8> [8, 9, 16]) [8, 9, 11, 16]

P S-adenosyl-L-homocysteine + m^7G(5')pppG <5, 8> [8, 16]

S S-adenosyl-L-methionine + G(5')pppR-RNA <1-13> (<1> acceptor substrate specificity, overview [9]; <5> active also on E. coli tRNA [8]; <1,5> R may be guanosine or adenosine, methyl acceptors are unmethylated vaccinia virus mRNAs [5,8,9,15]; <3> unmethylated mRNA of reovirus and poly A(+) RNA [11]; <4> methylates also undermethylated poly(A)-rich mRNA [13]; <1,5> mRNA, synthetic polyribonucleotides terminating in a dinucleoside triphosphate [8,9]; <1,5> specific for mRNAs with 5'-terminal sequence G(5')pppA- and G(5')pppG- [4,8,9]; <1> transfers methyl group from S-adenosyl-L-methionine specifically to position 7 of terminal guanosine [3,4,8]; <5> free G(5')pppG is utilized, G(5')ppG is a poor substrate, while GTP, GDP, and G(5')ppppG are no substrates [8]; <1> utilizes GMP, GDP, guanosine, ITP as substrates [9]) (Reversibility: ? <1-13> [1-23]) [1-23]

P S-adenosyl-L-homocysteine + m^7G(5')pppR-RNA <1-13> (<1-7> R may be guanosine or adenosine, mRNA containing an N^7-methylguanine cap [1-14]) [1-23]

S S-adenosyl-L-methionine + GTP <1, 7, 14> (<7> viral, not host enzyme [2]) (Reversibility: ? <1, 7, 14> [2, 9, 18, 24]) [2, 9, 18, 24]

P S-adenosyl-L-homocysteine + m^7GTP <7, 14> [2, 24]

S S-adenosyl-L-methionine + guanylimidodiphosphate <14> (Reversibility: ? <14> [24]) [24]

P ?

S S-adenosyl-L-methionine + pp(A)n <1> (Reversibility: ? <1> [4]) [4]

P ?

S Additional information <1, 5> (<1> isolated methyltransferase domain shows a lower but kinetically equivalent activity than the complete enzyme, which is highly enhanced by association with the D12 subunit [18,19]; <1> no methylation of internal or conventional 5'-terminal guanosine residues [4]; <1,5> no substrates are poly-(A) terminating in a single phosphate or a diphosphate [5,8]; <5> GDP, GTP, G(5')ppppG [8]; <1> ATP, XTP, CTP, UTP, and guanosine containing compounds with phospho groups in either position 2' or 3' or in 3'-5'phosphodiester linkage [9]) [4, 5, 8, 9, 18, 19]

P ?

Inhibitors

3-[(3-cholamidopropyl) dimethylammonio]-1-propanesulfonate <8> (<8> i.e. CHAPS, strong inhibition [16]) [16]

5',N^8-adenosyl-α,β-diaminobutyric acid <1> (<1> moderate [1]) [1]

Ca^{2+} <1, 4> (<1,4> strong [4,13]) [4, 13]

Cu^{2+} <1> (<1> strong [4]) [4]

GpppG <5> (<5> inhibits binding of the enzyme to RNA [22]) [22]

Mg^{2+} <1, 4> (<1> weak [4]; <1,4> strong [13,18]; <1> activation [5]) [4, 13, 18]

Mn^{2+} <1, 4> (<1> weak [4]; <4> strong [13]; <1> activation [5]) [4, 13]

N,N-bis-(3-D-gluconamidopropyl)-deoxycholamide <8> (<8> strong inhibition [16]) [16]

N-tetradecyl-N,N-dimethyl-3-ammonio-1-propane sulfonate <8> (<8> i.e. Zwittergent 3-14, strong inhibition [16]) [16]

S-(2-azaadenosyl)-L-homocysteine <1> (<1> moderate [1]) [1]

S-(3-aminoadenosyl)-L-homocysteine <1> (<1> weak [1]) [1]

S-(3-deazaadenosyl)-L-homocysteine <1, 2> (<1,2> strong [1,10]) [1, 10]

S-(8-azaadenosyl)-L-homocysteine <1, 2> (<1> weak [1]; <2> moderate [10]) [1, 10]

S-(N^6-dimethyl-3-deazaadenosyl)-L-homocysteine <1> (<1> weak [1]) [1]

S-(N^6-methyladenosyl)-L-homocysteine <1> (<1> moderate [1]) [1]

S-adenosyl-D-homocysteine <1, 2> (<1> weak [1]; <2> moderate [10]) [1, 10]
S-adenosyl-L-cysteine <1> (<1> moderate [1]) [1]
S-adenosyl-L-homocysteine <1, 2, 4> (<1,2,4> product inhibition, competitive [1,4,10,13]) [1, 4, 10, 13]
S-adenosyl-L-homocysteine structural analogues <1, 2> [1, 10]
S-adenosyl-L-homocysteine sulfone <1, 2> (<1> weak [1]; <2> strong [10]) [1, 10]
S-adenosyl-L-homocysteine sulfoxide <1, 2> (<1> weak [1]; <2> strong [10]) [1, 10]
S-aristeromycinyl-L-homocysteine <1, 2> [1, 10]
S-inosyl-L-homocysteine <1> (<1> weak [1]) [1]
S-tubercidinyl-L-homocysteine <1, 2> (<1> weak [1]; <2> strong [10]) [1, 10]
Thesit <8> (<8> strong inhibition [16]) [16]
Triton X-100 <8> (<8> inactivation, reversible by addition of lipids: cardiolipin, phosphatidylglycerol, and especially phosphatidylserine [16]) [16]
Zn^{2+} <1> (<1> strong [4]) [4]
amino acid modified structure analogue of adenosyl-L-homocysteine <1> (<1> i.e. A9145C, strong [1]) [1]
cycloheximide <7> (<7> inhibits protein synthesis in infected BHK cells [2]) [2]
deoxycholate <8> (<8> inactivation [16]) [16]
digitonin <8> (<8> strong inhibition [16]) [16]
n-octyl-β-D-gluconpyranoside <8> (<8> i.e. octylglucoside, strong inhibition [16]) [16]
oligoadenylic acid mono- and triphosphates <1, 6> (<1,6> 2'-5'-linked, with varying numbers of phosphate groups, methylated in the 3'-terminal hydroxy group or all three 3'-hydroxy groups, different types of inhibition of viral and L-cell enzyme [12]) [12]
sinefungin <1> (<1> i.e. A9145, strong [1]) [1]
Additional information <1, 2, 4, 6, 8> (<8> detergents prevent the association of S-adenosyl-L-methionine with the enzyme [16]; <4> inhibition by a natural low-molecular-weight inhibitor in the crude extract, removed during purification [13]; <1,2> no inhibition by S-uridinyl-L-homocysteine, S-cytidinyl-L-homocysteine [1,10]; <2> S-guanosyl-L-homocysteine [10]; <1,6> 2'-AMP, 3'-AMP [12]) [1, 10, 12, 13, 16]

Cofactors/prosthetic groups

Additional information <1, 5> (<1,5> no requirement of GTP [4,8]) [4, 8]

Activating compounds

importin-α <5> (<5> enhances specific binding of the enzyme to RNA, enhances RNA modification activity in vivo and in vitro [22]) [22]

Metals, ions

KCl <5> (<4> no activation [13]) [8]
Mg^{2+} <1, 4, 5> (<1,5> activation [5,8]; <1> not required [9]; <4> inactivation [13]) [5, 8, 13]

Mn^{2+} <1, 4> (<1> activation [5]; <4> inhibition [13]) [5, 13]
Additional information <1, 4, 5> (<1> no requirement for divalent cations [4]; <4> no activation by NaCl [13]) [4, 8, 13]

Specific activity (U/mg)
0.00000235 <4> (<4> purified enzyme [13]) [13]
0.000593 <5> (<5> purified enzyme [8]) [8]
0.0196 <1> (<1> purified D1 fragment, residues 498-844 [18]) [18]
Additional information <1> [19]

K_m-Value (mM)
0.002 <4> (S-adenosyl-L-methionine) [13]
0.003 <1> (S-adenosyl-L-methionine, <1> intact mRNA capping enzyme and isolated methyltransferase domain [18]) [18]
0.033 <1> (GTP, <1> intact mRNA capping enzyme and isolated methyltransferase domain [18]) [18]
0.036 <1> (GTP, <1> isolated methyltransferase domain [18]) [18]
0.085 <1> (G(5')pppG, <1> intact mRNA capping enzyme [18]) [18]
0.101 <1> (G(5')pppG, <1> isolated methyltransferase domain [18]) [18]

K_i-Value (mM)
0.00000018 <1> (amino acid modified structure analogue of S-adenosyl-L-homocysteine) [6]
0.00000134 <1> (amino acid modified structure analogue of S-adenosyl-L-homocysteine) [3]
0.0000075 <1> (sinefungin) [6]
0.0000194 <1> (sinefungin) [3]
0.00053 <1> (S-adenosyl-L-homocysteine) [6]
0.001 <1> (S-adenosyl-L-homocysteine) [3]
0.0012 <1> (S-adenosyl-L-homocysteine sulfone) [6]
0.0012 <1> (S-adenosyl-L-homocysteine sulfoxide) [6]
0.0012 <1> (S-tubercidinyl-L-homocysteine) [6]
0.0021 <1> (S-(3-deazaadenosyl)-L-homocysteine) [6]
0.0026 <1> (S-aristeromycinyl-L-homocysteine) [6]
0.0046 <1> (S-(3'-aminoadenosyl)-L-homocysteine) [6]
0.009 <1> (S-(N^6-methyladenosyl)-L-homocysteine) [6]
0.014 <1> (S-adenosyl-D-homocysteine) [6]
0.058 <1> (S-adenosyl-L-cysteine) [6]

pH-Optimum
7 <4> [13]
7.2-7.8 <5> [8]
7.5 <1, 8, 14> (<1,8,14> assay at [16,19,23,24]) [16, 19, 23, 24]
7.8 <1, 3> (<1,3> assay at [1,3,9,11]) [1, 3, 9, 11]
8 <1, 9, 10> (<1,9,10> assay at [7,18,20]) [7, 18, 20]

pH-Range
5-8 <4> (<4> half-maximal activity at pH 5.0, unstable above pH 8.0 [13]) [13]

Temperature optimum (°C)

30 <4, 7, 8, 14> (<7,8,14> assay at [2,16,24]; <4> around [13]) [2, 13, 16, 24]
37 <1, 3, 6, 9, 10> (<1,3,6,9,10> assay at [1,3-7,9,11,12,18-20,23]) [1, 3-7, 9, 11, 12, 18-20, 23]

4 Enzyme Structure

Molecular weight

33000 <1> (<1> small subunit of mRNA capping enzyme , gel filtration, SDS-PAGE [14,18]) [14, 18]
47000 <13> (<13> western blot [21]) [21]
49000 <4> (<4> gel filtration [13]) [13]
56000 <5> (<5> gel filtration, sucrose density gradient centrifugation [8]) [8]
58660-60000 <12> (<12> DNA sequence determination, western blot [21]) [21]
68820-70000 <11> (<11> DNA sequence determination, western blot [21]) [21]
97000 <1> (<1> large subunit of mRNA capping enzyme, gel filtration, SDS-PAGE [14,18]) [14, 18]
120000-129000 <1> (<1> glycerol gradient centrifugation [15]; <1> gel filtration, sucrose density gradient centrifugation [3]) [3, 15]
Additional information <1> (<1> components of the mRNA capping enzyme [14]; <1> mutants [6]; <1> Stokes radius determination [3]) [3, 6, 14]

Subunits

dimer <1> (<1> 1 * 31000 + 1 * 95000, SDS-PAGE [6]) [6]
monomer <4> (<4> 1 * 49000, SDS-PAGE [13]) [13]
Additional information <1> (<1> isolated methyltransferase domain shows a lower but kinetically equivalent activity than the complete enzyme, which is highly enhanced by association with the D12 subunit [18,19]; <1> subunit organisation of the mRNA capping enzyme complex [6]; <1> the methyl/guanylyltransferase complex on vaccinia virus dissociates into two subunits of MW 31400 and 95000 by action of SDS-PAGE [3]) [3, 6, 18, 19]

5 Isolation/Preparation/Mutation/Application

Source/tissue

cell suspension culture <3-6> [8, 11-13]
mycelium <3> [11]
zygote <11, 13> [21]

Localization

cytoplasm <5> (<5> 70% of total activity, 1/3 of which is losely associated to ribosomes [8]) [8]
membrane <8> (<8> lipid-binding site, membrane binding activity of wild-type and diverse mutants [16]) [16]

mitochondrion <7> (<7> of infected BHK cells [2]) [2]
nucleus <5> (<5> 30% of total activity [8]; <5> GFP-fusion protein analysis [22]) [8, 22]
viral nucleocapsid <1, 2, 8> [1, 3-6, 9, 10, 12, 15, 16]

Purification

<1> (from infected HeLa cells [9]; methyltransferase domain, and small subunit of mRNA capping enzyme, recombinant from E. coli [18,19]; small and large subunit of mRNA capping enzyme , recombinant from E. coli [7,14]; copurifies with mRNA-guanylyltransferase [3,15]) [3, 7, 9, 12, 14, 15, 18, 19]
<3> (DNA-agarose affinity chromatography [11]) [11]
<4> [13]
<5> (DNA-agarose affinity chromatography [8]) [8]
<6> [12]
<7> (recombinant from E. coli and recombinant from in vitro transcription and translation [17]) [17]
<8> (recombinant from Spodoptera frugiperda cells [16]) [16]
<14> (enzyme and truncated mutants recombinant from Saccharomyces cerevisiae [24]) [24]

Cloning

<1> (expression of D1 fragment mutants in Escherichia coli [23]; individual and coexpression of His-tagged D1 large subunit fragment, being the methyltransferase domain of the enzyme, and of D12 small subunit in Escherichia coli [18,19,23]; D1 fragment, amino acids 520-844 [19]; D1 fragment, amino acids 498 to 844 [18,23]; vaccinia virus genes D1 and D12 encoding the large and the small subunit of the mRNA capping enzyme, respectively, individual overexpression in Escherichia coli BL21 [6,7,14]; 2 carboxyl-truncated versions of the large subunit of the mRNA capping enzyme, expression in Escherichia coli [6,14]; <1> coexpression of genes D1 and D12 from 1 plasmid [14]) [6, 7, 14, 18, 23]
<5> (expression of full length enzyme and truncated enzyme in yeast and in HeLa cells, His-tagged, as GFP-fusion protein, or cMyc-tagged [22]) [22]
<7> (in vitro transcription and translation in presence of liposomes, enzyme is attached to the liposomes surface [17]; expression in Escherichia coli [17]) [17]
<8> (expression of VP4 protein in Spodoptera frugiperda cells [16]) [16]
<9> (expression of 2 different variants, each with and without His-tag, in Escherichia coli, amino acid sequence analysis [20]) [20]
<10> (expression in Saccharomyces cerevisiae, His-tagged, amino acid sequence analysis [20]) [20]
<11> (expression in Escherichia coli, His-tagged, amino acid sequence analysis [21]) [21]
<12> (expression in Escherichia coli, His-tagged, amino acid sequence analysis [21]) [21]
<13> (expression in Escherichia coli, His-tagged [21]) [21]
<14> (expression of full length and 2 truncated genes in Saccharomyces cerevisiae and Escherichia coli, location of methyltransferase activity [24]) [24]

Engineering

G600A <1> (<1> in D1 fragment, amino acids 498-844, 4% activity remaining [23]) [23]

H682A <1> (<1> in D1 fragment, amino acids 498-844, 4% activity remaining, unable to crosslink the cap in presence of S-adenosyl-L-homocysteine [23]) [23]

Y683A <1> (<1> in D1 fragment, amino acids 498-844, 0.05% activity remaining, unable to crosslink the cap in presence of S-adenosyl-L-homocysteine [23]) [23]

Y683F <1> (<1> in D1 fragment, amino acids 498-844, 4% activity remaining [23]) [23]

Y683S <1> (<1> in D1 fragment, amino acids 498-844, catalytically defective, unable to crosslink the cap in presence of S-adenosyl-L-homocysteine [23]) [23]

Additional information <1, 8> (<1> exchange of 8 residues for alanine in the D1 fragment, amino acids 498-844 [23]; <1> exchange for alanine at single sites within the methyltransferase domain fragment of D1, loss of activity when exchanged hits His682 and Tyr683, active site [19]; <8> mapping of lipid-binding site, membrane binding activity of wild-type and diverse constructed mutants [16]; <1> construction of carboxyl-truncated D1 genes, amino acid 1-666 and 1-425: 347 residues are sufficient for activity [6]) [6, 16, 19]

6 Stability

pH-Stability

4.5 <1> (<1> inactivation after 16 h at 0°C [15]) [15]

7.5 <1> (<1> 16 h stable at 0°C [15]) [15]

General stability information

<1>, gel electrophoresis inactivates [15]

<1>, small and large subunit of the mRNA capping enzyme are not stable and active when isolated, respectively [14]

<7>, expression and enzyme activity stable for 3 h in infected BHK cells [2]

<8>, cardiolipin, phosphatidylglycerol, and phosphatidylserine [16]

Storage stability

<1>, 0°C, 16 h at pH 7.5 [15]

<1>, 4°C or -20°C, several months, stable [3]

<3>, -20°C, in 10% v/v glycerol, at least several months [11]

<5>, 4°C, $t_{1/2}$: less than 36 h [8]

<4, 5>, -80°C, several months [8, 13]

References

[1] Pugh, C.S.G.; Borchardt, R.T.: Effects of S-adenosylhomocysteine analogues on vaccinia viral messenger ribonucleic acid synthesis and methylation. Biochemistry, **21**, 1535-1541 (1982)
[2] Cross, R.K.: Identification of a unique guanine-7-methyltransferase in Semliki Forest virus (SFV) infected cell extracts. Virology, **130**, 452-463 (1983)
[3] Martin, S.A.; Paoletti, E.; Moss, B.: Purification of mRNA guanylyltransferase and mRNA (guanine-7-) methyltransferase from vaccinia virions. J. Biol. Chem., **250**, 9322-9329 (1975)
[4] Martin, S.A.; Moss, B.: Modification of RNA by mRNA guanylyltransferase and mRNA (guanine-7-)methyltransferase from vaccinia virions. J. Biol. Chem., **250**, 9330-9335 (1975)
[5] Ensinger, M.J.; Martin, S.A.; Paoletti, E.; Moss, B.: Modification of the 5-terminus of mRNA by soluble guanylyl and methyl transferases from vaccinia virus. Proc. Natl. Acad. Sci. USA, **72**, 2525-2529 (1975)
[6] Cong, P.; Shuman, S.: Methyltransferase and subunit association domains of vaccinia virus mRNA capping enzyme. J. Biol. Chem., **267**, 16424-16429 (1992)
[7] Higman, M.A.; Bourgeois, N.; Niles, E.G.: The vaccinia virus mRNA (guanine-N^7-)-methyltransferase requires both subunits of the mRNA capping enzyme for activity. J. Biol. Chem., **267**, 16430-16437 (1992)
[8] Ensinger, M.J.; Moss, B.: Modification of the 5 terminus of mRNA by an RNA (guanine-7-)-methyltransferase from HeLa cells. J. Biol. Chem., **251**, 5283-5291 (1976)
[9] Martin, S.A.; Moss, B.: mRNA guanylyltransferase and mRNA (guanine-7-)-methyltransferase from vaccinia virions. Donor and acceptor substrate specificites. J. Biol. Chem., **251**, 7313-7321 (1976)
[10] Pugh, C.S.; Borchardt, R.T.; Stone, H.O.: Inhibition of Newcastle disease virion messenger RNA (guanine-7-)-methyltransferase by analogues of S-adenosylhomocysteine. Biochemistry, **16**, 3928-3932 (1977)
[11] Germershausen, J.; Goodman, D.; Somberg, E.W.: 5 Cap methylation of homologous poly A(+) RNA by a RNA (guanine-7) methyltransferase from Neurospora crassa. Biochem. Biophys. Res. Commun., **82**, 871-878 (1978)
[12] Goswami, B.B.; Crea, R.; Van Boom, J.H.; Sharma, O.K.: 2-5-Linked oligo(adenylic acid) and its analogs. A new class of inhibitors of mRNA methylation. J. Biol. Chem., **257**, 6867-6870 (1982)
[13] Locht, C.; Beaudart, J.L.; Delcour, J.: Partial purification and characterization of mRNA (guanine-7-) methyltransferase from the yeast Saccharomyces cerevisiae. Eur. J. Biochem., **134**, 117-121 (1983)
[14] Guo, P.; Moss, B.: Interaction and mutual stabilization of the two subunits of vaccinia virus mRNA capping enzyme coexpressed in Escherichia coli. Proc. Natl. Acad. Sci. USA, **87**, 4023-4027 (1990)
[15] Monroy, G.; Spencer, E.; Hurwitz, J.: Purification of mRNA guanylyltransferase from vaccinia virions. J. Biol. Chem., **253**, 4481-4489 (1978)

[16] Ramadevi, N.; Burroughs, N.J.; Mertens, P.P.C.; Jones, I.M.: Capping and methylation of mRNA by purified recombinant VP4 protein of bluetongue virus. Proc. Natl. Acad. Sci. USA, **95**, 13537-13542 (1998)
[17] Ahola, T.; Lampio, A.; Auvinen, P.; Kaariainen, L.: Semliki forest virus mRNA capping enzyme requires association with anionic membrane phospholipids for activity. EMBO J., **18**, 3164-3172 (1999)
[18] Higman, M.A.; Christen, L.A.; Niles, E.G.: The mRNA (guanine-7-)methyltransferase domain of the vaccinia virus mRNA capping enzyme. Expression in Escherichia coli and structural and kinetic comparison to the intact capping enzyme. J. Biol. Chem., **269**, 14974-14981 (1994)
[19] Mao, X.; Shuman, S.: Intrinsic RNA (guanine-7) methyltransferase activity of the vaccinia virus capping enzyme D1 subunit is stimulated by the D12 subunit. Identification of amino acid residues in the D1 protein required for subunit association and methyl group transfer. J. Biol. Chem., **269**, 24472-24479 (1994)
[20] Tsukamoto, T.; Shibagaki, Y.; Niikura, Y.; Mizumoto, K.: Cloning and characterization of three human cDNAs encoding mRNA (guanine-7-)-methyltransferase, an mRNA cap methylase. Biochem. Biophys. Res. Commun., **251**, 27-34 (1998)
[21] Yokoska, J.i.; Tsukamoto, T.; Miura, K.i.; Shiokawa, K.; Mizumoto, K.: Cloning and characterization of mRNA capping enzyme and mRNA (guanine-7-)-methyltransferase cDNAs from Xenopus laevis. Biochem. Biophys. Res. Commun., **268**, 617-624 (2000)
[22] Wen, Y.; Shatkin, A.J.: Cap methyltransferase selective binding and methylation of GpppG-RNA are stimulated by importin-α. Genes Dev., **14**, 2944-2949 (2000)
[23] Mao, X.; Shuman, S.: Vaccinia virus mRNA (guanine-7-)methyltransferase: mutational effects on cap methylation and AdoHcy-dependent photo-cross-linking of the cap to the methyl acceptor site. Biochemistry, **35**, 6900-6910 (1996)
[24] Li, Y.I.; Chen, Y.J.; Hsu, Y.H.; Meng, M.: Characterization of the Adomet-dependent guanylyltransferase activity that is associated with the N terminus of bamboo mosaic virus replicase. J. Virol., **75**, 782-788 (2001)

mRNA (nucleoside-2'-O-)-methyltransferase 2.1.1.57

1 Nomenclature

EC number
2.1.1.57

Systematic name
S-adenosyl-L-methionine:mRNA (nucleoside-2'-O-)-methyltransferase

Recommended name
mRNA (nucleoside-2'-O-)-methyltransferase

Synonyms
EC 2.1.1.58 (formerly, formation of 2'-O-methyladenosine cap)
mRNA 5' cap-specific (nucleoside-2'-O)-methyltransferase <1> (<1> i.e. VP39 [12]) [12]
mRNA cap (nucleoside-2'-O)-methyltransferase <4> (<4> i.e. NS5 [11]) [11]
mRNA cap-specific 2'-O-methyltransferase <1> (<1> i.e. VP39 [9]) [9]
messenger RNA (nucleoside-2'-)-methyltransferase
messenger ribonucleate nucleoside 2'-methyltransferase
methyltransferase, messenger ribonucleate nucleoside 2'-

CAS registry number
61970-02-3

2 Source Organism

<1> *Vaccinia virus* (J3 protein [13]; VP39 protein [9,12]; strain WR [1-3,5-7]) [1-3, 5-7, 9, 12, 13]
<2> *Homo sapiens* (HeLa S-3 cells infected with vaccinia virus, no activity in uninfected cells [8]) [4, 8]
<3> *bluetongue virus* (enzyme activity belongs to VP4 protein [10]) [10]
<4> *Dengue virus* (type 2, NS5 protein [11]) [11]

3 Reaction and Specificity

Catalyzed reaction
S-adenosyl-L-methionine + m^7G(5')pppR-RNA = S-adenosyl-L-homocysteine + m^7G(5')pppRm-RNA (mRNA containing a 2'-O-methylpurine cap) (<1> binding sites of the enzyme for the terminal cap nucleotide, m^7G, the cap-proximal RNA, and for RNA, immediately downstream, are characterized [9];

<1> R may be guanine or adenosine, mechanism [2]; <4> nucleotide binding analysis based on crystal structure [11])

Reaction type

methyl group transfer

Natural substrates and products

S S-adenosyl-L-methionine + m^{7}G(5')pppR-RNA <1> (<1> final step in formation of 7-methylguanosine(5')pppNm-cap structure of viral RNA [2,6]; <1> RNA-polymerase holoenzyme has methyltransferase activity: the same protein acts as methyltransferase and poly(A)-polymerase stimulatory factor VP39 [5]) (Reversibility: ? <1> [2, 5, 6]) [2, 5, 6]

P S-adenosyl-L-homocysteine + m^{7}G(5')pppRm-RNA

Substrates and products

S S-adenosyl-L-methionine + ApppGR-RNA <1> (<1> synthetic substrate with unusual cap structure, 20% activity compared to m^{7}G-terminated substrates [9]; <1> adenine-capped RNA, wild-type and mutants, indicating the redundancy of m^{7}G-contact residues able to confer cap-type specificity [9]) (Reversibility: ? <1> [9]) [9]

P ?

S S-adenosyl-L-methionine + GpppACCCCC <4> (Reversibility: ? <4> [11]) [11]

P S-adenosyl-L-homocysteine + GpppAmCCCCC <4> [11]

S S-adenosyl-L-methionine + m^{7}G(5')pppNmpN-RNA <2> (<2> N: purin and pyrimidine nucleotides are methylated, enzyme: cap II-methyltransferase [4]) (Reversibility: ? <2> [4]) [4]

P S-adenosyl-L-homocysteine + m^{7}G(5')pppNmpNm-RNA <2> (<2> i.e. cap II [4]) [4]

S S-adenosyl-L-methionine + m^{7}G(5')pppNpN-RNA <2> (<2> i.e. cap O, N: purine and pyrimidine nucleotides are methylated, enzyme: cap I-methyltransferase, substrates are RNAs with a capped terminus with at least 2 additional nucleotides, G(5')pppNpNp [4]) (Reversibility: ? <2> [4]) [4]

P S-adenosyl-L-homocysteine + m^{7}G(5')pppNmpN-RNA <2> (<2> i.e. cap I [4]) [4]

S S-adenosyl-L-methionine + m^{7}G(5')pppR-RNA <1-4> (<4> His-tagged N-terminal part of NS5 protein, termed NS5MTaseDV, amino acids 1-296, utilizes methylated and unmethylated capped RNA, but not uncapped RNA [11]; <1> adenine-capped RNA, wild-type and mutants, indicating the redundancy of m^{7}G-contact residues able to confer cap-type specificity [9]; <1,2> specific, in vitro methyl group acceptors are polyribonucleosides containing 7-methylguanine(5')pppG-terminals [1,2,8]; <1,2> brome mosaic virus RNA ending in m^{7}G(5')pppG as substrate [1,2,5,8]; <1,2> unmethylated vaccinia virus mRNA [2,4,7,8]; <2> satellite tobacco necrosis virus [4]; <1,2> best substrates are poly(A)-nucleosides and poly-(I)-nucleosides with m^{7}G(5')pppN-ends [2,4]; <1> much less activity with poly(G), poly(U) and poly(C) [2]; <1> poor substrates are

m^7GpppN-dinucleotides [2]; <1> methylates 2'-O-position of penultimate nucleoside of RNA, no methylation of internal sites [2,7]) (Reversibility: ? <1-4> [1,2,4,5,7-12]) [1, 2, 4, 5-12]

P S-adenosyl-L-homocysteine + m^7G(5')pppRm-RNA <1, 2, 4> (<1,2> mRNA containing a 2'-O-methylpurine cap, R may be guanine or adenosine, prereplicative viral product [2,8]) [2, 8, 10-12]

S S-adenosyl-L-methionine + m^7GpppACCCCC <4> (Reversibility: ? <4> [11]) [11]

P S-adenosyl-L-homocysteine + m^7GpppAmCCCCC <4> [11]

S S-adenosyl-L-methionine + m^7GpppApGp <2> (<2> i.e. capped dinucleotide, cap I-methyltransferase [4]) (Reversibility: ? <2> [4]) [4]

P S-adenosyl-L-homocysteine + m^7GpppAmpGp <2> [4]

S S-adenosyl-L-methionine + m^7GpppApGpUp <2> (<2> i.e. capped trinucleotide, cap I-methyltransferase [4]) (Reversibility: ? <2> [4]) [4]

P S-adenosyl-L-homocysteine + m^7GpppAmpGpUp <2> [4]

S S-adenosyl-L-methionine + m^7GpppG <3> (Reversibility: ? <3> [10]) [10]

P S-adenosyl-L-homocysteine + m^7GpppGm <3> [10]

S S-adenosyl-L-methionine + m^7GpppGp <1> (<1> substrate of minimal chain length for the cap-specific enzyme, m^7GpppG is no substrate [9]) (Reversibility: ? <1> [9]) [9]

P S-adenosyl-L-homocysteine + m^7GpppGmp

S S-adenosyl-L-methionine + m^7GpppGpN <1> (Reversibility: ? <1> [9]) [9]

P S-adenosyl-L-homocysteine + m^7GpppGmpN

S S-adenosyl-L-methionine + m^7GpppGpNp <1> (Reversibility: ? <1> [9]) [9]

P ?

S S-adenosyl-L-methionine + m^7GpppGpUbiotin-p <1> (Reversibility: ? <1> [9]) [9]

P ?

S S-adenosyl-L-methionine + mononucleotide <2> (<2> such as m^7GpppA or GpppA, poor substrates for cap I-methyltransferase [4]) (Reversibility: ? <2> [4]) [4]

P ?

S Additional information <1> (<1> no substrates are RNAs with pN-, ppN-, guanine(5')pppN-terminals [2]; <1> poly(A) and internal nucleotides of RNA [7]) [2, 7]

P ?

Inhibitors

5',N^8-adenosyl-α,β-diaminobutyric acid <1> (<1> moderate [6]) [6]

7-methylguanosine(5')pppN-dinucleotides <1> (<1> weak, competitive [2]) [2]

A9145C <1> (<1> i.e. amino acid modified structure analogue of adenosyl-L-homocysteine, competitive [3]; <1> strong [6]) [3, 6]

NaCl <1> (<1> above 0.05 M [2]) [2]

S-(2-azaadenosyl)-L-homocysteine <1> [6]

S-(3'-aminoadenosyl)-L-homocysteine <1> [6]
S-(3-deazaadenosyl)-L-homocysteine <1> (<1> strong [6]) [6]
S-(8-azaadenosyl)-L-homocysteine <1> (<1> weak [6]) [6]
S-(N^6-dimethyl-3-deazaadenosyl)-L-homocysteine <1> (<1> weak [6]) [6]
S-(N^6-methyladenosyl)-L-homocysteine <1> [6]
S-adenosyl-D-homocysteine <1> [6]
S-adenosyl-L-cysteine <1> (<1> moderate [6]) [6]
S-adenosyl-L-homocysteine <1, 2> (<1> product inhibition, competitive [3]; <1,2> strong, [4,6]; <2> cap I-methyltransferase [4]) [2-4, 6]
S-adenosyl-L-homocysteine sulfone <1> (<1> strong [6]) [6]
S-adenosyl-L-homocysteine sulfoxide <1> (<1> strong [6]) [6]
S-aristeromycinyl-L-homocysteine <1> [6]
S-cytidyl-L-homocysteine <1> (<1> weak [6]) [6]
S-inosyl-L-homocysteine <1> (<1> weak [6]) [6]
S-tubercidinyl-L-homocysteine <1> (<1> strong [6]) [6]
S-uridyl-L-homocysteine <1> (<1> weak [6]) [6]
sinefungin <1> (<1> i.e. A9145, competitive [3]; <1> strong [6]) [3, 6]
Additional information <1> (<1> no inhibition by poly(A), poly(U), poly(C) or poly(G) [2]; <1> EDTA [7]) [2, 7]

Metals, ions

KCl <2> (<2> activation, 140-180 mM, cap I-methyltransferase [4]) [4]
Additional information <1> (<1> no divalent cations required [2]) [2]

Specific activity (U/mg)

0.00000233 <1> (<1> purified enzyme [1]) [1]
Additional information <1, 2> [4, 7]

K_m-Value (mM)

0.0000014 <2> (G(5')pppA-poly(A)) [4]
0.0000016 <2> (m^7G(5')pppA-poly(A)) [4]
0.000005 <1> (brome mosaic virus RNA) [2]
0.000005-0.000015 <1> (RNA chain length 20-50 nt) [9]
0.00016-0.00023 <1> (RNA chain length 2-6 nt) [9]
0.0006 <2> (S-adenosyl-L-methionine) [4]
0.002 <1> (S-adenosyl-L-methionine) [2]
Additional information <1> [9]

K_i-Value (mM)

0.00000018 <1> (A9145C) [6]
0.00000134 <1> (A9145C) [3]
0.0000075 <1> (sinefungin) [6]
0.0000194 <1> (sinefungin) [3]
0.00053 <1> (S-adenosyl-L-homocysteine) [6]
0.001 <1> (S-adenosyl-L-homocysteine) [3]
0.0012 <1> (S-adenosyl-L-homocysteine sulfone) [6]
0.0012 <1> (S-adenosyl-L-homocysteine sulfoxide) [6]
0.0012 <1> (S-tubercidinyl-L-homocysteine) [6]
0.0021 <1> (S-(3-deazaadenosyl)-L-homocysteine) [6]

0.0026 <1> (S-aristeromycinyl-L-homocysteine) [6]
0.0046 <1> (S-(3'-aminoadenosyl)-L-homocysteine) [6]
0.009 <1> (S-(N^6-methyladenosyl)-L-homocysteine) [6]
0.014 <1> (S-adenosyl-D-homocysteine) [6]
0.058 <1> (S-adenosyl-L-cysteine) [6]

pH-Optimum
7.5 <1, 3> (<3> assay at [10]) [2, 10]
7.6 <1> (<1> assay at [1,7,9]) [1, 7, 9]
Additional information <1> (<1> pI: 8.4 [1]) [1]

pH-Range
6.5-8 <1> (<1> m^7G- and A-capped RNA substrates [9]) [9]
6.6-8.1 <1> (<1> about half-maximal activity at pH 6.6 and 8.1 [2]) [2]
Additional information <2> (<2> beginning from pH 6.9 continous increase of activity [4]) [4]

Temperature optimum (°C)
30 <3> (<3> assay at [10]) [10]
37 <1> (<1> assay at [1,7,9]) [1, 2, 7, 9]

4 Enzyme Structure

Molecular weight
33000 <1> (<1> gel filtration [1]) [1]
33000 <4> (<4> recombinant truncated N-terminal protein, gel filtration [11]) [11]
36000 <1> (<1> sucrose density gradient centrifugation [1]) [1]
38000 <1> (<1> gel filtration [1,5]) [1, 5]

Subunits
monomer <1, 4> (<4> 1 * 33000, recombinant truncated N-terminal protein, gel filtration, SDS-PAGE [11]; <1> 1 * 38000, SDS-PAGE [1]) [1, 11]

5 Isolation/Preparation/Mutation/Application

Source/tissue
cell suspension culture <2> (<2> HeLa S-3 cells [8]) [4, 8]

Localization
cytoplasm <2> (<2> cap I- and cap II-methyltransferase [4]) [4]
nucleus <2> (<2> cap I-methyltransferase [4]) [4]
viral nucleocapsid <1, 3> [1-3, 5-7, 10]

Purification
<1> (fragment mutant and exchange mutants from recombinant E. coli [12]; mutants recombinant from E. coli [9]; affinity chromatography on ADP-Sepharose [1]; solubilized from viral core [7]) [1, 5, 7, 9, 12]

<2> (HeLa-cells, phosphocellulose chromatography [4]) [4, 8]
<3> (recombinant from Spodoptera frugiperda cells [10]) [10]
<4> (recombinant His-tagged N-terminal part of NS5 protein from E. coli [11]) [11]

Crystallization

<4> (hanging drop method at room temperature, 0.1 M sodium citrate pH 5.8, 1.2 M lithium sulfate, 0.5 M ammonium sulfate, vapour diffusion for 1 week, structure analysis with and without bound nucleotide, model [11]) [11]

Cloning

<1> (expression of wild-type, exchange mutants and open-reading-frame mutant in Escherichia coli, all glutathione-S-transferase-tagged [12]; mutant enzymes are expressed in Escherichia coli and the tag is cleaved off [9]) [9, 12]
<3> (expression of VP4 protein in Spodoptera frugiperda cells [10]) [10]
<4> (cloning of His-tagged N-terminal part of NS5 protein, termed NS5MTaseDV, amino acids 1-296, overexpression in Escherichia coli [11]) [11]

Engineering

C178S <1> (<1> point mutations, glutathione-S-transferase-tagged, exchange shows no effect [12]) [12]
C272S <1> (<1> point mutations, glutathione-S-transferase-tagged, exchange shows no effect [12]) [12]
D182A <1> (<1> m^7G binding pocket mutant, mutagenesis, C-terminal truncated by 26 amino acids, lacks specific m^7G-contact side chains and shows reduced activity, but remains cap-dependent [9]) [9]
D182A/E233A <1> (<1> m^7G binding pocket double mutant, mutagenesis, N-terminal glutathione-S-transferase-tagged, C-terminal truncated by 26 amino acids, lacks specific m^7G-contact side chains and shows reduced activity, but remains cap-dependent [9]) [9]
E233A <1> (<1> m^7G binding pocket mutant, mutagenesis, N-terminal glutathione-S-transferase-tagged, C-terminal truncated by 26 amino acids, lacks specific m^7G-contact side chains and shows reduced activity, but remains cap-dependent [9]) [9]
F180A <1> (<1> m^7G binding pocket mutant, mutagenesis, C-terminal truncated by 26 amino acids, lacks specific m^7G-contact side chains and shows reduced activity, but remains cap-dependent [9]) [9]
G96D <1> (<1> mutation in J3 protein with nucleoside-2-O'-methyltransferase activity evokes a phenotype with abnormally long RNA transcripts analogously to A18 gene-mutation Cts23 [13]; <1> double mutant G96D plus Cts23 mutation in gene A18 shows that the J3-mutants serves as an extragenic suppressor of A18-Cts23 mutant [13]) [13]
Y22A <1> (<1> m^7G binding pocket mutant, mutagenesis, N-terminal glutathione-S-transferase-tagged, C-terminal truncated by 26 amino acids, lacks specific m^7G-contact side chains and shows reduced activity, but remains cap-dependent [9]) [9]

Additional information <1> (<1> 333-codon open-reading frame is cut out and expressed in E. coli, reduced methyltransferase and RNA binding activity [12]; <1> charged amino acids Asp, Glu, His, Lys, Arg are exchanged for Ala by oligonucleotide-directed mutagenesis, glutathione-S-transferase-tagged, 11 of 21 Ala-mutants show reduced activity, 4 show increased activity [12]) [12]

6 Stability

pH-Stability

Additional information <1> (<1> binding site of the enzyme for RNA, immediately downstream, is pH-sensitive, above pH 7.5: dramatic increase of K_m [9]) [9]

Temperature stability

37 <2> (<2> 10 min stable [4]) [4]
44 <2> (<2> and above, inactivation after 20 min [4]) [4]
50 <1> (<1> and below, stable in the presence of 10% v/v glycerol and 250 mg/l bovine serum albumin [2]) [2]
59 <1> (<1> rapid inactivation [2]) [2]

General stability information

<1>, bovine serum albumin stabilizes [2]
<1>, glycerol, 10% v/v stabilizes [2]

Storage stability

<1>, -20°C, several months [1]
<1>, 4°C, several months [1]

References

[1] Barbosa, E.; Moss, B.: mRNA(nucleoside-2-)-methyltransferase from vaccinia virus. Purification and physical properties. J. Biol. Chem., **253**, 7692-7697 (1978)
[2] Barbosa, E.; Moss, B.: mRNA(nucleoside-2-)-methyltransferase from vaccinia virus. Characteristics and substrate specificity. J. Biol. Chem., **253**, 7698-7702 (1978)
[3] Pugh, C.S.G.; Borchardt, R.T.; Stone, H.O.: Sinefungin, a potent inhibitor of virion mRNA(guanine-7-)-methyltransferase, mRNA(nucleoside-2-)-methyltransferase, and viral multiplication. J. Biol. Chem., **253**, 4075-4077 (1978)
[4] Langberg, S.R.; Moss, B.: Post-transcriptional modifications of mRNA. Purification and characterization of cap I and cap II RNA (nucleoside-2-)-methyltransferases from HeLa cells. J. Biol. Chem., **256**, 10054-10060 (1981)
[5] Schnierle, B.S.; Gershon, P.; Moss, B.: Cap-specific mRNA (nucleoside-$O^{2'}$)-methyltransferase and poly(A) polymerase stimulatory activities of vaccinia virus are mediated by a single protein. Proc. Natl. Acad. Sci. USA, **89**, 2897-2901 (1992)

[6] Pugh, C.S.G.; Borchardt, R.T.: Effects of S-adenosylhomocysteine analogues on vaccinia viral messenger ribonucleic acid synthesis and methylation. Biochemistry, **21**, 1535-1541 (1982)

[7] Ensinger, M.J.; Martin, S.A.; Paoletti, E.; Moss, B.: Modification of the 5-terminus of mRNA by soluble guanylyl and methyl transferases from vaccinia virus. Proc. Natl. Acad. Sci. USA, **72**, 2525-2529 (1975)

[8] Boone, R.F.; Ensinger, M.J.; Moss, B.: Synthesis of mRNA guanylyltransferase and mRNA methyltransferases in cells infected with vaccinia virus. J. Virol., **21**, 475-483 (1977)

[9] Lockless, S.W.; Cheng, H.-T.; Hodel, A.E.; Quiocho, F.A.; Gershon, P.D.: Recognition of capped RNA substrates by VP39, the vaccinia virus-encoded mRNA cap-specific 2'-O-methyltransferase. Biochemistry, **37**, 8564-8574 (1998)

[10] Ramadevi, N.; Burroughs, N.J.; Mertens, P.P.C.; Jones, I.M.: Capping and methylation of mRNA by purified recombinant VP4 protein of bluetongue virus. Proc. Natl. Acad. Sci. USA, **95**, 13537-13542 (1998)

[11] Egloff, M.-P.; Benarroch, D.; Selisko, B.; Romette, J.-L.; Canard, B.: An RNA cap (nucleoside-2'-O-)-methyltransferase in the flavivirus RNA polymerase NS5: crystal structure and functional characterization. EMBO J., **21**, 2757-2768 (2002)

[12] Schnierle, B.S.; Gershon, P.D.; Moss, B.: Mutational analysis of a multifunctional protein, with mRNA 5' cap-specific (nucleoside-2'-O-)-methyltransferase and 3'-adenylyltransferase stimulatory activities, encoded by vaccinia virus. J. Biol. Chem., **269**, 20700-20706 (1994)

[13] Latner, D.R.; Xiang, Y.; Lewis, J.I.; Condit, J.; Condit, R.C.: The vaccinia virus bifunctional gene J3 (nucleoside-2'-O-)-methyltransferase and poly(A) polymerase stimulatory factor is implicated as a positive transcription elongation factor by two genetic approaches. Virology, **269**, 345-355 (2000)

mRNA (adenosine-2′-O-)-methyltransferase 2.1.1.58

1 Nomenclature

EC number

2.1.1.58 (deleted, included in EC 2.1.1.57)

Recommended name

mRNA (adenosine-2'-O-)-methyltransferase

[Cytochrome c]-lysine N-methyltransferase 2.1.1.59

1 Nomenclature

EC number
2.1.1.59

Systematic name
S-adenosyl-L-methionine:[cytochrome c]-L-lysine N^6-methyltransferase

Recommended name
[cytochrome c]-lysine N-methyltransferase

Synonyms
cytochrome c (lysine) methyltransferase
cytochrome c methyltransferase
cytochrome c-specific protein methylase III
cytochrome c-specific protein-lysine methyltransferase
methyltransferase, cytochrome c (lysine)

CAS registry number
82047-78-7

2 Source Organism

<1> *Triticum aestivum* [8]
<2> *Crithidia oncopelti* [1]
<3> *Neurospora crassa* [2, 3, 6, 7]
<4> *Saccharomyces cerevisiae* (encoded by Ctm1p, strain YHR109w [5]) [4, 5, 9]

3 Reaction and Specificity

Catalyzed reaction
S-adenosyl-L-methionine + cytochrome c L-lysine = S-adenosyl-L-homocysteine + cytochrome c N^6-methyl-L-lysine (<3> hybrid ping-pong mechanism [6]; <2,3> specific methylation site, 1 lysine in cytochrome c, comparison [1-3])

Reaction type
methyl group transfer

Natural substrates and products

S S-adenosyl-L-methionine + cytochrome c L-lysine <1-4> (<4> regulation of enzyme and cytochrome c substrate is not in parallel manner [9]; <4> in vivo methylation of cytochrome c in yeast occurs probably before cytochrome c is attached to the mitochondria [9]) (Reversibility: ? <1-4> [1-4, 6-9]) [1-4, 6-9]

P S-adenosyl-L-homocysteine + cytochrome c N^6-methyl-L-lysine <1-4> [1-4, 6-9]

Substrates and products

S S-adenosyl-L-methionine + cytochrome c L-lysine <1-4> (<4> specific for unmethylated cytochrome c [5,9]; <1,3> also CNBr peptides of horse heart cytochrome c can serve as substrates, overview [3,7-9]; <1,3,4> horse heart cytochrome c-72 [2,3,5-9]; <4> trimethylation of recombinant iso-1-cytochrome c of lysine at position 72 [5]; <4> recombinant substrate iso-1-apocytochrome c [4]; <2> Crithidia cytochrome c-557 is far the best substrate [1]; <1> avian cytochromes c are much better substrates than those from mammalian sources [8]; <2> lysine-72 of horse cytochrome c is a poor acceptor [1]; <3> important features in determining methyl-accepting activity: 1. reactive site such as-X-Lys-Lys-Y-, 2. length of polypeptide chain, 3. tertiary structure [3]; <1,3,4> overview: substrate activity of various cytochromes c [3,8,9]; <1,3> does not randomly methylate cytochrome c, but shows absolute specificity for some amino acid sequences [2,3,8]) (Reversibility: ? <1-4> [1-9]) [1-9]

P S-adenosyl-L-homocysteine + cytochrome c N^6-methyl-L-lysine <1-4> (<2> uptake of methyl groups occurs only at lysine residue 8 and the terminal proline residue [1]; <3,4> mixture of ε-N-monomethyllysine, ε-N-dimethyllysine and ε-N-trimethyllysine in the ratio 1:3:4 [2-4,7-9]; <4> 9:17:74 [9]; <3,4> a single lysine residue at position 72 of horse heart cytochrome c is methylated [3,7,9]) [1-4, 6-9]

S S-adenosyl-L-methionine + myoglobin L-lysine <3> (<3> 10% activity compared to cytochrome c [2]) (Reversibility: ? <3> [2]) [2]

P S-adenosyl-L-homocysteine + myoglobin N^6-methyl-L-lysine <3> [2]

Inhibitors

S-adenosyl-D-homocysteine <1> [8]
S-adenosyl-L-homocysteine <1, 3, 4> (<3> competitive versus S-adenosyl-L-methionine [3,6,7]) [3, 4, 6-9]
puromycin <4> [4]

Specific activity (U/mg)

0.000117 <4> (<4> partially purified enzyme [9]) [9]
0.000289 <3> (<3> partially purified enzyme [2]) [2]
0.000508 <1> (<1> purified enzyme [8]) [8]
0.0295 <3> (<3> purified enzyme [3,7]) [3, 7]
Additional information <1, 2, 4> (<2> development of a radioactive assay method [1]) [1, 5, 8]

K_m-Value (mM)

0.000121 <1> (apocytochrome c) [8]
0.007 <3> (CNBr peptide, <3> residue 1-80 of horse heart cytochrome c [3,7]) [3, 7]
0.0095 <3> (S-adenosyl-L-methionine) [2]
0.03 <3> (cytochrome c, <3> of horse heart, heme-free [3,7]) [3, 7]
0.04 <3> (CNBr peptide, <3> residue 1-65 or residue 66-104 of horse heart cytochrome c [3,7]) [3, 7]
0.04 <4> (S-adenosyl-L-methionine) [9]
0.12 <3> (cytochrome c) [2]
0.2 <3> (cytochrome c, <3> of horse heart, ethanol-denatured [3,7]) [3, 7]
0.32 <3> (cytochrome c, <3> of horse heart, native [3,7]) [3, 7]
1.33 <4> (cytochrome c, <4> of horse heart [9]) [9]
1.7 <3> (cytochrome c) [3, 6, 7]
19 <3> (S-adenosyl-L-methionine) [3, 6, 7]
Additional information <1> [8]

K_i-Value (mM)

0.002 <3> (S-adenosyl-L-methionine) [3, 6, 7]

pH-Optimum

9 <3, 4> (<3> assay at [7]) [2, 3, 7, 9]

Temperature optimum (°C)

37 <2, 3> (<2,3> assay at [1,3,7]) [1, 3, 7]

4 Enzyme Structure

Molecular weight

97000 <4> (<4> gel filtration [9]) [9]
120000 <3> (<3> gel filtration [3,7]) [3, 7]

5 Isolation/Preparation/Mutation/Application

Source/tissue

germ <1> [8]
mycelium <3> [2, 7]

Localization

cytosol <3> (<3> 74.3% of the enzyme activity [2]) [2, 3, 7]
lipid bilayer <3> (<3> 13.6% of the enzyme activity [2]) [2]
membrane <3> [2]
Additional information <3> (<3> overview [7]) [7]

Purification

<1> [8]
<3> (partially [2]) [2, 3, 6, 7]
<4> (partially [9]) [9]

Cloning

<4> (expression of gluthathione-S-transferase-tagged Cmt1p in Saccharomyces cerevisiae and of His-tagged in Escherichia coli, Myc-tagged [5]; in vitro transcription and translation of full length iso-1-cytochrome c gene, serves as substrate [4]) [4, 5]

6 Stability

Storage stability

<3>, 0-4°C, 50% saturated ammonium sulfate, stable for several months [7]
<4>, -20°C, pH 7.0, crude extract supernatant, stable [9]
<4>, purified enzyme is stabilized by addition of 66% ammonium sulfate during storage [9]

References

[1] Valentine, J.; Pettigrew, G.W.: A cytochrome c methyltransferase from Crithidia oncopelti. Biochem. J., **201**, 329-338 (1982)
[2] Nochumson, S.; Durban, E.; Kim, S.; Paik, W.K.: Cytochrome c-specific protein methylase III from Neurospora crassa. Biochem. J., **165**, 11-18 (1977)
[3] Durban, E.; Nochumson, S.; Kim, S.; Ki Paik, W.: Cytochrome c-specific protein-lysine methyltransferase from Neurospora crassa. Purification, characterization, and substrate requirements. J. Biol. Chem., **253**, 1427-1435 (1978)
[4] Park, K.S.; Frost, B.; Tuck, M.; Ho, L.L.; Kim, S.; Paik, W.K.: Enzymatic methylation of in vitro synthesized apocytochrome c enhances its transport into mitochondria. J. Biol. Chem., **262**, 14702-14708 (1987)
[5] Polevoda, B.; Martzen, M.R.; Das, B.; Phizicky, E.M.; Sherman, F.: Cytochrome c methyltransferase, Ctm1p, of yeast. J. Biol. Chem., **275**, 20508-20513 (2000)
[6] Durban, E.; Kim, S.; Jun, G.J.; Paik, W.K.: Cytochrome c-specific protein-lysine methyltransferase from Neurospora crassa: kinetic mechanism. Korean J. Biochem., **15**, 19-24 (1983)
[7] Paik, W.K.; DiMaria, P.: Enzymatic methylation and demethylation of protein-bound lysine residues. Methods Enzymol., **106**, 274-287 (1984)
[8] DiMaria, P.; Kim, S.; Paik, W.K.: Cytochrome c specific methylase from wheat germ. Biochemistry, **21**, 1036-1044 (1982)
[9] DiMaria, P.; Polastro, E.; DeLange, R.J.; Kim, S.; Paik, W.K.: Studies on cytochrome c methylation in yeast. J. Biol. Chem., **254**, 4645-4652 (1979)

Calmodulin-lysine N-methyltransferase 2.1.1.60

1 Nomenclature

EC number
2.1.1.60

Systematic name
S-adenosyl-L-methionine:calmodulin-L-lysine N^6-methyltransferase

Recommended name
calmodulin-lysine N-methyltransferase

Synonyms
S-adenosylmethionine:calmodulin N-methyltransferase (lysine)
calmodulin N-methyltransferase
calmodulin lysine N-methyltransferase
methyltransferase, calmodulin (lysine)

CAS registry number
75603-20-2

2 Source Organism

<1> *Rattus norvegicus* [1-3, 5, 8]
<2> *Ovis aries* [4]
<3> *Escherichia coli* [6, 8, 9]
<4> *Paramecium tetraurelia* [7]

3 Reaction and Specificity

Catalyzed reaction
S-adenosyl-L-methionine + calmodulin L-lysine = S-adenosyl-L-homocysteine + calmodulin N^6-methyl-L-lysine (<1> Bi Bi sequential mechanism [8])

Reaction type
methyl group transfer

Natural substrates and products
S S-adenosyl-L-methionine + calmodulin L-lysine <1, 2> (<1,2> posttranslational modification of calmodulin [2,4]) [2, 4]
P S-adenosyl-L-homocysteine + calmodulin N^6-methyl-L-lysine

Substrates and products

- **S** S-adenosyl-L-methionine + calmodulin L-lysine <1-4> (<1> Dictyostelium discoideum calmodulin [3]; <2> Pisum sativum calmodulin [4]; <1> octopus calmodulin [5]; <1> mushroom demthylcalmodulin [8]; <1> highly specific [1-3]) (Reversibility: ? <1-4> [1-5, 7, 8]) [1-5, 7, 8]
- **P** S-adenosyl-L-homocysteine + calmodulin N^6-methyl-L-lysine <1, 2, 4> (<1,4> trimethyllysine calmodulin, major product, + dimethyllysine + monomethyllysine [1,7,8]; <1> 3 mol of methyl per mol of calmodulin are incorporated into lysine 115 of des(methyl) calmodulin from Dictyostelium discoideum resulting in formation of 1 mol of trimethyllysine at the site normally methylated [3]; <1> formation of ε-N-mono-, ε-N-di- and ε-N-trimethyllysine, the labelled N-methyllysine lies in the 107-126 peptide [5]; <1> enzyme methylates a specific lysine residue of octopus calmodulin [5]) [1-5, 7, 8]
- **S** Additional information <1, 3> (<1> troponin C has low activity at very high troponin concentration [2]; <3> structural requirements of calmodulin [9]) [2, 9]
- **P** ?

Inhibitors

EGTA <1> [1]
S-adenosylhomocysteine <1, 4> [7, 8]
VU-3 calmodulin <1, 2> (<2> site-specific mutant of calmodulin, Lys to Arg-115 [4]) [4, 8]
VU-4 calmodulin <2> (<2> site-specific mutant of calmodulin, Lys to Ile-115 [4]) [4]
calmidazolium <1, 4> [7, 8]
mellitin <4> [7]
p-chloromercuriphenylsulfonic acid <1> [1]
trimethyllysine calmodulin <1> [8]
Additional information <4> (<4> not inhibitory: sinefungin, tubercidin, calmodulin antagonist W7 [7]) [7]

Activating compounds

Ca^{2+} <1> (<1> stimulation [5]) [5]
Mg^{2+} <1> [2]
Mn^{2+} <1> (<1> stimulation by divalent cations, Mn^{2+} giving highest activity [1]; <1> stimulation [2,5]) [1, 2, 5]
Triton <1> [3]
[Thr136]calmodulin <3> (<3> defective in Ca-binding, cannot be methylated by the enzyme [6]) [6]
calmodulin <3> [6]
dithiothreitol <1> (<1> stimulates [1,3]; <4> absolute requirement [7]) [1, 3]

Metals, ions

Ca^{2+} <1, 4> (<1,4> stimulation [5,7]) [5, 7]
Mg^{2+} <1, 4> (<1,4> activation [2,7]) [2, 7]

Mn^{2+} <1, 4> (<1> stimulation by divalent cations, Mn^{2+} giving highest activity [1]; <1> stimulation [2,5]; <4> stimulation [7]) [1, 2, 5, 7]
Ni^{2+} <4> (<4> stimulation [7]) [7]
Additional information <4> (<4> not stimulating: Zn^{2+} [7]) [7]

Specific activity (U/mg)
Additional information <1, 2> [3, 4]

K_m-Value (mM)
0.00001 <1> (calmodulin) [2]
0.00002 <1> (octopus calmodulin) [5]
0.00005 <2> (VU-1 calmodulin) [4]
0.00023 <1> (mushroom demethylcalmodulin) [8]
0.0008 <1> (S-adenosyl-L-methionine) [5]
0.001 <1> (S-adenosyl-L-methionine) [2]
0.001 <4> (S-adenosyl-L-methionine) [7]
0.002 <1> (S-adenosyl-L-methionine) [8]
Additional information <1> [1]

K_i-Value (mM)
0.00031 <1> (trimethyllysine calmodulin) [8]
0.0004 <1> (VU-3 calmodulin) [8]

pH-Optimum
7.5 <4> [7]
7.5-8.5 <1> [5]
8 <1> [2]
Additional information <1> (<1> pH 8.0, 7.8 or 6.0 depending on assay method [1]) [1]

4 Enzyme Structure

Molecular weight
33000 <4> (<4> gel filtration [7]) [7]
55000 <1> (<1> gel filtration [3]) [3]
57000 <1> (<1> gel filtration [5]) [5]
65000 <1> (<1> gel filtration [2]) [2]

Subunits
monomer <4> (<4> 1 * 37000, SDS-PAGE [7]) [7]

5 Isolation/Preparation/Mutation/Application

Source/tissue
brain <1, 2> [1-5]
kidney <1> [3]
spleen <1> [3]

testis <1> [3, 8]
Additional information <1> (<1> found at high levels in tissues with high levels of calmodulin, at low levels in tissues with low levels of calmodulin like liver, heart, skeletal muscle [3]) [3]

Localization
cytosol <1> [1, 3, 5, 7]

Purification
<1> (partial [1,2]) [1-3, 5]
<2> [4]
<4> [7]

6 Stability

General stability information
<1>, Triton stabilizes during purification [3]

Storage stability
<1>, -20°C [2]
<1>, -20°C, up to 2 weeks [3]

References

[1] Sitaramayya, A.; Wright, L.S.; Siegel, F.L.: Enzymatic methylation of calmodulin in rat brain cytosol. J. Biol. Chem., **255**, 8894-8900 (1980)
[2] Murtaugh, T.J.; Rowe, P.M.; Vincent, P.L.; Wright, L. S.; Siegel, F.L.: Posttranslational modification of calmodulin. Methods Enzymol., **102**, 158-170 (1983)
[3] Rowe, P.M.; Wright, L.S.; Siegel, F.L.: Calmodulin N-methyltransferase. Partial purification and characterization. J. Biol. Chem., **261**, 7060-7069 (1986)
[4] Oh, S.H.; Roberts, D.M.: Analysis of the state of posttranslational calmodulin methylation in developing pea plants. Plant Physiol., **93**, 880-887 (1990)
[5] Morino, H.; Kawamoto, T.; Miyake, M.; Kakimoto, Y.: Purification and properties of calmodulin-lysine N-methyltransferase from rat brain cytosol. J. Neurochem., **48**, 1201-1208 (1987)
[6] Han, C.H.; Roberts, D.M.: Altered methylation substrate kinetics and calcium binding of a calmodulin with a Val136 → Thr substitution. Eur. J. Biochem., **244**, 904-912 (1997)
[7] Pech, L.L.; Nelson, D.L.: Purification and characterization of calmodulin (lysine 115) N-methyltransferase from Paramecium tetraurelia. Biochim. Biophys. Acta, **1199**, 183-194 (1994)
[8] Wright, L.S.; Bertics, P.J.; Siegel, F.L.: Calmodulin N-methyltransferase. Kinetics, mechanism, and inhibitors. J. Biol. Chem., **271**, 12737-12743 (1996)
[9] Cobb, J.A.; Roberts, D.M.: Structural requirements for N-trimethylation of lysine 115 of calmodulin. J. Biol. Chem., **275**, 18969-18975 (2000)

tRNA (5-methylaminomethyl-2-thiouridylate)-methyltransferase 2.1.1.61

1 Nomenclature

EC number
2.1.1.61

Systematic name
S-adenosyl-L-methionine:tRNA (5-methylaminomethyl-2-thio-uridylate)-methyltransferase

Recommended name
tRNA (5-methylaminomethyl-2-thiouridylate)-methyltransferase

Synonyms
S-adenosyl-L-methionine:tRNA (5-methylaminomethyl-2-thiouridylate)-methyltransferase
methyltransferase, transfer ribonucleate 5-methylaminomethyl-2-thiouridylate 5-
tRNA 5-methylaminomethyl-2-thiouridylate 5'-methyltransferase

CAS registry number
39391-17-8

2 Source Organism

<1> *Escherichia coli* (B [1]; K12 strain pTH32/1100, harbouring multi-copy plasmid [2, 3]) [1-3]

3 Reaction and Specificity

Catalyzed reaction
S-adenosyl-L-methionine + tRNA = S-adenosyl-L-homocysteine + tRNA containing 5-methylaminomethyl-2-thiouridylate (<1> mechanismn [2])

Reaction type
methyl group transfer

Natural substrates and products
S S-adenosyl-L-methionine + tRNA <1> (Reversibility: ? <1> [1, 2]) [1, 2]
P S-adenosyl-L-homocysteine + tRNA containing 5-methylaminomethyl-2-thiouridylate

Substrates and products

S S-adenosyl-L-methionine + tRNA <1> (<1> specific, methyl group acceptor is methyl deficient $tRNA^{Glu}$ from E. coli [1]; <1> tRNA from E. coli mutants [2]) (Reversibility: ? <1> [1, 2]) [1, 2]

P S-adenosyl-L-homocysteine + tRNA containing 5-methylaminomethyl-2-thiouridylate <1> [1, 2]

S Additional information <1> (<1> the enzyme exhibits two activities: enzyme from trmC1 mutant reacts with 5-carboxymethylaminomethyl-2-thiouridine to form 5-aminomethyl-2-thiouridine without S-adenosyl-L-methionine, enzyme from trmC2 mutant methylates 5-aminomethyl-2-thiouridine in the presence of S-adenosyl-L-methionine [2]) [2]

P ?

Inhibitors

Mg^{2+} <1> (<1> strong, 5 mM [2]) [2]
NaCl <1> (<1> mutant trmC1: 100 mM, mutant trmC2: 200 mM [2]) [2]
p-chloromercuribenzoate <1> [2]
Additional information <1> (<1> no inhibition by putrescine or spermidine [2]) [2]

Activating compounds

NH_4^+ <1> [2]
Additional information <1> (<1> no activation by putrescine or spermidine [2]) [2]

Metals, ions

NH_4^+ <1> (<1> activation [2]) [2]

Turnover number (min^{-1})

74 <1> (5-aminomethyl-2-thiouridine) [2]

Specific activity (U/mg)

0.1686 <1> (<1> mutant trmC1 [2]) [2]
0.9252 <1> (<1> mutant trmC2 [2]) [2]

pH-Optimum

8-8.5 <1> [2]

Temperature optimum (°C)

37 <1> (<1> assay at [1,2]) [1, 2]

4 Enzyme Structure

Subunits

monomer <1> (<1> 1 * 79000, SDS-PAGE [2]) [2]

5 Isolation/Preparation/Mutation/Application

Localization

soluble <1> [1]

Purification

<1> (affinity chromatography [1, 2]; on Heparin-Sepharose [2]) [1, 2]

Cloning

<1> (cloning and mapping of trmC gene [3]) [3]

References

[1] Taya, Y.; Nishimura, S.: Biosynthesis of 5-methylaminomethyl-2-thiouridylate. I. Isolation of a new tRNA-methylase specific for 5-methylaminomethyl-2-thiouridylate. Biochem. Biophys. Res. Commun., **51**, 1062-1068 (1973)

[2] Hagervall, T.G.; Edmonds, C.G.; McCloskey, J.A.; Björk, G.R.: Transfer RNA(5-methylaminomethyl-2-thiouridine)-methyltransferase from Escherichia coli K-12 has two enzymatic activities. J. Biol. Chem., **262**, 8488-8495 (1987)

[3] Hagervall, T.G.; Björk, G.R.: Genetic mapping and cloning of the gene (trmC) responsible for the synthesis of tRNA (mnm5s2U)methyltransferase in Escherichia coli K12. Mol. Gen. Genet., **196**, 201-207 (1984)

mRNA (2'-O-methyladenosine-N^6-)-methyltransferase 2.1.1.62

1 Nomenclature

EC number
2.1.1.62

Systematic name
S-adenosyl-L-methionine:mRNA (2'-O-methyladenosine-N^6-)-methyltransferase

Recommended name
mRNA (2'-O-methyladenosine-N^6-)-methyltransferase

Synonyms
VP39 <3> [3]
cap-dependent 2'-O-methyltransferase <3> [3]
messenger ribonucleate 2'-O-methyladenosine N^g-methyltransferase

CAS registry number
68009-87-0

2 Source Organism

<1> *Homo sapiens* (human, HeLa S3 cells [1]) [1]
<2> *Staphylococcus aureus* [2]
<3> *Variola sp.* (vaccinia virus [3]) [3]

3 Reaction and Specificity

Catalyzed reaction
S-adenosyl-L-methionine + m^7G(5')pppAm = S-adenosyl-L-homocysteine + m^7G(5')pppm^{6}Am (mRNA containing an N^6,2'-O-dimethyladenosine cap)

Reaction type
methyl group transfer

Natural substrates and products
S S-adenosyl-L-methionine + m^7G(5')pppAm-mRNA <1> (<1> involved in post-transcriptional modification of 5'-end of messenger-RNA [1]) (Reversibility: ? <1> [1]) [1]
P S-adenosyl-L-homocysteine + m^7G(5')pppm^{6}Am-mRNA <1> (<1> mRNA containing an N^6,2'-O-dimethyladenosine cap [1]) [1]

Substrates and products

S S-adenosyl-L-methionine + G(5')pppAm-mRNA <1> (<1> poor substrate [1]) [1]

P S-adenosyl-L-homocysteine + G(5')pppm^{6}Am-mRNA

S S-adenosyl-L-methionine + m^{7}G(5')pppA-mRNA <1> (Reversibility: ? <1> [1]) [1]

P S-adenosyl-L-homocysteine + m^{7}G(5')pppm^{6}A-mRNA

S S-adenosyl-L-methionine + m^{7}G(5')pppAm-mRNA <1, 3> (<1> methyl group acceptor is methylated or unmethylated vaccinia virus RNA [1]; <3> also acts as poly(A) polymerase progressivity factor [3]) (Reversibility: ? <1> [1]) [1, 3]

P S-adenosyl-L-homocysteine + m^{7}G(5')pppm^{6}Am-mRNA <1> (<1> mRNA containing an N^6,2'-O-dimethyladenosine cap [1]) [1]

S S-adenosyl-L-methionine + pppA-mRNA <1> (Reversibility: ? <1> [1]) [1]

P S-adenosyl-L-homocysteine + pppm^{6}A-mRNA

S Additional information <1> (<1> no substrates are the oligonucleotides m^{7}G(5')pppAm, m^{7}G(5')pppA, m^{7}G(5')pppAmpN, no methyl group acceptors are brome mosaic virus RNA, E. coli tRNA and rRNA and yeast tRNA [1]) [1]

P ?

Activating compounds

erythromycin <2> (<2> enzyme synthesis induced by [2]) [2]

Specific activity (U/mg)

0.000418 <1> [1]

4 Enzyme Structure

Molecular weight

29000 <2> (<2> PAGE [2]) [2]
65000 <1> (<1> sucrose gradient [1]) [1]

Subunits

monomer <1> (<1> 1 * 65000, SDS-PAGE [1]) [1]

5 Isolation/Preparation/Mutation/Application

Localization

cytoplasm <1> [1]

Purification

<1> (partially [1]) [1]
<2> [2]

Cloning

<2> (ermC gene of pE194 specifies resistance to macrolide-lincosamide-streptogramin B antibiotics [2]) [2]

6 Stability

Storage stability

<1>, -70°C, 0.21 M $(NH_4)_2SO_4$, stable for longer than 5 months [1]
<1>, 0°C, 0.21 M $(NH_4)_2SO_4$, stable for 4-5 months [1]

References

[1] Keith, J.M.; Ensinger, M.J.; Moss, B.: HeLa cell RNA (2-O-methyladenosine-N^6-)-methyltransferase specific for the capped 5-end of messenger RNA. J. Biol. Chem., **253**, 5033-5041 (1978)
[2] Shivakumar, A.G.; Dubnau, D.: Characterization of a plasmid-specified ribosome methylase associated with macrolide resistance. Nucleic Acids Res., **9**, 2549-2562 (1981)
[3] Shi, X.; Yao, P.; Jose, T.; Gershon, P.D.: Methyltransferase-specific domains within VP39, a bifunctional protein that participates in the modification of both mRNA ends. RNA, **2**, 88-101 (1996)

Methylated-DNA-[protein]-cysteine S-methyltransferase 2.1.1.63

1 Nomenclature

EC number
2.1.1.63

Systematic name
DNA-6-O-methylguanine:[protein]-L-cysteine S-methyltransferase

Recommended name
methylated-DNA-[protein]-cysteine S-methyltransferase

Synonyms
Ada protein
MGMT
O6-MGMT
O6-methylguanine-DNA methyltransferase
Pk-MGMT <2> [3]

CAS registry number
77271-19-3

2 Source Organism

<1> *Escherichia coli* [1, 2, 11, 12, 19]
<2> *Pyrococcus kodakaraensis* (strain KOD1 [3]) [3]
<3> *Homo sapiens* (three allelic variants: V1, V2, and V3 [7]) [4, 5, 6, 7, 11, 13, 14, 15, 16, 17, 18, 20, 21]
<4> *Rattus norvegicus* (wild-type and mutant with tetrapeptide deleted [15]) [8, 11, 13]
<5> *Xiphophorus maculatus* (Jp 163 A [9]) [9]
<6> *Xiphophorus couchianus* [9]
<7> *Xiphophorus helleri* [9]
<8> *Prococcus sp.* (KOD1 [10]) [10]
<9> *Mus musculus* [15]

3 Reaction and Specificity

Catalyzed reaction
DNA (containing 6-O-methylguanine) + protein L-cysteine = DNA (without 6-O-methylguanine) + protein S-methyl-L-cysteine

Reaction type

methyl group transfer

Natural substrates and products

S DNA containing 6-O-methylguanine + protein L-cysteine <1, 3, 4, 9> (<3>, the enzyme repairs DNA by transferring alkyl (R)-adducts from 6-O-alkylguanine in DNA to its own cysteine residue at codon 145 [4]; <3>, the enzyme is a critical defense against alkylation-induced mutagenesis and carcinogenesis [7]; <4>, the DNA repair enzyme can specifically remove methyl groups from 6-O-methylguanine, which is a major mutagenic and carcinogenic DNA lesion leading to the G-A transition [8]; <1>, 6-O-methylguanine is an important adduct formed by methylating agents, that, if not repaired, can lead to mutations and death. Its repair is carried out by 6-O-methylguanine DNA-methyltransferase. Exposure of E. coli cells to sublethal concentrations of methylating agent triggers the expression of the gene [12]; <3,9>, the enzyme repairs mutagenic and carcinogenic 6-O-alkylguanine in DNA [15]; <1>, the enzyme functions in DNA repair by direct dealkylation of mutagenic 6-O-alkylguanine. The protein methylated at Cys69 becomes a transcriptional activator of the genes in the ada regulon, including its own [19]) (Reversibility: ? <1, 3, 4, 9> [4, 7, 8, 12, 15, 19]) [4, 7, 8, 12, 15, 19]

P DNA lacking 6-O-methylguanine + protein S-methyl-L-cysteine

Substrates and products

S DNA containing 4-O-methylthymine + protein L-cysteine <3> (Reversibility: ? <3> [20]) [20]

P DNA lacking 4-O-methylthymine + protein S-methyl-L-cysteine

S DNA containing 6-O-ethylguanine + protein L-cysteine <1, 3, 4> (<1,3,4> CGC(e6G)AGCTCGCG [11]) (Reversibility: ? <1, 3, 4> [11]) [11]

P DNA lacking 6-O-ethylguanine + protein S-ethyl-L-cysteine <1, 3, 4> [11]

S DNA containing 6-O-methylguanine + protein L-cysteine <1, 2, 3, 4, 5, 6, 7, 8, 9> (<1>, synthetic DNA polymer poly(dC,dG,m^{6}dG) [2]; <1>, transferase activity methylates itself on removal of the methyl group from the 6-O position of guanine [1]; <2>, the enzyme repairs alkylated DNA by suicidal alkyl transfer from guanine 6-O to its own cysteine residue [3]; <3>, the enzyme repairs DNA by transferring alkyl (R)-adducts from 6-O-alkylguanine in DNA to its own cysteine residue at codon 145 [4]; <3>, the alkyl group is transferred without a cofactor to Cys145 residue of the enzyme and thereby inactivates the protein [5]; <3>, oligonucleotides containing an 6-O-(2-fluorobenzyl)guanine, 6-O-(3-fluorobenzyl)-guanine, 6-O-benzylhypoxanthine or 6-O-methylguanine are all good substrates [6]; <5, 6, 7>, 6-O-methylguanine oligonucleotide [9]; <1, 4>, CGC(m6G)CG and CGC(m6G)AGCTCGCG [11]; <3>, the binding to DNA is the rate determining step in the repair process. Approximately eight base pairs of the DNA substrate are covered by the human enzyme. Binding affinity to methylated DNA is two times higher than that to unmodified DNA. The interaction with DNA induces a conformational change in the enzyme [14]; <3,9>, enzyme reacts with the alkylated base in a synthetic DNA substrate poly(dC, dG, m6dG) [15]; <1>, the enzyme functions in DNA repair by direct dealkylation of mutagenic 6-O-alkylguanine. The protein methylated at Cys69 becomes a transcriptional activator of the genes in the ada regulon, including its own [19]; <3>, the methyltransferase protein itself accepts methyl groups from methylated DNA, one enzyme molecule accepts one methyl group [20]) (Reversibility: ? <1-9> [1-21]) [1, 2, 3, 4, 5, 6, 7, 8, 9, 10, 11, 12, 13, 14, 15, 16, 17, 18, 19, 20, 21]

P DNA lacking 6-O-methylguanine + protein S-methyl-L-cysteine

S Additional information <3> (<3>, the 28-amino acid carboxy-terminal tail of the enzyme is not required for activity and modulates the rate of 6-O-methylguanine DNA methyltransferase repair at reduced temperatures and plays a role in substrate specificity [21]) [21]

P ?

Inhibitors

6-O-benzylguanine <5, 6, 7> (<5,6,7>, competitive [9]) [9]

DNA <3, 9> [15]

Na_2SO_4 <3, 9> [15]

Na_3 citrate <3, 9> [15]

NaCl <3, 9> (<3>, 0.2 M, 80% inhibition [15]; <9>, 0.2 M, wild-type and mutant enzyme, 55% inhibition [15]) [15]

RNA <3, 9> [15]

double-stranded oligonucleotides <3> (<3>, 6-O-methylguanine, 6-O-(4-fluorobenzyl)-guanine, 6-O-(3-fluorobenzyl)-guanine, 6-O-(2-fluorobenzyl)-guanine, 6-O-benzylguanine, 6-O-benzylhypoxanthine. IC50: 1.4-3.0 nM [6]) [6]

methyl iodide <3> (<3>, can directly alkylate the active site of the enzyme, the agent can increase the effectiveness of environmental and endogenously produced alkylating carcinogens in producing the mutagenic 6-O-alkylguanine residue in DNA in vivo [4]) [4]

Additional information <1, 2, 3> (<1>, transferase activity methylates itself on removal of the methyl group from the 6-O position of guanine. Modification of a reactive sulfhydryl group can account for the enzyme inactivation accompanying the reaction [1]; <2>, the enzyme repairs alkylated DNA by suicidal alkyl transfer from guanine 6-O to its own Cys residue [3]; <3>, the alkyl group is transferred without a cofactor to Cys145 residue of the enzyme and thereby inactivates the protein [5]; <3>, high level of enzyme in tumors and relative resistance to cyclophosphamide in lung cancer indicates that 6-O-methylguanine-DNA methyltransferase may be a predictive factor of resistance to cyclophosphamide [16]) [1, 3, 5, 16]

Turnover number (min^{-1})

Additional information <1, 3, 4> (<1,3,4>, second order repair rate constants of 6-O-methyl and 6-O-ethylguanine containing substrates [11]) [11]

Specific activity (U/mg)

Additional information <3> (<3>, developmemt of a novel fluorometric oligonucleotide assay to measure enzyme activity in cellular and tissue extracts [18]) [14, 18, 20]

K_m-Value (mM)

Additional information <3> [7]

Temperature optimum (°C)

23 <5> [9]

Temperature range (°C)

15-40 <5> (<5>, about 75% of maximal activity at 15° or at 40°C [9]) [9]

4 Enzyme Structure

Molecular weight

19000 <3> (<3>, gel filtration, glycerol gradient sedimentation [20]) [20]

Subunits

? <1, 3, 9> (<3>, x * 20900, SDS-PAGE [15]; <9>, x * 22500, SDS-PAGE [15]; <1>, x * 39356, inactive enzyme, electrospray mass spectrometry [19]; <1>, x * 39377, active enzyme, electrospray mass spectrometry [19]) [15, 19]
monomer <3> (<3>, 1 * 25000, SDS-PAGE [20]) [20]

5 Isolation/Preparation/Mutation/Application

Source/tissue

JURKAT cell <3> [17]
SF-767 cell <3> [18]
astrocyte <3> (<3>, low activity [13]) [13]
bone marrow <3> [18]
brain <3, 5, 6, 7> [9, 18]
cell culture <1> (<1>, activity in cells adapted by exposure to N-methyl-N^1-nitroguanidine, no activity in extracts of nonadapted cells [2]) [2]
fibroblast <3> [7]
gill <5, 6, 7> [9]
kidney <3> [18]
liver <3, 5, 6, 7> [9, 18]
lung <3> [18]
lung cancer cell <3> [16]
mammary tumor cell <4> (<4>, no statistically significant difference in the amount of enzyme between tumors with and without activated H1-ras oncogene [8]) [8]
oligodendrocyte <4> [13]

oligodendroglioma cell <3> (<3>, low 6-O-methylguanine-DNA methyltransferase activity may contribute to the chemosensitivity of some human oligodendrogliomas [13]) [13]
spleen <3> [18]
testis <5, 6, 7> [9]

Purification

<3> (recombinant enzyme [6,7,14]) [6, 7, 14, 19, 20]
<8> (recombinant enzyme [10]) [10]

Renaturation

<1> (renaturation by dilution with buffer from 8 M urea or 6 M guanidine HCl instantly restores the secondary structure of the Ada protein to a state that is almost indistinguishable from the native state. Kinetics of renaturation is fast, at 0°C the restoration is quantitative in less than 30 s. The heat-coagulated protein can be restored to full activity by cycling it through treatment with 8 M urea or 6 M guanidine [19]) [19]

Crystallization

<2> (crystals formed at 20°C by conventional hanging drop-vapor diffusion method [3]) [3]

Cloning

<3> (expressed in Escherichia coliKT233 and in HeLa cells [7]; cDNA is cloned into a pET-11a plasmid and expressed in Escherichia coli [14]) [7, 14]
<8> (the gene is placed under the control of the lac promoter and overexpressed in the methyltransferase-deficient Escherichia coli strain KT233 [10]) [10]

Engineering

C145A <3> (<3>, inactive mutant enzyme forms a specific and stable complex with a 6-O-methylguanine-containing oligonucleotide substrate [5]) [5]
C145S <3> (<3>, inactive mutant enzyme forms a specific and stable complex with a 6-O-methylguanine-containing oligonucleotide substrate [5]) [5]
K107L <3> (<3>, mutant is deficient in DNA repair [4]) [4]
Additional information <3> (<3>, deletion of more than 8 or 31 residues from the amino or carboxyl terminus, respectively, leads to the loss of both activity and substrate binding. Removal of Arg9 or Leu176 and distal residues inactivates the protein [5]; <3>, three allelic variants: V1 with amino acid substitution Leu84Phe, variant V2 with amino acid substitution Trp65Cys and variant V3 with a silent mutation. Wild-type and V1 variant have similar enzymatic and physicochemical properties, while variant V2 is considered to be unstable and rare [7]; <3>, the 28-amino acid carboxy-terminal tail of the enzyme is not required for activity and modulates the rate of 6-O-methylguanine DNA methyltransferase repair at reduced temperatures and plays a role in substrate specificity [21]) [5, 7, 21]

Application

medicine <3> (<3>, high level of enzyme in tumors and relative resistance to cyclophosphamide in lung cancer indicates that 6-O-methylguanine-DNA methyltransferase may be a predictive factor of resistance to cyclophosphamide [16]; <3>, the enzyme is important in cellular resistance to certain alkylating antitumor agents such as the methylating drug temozolomide. Cisplatin is able to decrease enzyme levels in Jurkat cells, probably via the inhibition of gene transcription. The clinical efficiacy of triazene compounds might be improved by combination with cisplatin using appropriate doses and schedules of administration [17]; <3>, the efficiacy of 6-O-benzylguanine as a chemomodulator depends on the extent of depletion of 6-O-methylguanine DNA methyltransferase in normal tissues and the optimal therapeutic index for combination of 6-O-benzylguanine and 1,3-bis(2-chloroethyl)-1-nitrosourea therapy should be achieved by depleting 6-O-methylguanine DNA methyltransferase in the target tumor for 24 h with minimal depletion in normal tissues [18]) [16, 17, 18]

6 Stability

Temperature stability

37 <1> (<1>, spontaneously loses about 30% of its secondary structure in less than 30 min concomitant with a complete loss of activity [19]) [19]
50 <3, 9> (<3,9>, 20 min, 55-65% loss of activity in absence of DNA, 25-35% loss of activity in presence of DNA, 0.2 mg/ml [15]) [15]
65 <5> (<5>, 5 min, almost complete inactivation [9]) [9]
70 <1> (<1>, protein immediately coagulates [19]) [19]
75 <8> (<8>, 60 min, 25% loss of activity [10]) [10]
90 <8> (<8>, stable for at least 30 min. 66% loss of activity after 60 min [10]) [10]
Additional information <1, 2> (<2>, enzyme is extremely thermostable [3]; <1>, temperature-dependent inactivation of the enzyme is inhibited by DNA [19]) [3, 19]

General stability information

<1>, enzyme is denatured by 8 M urea at pH 7.5 for 30 s or by 6 M guanidine hydrochloride at pH 7.5 for 16-24 h [19]

References

[1] Olsson, M.; Lindahl, T.: Repair of alkylated DNA in Escherichia coli. Methyl group transfer from O^6-methylguanine to a protein cysteine residue. J. Biol. Chem., **255**, 10569-10571 (1980)
[2] Foote, R.S.; Mitra, S.; Pal, B.C.: Demethylation of O^6-methylguanine in a synthetic DNA polymer by an inducible activity in Escherichia coli. Biochem. Biophys. Res. Commun., **97**, 654-659 (1980)
[3] Hashimoto, H.; Inoue, T.; Nishioka, M.; Fujiwara, S.; Takagi, M.; Imanaka, T.; Kai, Y.: Hyperthermostable protein structure maintained by intra and

inter-helix ion-pairs in archaeal O^6-methylguanine-DNA methyltransferase. J. Mol. Biol., **292**, 707-716 (1999)
[4] Oh, H.K.; Teo, A.K.C.; Ali, R.B.; Lim, A.; Ayi, T.C.; Yarosh, D.B.; Li, B.F.L.: Conformational change in human DNA repair enzyme O^6-methylguanine-DNA methyltransferase upon alkylation of its active site by SN1 (indirect-acting) and SN2 (direct-acting) alkylating agents: breaking a "salt-link"?. Biochemistry, **35**, 12259-12266 (1996)
[5] Hazra, T.K.; Roy, R.; Biswas, T.; Grabowski, D.T.; Pegg, A.E.; Mitra, S.: Specific recognition of O^6-methylguanine in DNA by active site mutants of human O^6-methylguanine-DNA methyltransferase. Biochemistry, **36**, 5769-5776 (1997)
[6] Terashima, I.; Kawate, H.; Sakumi, K.; Sekiguchi, M.; Kohda, K.: Substrate specificity of human O^6-methylguanine-DNA methyltransferase for O^6-benzylguanine derivatives in oligodeoxynucleotides. Chem. Res. Toxicol., **10**, 1234-1239 (1997)
[7] Inoue, R.; Abe, M.; Nakabeppu, Y.; Sekiguchi, M.; Mori, T.; Suzuki, T.: Characterization of human polymorphic DNA repair methyltransferase. Pharmacogenetics, **10**, 59-66 (2000)
[8] Inui, H.; Watatani, M.; Yasutomi, M.; Nishisho, I.: Comparison of O^6-methylguanine-DNA methyltransferase mRNA levels in Ha-ras mutated and non-mutated rat mammary tumors induced by N-methyl-N-nitrosourea. Carcinogenesis, **15**, 459-463 (1994)
[9] Walter, R.B.; Sung, H.M.; Intano, G.W.; Walter, C.A.: Characterization of O^6-methylguanine-DNA-methyltransferase (O^6-MGMT) activity in Xiphophorus fishes. Mutat. Res., **493**, 11-22 (2001)
[10] Leclere, M.M.; Nishioka, M.; Yuasa, T.; Fujiwara, S.; Takagi, M.; Imanaka, T.: The O^6-methylguanine-DNA methyltransferase from the hyperthermophilic archaeon Pyrococcus sp. KOD1: a thermostable repair enzyme. Mol. Gen. Genet., **258**, 69-77 (1998)
[11] Liem, L.K.; Lim, A.; Li, B.F.L.: Specificities of human, rat and E. coli O^6-methylguanine-DNA methyltransferases towards the repair of O6-methyl and O^6-ethylguanine in DNA. Nucleic Acids Res., **22**, 1613-1619 (1994)
[12] Kleibl, K.: Molecular mechanisms of adaptive response to alkylating agents in Escherichia coli and some remarks on O^6-methylguanine DNA-methyltransferase in other organisms. Mutat. Res., **512**, 67-84 (2002)
[13] Nutt, C.L.; Costello, J.F.; Bambrick, L.L.; Yarosh, D.B.; Swinnen, L.J.; Chambers, A.F.; Cairncross, J.G.: O^6-Methylguanine-DNA methyltransferase in tumors and cells of the oligodendrocyte lineage. Can. J. Neurol. Sci., **22**, 111-115 (1995)
[14] Chan, C.L.; Wu, Z.; Ciardelli, T.; Eastman, A.; Bresnick, E.: Kinetic and DNA-binding properties of recombinant human O^6-methylguanine-DNA methyltransferase. Arch. Biochem. Biophys., **300**, 193-200 (1993)
[15] Roy, R.; Shiota, S.; Kennel, S.J.; Raha, R.; von Wronski, M.; Brent, T.P.; Mitra, S.: A comparative study of the biochemical properties of human and mouse recombinant O^6-methylguanine-DNA methyltransferases. Carcinogenesis, **16**, 405-411 (1995)

[16] Mattern, J.; Eichhorn, U.; Kaina, B.; Volm, M.: O^6-Methylguanine-DNA methyltransferase activity and sensitivity to cyclophosphamide and cisplatin in human lung tumor xenografts. Int. J. Cancer, **77**, 919-922 (1998)
[17] D'Atri, S.; Graziani, G.; Lacal, P.M.; Nistico, V.; Gilberti, S.; Faraoni, I.; Watson, A.J.; Bonmassar, E.; Margison, G.P.: Attenuation of O(6)-methylguanine-DNA methyltransferase activity and mRNA levels by cisplatin and temozolomide in jurkat cells. J. Pharmacol. Exp. Ther., **294**, 664-671 (2000)
[18] Kreklau, E.L.; Liu, N.; Li, Z.; Cornetta, K.; Erickson, L.C.: Comparison of single- versus double-bolus treatments of O^6-benzylguanine for depletion of O^6-methylguanine DNA methyltransferase (MGMT) activity in vivo: development of a novel fluorometric oligonucleotide assay for measurement of MGMT activity. J. Pharmacol. Exp. Ther., **297**, 524-530 (2001)
[19] Bhattacharyya, D.; Hazra, T.K.; Behnke, W.D.; Chong, P.L.; Kurosky, A.; Lee, J.C.; Mitra, S.: Reversible folding of Ada protein (O^6-methylguanine-DNA methyltransferase) of Escherichia coli. Biochemistry, **37**, 1722-1730 (1998)
[20] Koike, G.; Maki, H.; Takeya, H.; Hayakawa, H.; Sekiguchi, M.: Purification, structure, and biochemical properties of human O^6-methylguanine-DNA methyltransferase. J. Biol. Chem., **265**, 14754-14762 (1990)
[21] Morgan, S.E.; Kelley, M.R.; Pieper, R.O.: The role of the carboxyl-terminal tail in human O^6-methylguanine DNA methyltransferase substrate specificity and temperature sensitivity. J. Biol. Chem., **268**, 19802-19809 (1993)

3-Demethylubiquinone-9 3-O-methyltransferase 2.1.1.64

1 Nomenclature

EC number
2.1.1.64

Systematic name
S-adenosyl-L-methionine:2-nonaprenyl-3-methyl-5-hydroxy-6-methoxy-1,4-benzoquinone 3-O-methyltransferase

Recommended name
3-demethylubiquinone-9 3-O-methyltransferase

Synonyms
2-octaprenyl-3-methyl-5-hydroxy-6-methoxy-1,4-benzoquinone methyltransferase
5-demethylubiquinone-9 methyltransferase
OMHMB-methyltransferase
S-adenosyl-L-methionine:2-octaprenyl-3-methyl-5-hydroxy-6-methoxy-1,4-benzoquinone-O-methyltransferase

CAS registry number
63774-48-1

2 Source Organism

<1> *Escherichia coli* (strain K12 and derivatives [1]) [1]
<2> *Rattus norvegicus* (Wistar albino [2]) [2]

3 Reaction and Specificity

Catalyzed reaction
S-adenosyl-L-methionine + 3-demethylubiquinone-9 = S-adenosyl-L-homocysteine + ubiquinone-9

Reaction type
methyl group transfer

Natural substrates and products
S 2-octaprenyl-3-methyl-5-hydroxy-6-methoxy-1,4-benzoquinone + S-adenosyl-L-methionine + NADH <1> (<1> final reaction in the biosynthesis of ubiquinone [1]) (Reversibility: ? <1> [1]) [1]
P ubiquinone-8 + S-adenosyl-L-homocysteine + NAD^+ <1> [1]

Substrates and products

S 2-octaprenyl-3-methyl-5-hydroxy-6-methoxy-1,4-benzoquinone + S-adenosyl-L-methionine + NADH <1> (Reversibility: ? <1> [1]) [1]
P ubiquinone-8 + S-adenosyl-L-homocysteine + NAD^+ <1> [1]
S 5-demethylubiquinone-9 + S-adenosyl-L-methionine + NADH <2> (Reversibility: ? <2> [2]) [2]
P ubiquinone-9 + S-adenosyl-L-homocysteine + NAD^+ <2> [2]

Inhibitors

EDTA <2> [2]
fumigatol <2> [2]
phospholipase A <2> (<2> preincubation causes complete inhibition showing a strong requirement for phospholipids [2]) [2]
pyrogallol <2> (<2> poor inhibitor [2]) [2]

Cofactors/prosthetic groups

NADH <1, 2> (<2> required to generate the hydroquinone, the immediate substrate for methylation [2]) [1, 2]
NADPH <2> [2]

Activating compounds

ascorbic acid <2> [2]
dithionite <2> (<2> partially substitute for NADH [2]) [2]
dithiothreitol <2> [2]
mercaptoethanol <2> [2]

Metals, ions

Ca^{2+} <1> [1]
Cd^{2+} <1> [1]
Co^{2+} <1> [1]
Cu^{2+} <1> [1]
Fe^{2+} <1> [1]
Mg^{2+} <1, 2> [1, 2]
Mn^{2+} <1> [1]
Ni^{2+} <1> [1]
Zn^{2+} <1> (<1> absolute requirement for a divalent metal ion, 1 mM $ZnSO_4$ stimulates activity over 100fold [1]) [1]

Specific activity (U/mg)

0.00068 <2> (<2> 5-demethylubiquinone-9 as substrate [2]) [2]

K_m-Value (mM)

0.00006-0.00008 <2> (5-demethylubiquinone-9) [2]
0.022 <2> (S-adenosylmethionine) [2]

4 Enzyme Structure

Molecular weight

50000 <1> (<1> gel filtration [1]) [1]

5 Isolation/Preparation/Mutation/Application

Source/tissue

liver <2> [2]

Localization

cytoplasm <1> [1]
mitochondrial inner membrane <2> [2]
mitochondrion <2> [2]
plasma membrane <1> (<1> internal surface [1]) [1]

Purification

<1> (partially) [1]
<2> [2]

Cloning

<1> (gene designated ubiG located at about minute 42 on the Escherichia coli chromosome [1]) [1]

6 Stability

Oxidation stability

<1>, dithiothreitol absolutely necessary for stabilization [1]

General stability information

<2>, freezing and thawing causes inactivation [2]
<2>, solubilization by Triton X-100 inactivates the enzyme [2]

References

[1] Leppik, R.A.; Stoobant, P.; Shineberg, B.; Young, I.G.; Gibson, F.: Membrane-associated reactions in ubiquinone biosynthesis. 2-Octaprenyl-3-methyl-5-hydroxy-6-methoxy-1,4-benzoquinone methyltransferase. Biochim. Biophys. Acta, **428**, 146-156 (1976)

[2] Houser, R.M.; Olson, E.: 5-Demethylubiquinone-9-methyltransferase from rat liver mitochondria. Characterization, localization, and solubilization. J. Biol. Chem., **252**, 4017-4021 (1977)

Licodione 2'-O-methyltransferase 2.1.1.65

1 Nomenclature

EC number
2.1.1.65

Systematic name
S-adenosyl-L-methionine:licodione 2'-O-methyltransferase

Recommended name
licodione 2'-O-methyltransferase

Synonyms
methyltransferase, licodione 2'-O-

CAS registry number
77000-07-8

2 Source Organism

<1> *Glycyrrhiza echinata* (strains W, KN, MS, DK [2]) [1-3]
<2> *Medicago sativa* [3]

3 Reaction and Specificity

Catalyzed reaction
S-adenosyl-L-methionine + licodione = S-adenosyl-L-homocysteine + 2'-O-methyllicodione

Reaction type
methyl group transfer

Natural substrates and products
S S-adenosyl-L-methionine + licodione <1, 2> (<1,2> involved in retrochalcone echinatin biosynthesis, selective induction in response to stress [1-3]; <1,2> biosynthesis pathway [3]) (Reversibility: ? <1, 2> [1-3]) [1-3]
P S-adenosyl-L-homocysteine + 2'-O-methyllicodione <1> [2, 3]

Substrates and products
S S-adenosyl-L-methionine + 2'-deoxylicodione <1> (<1> very poor substrate [2]) (Reversibility: ? <1> [2]) [2]
P S-adenosyl-L-homocysteine + ?

S S-adenosyl-L-methionine + 2'-hydroxylicodione <1> (<1> methylated at 38% of licodione methylation [2]) (Reversibility: ? <1> [2]) [2]
P S-adenosyl-L-homocysteine + ?
S S-adenosyl-L-methionine + 2,4-dihydroxybenzoic acid <1> (<1> methylated at 4.5% of licodione methylation [2]) (Reversibility: ? <1> [2]) [2]
P S-adenosyl-L-homocysteine + 2-methoxy-4-hydroxybenzoic acid
S S-adenosyl-L-methionine + dihydroisoliquiritigenin <1> (<1> methylated at 7.2% of licodione methylation [2]) (Reversibility: ? <1> [2]) [2]
P S-adenosyl-L-homocysteine + 2'-O-methyldihydroisoliquiritigenin
S S-adenosyl-L-methionine + echinatin <1> (<1> methylated at 4% of licodione methylation [2]) (Reversibility: ? <1> [2]) [2]
P S-adenosyl-L-homocysteine + ?
S S-adenosyl-L-methionine + isoliquiritigenin <1> (<1> methylated at 10.7% of licodione methylation [2]) (Reversibility: ? <1> [2]) [2]
P S-adenosyl-L-homocysteine + 2'-O-methylisoliquiritigenin
S S-adenosyl-L-methionine + licodione <1, 2> (<1,2> i.e. 1-(2,4-dihydroxyphenyl)-3-(4-hydroxyphenyl)-1,3-propanedione [2,3]; <1,2> hydroxygroup ortho to C3-linkage in licodione is methylated [2,3]) (Reversibility: ? <1, 2> [1-3]) [1-3]
P S-adenosyl-L-homocysteine + 2'-O-methyllicodione <1> [2, 3]
S S-adenosyl-L-methionine + resacetophenone <1> (<1> methylated at 2.6% of licodione methylation [2]) (Reversibility: ? <1> [2]) [2]
P S-adenosyl-L-homocysteine + ?

Inhibitors

Cu^{2+} <1> (<1> 80% loss of activity [2]) [2]
Zn^{2+} <1> (<1> 30% loss of activity [2]) [2]

Activating compounds

EDTA <1> (<1> slight increase of activity [2]) [2]
Additional information <1, 2> (<2> treatment with yeast extracts increases activity with the maximum at 24 h [3]; <1> treatment with yeast extracts increases activity with the maximum at 16 to 24 h [3]) [3]

Metals, ions

Mg^{2+} <1> (<1> requirement [1]; <1> not requirement [2]) [1, 2]

Specific activity (U/mg)

Additional information <1> [2]

K_m-Value (mM)

0.0028 <1> (S-adenosyl-L-methionine) [2]
0.011 <1> (licodione) [2]

pH-Optimum

8 <1, 2> [2, 3]

Temperature optimum (°C)

30 <1> (<1> assay at [2]) [2]

5 Isolation/Preparation/Mutation/Application

Source/tissue

callus <1> (<1> cultured [1]) [1]
cell culture <1, 2> [2, 3]
root <2> [3]
shoot <2> [3]

Purification

<1> [2]

Cloning

<2> (expressed in Escherichia coli [3]) [3]

References

[1] Ayabe, S.; Udagawa, A.; Iida, K.; Yoshikawa, T.; Furuya, T.: Studies on plant tissue cultures, Part 47, Regulation of retrochalcone biosynthesis: activity changes of O-methyltransferases in the yeast extract-induced Glycyrrhiza echinata cells. Plant Cell Rep., **6**, 16-19 (1987)
[2] Ayabe, S.; Yoshikawa, T.; Kobayashi, M.; Furuya, T.: Biosynthesis of a retrochalcone echinatin: involvement of *o*-methyltransferase to licodione. Phytochemistry, **19**, 2331-2336 (1980)
[3] Ichimura, M.; Furuno, T.; Takahashi, T.; Dixon, R.A.; Ayabe, S.: Enzymic O-methylation of isoliquiritigenin and licodione in alfalfa and licorice cultures. Phytochemistry, **44**, 991-995 (1997)

rRNA (adenosine-2'-O-)-methyltransferase 2.1.1.66

1 Nomenclature

EC number
2.1.1.66

Systematic name
S-adenosyl-L-methionine:rRNA (adenosine-2'-O)-methyltransferase

Recommended name
rRNA (adenosine-2'-O-)-methyltransferase

Synonyms
23S rRNA A1067 methyltransferase <1> [3]
RNA-pentose methylase
rRNA adenosine 2'-methylase
ribosomal ribonucleate adenosine 2'-methyltransferase
thiostrepton-resistance methylase

CAS registry number
80146-87-8

2 Source Organism

<1> *Streptomyces azureus* (ATCC 14921 [1]) [1-3]

3 Reaction and Specificity

Catalyzed reaction
S-adenosyl-L-methionine + rRNA = S-adenosyl-L-homocysteine + rRNA containing a single residue of 2'-O-methyladenosine

Reaction type
methyl group transfer

Natural substrates and products
S S-adenosyl-L-methionine + rRNA <1> (Reversibility: ? <1> [1-3]) [1-3]
P S-adenosyl-L-homocysteine + rRNA containing a single residue of 2'-O-methyladenosine

Substrates and products
S S-adenosyl-L-methionine + E. coli 23SrRNA <1> (Reversibility: ? <1> [1, 3]) [1, 3]

P S-adenosyl-L-homocysteine + E. coli rRNA containing a single residue of 2'-O-methyladenosine
S S-adenosyl-L-methionine + rRNA <1> (Reversibility: ? <1> [1-3]) [1-3]
P S-adenosyl-L-homocysteine + rRNA containing a single residue of 2'-O-methyladenosine
S Additional information <1> (<1> preferred substrate is phenol-extracted 23S RNA, mature 50S ribosomal subunits are unaffected by the enzyme, the methylase is inactive on the complex of 23S RNA with ribosomal protein L11 [1]) [1]
P ?

Inhibitors
AMP <1> (<1> weak inhibition [1]) [1]
ATP <1> (<1> moderate inhibition [1]) [1]
KH_2PO_4 <1> (<1> 50% inhibition [1]) [1]
S-adenosyl-L-homocysteine <1> [1]
adenosine <1> (<1> moderate inhibition [1]) [1]
protein L11 <1> [1, 3]
thiostrepton <1> (<1> inhibits methylation [3]) [3]

Cofactors/prosthetic groups
S-adenosyl-L-methionine <1> [1]

Activating compounds
NH_4^+ <1> [1]

Metals, ions
Mg^{2+} <1> [1]

Specific activity (U/mg)
0.0000608 <1> [1]
0.0002529 <1> [1]

K_m-Value (mM)
0.001 <1> (E. coli RNA fragment 1029-1122 23S rRNA) [3]
0.1 <1> (S-adenosyl-L-methionine) [3]
0.15 <1> (S-adenosyl-L-methionine) [1]

K_i-Value (mM)
0.0005 <1> (S-adenosyl-L-homocysteine) [1]
0.013 <1> (S-adenosyl-L-homocysteine) [3]

pH-Optimum
7.5-7.6 <1> [1]

pH-Range
6.5-8 <1> [1]

Temperature range (°C)
30-45 <1> [1]

4 Enzyme Structure

Molecular weight

35000-38000 <1> (<1> gel filtration [1]) [1]

5 Isolation/Preparation/Mutation/Application

Localization

ribosome <1> [1]

Purification

<1> (recombinant protein, expressed in E. coli [3]) [1, 3]

Cloning

<1> (gene is cloned into Streptomyces lividans, enzyme confers resistance to thiostrepton in vivo [1,2]; thiostrepton resistance gene overexpressed in Escherichia coli BL21(DE3)xplysS using a T7-RNA-polymerase-dependent expression vector [3]) [1-3]

Application

medicine <1> (<1> peptide antibiotic thiostrepton inhibits protein synthesis in bacterial extracts [1-3]) [1-3]

6 Stability

pH-Stability

6 <1> (<1> unstable below [1]) [1]

References

[1] Thompson, J.; Cundliffe, E.: Purification and properties of an RNA methylase produced by Streptomyces azureus and involved in resistance to thiostrepton. J. Gen. Microbiol., **124**, 291-297 (1981)

[2] Thompson, J.; Schmidt, F.; Cundliffe, E.: Site of action of a ribosomal RNA methylase conferring resistance to thiostrepton. J. Biol. Chem., **257**, 7915-7917 (1982)

[3] Bechthold, A.; Floss, H.G.: Overexpression of the thiostrepton-resistance gene from Streptomyces azureus in Escherichia coli and characterization of recognition sites of the 23S rRNA A1067 2'-methyltransferase in the guanosine triphosphatase center of 23S ribosomal RNA. Eur. J. Biochem., **224**, 431-437 (1994)

Thiopurine S-methyltransferase 2.1.1.67

1 Nomenclature

EC number
2.1.1.67

Systematic name
S-adenosyl-L-methionine:thiopurine S-methyltransferase

Recommended name
thiopurine S-methyltransferase

Synonyms
6-thiopurine transmethylase
TPMT
mercaptopurine methyltransferase
thiopurine methyltransferase

CAS registry number
67339-09-7

2 Source Organism

<1> *Homo sapiens* (human [2,4-15]; 293T cells [14]; HeLa cells [14]) [2, 4-15]
<2> *Mus musculus* (mouse, strain C57/BL6, GenBank accession number AA109870 [12]) [12]
<3> *Mus musculus* (mouse, strain C57/BL6, GenBank accession number AA637718 [12]) [12]
<4> *Mus musculus* (mouse [1,5,6,12]; Swiss-Webster [1]; NIH-3T3 cells [14]) [1, 5, 6, 12, 14]
<5> *Rattus norvegicus* (rat, Sprague-Dawley [1,3]) [1, 3]

3 Reaction and Specificity

Catalyzed reaction
S-adenosyl-L-methionine + a thiopurine = S-adenosyl-L-homocysteine + a thiopurine S-methylether (also acts, more slowly, on thiopyrimidines and aromatic thiols. Not identical with EC 2.1.1.9 thiol S-methyltransferase)

Reaction type
methyl group transfer

Natural substrates and products

- **S** S-adenosyl-L-methionine + a thiopurine <1-5> (<4,5> specific for S-adenosyl-L-methionine [1]; <1> enzyme plays an important role in metabolism of heterocyclic sulfhydryl drugs such as 6-thiopurine and 6-thioguanine [4]; <1> S-methylation, enzyme of major catabolic pathway of thiopurines [2,10]) (Reversibility: ? <1-5> [1-15]) [1-15]
- **P** S-adenosyl-L-homocysteine + a thiopurine S-methylether

Substrates and products

- **S** S-adenosyl-L-methionine + 2,8-dihydroxy-6-thiopurine <4, 5> (Reversibility: ? <4, 5> [1]) [1]
- **P** S-adenosyl-L-homocysteine + 2,8-dihydroxy-6-methylthiopurine
- **S** S-adenosyl-L-methionine + 2-amino-6-thiopurine <4, 5> (Reversibility: ? <4, 5> [1]) [1]
- **P** S-adenosyl-L-homocysteine + 2-amino-6-methylthiopurine
- **S** S-adenosyl-L-methionine + 2-aminothiophenol <1> (Reversibility: ? <1> [4]) [4]
- **P** S-adenosyl-L-homocysteine + 2-methylsulfanylphenylamine
- **S** S-adenosyl-L-methionine + 2-bromothiophenol <1> (Reversibility: ? <1> [4]) [4]
- **P** S-adenosyl-L-homocysteine + 1-bromo-2-methylsulfanylbenzene
- **S** S-adenosyl-L-methionine + 2-mercaptoethanol <1> (Reversibility: ? <1> [2]) [2]
- **P** S-adenosyl-L-homocysteine + 2-methylmercaptoethanol
- **S** S-adenosyl-L-methionine + 2-methoxythiophenol <1> (Reversibility: ? <1> [4]) [4]
- **P** S-adenosyl-L-homocysteine + 1-methoxy-2-methylsulfanylbenzene
- **S** S-adenosyl-L-methionine + 2-thiothymine <4, 5> (Reversibility: ? <4, 5> [1]) [1]
- **P** S-adenosyl-L-homocysteine + 2-methylthiothymine
- **S** S-adenosyl-L-methionine + 2-thiouracil <1, 4, 5> (Reversibility: ? <1, 4, 5> [1, 2]) [1, 2]
- **P** S-adenosyl-L-homocysteine + thiouracil-2-S-methylether
- **S** S-adenosyl-L-methionine + 3-methoxythiophenol <1> (Reversibility: ? <1> [4]) [4]
- **P** S-adenosyl-L-homocysteine + 1-methylsulfanyl-3-methoxybenzene
- **S** S-adenosyl-L-methionine + 4-(aminomethoxy)thiophenol <1> (Reversibility: ? <1> [4]) [4]
- **P** S-adenosyl-L-homocysteine + 1-aminomethoxy-4-methylsulfanylbenzene
- **S** S-adenosyl-L-methionine + 4-bromothiophenol <1> (Reversibility: ? <1> [4]) [4]
- **P** S-adenosyl-L-homocysteine + 1-bromo-4-methylsulfanylbenzene
- **S** S-adenosyl-L-methionine + 4-chlorothiophenol <1> (Reversibility: ? <1> [4]) [4]
- **P** S-adenosyl-L-homocysteine + 1-chloro-4-methylsulfanylbenzene
- **S** S-adenosyl-L-methionine + 4-fluorothiophenol <1> (Reversibility: ? <1> [4]) [4]

P S-adenosyl-L-homocysteine + 1-fluoro-4-methylsulfanylbenzene
S S-adenosyl-L-methionine + 4-methoxythiophenol <1> (Reversibility: ? <1> [4]) [4]
P S-adenosyl-L-homocysteine + 1-methoxy-4-methylsulfanylbenzene
S S-adenosyl-L-methionine + 4-methylthiophenol <1> (Reversibility: ? <1> [4]) [4]
P S-adenosyl-L-homocysteine + 1-methyl-4-methylsulfanylbenzene
S S-adenosyl-L-methionine + 4-nitrothiophenol <1> (Reversibility: ? <1> [4]) [4]
P S-adenosyl-L-homocysteine + 1-methylsulfanyl-4-nitrobenzene
S S-adenosyl-L-methionine + 4-thiobenzoate <1> (Reversibility: ? <1> [4]) [4]
P S-adenosyl-L-homocysteine + 4-methylsulfanyl benzoate
S S-adenosyl-L-methionine + 6-hydroxy-8-mercaptopurine <1> (Reversibility: ? <1> [7]) [7]
P S-adenosyl-L-homocysteine + 6-hydroxy-8-methylmercaptopurine
S S-adenosyl-L-methionine + 6-mercaptopurine <1, 4> (<1> immunosuppressant medication [5]) (Reversibility: ? <1, 4> [2, 5-8, 10-15]) [2, 5-8, 10-15]
P S-adenosyl-L-homocysteine + 6-methylmercaptopurine
S S-adenosyl-L-methionine + 6-mercaptopurine-riboside <1> (Reversibility: ? <1> [7]) [7]
P S-adenosyl-L-homocysteine + 6-methylmercaptopurine-riboside
S S-adenosyl-L-methionine + 6-mercaptopurine-riboside-5'-monophosphate <1> (Reversibility: ? <1> [7]) [7]
P S-adenosyl-L-homocysteine + 6-methylmercaptopurine-riboside-5'-monophosphate
S S-adenosyl-L-methionine + 6-mercaptopurine-riboside-5'-triphosphate <1> (Reversibility: ? <1> [7]) [7]
P S-adenosyl-L-homocysteine + 6-methylmercaptopurine-riboside-5'-triphosphate
S S-adenosyl-L-methionine + 6-selenoguanine-riboside <1> (Reversibility: ? <1> [7]) [7]
P S-adenosyl-L-homocysteine + 6-methylselenoguanine-riboside
S S-adenosyl-L-methionine + 6-selenopurine <1> (Reversibility: ? <1> [7]) [7]
P S-adenosyl-L-homocysteine + 6-methylselenopurine
S S-adenosyl-L-methionine + 6-selenopurine-riboside <1> (Reversibility: ? <1> [7]) [7]
P S-adenosyl-L-homocysteine + 6-methylselenopurine-riboside
S S-adenosyl-L-methionine + 6-thiodeoxyguanosine monophosphate <1> (Reversibility: ? <1> [8]) [8]
P S-adenosyl-L-homocysteine + 6-methylthiodeoxyguanosine monophosphate
S S-adenosyl-L-methionine + 6-thiodeoxyinosine monophosphate <1> (Reversibility: ? <1> [8]) [8]
P S-adenosyl-L-homocysteine + 6-methyl thiodeoxyinosine monophosphate

S S-adenosyl-L-methionine + 6-thioguanine <1, 4, 5> (Reversibility: ? <1, 4, 5> [1, 2, 5-9, 12-15]) [1, 2, 5-9, 12-15]
P S-adenosyl-L-homocysteine + 6-methylthioguanine
S S-adenosyl-L-methionine + 6-thioguanine-riboside <1> (Reversibility: ? <1> [7]) [7]
P S-adenosyl-L-homocysteine + 6-methylthioguanine-riboside
S S-adenosyl-L-methionine + 6-thioguanine-riboside-5'-monophosphate <1> (Reversibility: ? <1> [7]) [7]
P S-adenosyl-L-homocysteine + 6-methylthioguanine-riboside-5'-monophosphate
S S-adenosyl-L-methionine + 6-thioguanosine monophosphate <1> (Reversibility: ? <1> [8]) [8]
P S-adenosyl-L-homocysteine + 6-methyl thioguanosine monophosphate
S S-adenosyl-L-methionine + 6-thioinosine monophosphate <1> (Reversibility: ? <1> [8]) [8]
P S-adenosyl-L-homocysteine + 6-methyl thioinosine monophosphate
S S-adenosyl-L-methionine + 6-thiopurine <1, 4, 5> (<4,5> preferred substrate [1]) (Reversibility: ? <1, 4, 5> [1-4]) [1-4]
P S-adenosyl-L-homocysteine + 6-methylthiopurine
S S-adenosyl-L-methionine + 6-thiouric acid <4, 5> (Reversibility: ? <4, 5> [1]) [1]
P S-adenosyl-L-homocysteine + 6-methylthiouric acid
S S-adenosyl-L-methionine + 7-methyl-6-mercaptopurine <1> (Reversibility: ? <1> [7]) [7]
P S-adenosyl-L-homocysteine + ?
S S-adenosyl-L-methionine + 8-hydroxy-6-mercaptopurine <1> (Reversibility: ? <1> [7]) [7]
P S-adenosyl-L-homocysteine + 8-hydroxy-6-methylmercaptopurine
S S-adenosyl-L-methionine + 9(n-propyl)6-thioguanine <1> (Reversibility: ? <1> [7]) [7]
P S-adenosyl-L-homocysteine + 9(n-propyl)6-methylthioguanine
S S-adenosyl-L-methionine + 9-(n-butyl)-6-mercaptopurine <1> (Reversibility: ? <1> [7]) [7]
P S-adenosyl-L-homocysteine + 9-(n-butyl)-6-methylmercaptopurine
S S-adenosyl-L-methionine + 9-ethyl-6-mercaptopurine <1> (Reversibility: ? <1> [7]) [7]
P S-adenosyl-L-homocysteine + 9-ethyl-6-methylmercaptopurine
S S-adenosyl-L-methionine + a thiopurine <1-5> (<4,5> specific for S-adenosyl-L-methionine [1]) (Reversibility: ? <1-5> [1-15]) [1-15]
P S-adenosyl-L-homocysteine + a thiopurine S-methylether
S S-adenosyl-L-methionine + azathioprine <1, 4> (<1> antineoplastic medication [5-7,12]) (Reversibility: ? <1, 4> [5-7, 11, 12, 15]) [5-7, 11, 12, 15]
P S-adenosyl-L-homocysteine + methylazathioprine
S Additional information <1, 5> (<5> glutathione is no substrate [3]; <1> aliphatic thiol compounds are poor substrates or not methylated, dithiothreitol is no substrate [2]) [2, 3]
P ?

Inhibitors

2,3-dithiopropanol <4, 5> [1]
2,4-dithiopyrimidine <4, 5> [1]
2,8-dihydroxy-6-mercaptopurine <1> [7, 11]
2-amino-6-thiopurine <4, 5> [1]
2-hydroxy-6-mercaptopurine <1> [7, 11]
2-mercaptoethanol <4, 5> [1]
2-methylthio-4-hydroxypyrimidine <4, 5> (<4,5> 1.8 mM, 45% inhibition [1]) [1]
2-methylthiouracil <4, 5> [1]
3,4,5-triodobenzoic acid <1> [4]
3,4-dimethoxy-5-hydroxybenzoic acid <1> [5]
3,4-dimethoxybenzoic acid <1> [4]
3,5-dichlorobenzoic acid <1> [4]
3,5-dihydroxy-4-methoxybenzoic acid <1> [4]
3,5-dimethylbenzoic acid <1> [4]
3,5-dinitrobenzoic acid <1> [4]
3-bromo-5-hydroxy-4-methoxybenzoic acid <1> [4]
3-chloro-5-hydroxy-4-methoxybenzoic acid <1> [4]
3-hydroxy-4,5-dimethoxybenzoic acid <1> [4]
3-hydroxy-4-methoxy-5-(dimethylamino)benzoic acid <1> [4]
3-hydroxy-4-methoxy-5-nitrobenzoic acid <1> [4]
3-hydroxy-4-methoxybenzoic acid <1> [4]
3-hydroxy-5-iodo-4-methoxybenzoic acid <1> [4]
4-hydroxy-3-methoxybenzoic acid <1> [4]
5-amino-salicylic acid <1> [10]
5-methyl-2,4-dithiopyrimidine <4, 5> [1]
6-hydroxy-2-thiopurine <4, 5> [1]
6-methylmercaptopurine <1> [2]
6-methylthiopurine <4, 5> [1]
6-methylthiopurine <4, 5> [1]
6-thioguanine <1> (<1> mixed inhibitor of 6-mercaptopurine methylation [11]) [11]
L-cysteine <4, 5> (<4,5> 1.8 mM, 50% inhibition [1]) [1]
L-homocysteine <4, 5> [1]
L-homocysteine thiolactone <4, 5> [1]
N-ethylmaleimide <1> [2]
S-adenosyl-L-homocysteine <1> [2, 6]
S-adenosyl-L-methionine <4, 5> [1]
S-methylglutathione <5> [3]
bendroflumethiazide <1> [10]
benzoic acid <1> (<1> substituted benzoic acids, benzoic acid compounds, noncompetitive inhibition [4]; <1> benzoic acid derivatives [5,10]) [4, 5, 10, 15]
ethacrynic acid <1> [2]
furosemide <1> [10]
glutathione <4, 5> [1, 3]
sulphasalazine <1> [10]

trichlormethiazide <1> [10]
tropolone <1> [2, 6]

Activating compounds
SKF-525A <1> [2]

Metals, ions
Mg^{2+} <1> (<1> baseline activity 102% [2]) [2]

Specific activity (U/mg)
1.03 <1> [2]

K_m-Value (mM)
0.0024 <1> (S-adenosyl-L-methionine, <1> cell lysate [6]) [6]
0.0027 <1> (S-adenosyl-L-methionine) [2, 6]
0.00637 <1> (S-adenosyl-L-methionine, <1> isoenzyme II [5]) [5]
0.00722 <1> (S-adenosyl-L-methionine, <1> isoenzyme I [5]) [5]
0.0106 <1> (6-mercaptopurine, <1> recombinant enzyme, expressed in yeast [8]) [8]
0.0127 <1> (6-thiodeoxyinosine, <1> recombinant enzyme, expressed in yeast [8]) [8]
0.0181 <1> (6-thioguanine, <1> recombinant enzyme, expressed in yeast [8]) [8]
0.0257 <1> (6-thioinosine 5'-monophosphate, <1> recombinant enzyme, expressed in yeast [8]) [8]
0.0261 <1> (6-thioguanosine, <1> recombinant enzyme, expressed in yeast [8]) [8]
0.0271 <1> (6-thioguanosine 5'-monophosphate, <1> recombinant enzyme, expressed in yeast [8]) [8]
0.0291 <1> (6-selenopurine) [7]
0.0518 <1> (6-selenopurine-riboside) [7]
0.0551 <1> (6-thioinosine, <1> recombinant enzyme, expressed in yeast [8]) [8]
0.0961 <1> (8-hydroxy-6-mercaptopurine) [7]
0.13 <1> (6-thioguanine) [13]
0.1314 <1> (6-thiodeoxyguanosine, <1> recombinant enzyme, expressed in yeast [8]) [8]
0.138 <1> (6-hydroxy-8-mercaptopurine) [7]
0.139 <1> (6-selenoguanine-riboside) [7]
0.156 <1> (6-mercaptopurine) [13]
0.159 <1> (9-(n-propyl)-6-thioguanine) [7]
0.2 <4> (6-thiopurine) [1]
0.224 <1> (6-mercaptopurine, <1> isoenzyme II [5]) [5]
0.231 <1> (7-methyl-6-mercaptourine) [7]
0.29 <1> (6-mercaptopurine, <1> cell lysate [6]) [6]
0.292 <1> (9-(n-butyl)6-mercaptopurine) [7]
0.3 <1> (6-mercaptopurine) [2, 6]
0.307 <1> (6-mercaptopurine, <1> isoenzyme I [5]) [5]
0.32 <1> (6-mercaptopurine) [6]

0.372 <1> (9-ethyl-6-mercaptopurine) [7]
0.38 <1> (6-mercaptopurine, <1> kidney enzyme [11]) [11]
0.383 <1> (6-mercaptopurine) [7]
0.55 <1> (6-thioguanine) [2]
0.557 <1> (6-thioguanine) [7]
0.58 <1> (6-mercaptopurine, <1> liver enzyme [11]) [6, 11]
0.761 <1> (6-thioguanine-riboside) [7]
0.89 <1> (6-mercaptopurine-riboside-5'-triphosphate) [7]
0.99 <1> (6-mercaptopurine, <1> recombinant enzyme, baculovirus expressed [11]) [11]
1.04 <1> (6-thioguanine-riboside-5'-monophosphate) [7]
1.17 <1> (6-mercaptopurine-riboside) [7]
1.27 <1> (6-mercaptopurine-riboside-5'-monophosphate) [7]
1.7 <4> (2-thiouracil) [1]
2 <1> (2-thiouracil) [2]
168 <1> (2-mercaptoethanol) [2]
Additional information <1> (<1> various thiophenols as substrates, K_m 0.0008-0.0078 [4]) [4]

K_i-Value (mM)
0.00075 <1> (S-adenosyl-L-homocysteine) [2]
0.00802 <1> (6-mercaptopurine, <1> isoenzyme II [5]) [5]
0.01 <1> (3,4-dimethoxy-5-hydroxybenzoic acid) [7]
0.0107 <1> (3,4-dimethoxy-5-hydroxybenzoic acid, <1> isoenzyme II [5]) [5]
0.0113 <1> (3,4-dimethoxy-5-hydroxybenzoic acid, <1> isoenzyme I [5]) [5]
0.0113 <1> (6-mercaptopurine, <1> isoenzyme I [5]) [5]
0.027 <1> (2-hydroxy-6-mercaptopurine) [7]
0.45 <1> (6-thioguanine) [11]
0.85 <1> (tropolone) [2]
0.9 <1> (6-thioguanine) [11]

pH-Optimum
6.5 <1> [5]
6.6 <1> [6]
6.7 <1> [2]

pH-Range
3.5-8.6 <1> [6]

4 Enzyme Structure

Molecular weight
28000 <1> [8, 12]
30000 <1> (<1> 2 isozymes, gel filtration [5]) [5]
36000 <1> (<1> gel filtration [2]) [2]

5 Isolation/Preparation/Mutation/Application

Source/tissue

blood <1> [2, 6, 9, 10, 12, 13]
erythrocyte <1> [2, 4, 6, 8-13]
fibroblast <4> [14]
intestine <1> [8]
kidney <1, 4, 5> [1-10, 12, 13]
leukocyte <1> [8]
liver <1, 4, 5> [1, 5-10, 12]
lymphocyte <1> [2, 5, 6, 9, 10]

Localization

cytoplasm <1, 4> [5, 6, 9, 12]
cytosol <1, 4, 5> [3, 8, 11, 12, 14]

Purification

<1> (partially [2,4]; 2 isozymes [5,7]) [2, 4-7, 13]

Cloning

<1> (cDNA cloning, heterologous expression in yeast [8,9]; baculovirus expressed [11]; retroviral gene transfer [14]) [8-11, 14, 15]
<4> (cDNA cloned and heterologously expressed in rabbit reticulocytes and wheat germ lysate [12]) [12]

Application

medicine <1> (<1> enzyme activity might represent one factor responsible for variations in the therapeutic or toxic effects of thiopurine and thiopyrimidine drugs used clinically to treat recipients of transplanted kidneys and patients with renal disease such as glomerulonephritis [4]; <1> enzyme may be a target for clinically significant drug interactions, the common genetic polymorphism might be a risk factor for the occurence of therapy-dependent secondary leukemia [15]) [4, 8-12, 15]

6 Stability

Temperature stability

45-56 <4, 5> (<4,5> heat inactivation, 23% of the activity is lost after 20 min at 45°C, 95% activity is lost after 6 min at 56°C [1]) [1]

Storage stability

<1>, 80°C, activity in cell lysate is stable during storage [6]

References

[1] Remy, C.N.: Metabolism of thiopyrimidines and thiopurines. J. Biol. Chem., **238**, 1078-1084 (1963)

[2] Woodson, L.C.; Weinshilboum, R.M.: Human kidney thiopurine methyltransferase. Purification and biochemical properties. Biochem. Pharmacol., **32**, 819-826 (1983)
[3] Loo, G.; Smith, J.T.: Glutathione: an endogenous substrate for thiopurine methyltransferase?. Biochem. Biophys. Res. Commun., **126**, 1201-1207 (1985)
[4] Ames, M.M.; Selassie, C.D.; Woodson, L.C.; Van Loon, J.A.; Hansch, C.; Weinshilboum, R.M.: Thiopurine methyltransferase: structure-activity relationships for benzoic acid inhibitors and thiophenol substrates. J. Med. Chem., **29**, 354-358 (1986)
[5] Van Loon, J.A.; Weinshilboum, R.M.: Thiopurine methyltransferase isozymes in human renal tissue. Drug Metab. Dispos., **18**, 632-638 (1990)
[6] Szumlanski, C.L.; Honchel, R.; Scott, M.C.; Weinshilboum, R.M.: Human liver thiopurine methyltransferase pharmacogenetics: biochemical properties, liver-erythrocyte correlation and presence of isozymes. Pharmacogenetics, **2**, 148-159 (1992)
[7] Deininger, M.; Szumlanski, C.L.; Otterness, D.M.; Van Loon, J.; Ferber, W.; Weinshilboum, R.M.: Purine substrates for human thiopurine methyltransferase. Biochem. Pharmacol., **48**, 2135-2138 (1994)
[8] Krynetski, E.Y.; Krynetskaia, N.F.; Yanishevski, Y.; Evans, W.E.: Methylation of mercaptopurine, thioguanine, and their nucleotide metabolites by heterologously expressed human thiopurine S-methyltransferase. Mol. Pharmacol., **47**, 1141-1147 (1995)
[9] Krynetski, E.Y.; Schuetz, J.D.; Galpin, A.J.; Pui, C.H.; Relling, M.V.; Evans, W.E.: A single point mutation leading to loss of catalytic activity in human thiopurine S-methyltransferase. Proc. Natl. Acad. Sci. USA, **92**, 949-953 (1995)
[10] Lysaa, R.A.; Giverhaug, T.; Wold, H.L.; Aarbakke, J.: Inhibition of human thiopurine methyltransferase by furosemide, bendroflumethiazide and trichlormethiazide. Eur. J. Clin. Pharmacol., **49**, 393-396 (1996)
[11] Stupans, I.; Kirlich, A.; McKinnon, R.A.: Human thiopurine methyltransferase: no evidence of activation by its substrates. Life Sci., **62**, 343-350 (1997)
[12] Fessing, M.Y.; Belkov, V.M.; Krynetski, E.Y.; Evans, W.E.: Molecular cloning and functional characterization of the cDNA encoding the murine thiopurine S-methyltransferase (TPMT). FEBS Lett., **424**, 143-145 (1998)
[13] Kroeplin, T.; Iven, H.: Methylation of 6-mercaptopurine and 6-thioguanine by thiopurine S-methyltransferase. A comparison of activity in red blood cell samples of 199 blood donors. Eur. J. Clin. Pharmacol., **56**, 343-345 (2000)
[14] Dervieux, T.; Blanco, J.G.; Krynetski, E.Y.; Vanin, E.F.; Roussel, M.F.; Relling, M.V.: Differing contribution of thiopurine methyltransferase to mercaptopurine versus thioguanine effects in human leukemic cells. Cancer Res., **61**, 5810-5816 (2001)
[15] Weinshilboum, R.: Thiopurine pharmacogenetics: Clinical and molecular studies of thiopurine methyltransferase. Drug Metab. Dispos., **29**, 601-605 (2001)

Caffeate O-methyltransferase 2.1.1.68

1 Nomenclature

EC number
2.1.1.68

Systematic name
S-adenosyl-L-methionine:3,4-dihydroxy-trans-cinnamate 3-O-methyltransferase

Recommended name
caffeate O-methyltransferase

Synonyms
COMT <3, 5, 11> [1, 2, 3, 4, 7, 8, 12]
S-adenosyl-L-methionine:caffeic acid-O-methyltransferase
caffeate 3-O-methyltransferase
caffeate methyltransferase

CAS registry number
50936-45-3

2 Source Organism

<1> *Beta vulgaris* (ssp. vulgaris, spinach beet [17]) [17]
<2> *Bambusa sp.* (bamboo [18]) [18]
<3> *Medicago sativa* (strain Saranac, alfalfa [13]; strain cv Apollo [4]; transgenic line with downregulated caffeate O-methyltransferase activity, cv Regen SY [4]; elicitor-treated, elicitors are fungal macromolecules from the cell walls of Colletotrichum lindemuthianum, two isoforms [12]) [1, 2, 4, 12, 13]
<4> *Glycine max* (var. Mandarin, soybean [16]) [16]
<5> *Nicotiana tabacum* (tobacco, strain Samsun NN, three isozymes isozyme I, isozyme II and isozyme III, only isozyme I has the characteristics of caffeate O-methyltransferase [7,14]) [7, 10, 14]
<6> *Populus tremuloides* (aspen [6,9,11]) [6, 9, 10, 11]
<7> *Daucus carota* [15]
<8> *Phaseolus vulgaris* (french bean [17]) [17]
<9> *Spinacia oleracea* (spinach [17]) [17]
<10> *Pisum sativum* (pea [17]) [17]
<11> *Linum usitatissimum* (flax, strain cv Barbara [3]) [3]

<12> *Vanilla planifolia* (has a low caffeic acid 4-O-methyltransferase activity, 2% of 3-O-methyltransferase activity [5]) [5]
<13> *Petroselinum crispum* (parsley [8]) [8]
<14> *angiosperms* [10]

3 Reaction and Specificity

Catalyzed reaction
S-adenosyl-L-methionine + 3,4-dihydroxy-trans-cinnamate = S-adenosyl-L-homocysteine + 3-methoxy-4-hydroxy-trans-cinnamate (<4> mechanism [16])

Reaction type
methyl group transfer

Natural substrates and products
S S-adenosyl-L-methionine + 3,4-dihydroxy-trans-cinnamate <1, 3, 5, 8, 9, 10, 12> (<3,5> i.e. caffeic acid, involved in the biosynthesis of lignin [1,2,13,14]; <1,8,9,10> i.e. caffeic acid, involved in the biosynthesis of lignin and of some flavonoids in plants [17]; <12> the enzyme is co-induced with other enzymes of the general phenylpropanoid pathway and is involved in the biosynthesis of lignin precursors [5]; <14> the enzyme does not use caffeic acid as a substrate in the presence of 5-hydroxyconiferaldehyde [10]; <5,6> enzyme essentially controls the biosynthesis of syringyl lignin units [10]) (Reversibility: ? <1, 3, 5, 8, 9, 10, 12> [1, 13, 14, 17]) [1, 5, 10, 13, 14, 17]
P S-adenosyl-L-homocysteine + 3-methoxy-4-hydroxy-trans-cinnamate
S S-adenosyl-L-methionine + caffeoyl alcohol <3> (<3> alternative pathway to monolignols involving methylation of caffeoyl aldehyde and/or caffeoyl alcohol [4]) (Reversibility: ? <3> [4]) [4]
P ?
S S-adenosyl-L-methionine + caffeoyl aldehyde <3> (<3> alternative pathway to monolignols involving methylation of caffeoyl aldehyde and/or caffeoyl alcohol [4]) (Reversibility: ? <3> [4]) [4]
P ?

Substrates and products
S S-adenosyl-L-methionine + 3,4,5-trihydroxycinnamic acid <4> (Reversibility: ? <4> [16]) [16]
P S-adenosyl-L-homocysteine + 4,5-dihydroxy-3-methoxycinnamic acid
S S-adenosyl-L-methionine + 3,4-dihydroxy-hydrocinnamic acid <3> (Reversibility: ? <3> [12]) [12]
P S-adenosyl-L-homocysteine + 4-hydroxy-3-methoxyhydrocinnamic acid
S S-adenosyl-L-methionine + 3,4-dihydroxy-trans-cinnamate <1, 2, 3, 4, 5, 6, 8, 9, 10, 11, 13> (<5,6> i.e. caffeate, *m*-specific methylation [9,14]) (Reversibility: ? <1, 2, 3, 4, 5, 6, 8, 9, 10, 11, 13> [1, 2, 3, 4, 6, 7, 8, 9, 12, 13, 14, 16, 17, 18]) [1, 2, 3, 4, 6, 7, 8, 9, 12, 13, 14, 16, 17, 18]

P S-adenosyl-L-homocysteine + 3-methoxy-4-hydroxy-trans-cinnamate <1, 2, 3, 4> (<1,2,3,4> i.e. ferulate [1,12,13,16,17,18]) [1, 12, 13, 16, 17, 18]
S S-adenosyl-L-methionine + 5-hydroxyconiferaldehyde <3, 14> (Reversibility: ? <3, 14> [2, 4, 10]) [2, 4, 10]
P S-adenosyl-L-homocysteine + 5-hydroxy-3-methoxyconiferyl aldehyde
S S-adenosyl-L-methionine + 5-hydroxyconiferyl alcohol <3, 14> (Reversibility: ? <3, 14> [2, 4, 10]) [2, 4, 10]
P S-adenosyl-L-homocysteine + 5-hydroxy-3-methoxyconiferyl alcohol
S S-adenosyl-L-methionine + 5-hydroxyferulic acid <2, 3, 4, 5, 6, 13> (<5,6> *m*-specific methylation [9,14]; <3> at 70% the rate of caffeic acid methylation [13]; <5> best substrate [14]) (Reversibility: ? <2, 3, 4, 5, 6, 13> [1, 2, 4, 6, 7, 8, 9, 13, 14, 16, 18]) [1, 2, 4, 6, 7, 8, 9, 13, 14, 16, 18]
P S-adenosyl-L-homocysteine + sinapic acid <4> [16]
S S-adenosyl-L-methionine + caffeoyl alcohol <3> (Reversibility: ? <3> [2, 4]) [2, 4]
P S-adenosyl-L-homocysteine + 3-methoxycaffeoyl alcohol
S S-adenosyl-L-methionine + caffeoyl aldehyde <3> (Reversibility: ? <3> [2, 4]) [2, 4]
P S-adenosyl-L-homocysteine + 3-methoxycaffeoyl aldehyde
S S-adenosyl-L-methionine + catechol <1, 4, 5> (<4> poor substrate [16]; <5> isozyme II and III, no substrate for isozyme I [14]) (Reversibility: ? <1, 4, 5> [14, 16, 17]) [14, 16, 17]
P S-adenosyl-L-homocysteine + 2-methoxyphenol
S S-adenosyl-L-methionine + chlorogenic acid <5> (Reversibility: ? <5> [14]) [14]
P ?
S S-adenosyl-L-methionine + daidzein <3> (<3> poor substrate [13]) (Reversibility: ? <3> [13]) [13]
P ?
S S-adenosyl-L-methionine + esculetin <5> (Reversibility: ? <5> [14]) [14]
P ?
S S-adenosyl-L-methionine + genistein <3> (<3> poor substrate [13]) (Reversibility: ? <3> [13]) [13]
P ?
S S-adenosyl-L-methionine + homocatechol <5> (Reversibility: ? <5> [14]) [14]
P S-adenosyl-L-homocysteine + 2-methoxy-4-methylphenol
S S-adenosyl-L-methionine + hydroxycaffeic acid <5> (Reversibility: ? <5> [14]) [14]
P ?
S S-adenosyl-L-methionine + protocatechuic acid <1, 3, 4, 5> (<3,4> poor substrate [13,16]) (Reversibility: ? <1, 3, 4, 5> [13, 14, 16, 17]) [13, 14, 16, 17]
P S-adenosyl-L-homocysteine + 4-hydroxy-3-methoxybenzoate
S S-adenosyl-L-methionine + protocatechuic aldehyde <1, 4, 5> (Reversibility: ? <1, 4, 5> [14, 16, 17]) [14, 16, 17]
P S-adenosyl-L-homocysteine + 4-hydroxy-3-methoxybenzaldehyde

S S-adenosyl-L-methionine + pyrogallol <3> (<3> poor substrate [13]) (Reversibility: ? <3> [13]) [13]

P ?

S Additional information <3, 4, 5, 6> (<3> bifunctional enzyme, methylation of 3-hydroxy and 5-hydroxy positions on the aromatic ring [1]; <4,6> methylation only in *meta*-position [9,16]; <5> *p*- and *m*-methylation of protocatechuic aldehyde and acid, as well as esculetin [14]) [1, 9, 14, 16]

P ?

Inhibitors

2-methoxyethanol <1> [17]
5-hydroxyferulic acid <2> (<2> competitive [18]) [18]
Mg^{2+} <1> (<1> above 4 mM [17]) [17]
S-adenosyl-L-homocysteine <1, 3, 4> (<1,3,4> strong, product inhibition, competitive [12,13,16,17]) [12, 13, 16, 17]
ammonium sulfate <1> [17]
ethanol <1> [17]
luteolin <4> [16]
quercetin <4> [16]
rabbit anti-O-methyltransferase immunoglobulin G <6> (<6> strong [9]) [9]
Additional information <4> (<4> no inhibition by EDTA [16]) [16]

Activating compounds

2-mercaptoethanol <3> (<3> 7% activity increase at 20 mM [12]) [12]
dithiothreitol <3> (<3> 7% activity increase at 20 mM [12]) [12]

Metals, ions

Additional information <1, 5> (<1,5> no Mg^{2+} requirement [14,17]) [14, 17]

Specific activity (U/mg)

0.0027 <1> [17]
0.00546 <4> [16]
0.017 <3> [13]
0.14 <6> (<6> with caffeoyl-CoA as substrate [6]) [6]
0.265 <6> (<6> with 5-hydroxyferuoyl-CoA as substrate [6]) [6]
0.48 <6> (<6> with caffeic acid as substrate [6]) [6]
1.077 <6> (<6> with 5-hydroxyferulic acid as substrate [6]) [6]
1.602 <6> [9]

K_m-Value (mM)

0.0018 <3> (5-hydroxyconiferaldehyde, <3> recombinant enzyme [4]) [4]
0.0038 <3> (5-hydroxyconiferyl alcohol, <3> recombinant enzyme [4]) [4]
0.0041 <3> (S-adenosyl-L-methionine) [13]
0.0045 <5> (S-adenosyl-L-methionine) [14]
0.005 <3> (5-hydroxyconiferaldehyde, <3> wild type [2]) [1, 2]
0.0068 <3> (5-hydroxyferulic acid, <3> recombinant enzyme [4]) [4]
0.0069 <3> (caffeoyl aldehyde, <3> recombinant enzyme [4]) [4]
0.007 <3> (S-adenosyl-L-methionine, <3> isozyme II [12]) [12]
0.0075 <3> (5-hydroxyconiferyl alcohol, <3> wild type [2]) [2]

0.01 <2> (5-hydroxyferulic acid) [18]
0.01 <3> (5-hydroxyferulic acid, <3> wild type [2]) [1, 2]
0.01 <3> (caffeoyl aldehyde, <3> wild type [2]) [2]
0.012 <3> (S-adenosyl-L-methionine, <3> isozyme I [12]) [12]
0.012 <3> (caffeoyl alcohol, <3> recombinant enzyme [4]) [4]
0.013 <1> (S-adenosyl-L-methionine) [17]
0.013 <3> (caffeoyl alcohol, <3> wild type [2]) [2]
0.015 <4> (S-adenosyl-L-methionine) [16]
0.03 <5> (esculetin, <5> isozyme I, *m*- and *p*-methylation [14]) [14]
0.04 <5> (homocatechol, <5> isozyme I [14]) [14]
0.043 <3> (caffeic acid, <3> wild type [2]) [1, 2]
0.05 <2> (caffeic acid) [18]
0.05 <4> (protocatechuic aldehyde) [16]
0.053 <3> (caffeic acid, <3> isozyme I [12]) [12]
0.055 <4> (5-hydroxyferulic acid) [16]
0.058 <3> (caffeic acid) [13]
0.059 <3> (caffeic acid, <3> isozyme II [12]) [12]
0.06 <3> (caffeic acid, <3> recombinant enzyme [4]) [4]
0.065 <5> (5-hydroxyferulic acid) [14]
0.068 <1> (caffeic acid) [17]
0.1 <4> (3,4,5-trihydroxycinnamic acid) [16]
0.1 <5> (caffeic acid, <5> isozyme I [14]) [14]
0.133 <4> (caffeic acid) [16]
0.23 <5> (hydrocaffeic acid, <5> isozyme I [14]) [14]
0.7 <5> (protocatechuic aldehyde, <5> isozyme I, *m*- and *p*-methylation [14]) [14]
2 <5> (catechol, <5> isozyme I [14]) [14]
2.5 <5> (chlorogenic acid, <5> isozyme I [14]) [14]
2.5 <5> (protocatechuic acid, <5> isozyme I, *m*- and *p*-methylation [14]) [14]

K_i-Value (mM)

0.00044 <3> (S-adenosyl-L-homocysteine) [13]
0.002 <3> (S-adenosyl-L-homocysteine, <3> isozyme II [12]) [12]
0.004 <1, 3> (S-adenosyl-L-homocysteine, <3> isozyme I [12]) [12, 17]
0.0069 <4> (S-adenosy-L-homocysteine) [16]

pH-Optimum

6.5 <1> (<1> around [17]) [17]
6.5-7 <4> (<4> methylation of caffeic and 5-hydroxyferulic acid [16]) [16]
7 <3> (<3> around [13]) [13]
7.2 <3> [12]
7.5 <6> (<6> assay at [6]) [6]

pH-Range

5-8.5 <4> (<4> more than half-maximal activity at pH 5.0 and 8.5 [16]) [16]
5.4-7.6 <1> (<1> about half-maximal activity at pH 5.4 and 7.6 [17]) [17]
6.5-8 <3> (<3> more than half-maximal activity at pH 6.5 and 8.0 [13]) [13]

Temperature optimum (°C)
30 <1> (<1> assay at [17]) [17]

4 Enzyme Structure

Molecular weight
39800 <6> (<6> amino acid analysis [11]) [11]
41000 <3> (<3> SDS-PAGE [12]) [12]
43000 <3> (<3> SDS-PAGE, recombinant enzyme [4]) [4]
60000 <6> (<6> HPLC gel filtration [9]) [9]
93000 <5> (<5> isozyme I, gel filtration [14]) [14]
103000 <3> (<3> gel filtration [13]) [13]

Subunits
? <6> (<6> x * 40000, SDS-PAGE [9]) [9]
monomer <3> (<3> 1 * 41000, SDS-PAGE [12]) [12]

5 Isolation/Preparation/Mutation/Application

Source/tissue
cell suspension culture <4, 7> [15, 16]
leaf <1, 5, 9> (<5> highly infected by tobacco mosaic virus [14]) [14, 17]
root nodule <3> [13]
shoot <2> [18]
stem <3, 11, 13> (<11> at different developmental stages, vegetative, flowering and seed maturation [3]; <3> first to tenth stem internodes of non-transgenic alfalfa [4]) [3, 4, 8]
xylem <6> (<6> developing secondary xylem [6,9]) [6, 9]

Localization
soluble <3> [12]

Purification
<1> [17]
<2> [18]
<3> (affinity chromatography [13]) [13]
<5> (three caffeic acid methylating activities: isozymes I, II, III, with an enzyme density of 1.277 kg/l, determined by isopycnic equilibrium centrifugation [14]) [14]
<6> (affinity chromatography on S-adenosyl-homocysteine-agarose [9]) [9]

Crystallization
<3> (crystallization from polyethylene glycol solution, 2.2 A resolution, complex with S-adenosyl-L-homocysteine and ferulic acid [2]; crystallization from polyethylene glycol solution, 2.4 A resolution, complex with S-adenosyl-L-homocysteine and 5-hydroxyconiferaldehyde [2]) [1, 2]

Cloning

<3> (expression in Escherichia coli BL21 [2,4]) [2, 4]
<6> (cDNA clone Ptomt1 in Escherichia coli [11]; expressed in Escherichia coli [6]) [6, 11]
<12> (expression in Escherichia coli XL-1 Blue [5]) [5]

Engineering

A162T <3> (<3> same reactivity as wild type [2]) [2]
F172Y <3> (<3> no reaction with caffeic acid, with other substrates same reactivity as wild type [2]) [2]
H183K <3> (<3> no reaction with caffeic acid, with other substrates same reactivity as wild type [2]) [2]
L136Y <3> (<3> same reactivity as wild type [2]) [2]
M130L <3> (<3> no reaction with caffeic acid and 5-hydroxyferulic acid, with other substrates same reactivity as wild type [2]) [2]
N131K <3> (<3> same reactivity as wild type [2]) [2]
N324Y <3> (<3> no reaction with caffeic acid, with other substrates same reactivity as wild type [2]) [2]

6 Stability

pH-Stability

7 <3> (<3> 37% of activity of pH 7.2 [12]) [12]
7.4 <3> (<3> 75% of activity of pH 7.2 [12]) [12]

Temperature stability

45 <5> (<5> isozyme I inactivates after 5 min, isozyme II loses 70% of activity after 2 h, isozyme III has 50% activation after 2 h [14]) [14]

General stability information

<1>, 2-mercaptoethanol or dithiothreitol or GSH reduces storage stability [17]
<5>, 2-mercaptoethanol stabilizes [14]
<5>, freezing causes complete inactivation [14]
<5>, gel filtration leads to severe loss of activity [14]

Storage stability

<1>, -20°C, at least a month in absence of thiol compounds [17]
<3>, -70°C, 2 months, stable [12]
<3>, 4°C, 4-5 d, assay buffer, 50% activity [12]
<4>, 4°C, 0.02% NaN_3, at least 2 months [16]
<5>, 0°C, 65%-85% of initial activity retained after 5 months in the presence of 2-mercaptoethanol [14]

References

[1] Eckardt, N.A.: Probing the mysteries of lignin biosynthesis: The crystal structure of caffeic acid/5-hydroxyferulic acid 3/5-O-methyltransferase provides new insights. Plant Cell, **14**, 1185-1189 (2002)

[2] Zubieta, C.; Kota, P.; Ferrer, J.L.; Dixon, R.A.; Noel, J.P.: Structural basis for the modulation of lignin monomer methylation by caffeic acid/5-hydroxyferulic acid 3/5-O-methyltransferase. Plant Cell, **14**, 1265-1277 (2002)

[3] Day, A.; Dehorter, B.; Neutelings, G.; Czeszak, X.; Chabbert, B.; Belingheri, L.; David, H.: Caffeoyl-coenzyme A 3-O-methyltransferase enzyme activity, protein and transcript accumulation in flax (Linum usitatissimum) stem during development. Physiol. Plant., **113**, 275-284 (2001)

[4] Parvathi, K.; Chen, F.; Guo, D.; Blount, J.W.; Dixon, R.A.: Substrate preferences of O-methyltransferases in alfalfa suggest new pathways for 3-O-methylation of monolignols. Plant J., **25**, 193-202 (2001)

[5] Xue, Z.T.; Brodelius, P.E.: Kinetin-induced caffeic acid O-methyltransferases in cell suspension cultures of Vanilla planifolia Andr. and isolation of caffeic acid O-methyltransferase cDNAs. Plant Physiol. Biochem., **36**, 779-788 (1998)

[6] Meng, H.; Campbell, W.H.: Substrate profiles and expression of caffeoyl coenzyme a and caffeic acid O-methyltransferases in secondary xylem of aspen during seasonal development. Plant Mol. Biol., **38**, 513-520 (1998)

[7] Martz, F.; Maury, S.; Pincon, G.; Legrand, M.: cDNA cloning, substrate specificity and expression study of tobacco caffeoyl-CoA 3-O-methyltransferase, a lignin biosynthetic enzyme. Plant Mol. Biol., **36**, 427-437 (1998)

[8] Grimmig, B.; Kneusel, R.E.; Junghanns, K.T.; Matern, U.: Expression of bifunctional caffeoyl-CoA 3-O-methyltransferase in stress compensation and lignification. Plant Biol., **1**, 299-310 (1999)

[9] Bugos, R.C.; Chiang, V.L.C.; Campbell, W.H.: Characterization of bispecific caffeic acid/5-hydroxyferulic acid O-methyltransferase from aspen. Phytochemistry, **31**, 1495-1498 (1992)

[10] Ye, Z.H.; Zhong, R.; Morrison, W.H.; Himmelsbach, D.S.: Caffeoyl coenzyme A O-methyltransferase and lignin biosynthesis. Phytochemistry, **57**, 1177-1185 (2001)

[11] Bugos, R.C.; Chiang, V.L.C.; Campbell, W.H.: cDNA cloning, sequence analysis and seasonal expression of lignin-bispecific caffeic acid/5-hydroxyferulic acid O-methyltransferase of aspen. Plant Mol. Biol., **17**, 1203-1215 (1991)

[12] Edwards, R.; Dixon, R.A.: Purification and characterization of S-adenosyl-L-methionine: caffeic acid 3-O-methyltransferase from suspension cultures of alfalfa (Medicago sativa L.). Arch. Biochem. Biophys., **287**, 372-379 (1991)

[13] Vance, C.P.; Bryan, J.W.: Purification and properties of caffeic acid O-methyltransferase from alfalfa root nodules. Phytochemistry, **20**, 41-43 (1981)

[14] Collendavelloo, J.; Legrand, M.; Geoffroy, P.; Barthelemy, J.; Fritig, B.: Purification and properties of the three *o*-diphenol-O-methyltransferases of tobacco leaves. Phytochemistry, **20**, 611-616 (1981)
[15] Sugano, N.; Ogawa, Y.: Effect of 2,4-dichlorophenoxyacetic acid on the activity of O-methyltransferase in carrot cell culture. Phytochemistry, **20**, 617-619 (1981)
[16] Poulton, J.E.; Hahlbrock, K.; Grisebach, H.: Enzymic synthesis of lignin precursors. Purification of properties of the S-adenosyl-l-methionine: caffeic acid 3-O-methyltransferase from soybean cell suspension cultures. Arch. Biochem. Biophys., **176**, 449-456 (1976)
[17] Poulton, J.E.; Butt, V.S.: Purification and properties of S-adenosyl-L-methionine: caffeic acid O-methyltransferase from leaves of spinach beet (Beta vulgaris L). Biochim. Biophys. Acta, **403**, 301-314 (1975)
[18] Shimada, M.; Kuroda, H.; Higuchi, T.: Evidence for the formation of methoxyl groups of ferulic acid and sinapic acids in Bambusa by the same O-methyltransferase. Phytochemistry, **12**, 2873-2875 (1973)

5-Hydroxyfuranocoumarin 5-O-methyltransferase 2.1.1.69

1 Nomenclature

EC number
2.1.1.69

Systematic name
S-adenosyl-L-methionine:5-hydroxyfuranocoumarin 5-O-methyltransferase

Recommended name
5-hydroxyfuranocoumarin 5-O-methyltransferase

Synonyms
furanocoumarin 5-O-methyltransferase
furanocoumarin 5-methyltransferase

CAS registry number
67339-12-2

2 Source Organism

<1> *Ruta graveolens* [1-3]
<2> *Heracleum lanatum* (Michx. [2]) [2]

3 Reaction and Specificity

Catalyzed reaction
S-adenosyl-L-methionine + 5-hydroxyfuranocoumarin = S-adenosyl-L-homocysteine + 5-methoxyfuranocoumarin (converts bergaptol into bergapten. Methylates the 5-hydroxyl of some hydroxy- and methyl- coumarins, but has little activity on non-coumarin phenols)

Reaction type
methyl group transfer

Natural substrates and products
S S-adenosyl-L-methionine + bergaptol <1> (<1> furanocoumarinbiosynthesis [1-3]) (Reversibility: ? <1> [1-3]) [1-3]
P S-adenosyl-L-homocysteine + bergapten

Substrates and products

S S-adenosyl-L-methionine + 5-hydroxyxanthotoxin <1, 2> (Reversibility: ? <1> [1-3]) [1-3]

P S-adenosyl-L-homocysteine + isopimpinellin

S S-adenosyl-L-methionine + bergaptol <1, 2> (Reversibility: ? <1> [1-3]) [1-3]

P S-adenosyl-L-homocysteine + bergapten

S S-adenosyl-L-methionine + caffeic acid <1> (Reversibility: ? <1> [2]) [2]

P S-adenosyl-L-homocysteine + ?

S Additional information <1> (<1> no activity against xanthoxol [1]; <1> activity against phenol, resorcinol, phloroglucinol, 3-hydroxy-4-methoxy-benzaldehyde, 2-hydroxybenzoic acid, 3-hydroxybenzoic acid, 2-hydroxy-phenylacetic acid, 3-hydroxycinnamic acid, and quercetin-7,4'-dimethyl ether lower than 10% of the activity observed with bergaptol as substrate [2]) [1, 2]

P ?

Inhibitors

Co^{2+} <1> [2]
Cu^{2+} <1> [2]
Fe^{2+} <1> [2]
Mg^{2+} <1> (<1> slightly inhibits the O-methylation of bergaptol [2]) [2]
Ni^{2+} <1> [2]
Zn^{2+} <1> [2]
kinetin <1> (<1> strong inhibition [2]) [2]
p-chloromercuribenzoic acid <1> [2]

Cofactors/prosthetic groups

S-adenosyl-L-methionine <1> [1]

Specific activity (U/mg)

0.000029 <1> (<1> bergaptol [2]) [2]
0.000056 <1> (<1> 5-hydroxyxanthotoxin [2]) [2]
0.000375 <1> [1]

pH-Optimum

7.3-8 <1> [2]
8.5-9 <1> (<1> 5-hydroxyanthotoxin as substrate [2]) [2]

4 Enzyme Structure

Molecular weight

110000 <1> (<1> gel filtration [2]) [2]

5 Isolation/Preparation/Mutation/Application

Source/tissue

cell culture <1> [3]
leaf <1, 2> (<1> sterile leaf cell culture [2]) [2]
shoot <1> (<1> young shoots of field-grown plants [2]) [2]

Purification

<1> (partial [1]) [1]

6 Stability

Storage stability

<1>, -20°C, stored and refrozen, activity against bergaptol is essentially unchanged after 18 days and is still 65% of the initial value after 34 days [2]
<1>, 0°C, stored below undesalted, 40% loss of activity after 5 weeks [2]

References

[1] Sharma, S.K.; Garrett, J.M.; Brown, S.A.: Separation of the S-adenosylmethionine: 5-and 8-hydroxyfuranocoumarin O-methyltransferases of Ruta graveolens L. by general ligand affinity chromatography. Z. Naturforsch. C, **34**, 387-391 (1979)
[2] Thompson, H.J.; Sharma, S.K.; Brown, S.A.: O-Methyltransferases of furanocoumarin biosynthesis. Arch. Biochem. Biophys., **188**, 272-281 (1978)
[3] Sharma, S.K.; Brown, S.A.: Affinity chromatography of Ruta graveolens L. O-methyltransferases. Studies demonstrating the potential of the technique in the mechanistic investigation of O-methyltransferases. Can. J. Biochem., **57**, 986-995 (1979)

8-Hydroxyfuranocoumarin 8-O-methyltransferase 2.1.1.70

1 Nomenclature

EC number
2.1.1.70

Systematic name
S-adenosyl-L-methionine:8-hydroxyfuranocoumarin 8-O-methyltransferase

Recommended name
8-hydroxyfuranocoumarin 8-O-methyltransferase

Synonyms
furanocoumarin 8-O-methyl-transferase
furanocoumarin 8-methyltransferase

CAS registry number
67339-13-3

2 Source Organism

<1> *Ruta graveolens* [1-3]
<2> *Heracleum lanatum* (Michx. [2]) [2]

3 Reaction and Specificity

Catalyzed reaction
S-adenosyl-L-methionine + 8-hydroxyfuranocoumarin = S-adenosyl-L-homocysteine + 8-methoxyfuranocoumarin

Reaction type
methyl group transfer

Natural substrates and products
S S-adenosyl-L-methionine + xanthotoxol <1> (<1> furanocoumarin biosynthesis [1-3]) (Reversibility: ? <1> [1]) [1-3]
P S-adenosyl-L-homocysteine + xanthotoxin

Substrates and products
S S-adenosyl-L-methionine + 7,8-dihydroxycoumarin <1> [2]
P ?
S S-adenosyl-L-methionine + 8-hydroxybergapten <1, 2> (Reversibility: ? <1> [2, 3]) [2, 3]

P S-adenosyl-L-homocysteine + isopimpinellin
S S-adenosyl-L-methionine + xanthotoxol <1, 2> (Reversibility: ? <1> [1]) [1, 2]
P S-adenosyl-L-homocysteine + xanthotoxin

Inhibitors
Co^{2+} <1> [2]
Cu^{2+} <1> [2]
Fe^{2+} <1> [2]
Mg^{2+} <1> [2]
Ni^{2+} <1> [2]
Zn^{2+} <1> [2]
kinetin <1> (<1> strong inhibition [2]) [2]
p-chloromercuribenzoic acid <1> [2]

Cofactors/prosthetic groups
S-adenosyl-L-methionine <1> [1]

Specific activity (U/mg)
0.0000073 <1> (<1> xanthotoxol [2]) [2]
0.000033 <1> (<1> 8-hydroxybergapten [2]) [2]

pH-Optimum
7.3-8 <1> [2]

4 Enzyme Structure

Molecular weight
850000 <1> (<1> gel filtration [2]) [2]

5 Isolation/Preparation/Mutation/Application

Source/tissue
cell culture <1> [3]
leaf <1, 2> (<1> sterile leaf cell culture [2]) [2]
shoot <1> (<1> young shoots of field-grown plants [2]) [2]

Purification
<1> (partial [1]) [1]

6 Stability

General stability information
<1>, marked loss of activity against xanthotoxol after resalting [2]

Storage stability

<1>, -20°C, stored and refrozen, activity against xanthotoxol falls rapidly to less than half of the initial value in 1 week and to only 10% after 4 weeks [2]

<1>, 0°C, stored below undesalted, 40% loss of activity after 5 weeks [2]

References

[1] Sharma, S.K.; Garrett, J.M.; Brown, S.A.: Separation of the S-adenosylmethionine: 5-and 8-hydroxyfuranocoumarin O-methyltransferases of Ruta graveolens L. by general ligand affinity chromatography. Z. Naturforsch. C, **34**, 387-391 (1979)

[2] Thompson, H.J.; Sharma, S.K.; Brown, S.A.: O-Methyltransferases of furanocoumarin biosynthesis. Arch. Biochem. Biophys., **188**, 272-281 (1978)

[3] Sharma, S.K.; Brown, S.A.: Affinity chromatography of Ruta graveolens L. O-methyltransferases. Studies demonstrating the potential of the technique in the mechanistic investigation of O-methyltransferases. Can. J. Biochem., **57**, 986-995 (1979)

Phosphatidyl-N-methylethanolamine N-methyltransferase 2.1.1.71

1 Nomenclature

EC number
2.1.1.71

Systematic name
S-adenosyl-L-methionine:phosphatidyl-N-methylethanolamine N-methyltransferase

Recommended name
phosphatidyl-N-methylethanolamine N-methyltransferase

Synonyms
PE N-MTase <2> [3]
PLMT
class I PLMT <4, 5> [4]
methyltransferase II
phosphatidyl-N-methylethanolamine methyltransferase
phosphatidyl-N-monomethylethanolamine methyltransferase
phosphatidylethanolamine methyltransferase I
phosphatidylmonomethylethanolamine methyltransferase
phospholipid methyltransferase

CAS registry number
67167-73-1

2 Source Organism

<1> *Rattus norvegicus* (rat [1-4,6]; strain Wistar [1]; Sprague-Dawley [6]; white rats [7]) [1-4, 6, 7]
<2> *Bos taurus* (bovine [2,3,4,7]) [2, 3, 4, 7]
<3> *Neurospora crassa* [2, 7]
<4> *Saccharomyces cerevisiae* (baker's yeast [4]) [4, 5]
<5> *Schizosaccharomyces pombe* (fission yeast [4]) [4]
<6> *Mus musculus* (mouse [4,8]) [4, 8]
<7> *Homo sapiens* (human [7]) [7, 10]
<8> *Oryctolagus cuniculus* (rabbit [9]) [9]

3 Reaction and Specificity

Catalyzed reaction

S-adenosyl-L-methionine + phosphatidyl-N-methylethanolamine = S-adenosyl-L-homocysteine + phosphatidyl-N-dimethylethanolamine (the enzyme also catalyzes the transfer of a further methyl group, producing phosphatidylcholine)

Reaction type

methyl group transfer

Substrates and products

S S-adenosyl-L-methionine + dipalmitoylphosphatidyl-N,N-dimethylethanolamine <1> (Reversibility: ? <1> [7]) [7]
P S-adenosyl-L-homocysteine + dipalmitoylphosphatidylcholine
S S-adenosyl-L-methionine + dipalmitoylphosphatidyl-N-methylethanolamine <1> (Reversibility: ? <1> [7]) [7]
P S-adenosyl-L-homocysteine + dipalmitoylphosphatidyl-N,N-dimethylethanolamine dipalmitoylphosphatidyl-N-methylethanolamine
S S-adenosyl-L-methionine + phosphatidyl-N,N-dimethylethanolamine <1, 2, 4> (Reversibility: ? <1, 2, 4> [1-3, 5-7]) [1-3, 5-7]
P S-adenosyl-L-homocysteine + phosphatidylcholine
S S-adenosyl-L-methionine + phosphatidyl-N-ethanolamine <1, 2> (Reversibility: ? <1, 2> [1, 3]) [1, 3]
P S-adenosyl-L-homocysteine + phosphatidyl-N-monomethylethanolamine
S S-adenosyl-L-methionine + phosphatidyl-N-monomethylethanolamine <1> (Reversibility: ? <1> [1]) [1]
P S-adenosyl-L-homocysteine + phosphatidylcholine
S S-adenosyl-L-methionine + phosphatidyl-N-monomethylethanolamine <1, 2, 4> (Reversibility: ? <1, 2, 4> [1, 2, 5]) [1, 2, 5]
P S-adenosyl-L-homocysteine + phosphatidyl-N-dimethylethanolamine
S Additional information <2> (<2> enzyme methylates only exogenously added mono-and dimethyl derivatives of phosphatidyl-N-ethanolamine, but not phosphatidyl-N-ethanolamine [2]) [2]
P ?

Inhibitors

$HgCl_2$ <4> [5]
S-adenosyl-L-homocysteine <1> (<1> competitive inhibition [1]) [1]
Triton X-100 <1, 4> [1, 5]
adriamycin <1> (<1> doxorubicin, antitumor drug [6]) [6]
octyl glucoside <4> [5]
p-chloromercuriphenylsulfonic acid <4> [5]
phosphatidylcholine <1> (<1> inhibits the incorporation of $[^3H]$-methyl groups from S-adenosyl-L-methionine into phosphatidyl-N-monomethylethanolamine and phosphatidyl-N,N-dimethylethanolamine [1]) [1]
sodium cholate <4> [5]

Metals, ions

Ca^{2+} <4> (<4> leads to a stimulation of activity [5]) [5]
Mg^{2+} <2, 4> (<2> requires magnesium ions for full activity [3]; <4> leads to a stimulation of activity [5]) [3, 5]

Specific activity (U/mg)

0.79 <1> [1]

K_m-Value (mM)

0.016 <1> (S-adenosyl-L-methionine, <1> substrate phosphatidyl-N,N-dimethylethanolamine [1]) [1]
0.022 <1> (S-adenosyl-L-methionine) [1]
0.037 <2> (S-adenosyl-L-methionine, <2> pH 8.5 [3]) [3]
0.038 <2> (S-adenosyl-L-methionine, <2> pH 10.0 [3]) [3]
0.054 <4> (phosphatidyl-N-monomethylethanolamine) [5]
0.059 <4> (phosphatidyl-N-dimethylethanolamine) [5]
0.067 <1> (phosphatidyl-N-monomethylethanolamine) [7]
0.08 <1> (phosphatidyl-N-monomethylethanolamine) [1]
0.1 <2> (S-adenosyl-L-methionine) [2, 7]
0.18 <4> (phosphatidyl-N-dimethylethanolamine) [5]
0.38 <4> (phosphatidyl-N-monomethylethanolamine) [5]

K_i-Value (mM)

0.0049 <1> (S-adenosyl-L-homocysteine) [1]
0.0067 <1> (S-adenosyl-L-homocysteine, <1> inhibition of the last step in the synthesis of phosphatidylcholine [1]) [1]
0.054 <4> (S-adenosyl-L-homocysteine, <4> phosphatidyl-N-dimethylethanolamine as substrate, competitive inhibition [5]) [5]
0.057 <4> (S-adenosyl-L-homocysteine, <4> phosphatidyl-N-monomethylethanolamine as substrate, competitive inhibition [5]) [5]
0.12 <4> (S-adenosyl-L-homocysteine, <4> phosphatidyl-N-dimethylethanolamine as substrate, noncompetitive inhibition [5]) [5]
0.32 <4> (S-adenosyl-L-homocysteine, <4> phosphatidyl-N-monomethylethanolamine as substrate, noncompetitive inhibition [5]) [5]

pH-Optimum

7.5 <4> [5]
8.5 <2> (<2> low substrate concentrations about 0.01 M S-adenosyl-L-methionine [3]) [3]
9.5 <1> [1]
10 <1, 2> (<2> high substrate concentrations about 0.2 M S-adenosyl-L-methionine [3]) [2, 3, 7]

pH-Range

5.5-9.5 <1> [1]
6.5-9.5 <4> [5]
6.5-10.5 <1> [7]
7-11 <2> [3]

Temperature optimum (°C)

40 <4> [5]

4 Enzyme Structure

Molecular weight

23150 <4> (<4> predicted from amino acid sequence [4]) [4]

5 Isolation/Preparation/Mutation/Application

Source/tissue

adrenal medulla <2> [2, 7]
aorta <1> [7]
brain <1> (<1> synaptosomes [7]) [7]
erythrocyte <1, 2> (<2> present in red blood cells on the interior side of the membrane [2]) [2, 7]
eye <2> [3]
heart <1> (<1> sarcolemma, sarcolemmal vesicles [6]) [6]
hepatocyte <1> [1]
liver <1, 6> [1-5, 7, 8]
muscle <8> (<8> skeletal muscle [9]) [9]
pancreas <1> (<1> islet cells [7]) [7]
placenta <7> (<7> microvilli [7]) [7]
retina <2> (<2> rod outer segments [3]) [3]

Localization

cytoplasm <1> [1]
cytosol <2> [3]
endoplasmic reticulum <4, 5> [4]
membrane <1> (<1> synaptosomal plasma membrane [3]) [3, 7]
microsome <1-3> [1-3, 5, 7]
mitochondrion <2> [2, 3, 7]
nucleus <1> [7]
sarcoplasmic reticulum <8> (<8> localized mainly in the extracytoplasmic luminal part of the membrane [9]) [9]

Purification

<1> (partial [1]) [1, 3, 5]

Cloning

<1> [4]
<4> (enzyme is encoded by PEM2/OPI3 gene, cloned by complementation [4,5]) [4, 5]
<5> [4]
<6> [4]

Application

medicine <1> (<1> adriamycin, an effective anticancer chemotherapeutic agent depresses enzyme activity in vivo and in vitro [6]) [6]

6 Stability

Temperature stability

30 <4> (<4> unstable above [5]) [5]

Storage stability

<1>, -70°C, not stable to freezing [1]
<1>, 2°C, loses 30% of its activity within 12 h [1]

References

[1] Schneider, W.J.; Vance, D.E.: Conversion of phosphatidylethanolamine to phosphatidylcholine in rat liver. Partial purification and characterization of the enzymatic activities. J. Biol. Chem., **254**, 3886-3891 (1979)
[2] Hirata, F.; Viveros, O.H.; Diliberto, E.J.; Axelrod, J.: Identification and properties of two methyltransferases in conversion of phosphatidylethanolamine to phosphatidylcholine. Proc. Natl. Acad. Sci. USA, **75**, 1718-1721 (1978)
[3] Roque, M.E.; Giusto, N.M.: Phosphatidylethanolamine N-methyltransferase activity in isolated rod outer segments from bovine retina. Exp. Eye Res., **60**, 631-643 (1995)
[4] Kanipes, M.I.; Henry, S.A.: The phospholipid methyltransferases in yeast. Biochim. Biophys. Acta, **1348**, 134-141 (1997)
[5] Gaynor, P.M.; Carman, G.M.: Phosphatidylethanolamine methyltransferase and phospholipid methyltransferase activities from Saccharomyces cerevisiae. Enzymological and kinetic properties. Biochim. Biophys. Acta, **1045**, 156-163 (1990)
[6] Iliskovic, N.; Panagia, V.; Slezak, J.; Kumar, D.; Li, T.; Singal, P.K.: Adriamycin depresses in vivo and in vitro phosphatidylethanolamine N-methylation in rat heart sarcolemma. Mol. Cell. Biochem., **176**, 235-240 (1997)
[7] Sastry, B.V.R.; Statham, C.N.; Axelrod, J.; Hirata, F.: Evidence for two methyltransferase involved in the conversion of phosphatidylethanolamine to phosphatidylcholine in the rat liver. Arch. Biochem. Biophys., **211**, 762-773 (1981)
[8] Tanaka, Y.; Amano, F.; Maeda, M.; Nishijima, M.; Akamatsu, Y.: Purification and properties of phosphatidyl-N-monomethylethanolamine N-methyltransferase, the enzyme catalyzing the second and the third steps in the phosphatidylethanolamine N-methyltransferase system, from mouse liver microsomes. Jpn. J. Med. Sci. Biol., **43**, 59-73 (1990)

[9] Pilarska, M.; Pikula, S.; Szymanska, G.; Sarzala, M.G.: Properties and topology of enzymes methylating phosphatidylethanolamine to phosphatidylcholine in sarcoplasmic reticulum. Int. J. Biochem., **19**, 705-711 (1987)
[10] Guan, Z.Z.; Wang, Y.N.; Xiao, K.Q.; Hu, P.S.; Liu, J.L.: Activity of phosphatidylethanolamine-N-methyltransferase in brain affected by Alzheimer's disease. Neurochem. Int., **34**, 41-47 (1999)

Site-specific DNA-methyltransferase (adenine-specific) 2.1.1.72

1 Nomenclature

EC number
2.1.1.72

Recommended name
site-specific DNA-methyltransferase (adenine-specific)

Synonyms
DNA-[N^6-adenine]-methyltransferase
N^6-adenine DNA -methyltransferase
T4 Dam (N^6-Ade)-MTase <2> [3]
T4DNA-(N^6-adenine)-methyltransferase <2> [5]
VspI methyltransferase <5> [13]
Wadmtase <1> [2]
adenine-N6 MTAse
modification methylase
restriction-modification system
Additional information <8> (<8>, this is a large group of enzymes most of which, with enzymes of similar site specificity listed as EC 3.1.21.3, 4 or 5, form so-called "restriction-modification systems". A complete listing of all these enzymes has been produced by R.J. Roberts, this list is updated annually [1]) [1]

CAS registry number
69553-52-2

2 Source Organism

<1> *Triticum aestivum* [2]
<2> *bacteriophage T4* [3, 4, 5, 6, 7, 10, 11, 12, 13, 14]
<3> *Caulobacter crescentus* [8]
<4> *Escherichia coli* (M.EcoRV, M.EcoRI, E. coli dam methyltransferase and M.FokI enzyme [16]) [9, 15, 16]
<5> *Vibrio sp.* [13, 17]
<6> *bacteriophage T2* [13, 14]
<7> *Flavobacterium okeanokoites* [16]
<8> *bacteria* [1]

3 Reaction and Specificity

Catalyzed reaction

S-adenosyl-L-methionine + DNA adenine = S-adenosyl-L-homocysteine + DNA 6-methylaminopurine (<2>, steady-state-ordered bi-bi mechanism in which the order of substrate binding and product release is S-adenosyl-L-methionine, DNA, DNAMe, S-adenosyl-L-homocysteine [6])

Reaction type

methyl group transfer

Natural substrates and products

S S-adenosyl-L-methionine + DNA adenine <1, 3> (<1>, the enzyme is responsible for mitochondrial DNA modification that might be involved in the regulation of replication of mitochondria in plants [2]; <3>, the enzyme is involved in cell cycle regulation of Caulobacter crescentus [8]) (Reversibility: ? <1, 3> [2, 8]) [2, 8]

P S-adenosyl-L-homocysteine + DNA 6-methylaminopurine <1, 3> [2, 8]

Substrates and products

S S-adenosyl-L-methionine + DNA adenine <1-8> (<1>, the enzyme de novo methylates the first adenine residue in the TGATCA sequence in the single-stranded or double-stranded DNA substrate, but it prefers single-stranded structures [2]; <1>, GGGTGATCAGGG [2]; <1>, CCCTGATCACCC [2]; <1>, GGGTGATCAGGG+CCCTGATCACCC, annealed [2]; <1>, GCGTGATCACGC [2]; <1>, GCGTGATCACGC [2]; <1> lambda phage DNA [2]; <2>, in unmethylated target (2A/A duplex), the enzyme alone randomly binds to the asymetric 2A/A duplex, S-adenosyl-L-methionine induces an allosteric T4 conformational change that promotes reorientation of the enzyme to the strand containing the native base. S-adenosyl-L-methionine increases enzyme binding-specificity, in addition to serving as the methyl donor, the enzyme recognizes the palindromic sequence GATC [3]; <2>, nonglucosylated unmethylated T4 gt- dam-DNA. The enzyme methylates the palindromic tetranucleotide, GATC, designated the canonical sequence. At high Mtase:DNA ratios, T4 Dam can methylate some noncanonical sequences belonging to GAY, where Y represents cytosine or thymine [4]; <2>, synthetic oligonucleotide substrates containing the native recognition site GATC or its modified variants [5]; <2>, the enzyme mediates methyl group transfer reaction from S-adenosyl-L-methionine to adenine in the palindromic recognition sequence, GATC, of a 20-mer oligonucleotide duplex [6]; <2>, oligonucleotide substrates containing the native or modified recognition site. The enzyme recognizes the palindromic sequence GATC and catalyzes transfer of the methyl group from S-adenosyl-L-methionine to the N^6-position of adenine. The release of product is the rate-limiting step in the reaction [7]; <3>, the enzyme preferentially binds DNA before S-adenosylmethionine. S-adenosylhomocysteine is preferentially released from the enzyme before fully methylated DNA. Binding of both substrates and methylation occurs first in a rapid step followed by regeneration of enzyme in a second rate-limiting step [8]; <4>, non-self-complementary tetradecanucleotide duplexes that contain the GATC target sequence. The enzyme is rather tolerant to base modification, binding of the enzyme is inversely proportional to the thermodynamic stability of the duplex in the ternary complex [9]; <2>, the enzyme catalyzes methyl group transfer from S-adenosyl-L-methionine to the N^6-position of adenine in the palindromic sequence GATC. The rate-limiting step is release of product methylated DNA from the enzyme [10]; <2>, canonical 12-mer and 20-mers various defective duplexes containing some defect in the DNA-target site, e.g. the absence of an internucleotide phosphate or a nucleotide within the recognition site, or a single-stranded region [10]; <2>, interaction with substrates containing defective recognition sites. Deoxyguanosine residues in both strands of the modified GATC are indispensable for complex formation [11]; <2>, various synthetic oligonucleotide substrates. Upon collision of an enzyme monomer with an oligonucleotide duplex, an asymmetrical complex forms in which the enzyme (randomly oriented relative to one of the strands of the specific recognition site) catalyzes a fast transfer of the methyl group from S-adenmosylmethionine to the adenosine residue. Simultaneously, a second T4MTAse subunit is added to the complex, providing for the continuation of the reaction [12]; <5>, the enzyme modifies the third adenine within the recognition sequence 5-ATTAAT-3' [13]; <2,6>, nonglucosylated, hydroxymethylcytosine-containing T2gt-virion DNA has a higher level of methylation than T4gt virion DNA does [14]; <4>, canonical 14mers and various substituted duplexes. Non-self-complementary tetradecamer duplex d[GCCGGATCTAGACG]*d[CGTCTAGATCCGGC] containing the hemimethylated GATC target sequence on one or the other strand and modifications in the GATC target sequence of the complementary strands. Large differences in DNA methylation of duplexes containing single dI or dG substitutions of the Dam recognition site are observed compared with the canonical substrate, if the substitution involves the top strand, on the G-C rich side. Substitution in either strand by uracil or 5-ethyluracil result in small perturbation of the methylation patterns. When 2,6-diamino-purine replaces the adenine to be methylated, small but significant methylation is observed [15]; <4>, methylation of DNA in a sequence specific manner, low substrate specificity with respect to the target base. Cytosine residues can be methylated if they are located in a C:T mismatch base pair at the target position of the enzyme, modification of cytosine residues at position N4 [16]) (Reversibility: ? <1-8> [1-17]) [1-17]

P S-adenosyl-L-homocysteine + DNA 6-methylaminopurine <1-8> [1-17]

Inhibitors

EDTA <1> [2]
NEM <2> [4]
S-adenosyl-L-homocysteine <2, 6> (<2>, competitive [4]; <2>, product inhibition [6]) [4, 6, 14]

methylated DNA <2> (<2>, competitive [4]) [4]
sinefungin <2> [6]
Additional information <2, 6> (<2>, inhibited at ionic strength greater than 0.2 M [4]; <2>, reduction of methylation rate at high concentrations of the substrate 20-mer DNA duplex [6]; <6>, enzyme is inhibited at ionic strength greater than 0.2 M [14]) [4, 6, 14]

Activating compounds

S-adenosyl-L-methionine <2> (<2>, stimulates at increasing concentrations [6]) [6]

Metals, ions

Ca^{2+} <1> (<1>, 2-3 mM required [2]) [2]
Mg^{2+} <1> (<1>, 2-3 mM required [2]; <2>, no requirement for Mg^{2+} [4]) [2]

Turnover number (min^{-1})

Additional information <2, 4, 6> (<2>, turnover-numbers of synthetic oligonucleotide substrates containing the native recognition site GATC or its modified variants [5]; <4>, turnover numbers for the canonical 14-mer duplex and various substituted duplexes [9]; <2>, turnover-numbers of the canonical 12-mer and 20-mers and for various defective duplexes [10]; <6>, residues at position 20 and 26 are responsible for the different turnover-number values of the two MTases bor both canonical and noncanonical sites [14]) [5, 9, 10, 14]

Specific activity (U/mg)

Additional information <2, 6> [4, 14]

K_m-Value (mM)

0.0000000011 <2, 6> (T4 gt- dam- DNA, <2,6>, nonglucosylated unmethylated T4 gt- dam- DNA [4,14]) [4, 14]
0.0001 <2, 6> (S-adenosyl-L-methionine) [4, 14]
Additional information <2, 4> (<2>, K_m-values for oligonucleotide substrates containing the native or modified recognition site: 5.3-12.9 nM [7]; <4>, turnover numbers for the canonical 14-mer duplex and various substituted duplexes [9]; <2>, the T4 Dam methylation reaction cannot be explained by a simple Michaelian scheme [10]; <4>, K_m-values fpr 14-mers and various substituted duplexes [15]) [7, 9, 10, 15]

K_i-Value (mM)

0.0000000046 <2> (methylated DNA) [4]
0.0024 <2, 6> (S-adenosyl-L-homocysteine) [4, 14]

pH-Optimum

7-8.5 <6> [14]
7.5-8 <1, 2> [2, 4]

pH-Range

2.8-9 <1> (<1>, pH 2.8: about 65% of maximal activity, pH 9.0: about 45% of maximal activity [2]) [2]

4 Enzyme Structure

Molecular weight

30690 <2> (<2>, gel filtration, glycerol gradient centrifugation [4]) [4]
Additional information <5> (<5>, sequence analysis of the gene for N^6-methyladenosine DNA methylase [17]) [17]

Subunits

? <1, 2> (<2>, x * 30400, SDS-PAGE [4]; <1>, x * 39000, SDS-PAGE [2]) [2, 4]

5 Isolation/Preparation/Mutation/Application

Source/tissue

coleoptile <1> [2]
seedling <1> [2]

Localization

vacuole <1> [2]

Purification

<1> [2]
<2> [4]
<3> [8]
<6> [14]

Cloning

<2> (the enzyme is subcloned into the plasmid expression vector pJW2 [4]) [4]
<3> (subcloned into pET21b overexpression in Escherichia coli [8]) [8]
<6> (cloning of T2 dam gene into expression vector pJW2, creation of T2-T4 hybrid genes in expression vector pJW2 [14]) [14]

6 Stability

Temperature stability

4 <3> (<3>, fully active [8]) [8]
30 <3> (<3>, thermally inactivated within 20 min [8]) [8]

References

[1] Roberts, R.J.; Vincze, T.; Posfai, J.; Macelis, D.: REBASE: restriction enzymes and methyltransferases. Nucleic Acids Res., **31**, 418-420 (2003)
[2] Fedoreyeva, L.I.; Vanyushin, B.F.: N^6-Adenine DNA-methyltransferase in wheat seedlings. FEBS Lett., **514**, 305-308 (2002)

[3] Malygin, E.G.; Evdokimov, A.A.; Zinoviev, V.V.; Ovechkina, L.G.; Lindstrom, W.M.; Reich, N.O.; Schlagman, S.L.; Hattman, S.: A dual role for substrate S-adenosyl-L-methionine in the methylation reaction with bacteriophage T4 Dam DNA-[N^6-adenine]-methyltransferase. Nucleic Acids Res., **29**, 2361-2369 (2001)

[4] Kossykh, V.G.; Schlagman, S.L.; Hattman, S.: Phage T4 DNA [N^6-adenine]-methyltransferase. Overexpression, purification, and characterization. J. Biol. Chem., **270**, 14389-14393 (1995)

[5] Malygin, E.G.; Ovechkina, L.G.; Evdokimov, A.A.; Zinoviev, V.V.: Single turnover kinetics of methylation by T4 DNA-(N^6-adenine)-methyltransferase. Mol. Biol., **35**, 56-68 (2001)

[6] Evdokimov, A.A.; Zinoviev, V.V.; Malygin, E.G.; Schlagman, S.L.; Hattman, S.: Bacteriophage T4 Dam DNA-[N^6-adenine]methyltransferase. Kinetic evidence for a catalytically essential conformational change in the ternary complex. J. Biol. Chem., **277**, 279-286 (2002)

[7] Malygin, E.G.; Lindstrom, W.M., Jr.; Schlagman, S.L.; Hattman, S.; Reich, N.O.: Pre-steady state kinetics of bacteriophage T4 Dam DNA-[N^6-adenine] methyltransferase: interaction with native (GATC) or modified sites. Nucleic Acids Res., **28**, 4207-4211 (2000)

[8] Berdis, A.J.; Lee, I.; Coward, J.K.; Stephens, C.; Wright, R.; Shapiro, L.; Benkovic, S.J.: A cell cycle-regulated adenine DNA methyltransferase from Caulobacter crescentus processively methylates GANTC sites on hemimethylated DNA. Proc. Natl. Acad. Sci. USA, **95**, 2874-2879 (1998)

[9] Thielking, V.; Du Bois, S.; Eritja, R.; Guschlbauer, W.: Dam methyltransferase from Escherichia coli. Kinetic studies using modified DNA oligomers. Nonmethylated substrates. Biol. Chem., **378**, 407-415 (1997)

[10] Zinoviev, V.V.; Evdokimov, A.A.; Gorbunov, Y.A.; Malygin, E.G.; Kossykh, V.G.; Hattman, S.: Phage T4 DNA [N^6-adenine] methyltransferase. Kinetic studies using oligonucleotides containing native or modified recognition sites. Biol. Chem., **379**, 481-488 (1998)

[11] Petrov, N.A.; Gorbunov, Y.A.; Malygin, E.G.: Interaction of T4 phage DNA-[N^6-adenine]-methyltransferase with substrates containing defective recognition sites. Mol. Biol., **31**, 827-831 (1997)

[12] Malygin, E.G.; Ovechkina, L.G.; Evdokimov, A.A.; Zinoviev, V.V.: Single turnover kinetics of methylation by T4 DNA-(N^6-adenine)-methyltransferase. Mol. Biol., **35**, 56-68 (2001)

[13] Degtyarev, S.K.; Prikhodko, E.A.; Rechkunova, N.I.; Prikhodko, G.G.; Krasnykh, V.N.: Biochemical characterization of VspI methyltransferase. Gene, **157**, 65-66 (1995)

[14] Kossykh, V.G.; Schlagman, S.L.; Hattman, S.: Comparative studies of the phage T2 and T4 DNA (N^6-adenine)methyltransferases: amino acid changes that affect catalytic activity. J. Bacteriol., **179**, 3239-3243 (1997)

[15] Marzabal, S.; DuBois, S.; Thielking, V.; Cano, A.; Eritja, R.; Guschlbauer, W.: Dam methylase from Escherichia coli: kinetic studies using modified DNA oligomers: hemimethylated substrates. Nucleic Acids Res., **23**, 3648-3655 (1995)

[16] Jeltsch, A.; Christ, F.; Fatemi, M.; Roth, M.: On the substrate specificity of DNA methyltransferases. Adenine-N6 DNA methyltransferases also modify cytosine residues at position N4. J. Biol. Chem., **274**, 19538-19544 (1999)
[17] Degtyarev, S.K.; Prikhodæko, E.A.; Prikhodæko, G.G.; Krasnykh, V.N.: VspI methylase belongs to m^6A-γ class of adenine methylase. Nucleic Acids Res., **21**, 2015 (1993)

Site-specific DNA-methyltransferase (cytosine-specific) 2.1.1.73

1 Nomenclature

EC number

2.1.1.73 (deleted, reaction is that of EC 2.1.1.37)

Recommended name

site-specific DNA-methyltransferase (cytosine-specific)

Methylenetetrahydrofolate-tRNA-(uracil-5-)-methyltransferase ($FADH_2$-oxidizing) 2.1.1.74

1 Nomenclature

EC number
2.1.1.74

Systematic name
5,10-methylenetetrahydrofolate:tRNA-UΨC (uracil-5-)-methyl-transferase

Recommended name
methylenetetrahydrofolate-tRNA-(uracil-5-)-methyltransferase ($FADH_2$-oxidizing)

Synonyms
EC 2.1.2.12 (formerly)
folate-dependent ribothymidyl synthase
methylenetetrahydrofolate-transfer ribonucleate uracil 5-methyltransferase,

CAS registry number
56831-74-4 (74665-78-4 deleted)

2 Source Organism

<-1> no activity in *Micrococcus luteus* (strains ATCC 4698 and ATCC12698, formerly Micrococcus lysodeikticus [1]) [1]
<1> *Bacillus cereus* [2]
<2> *Bacillus subtilis* [1-3]
<3> *Streptococcus faecalis* [1-3]

3 Reaction and Specificity

Catalyzed reaction
5,10-methylenetetrahydrofolate + tRNA UΨC + $FADH_2$ = tetrahydrofolate + tRNA TΨC + FAD

Reaction type
methyl group transfer
reductive methylation

Natural substrates and products
S 5,10-methylenetetrahydrofolate + tRNA UΨC + $FADH_2$ <1-3> (<2,3> post-transcriptional biosynthesis of ribothymidine in the tRNA [1,3]) (Reversibility: ? <1-3> [1-3]) [1-3]
P tetrahydrofolate + tRNA TΨC + FAD^+ <2, 3> [3]

Substrates and products

S 5,10-methylenetetrahydrofolate + tRNA UΨC + $FADH_2$ <1-3> (Reversibility: ? <1-3> [1-3]) [1-3]

P tetrahydrofolate + tRNA TΨC + FAD^+ <3> [3]

Cofactors/prosthetic groups

$FADH_2$ <2, 3> [2, 3]
NADPH <2, 3> [1]

Specific activity (U/mg)

0.1 <3> [2, 3]

K_m-Value (mM)

0.5 <3> (5,10-methylenetetrahydrofolate) [3]

pH-Optimum

9 <2, 3> [1]

5 Isolation/Preparation/Mutation/Application

Localization

cytoplasm <3> [1]

Purification

<3> [2, 3]

6 Stability

General stability information

<3>, enzyme requires FAD^+ for stability [2]

References

[1] Delk, A.S.; Romeo, J.M.; Nagle, D.P.; Rabinowitz, J.C.: Biosynthesis of ribothymidine in the transfer RNA of Streptococcus faecalis and Bacillus subtilis. A methylation of RNA involving 5,10-methylenetetrahydrofolate. J. Biol. Chem., **251**, 7649-7656 (1976)

[2] Delk, A.S.; Nagle, D.P.; Rabinowitz, J.C.: The methylenetetrahydrofolate-mediated biosynthesis of ribothymidine in the transfer-RNA of Streptococcus faecalis: incorporation of hydrogen from solvent into the methyl moiety. Biochem. Biophys. Res. Commun., **86**, 244-251 (1979)

[3] Delk, A.S.; Nagle, D.P.; Rabinowitz, J.C.: Methylenetetrahydrofolate-dependent biosynthesis of ribothymidine in transfer RNA of Streptococcus faecalis. Evidence for reduction of the 1-carbon unit by $FADH_2$. J. Biol. Chem., **255**, 4387-4390 (1980)

Apigenin 4'-O-methyltransferase 2.1.1.75

1 Nomenclature

EC number
2.1.1.75

Systematic name
S-adenosyl-L-methionine:5,7,4'-trihydroxyflavone 4'-O-methyltransferase

Recommended name
apigenin 4'-O-methyltransferase

Synonyms
flavonoid O-methyltransferase
methyltransferase, flavonoid

CAS registry number
118251-36-8

2 Source Organism

<1> *Robinia pseudoacacia* [1]

3 Reaction and Specificity

Catalyzed reaction
S-adenosyl-L-methionine + 5,7,4'-trihydroxyflavone = S-adenosyl-L-homocysteine + 4'-methoxy-5,7-dihydroxyflavone

Reaction type
methyl group transfer
para-O-methylation

Natural substrates and products
S S-adenosyl-L-methionine + 5,7,4'-trihydroxyflavone <1> (<1>, enzyme is involved in biosynthesis of 4'-methoxylated flavonoids [1]) [1]
P S-adenosyl-L-homocysteine + 4'-methoxy-5,7-dihydroxyflavone

Substrates and products
S S-adenosyl-L-methionine + 5,7,4'-trihydroxyflavone <1> (<1>, i.e. apigenin [1]) (Reversibility: ? <1> [1]) [1]
P S-adenosyl-L-homocysteine + 4'-methoxy-5,7-dihydroxyflavone <1> (<1>, i.e. acacetin [1]) [1]

Inhibitors

S-adenosyl-L-homocysteine <1> (<1>, 0.1 mM, 47% inhibition [1]) [1]

Activating compounds

2-mercaptoethanol <1> (<1>, 5 mM, required for optimum activity [1]) [1]

K_m-Value (mM)

0.031 <1> (S-adenosyl-L-methionine) [1]
0.05 <1> (apigenin) [1]

pH-Optimum

9 <1> (<1>, glycine-NaOH buffer [1]) [1]

pH-Range

7.5-10 <1> (<1>, about 50% of activity maximum at pH 7.5 and 10, glycine-NaOH buffer [1]) [1]

5 Isolation/Preparation/Mutation/Application

Source/tissue

callus <1> [1]
seedling <1> [1]
shoot <1> [1]

References

[1] Kuroki, G.; Poulton, J.E.: The *para*-O-methylation of apigenin to acacetin by cell-free extracts of Robinia pseudoacacia L.. Z. Naturforsch. C, **36c**, 916-920 (1981)

Quercetin 3-O-methyltransferase 2.1.1.76

1 Nomenclature

EC number
2.1.1.76

Systematic name
S-adenosyl-L-methionine:3,5,7,3',4'-pentahydroxyflavone 3-O-methyltransferase

Recommended name
quercetin 3-O-methyltransferase

Synonyms
3-OMT
S-adenosyl-L-methionine:quercetin 3-O-methyltransferase <3> [2]
flavonoid 3-methyltransferase
flavonol 3-O-methyltransferase
trimethylfavonol 3'/5'-OMT <3> [6]

CAS registry number
75603-21-3

2 Source Organism

<1> *Arabidopsis thaliana* [6]
<2> *Arabidopsis thaliana* (cDNA clone [6]) [6]
<3> *Chrysosplenium americanum* (Schwein ex Hooker [5]) [1-6]
<4> *Citrofortunella microcarpa* (calamondin orange, synonym Citrofortunella mitis [1,3]) [1, 3]
<5> *Nicotiana tabacum* (tobacco [1]) [1]

3 Reaction and Specificity

Catalyzed reaction
S-adenosyl-L-methionine + 3,5,7,3',4'-pentahydroxyflavone = S-adenosyl-L-homocysteine + 3-methoxy-5,7,3',4'-tetrahydroxyflavone (specific for quercetin, related enzymes bring about the 3-O-methylation of other flavonols, such as galangin and kaempferol)

Reaction type
methyl group transfer

Natural substrates and products

S S-adenosyl-L-methionine + 3,5,7,3',4'-pentahydroxyflavone <1-5> (<3> involved in metabolism of secondary plant products, especially formation of lignin and detoxification of flavonoids [2]; <3> enzymatic synthesis of polymethylated flavonol glucosides [5]; <2> involved in lignin biosynthesis [6]) (Reversibility: ? <1-5> [1-6]) [1-6]

P S-adenosyl-L-homocysteine + 3-methoxy-5,7,3',4'-tetrahydroxyflavone <3-5> (<3> 3-O-methylquercetin [2,3]) [1-3]

Substrates and products

S S-adenosyl-L-methionine + 3,5,7,3',4'-pentahydroxyflavone <1-5> (<4> quercetin, twice efficient methyl acceptor as galangin [1]; <2> flavonol quercetin is the preferred substrate [6]) (Reversibility: ? <1-5> [1-6]) [1-6]

P S-adenosyl-L-homocysteine + 3-methoxy-5,7,3',4'-tetrahydroxyflavone <3-5> [1]

S S-adenosyl-L-methionine + 7-methylquercetin <3, 3> (<3> rhamnetin, 30% of the activity with quercetin [3]) (Reversibility: ? <3> [3]) [3]

P ?

S S-adenosyl-L-methionine + galangin <4, 5> (<5> utilizes galangin more efficiently than quercetin [1]) (Reversibility: ? <4, 5> [1]) [1]

P S-adenosyl-L-homocysteine + 3'-O-methyl-galangin <4> [1]

S S-adenosyl-L-methionine + myricetin <2> (Reversibility: ? <2> [6]) [6]

P S-adenosyl-L-homocysteine + ?

Inhibitors

3,6,7,3'-tetramethylquercetagetin <3> (<3> noncompetitive inhibitor [4]) [4]
3-methylquercetin <3> (<3> noncompetitive inhibitor with respect to both S-adenosyl-L-methionine and quercetin [4]) [4]
N-ethylmaleimide <3> [3]
S-adenosyl-L-homocysteine <3> (<3> competitive inhibitor with respect to S-adenosyl-L-methionine [4]) [4]
p-chloromercuribenzoate <3> [3]

Activating compounds

2-mercaptoethanol <3> (<3> relative activity 105% [3]) [3]
iodoacetamide <3> (<3> relative activity 105% [3]) [3]

Metals, ions

K^+ <3> (<3> relative activity 115% [3]) [3]
Mg^{2+} <3> (<3> required for activity [1,2]) [1, 2]
Mn^{2+} <3> (<3> relative activity 110% [3]) [3]

Specific activity (U/mg)

Additional information <3> (<3> 53 pkat/mg [3]) [3]

K_m-Value (mM)

0.0014 <3> (quercetin) [1, 2]
0.07 <3> (S-adenosyl-L-methionine) [1, 2]

pH-Optimum
7.5-8.5 <3> [1, 2]
7.8-8.2 <3> [3]

pH-Range
7.2-9.3 <3> (<3> about 50% of activity maximum at pH 7.2 and 9.3 [3]) [3]

4 Enzyme Structure

Molecular weight
57000 <3> (<3> gel filtration [3]) [3]
65000 <3> (<3> gel filtration [1-3]) [1-3]

5 Isolation/Preparation/Mutation/Application

Source/tissue
cell culture <5> [1]
leaf <3> [5]
peel <4> [1]
root <4> [1]
shoot <3> [1]

Purification
<2> (recombinant protein, expressed in E. coli BL21 [6]) [6]
<3> (partially [2-4]) [1-5]
<4> (partially [3]) [3]

Cloning
<2> (pET-19b vector, cDNA clone overexpressed in Escherichia coli BL21) [6]

6 Stability

Storage stability
<3>, -15°C, purified enzyme, 0.2 M potassium phosphate, pH 8.0, 20% glycerol, 5 mM dithiothreitol, stable for several months [3]
<3>, 4°C, partially purified enzyme loses 80% of its activity over a period of 4-6 weeks [3]

References

[1] De Luca, V.; Brunet, G.; Khouri, H.; Ibrahim, R.: Flavonol 3-O-methyltransferase in plant tissues. Z. Naturforsch. C, **37c**, 134-135 (1982)
[2] Ibrahim, R.K.; De Luca, V.: Polymethylated flavonol synthesis is catalysed by distinct O-methyltranferases. Naturwissenschaften, **69**, 41-42 (1982)

[3] De Luca, V.; Ibrahim, R.K.: Enzymatic synthesis of polymethylated flavonols in Chrysosplenium americanum. I. Partial purification and some properties of S-adenosyl-L-methionine:flavonol 3-, 6-, 7-, and 4-O-methyltransferases. Arch. Biochem. Biophys., **238**, 596-605 (1985)
[4] De Luca, V.; Ibrahim, R.K.: Enzymatic synthesis of polymethylated flavonols in Chrysosplenium americanum. II. Substrate interaction and product inhibition studies of flavonol 3-, 6-, and 4-O-methyltransferases. Arch. Biochem. Biophys., **238**, 606-618 (1985)
[5] Khouri, H.E.; Ibrahim, R.K.: Resolution of five position-specific flavonoid O-methyltransferases by fast protein liquid chromatofocusing. J. Chromatogr., **407**, 291-297 (1987)
[6] Muzac, I.; Wang, J.; Anzellotti, D.; Zhang, H.; Ibrahim, R.K.: Functional expression of an Arabidopsis cDNA clone encoding a flavonol 3'-O-methyltransferase and characterization of the gene product. Arch. Biochem. Biophys., **375**, 385-388 (2000)

Protein-L-isoaspartate (D-aspartate) O-methyltransferase 2.1.1.77

1 Nomenclature

EC number
2.1.1.77

Systematic name
S-adenosyl-L-methionine:protein-L-isoaspartate O-methyltransferase

Recommended name
protein-L-isoaspartate (D-aspartate) O-methyltransferase

Synonyms
D-aspartyl/L-isoaspartyl methyltransferase
EC 2.1.1.24 (formerly)
HPIMT <13> [31]
IAMT <2, 10> [17]
L-aspartyl/L-isoaspartyl protein methyltransferase
L-isoaspartyl/D-aspartyl protein carboxyl methyltransferase
PCMT <5, 12, 13, 21> [1, 9, 15, 18, 21]
PCMT1 <13, 18> [23, 27]
PIMT <9, 13, 18, 21, 23> [16, 19, 22, 29, 31]
methyltransferase, protein (D-aspartate)
protein (L-isoaspartate) O-methyltransferase
protein D-aspartate methyltransferase
protein L-isoaspartate (D-aspartate) O-methyltransferase <6, 13> [23, 24]
protein L-isoaspartate methyltransferase
protein L-isoaspartyl methyltransferase
protein carboxyl methyltransferase type II <13> [3, 10]
protein repair L-isoaspartyl methyltransferase <2> [25]
protein-L-isoaspartate O-methyltransferase
protein-β-aspartate O-methyltransferase
type II methyltransferase <10, 13> [8]

CAS registry number
105638-50-4

2 Source Organism

<-8> no activity in *Bacillus subtilis* [20]
<-7> no activity in *Treponema pallidum* [20]
<-6> no activity in *Sinorhizobium meliloti* [20]

<-5> no activity in *Mycoplasma pneumoniae* [20]
<-4> no activity in *Mycoplasma genitalium* [20]
<-3> no activity in *Haemophilus influenzae* [20]
<-2> no activity in *Borellia burgdorferi* [20]
<-1> no activity in *Aquifex aeolicus* [20]
<1> *Agaricus bisporus* (mushroom [11]) [11]
<2> *Arabidopsis thaliana* (thale cress [17,25,28,32]; ecotype Columbia [17]) [17, 25, 28, 32]
<3> *Archaeoglobus fulgidus* [20]
<4> *Armadillidium pulchellum* (pill woodlouse) [11]
<5> *Bos taurus* (bovine [1,2,4-8,10,15,16]) [1, 2, 4-8, 10, 15, 16]
<6> *Caenorhabditis elegans* (strain N2 [28]) [20, 24, 25, 28, 32]
<7> *Cuniculus sp.* (rabbit [5,10]; New Zealand White [15]) [5, 10, 15]
<8> *Daucus carota* (carrot, var. Danvers Half Long [28]) [28]
<9> *Drosophila melanogaster* (fruit fly [22]) [10, 22, 25]
<10> *Escherichia coli* [4, 6, 8, 10, 16, 17, 20, 21, 24, 25, 28, 32]
<11> *Equus sp.* (horse [5]) [5]
<12> *Gallus gallus* (chicken [15,18]) [15, 18]
<13> *Homo sapiens* (human [1,3-8,10,12-15,17,20,21,23-25,27,28,30-32]; HeLa cells [27]) [1, 3-8, 10, 12-15, 17, 20, 21, 23-25, 27, 28, 30-32]
<14> *Limax maximus* (great slug [11]) [11]
<15> *Lycopersicon esculentum* (tomato, Mill. [25]) [25]
<16> *Methanobacterium thermoautotrophicum* [20]
<17> *Methanococcus jannaschii* [20]
<18> *Mus musculus* [2, 10, 17, 20, 23, 24, 27, 29, 32]
<19> *Oryza sativa* (rice [25,28]; strain M-101 [28]) [25, 28]
<20> *Pyrococcus furiosus* [26, 32]
<21> *Rattus norvegicus* (rat, pheochromocytoma PC-12 line [2,29]; Sprague-Dawley [9,15]) [2, 5, 6, 9, 10, 15, 16, 19, 20, 22, 25, 29]
<22> *Salmonella typhimurium* [2, 4, 6, 8, 10]
<23> *Thermotoga maritima* [20, 25, 31, 32]
<24> *Torpedo ocellata* [5, 15]
<25> *Triticum aestivum* (wheat [11,15,17,20,25,28]; winter wheat, cv Augusta [28]) [11, 15, 17, 20, 25, 28]
<26> *Xenopus laevis* [2, 5, 15, 25, 27]
<27> *Zea mays* (maize, corn [25,28]; hybrids White Knight, Kandy Korn and Silver Queen [28]) [25, 28]

3 Reaction and Specificity

Catalyzed reaction

S-adenosyl-L-methionine + protein L-isoaspartate = S-adenosyl-L-homocysteine + protein L-isoaspartate α-methyl ester (D-aspartate (but not L-aspartate) residues in proteins can also act as acceptors)

Reaction type

methyl group transfer

Natural substrates and products

S S-adenosyl-L-methionine + protein L-isoaspartate <1-10, 12-27> (<1,4,14,25> implicated in the repair or degradation of age-damaged proteins that contain atypical L-isoaspartyl residues [11]; <5,18,21,22,26> key role in degradation or repair of deamination damaged proteins [2]; <21> post-translational covalent modification [9]; <13> involved in the repair of age-damaged proteins, aspartyl and asparginyl residues that are susceptible to deamidation, isomerization and racemization [23]; <10> repair and/or degradation of spontaneously damaged proteins [8]; <9> function in protein repair [22]; <23> highly active protein repair enzyme [20]; <20> protein repair methyltransferase [26]; <2> repair enzyme, methylates abnormal L-isoaspartate residues in proteins which arise spontaneously as a result of aging [25]) (Reversibility: ? <1-4, 6-10, 12-20, 22-27> [2-4, 8, 10-15, 19, 20, 22-26, 28, 30, 31]; r <5, 21> [1, 9]) [1-32]

P S-adenosyl-L-homocysteine + protein L-isoaspartate α-methyl ester <1-10, 12-27> [2-4, 8, 10-15, 19, 20, 22-26, 28, 30, 31]

Substrates and products

S S-adenosyl-L-methionine + A-isoD-QLTEEQIAEFK <13> (Reversibility: ? <13> [12]) [12]

P S-adenosyl-L-homocysteine + ?

S S-adenosyl-L-methionine + A-isoD-T <13> (Reversibility: ? <13> [12]) [12]

P S-adenosyl-L-homocysteine + ?

S S-adenosyl-L-methionine + AA-isoD-F-amide <13> (Reversibility: ? <13> [12]) [12]

P S-adenosyl-L-homocysteine + ?

S S-adenosyl-L-methionine + AA-isoD-amide <13> (Reversibility: ? <13> [12]) [12]

P S-adenosyl-L-homocysteine + ?

S S-adenosyl-L-methionine + AL-isoD-T <13> (Reversibility: ? <13> [13]) [13]

P S-adenosyl-L-homocysteine + ?

S S-adenosyl-L-methionine + AP-isoD-amide <13> (Reversibility: ? <13> [12]) [12]

P S-adenosyl-L-homocysteine + ?

S S-adenosyl-L-methionine + Ala-Asp <13> (Reversibility: ? <13> [12]) [12]

P S-adenosyl-L-homocysteine + ?

S S-adenosyl-L-methionine + G-isoD-G <13> (Reversibility: ? <13> [12]) [12]

P S-adenosyl-L-homocysteine + ?

S S-adenosyl-L-methionine + GFDL-isoD-GGGVG <5, 13> (Reversibility: ? <5, 13> [7, 12]) [7, 12]
P S-adenosyl-L-homocysteine + ?
S S-adenosyl-L-methionine + KASA (D-Asp) LAKY <13, 20> (Reversibility: ? <13, 20> [13, 32]) [13, 32]
P S-adenosyl-L-homocysteine + ?
S S-adenosyl-L-methionine + KASA (isoD) LAKY <5, 6, 10, 13, 23, 25> (<6,10,13,23,25> synthetic methyl acceptor [20]) (Reversibility: ? <5, 6, 10, 13, 23, 25> [7, 20]) [7, 20]
P S-adenosyl-L-homocysteine + ?
S S-adenosyl-L-methionine + KASA-L-isoAsp-LAKY <2, 13, 19, 20, 25, 27> (Reversibility: ? <2, 13, 19, 20, 25, 27> [13, 25, 28, 32]) [13, 25, 28, 32]
P S-adenosyl-L-homocysteine + ?
S S-adenosyl-L-methionine + KMK-isoD-SEEEIR <13> (Reversibility: ? <13> [12]) [12]
P S-adenosyl-L-homocysteine + ?
S S-adenosyl-L-methionine + KMK-isoD-TDSEEEIR <13> (Reversibility: ? <13> [12, 14]) [12, 14]
P S-adenosyl-L-homocysteine + ?
S S-adenosyl-L-methionine + KQVV-isoD-SAYEVIK <5, 13> (Reversibility: ? <5, 13> [7, 12]) [7, 12]
P S-adenosyl-L-homocysteine + ?
S S-adenosyl-L-methionine + KVTCK-isoD-GQTNCYQSK <13> (Reversibility: ? <13> [12]) [12]
P S-adenosyl-L-homocysteine + ?
S S-adenosyl-L-methionine + LM-isoD-T <13> (Reversibility: ? <13> [12]) [12]
P S-adenosyl-L-homocysteine + ?
S S-adenosyl-L-methionine + LSH-isoD-GY-(3-sulfo)Ala-L <13> (Reversibility: ? <13> [7, 12]) [7, 12]
P S-adenosyl-L-homocysteine + ?
S S-adenosyl-L-methionine + LSH-isoD-GYCL <13> (Reversibility: ? <13> [7, 12]) [7, 12]
P S-adenosyl-L-homocysteine + ?
S S-adenosyl-L-methionine + N-acetyl-Asn <13> (Reversibility: ? <13> [12]) [12]
P S-adenosyl-L-homocysteine + ?
S S-adenosyl-L-methionine + N-acetyl-L-Asp <13> (Reversibility: ? <13> [12]) [12]
P S-adenosyl-L-homocysteine + ?
S S-adenosyl-L-methionine + N-succinyl-AAA-*p*-nitroanilide <20> (Reversibility: ? <20> [32]) [32]
P S-adenosyl-L-homocysteine + ?

S S-adenosyl-L-methionine + N-succinyl-AAPF-*p*-nitroanilide <20> (Reversibility: ? <20> [32]) [32]
P S-adenosyl-L-homocysteine + ?
S S-adenosyl-L-methionine + N-succinyl-AAPL-*p*-nitroanilide <20> (Reversibility: ? <20> [32]) [32]
P S-adenosyl-L-homocysteine + ?
S S-adenosyl-L-methionine + N-succinyl-AAV-*p*-nitroanilide <20> (Reversibility: ? <20> [32]) [32]
P S-adenosyl-L-homocysteine + ?
S S-adenosyl-L-methionine + N-succinyl-AAVA-*p*-nitroanilide <13, 20> (Reversibility: ? <13, 20> [32]) [32]
P S-adenosyl-L-homocysteine + ?
S S-adenosyl-L-methionine + T1-isoAsp-12 <13> (Reversibility: ? <13> [14]) [14]
P S-adenosyl-L-homocysteine + ?
S S-adenosyl-L-methionine + T4-isoAsp-38 <13> (Reversibility: ? <13> [14]) [14]
P S-adenosyl-L-homocysteine + ?
S S-adenosyl-L-methionine + TS-isoD-YSKY <13> (Reversibility: ? <13> [12]) [12]
P S-adenosyl-L-homocysteine + ?
S S-adenosyl-L-methionine + V2-isoAsp-12 <13> (Reversibility: ? <13> [14]) [14]
P S-adenosyl-L-homocysteine + ?
S S-adenosyl-L-methionine + V4-isoAsp-38 <13> (Reversibility: ? <13> [14]) [14]
P S-adenosyl-L-homocysteine + ?
S S-adenosyl-L-methionine + VYG-L-isoAsp-PA <13> (Reversibility: ? <13> [13]) [13]
P S-adenosyl-L-homocysteine + ?
S S-adenosyl-L-methionine + VYG-isoD-PA <13> (Reversibility: ? <13> [12]) [12]
P S-adenosyl-L-homocysteine + ?
S S-adenosyl-L-methionine + VYP (isoD) HA <6, 10, 13, 23, 25> (Reversibility: ? <6, 10, 13, 23, 25> [12, 20]) [12, 20]
P S-adenosyl-L-homocysteine + ?
S S-adenosyl-L-methionine + VYP-D-Asp-GA <13> (Reversibility: ? <13> [13]) [13]
P S-adenosyl-L-homocysteine + ?
S S-adenosyl-L-methionine + VYP-L-isoAsp-CA <2> (Reversibility: ? <2> [25]) [25]
P S-adenosyl-L-homocysteine + ?

S S-adenosyl-L-methionine + VYP-L-isoAsp-GA <13> (Reversibility: ? <13> [13, 14]) [13, 14]
P S-adenosyl-L-homocysteine + ?
S S-adenosyl-L-methionine + VYP-L-isoAsp-GHG <2> (Reversibility: ? <2> [25]) [25]
P S-adenosyl-L-homocysteine + ?
S S-adenosyl-L-methionine + VYP-L-isoAsp-HA <2, 13, 19, 20, 25, 27> (Reversibility: ? <2, 13, 19, 20, 25, 27> [13, 25, 28, 32]) [13, 25, 28, 32]
P S-adenosyl-L-homocysteine + ?
S S-adenosyl-L-methionine + VYP-L-isoAsp-RR <2> (Reversibility: ? <2> [25]) [25]
P S-adenosyl-L-homocysteine + ?
S S-adenosyl-L-methionine + VYP-isoD-AA <13> (Reversibility: ? <13> [12]) [12]
P S-adenosyl-L-homocysteine + ?
S S-adenosyl-L-methionine + VYP-isoD-CA <13> (Reversibility: ? <13> [7, 12]) [7, 12]
P S-adenosyl-L-homocysteine + ?
S S-adenosyl-L-methionine + VYP-isoD-Cya-G <13> (Reversibility: ? <13> [7]) [7]
P S-adenosyl-L-homocysteine + ?
S S-adenosyl-L-methionine + VYP-isoD-DA <13> (Reversibility: ? <13> [7, 12]) [7, 12]
P S-adenosyl-L-homocysteine + ?
S S-adenosyl-L-methionine + VYP-isoD-GA <5, 13> (Reversibility: ? <5, 13> [7, 12]) [7, 12]
P S-adenosyl-L-homocysteine + ?
S S-adenosyl-L-methionine + VYP-isoD-LA <13> (Reversibility: ? <13> [12]) [12]
P S-adenosyl-L-homocysteine + ?
S S-adenosyl-L-methionine + VYP-isoD-PA <13> (Reversibility: ? <13> [12]) [12]
P S-adenosyl-L-homocysteine + ?
S S-adenosyl-L-methionine + VYP-isoD-SA <13> (Reversibility: ? <13> [12]) [12]
P S-adenosyl-L-homocysteine + ?
S S-adenosyl-L-methionine + VYP-isoD-(3-sulfo)Ala <13> (Reversibility: ? <13> [12]) [12]
P S-adenosyl-L-homocysteine + ?
S S-adenosyl-L-methionine + VYR-isoD-RR <13> (Reversibility: ? <13> [12]) [12]
P S-adenosyl-L-homocysteine + ?

S S-adenosyl-L-methionine + WAGG-isoD-ASGE <13> (Reversibility: ? <13> [12]) [12]
P S-adenosyl-L-homocysteine + ?
S S-adenosyl-L-methionine + WM-D-Asp-F-amide <13> (Reversibility: ? <13> [13]) [13]
P S-adenosyl-L-homocysteine + ?
S S-adenosyl-L-methionine + WM-L-isoAsp-F-amide <13> (Reversibility: ? <13> [13]) [13]
P S-adenosyl-L-homocysteine + ?
S S-adenosyl-L-methionine + WM-isoD-F-NH_2 <13> (Reversibility: ? <13> [12]) [12]
P S-adenosyl-L-homocysteine + ?
S S-adenosyl-L-methionine + YVS-isoAsp-GHG <13> (Reversibility: ? <13> [12, 14]) [12, 14]
P S-adenosyl-L-homocysteine + ?
S S-adenosyl-L-methionine + YVS-isoD-G-(3-sulfo)Ala-G <13> (Reversibility: ? <13> [7]) [7]
P S-adenosyl-L-homocysteine + ?
S S-adenosyl-L-methionine + YVS-isoD-GCG <13> (Reversibility: ? <13> [7, 12]) [7, 12]
P S-adenosyl-L-homocysteine + ?
S S-adenosyl-L-methionine + YVS-isoD-GDG <13> (Reversibility: ? <13> [7, 12]) [7, 12]
P S-adenosyl-L-homocysteine + ?
S S-adenosyl-L-methionine + YVS-isoD-G-(3-sulfo)Ala-G <13> (Reversibility: ? <13> [12]) [12]
P S-adenosyl-L-homocysteine + ?
S S-adenosyl-L-methionine + acetyl-G-isoD-G <13> (Reversibility: ? <13> [12]) [12]
P S-adenosyl-L-homocysteine + ?
S S-adenosyl-L-methionine + acetyl-isoD-G <13> (Reversibility: ? <13> [12]) [12]
P S-adenosyl-L-homocysteine + ?
S S-adenosyl-L-methionine + β-101-lysozyme <13> (Reversibility: ? <13> [7, 12]) [7, 12]
P S-adenosyl-L-homocysteine + ?
S S-adenosyl-L-methionine + calbindin D9k <13> (Reversibility: ? <13> [12]) [12]
P S-adenosyl-L-homocysteine + ?
S S-adenosyl-L-methionine + calmodulin <13, 26> (Reversibility: ? <13, 26> [21, 27]) [21, 27]
P S-adenosyl-L-homocysteine + ?

S S-adenosyl-L-methionine + γ-globulin <2, 5, 21> (Reversibility: r <5> [1]; ? <2, 21> [16, 25]) [1, 16, 25]
P S-adenosyl-L-homocysteine + ?
S S-adenosyl-L-methionine + gelatin I <21> (Reversibility: ? <21> [9]) [9]
P S-adenosyl-L-homocysteine + ?
S S-adenosyl-L-methionine + histidine phosphocarrier protein (isoAsp-38) <13> (Reversibility: ? <13> [14]) [14]
P S-adenosyl-L-homocysteine + ?
S S-adenosyl-L-methionine + histidine phosphocarrier protein-2 (isoAsp-12,-38) <13> (Reversibility: ? <13> [14]) [14]
P S-adenosyl-L-homocysteine + ?
S S-adenosyl-L-methionine + isoACTH <5> (Reversibility: r <5> [1]) [1]
P S-adenosyl-L-homocysteine + ?
S S-adenosyl-L-methionine + isoAspDSIP <21> (Reversibility: ? <21> [16]) [16]
P S-adenosyl-L-homocysteine + ?
S S-adenosyl-L-methionine + isoD-56 calbindin D9k <13> (Reversibility: ? <13> [12]) [12]
P S-adenosyl-L-homocysteine + ?
S S-adenosyl-L-methionine + isoD-G <13> (Reversibility: ? <13> [12]) [12]
P S-adenosyl-L-homocysteine + ?
S S-adenosyl-L-methionine + native lysozyme <13> (Reversibility: ? <13> [7]) [7]
P S-adenosyl-L-homocysteine + ?
S S-adenosyl-L-methionine + ovalbumin <2, 5, 6, 8, 10, 13, 19-21, 23, 25, 27> (Reversibility: ? <2, 5, 6, 8, 10, 13, 19-21, 23, 25, 27> [7-9, 20, 25, 28, 32]) [7-9, 20, 25, 28, 32]
P S-adenosyl-L-homocysteine + ?
S S-adenosyl-L-methionine + protein D-aspartate <2, 5, 10, 13, 20> (Reversibility: ? <2, 5, 10, 13, 20> [3, 7, 8, 10, 12, 13, 25, 32]) [3, 7, 8, 10, 12, 13, 25, 32]
P S-adenosyl-L-homocysteine + protein D-aspartate β-methyl ester <2, 13> [3, 8, 10, 12, 13, 25]
S S-adenosyl-L-methionine + protein L-isoaspartate <1-10, 12-27> (Reversibility: ? <1-4, 6-10, 12-20, 22-27> [1-32]; r <5, 21> [1, 9]) [1-32]
P S-adenosyl-L-homocysteine + protein L-isoaspartate α-methyl ester <1-10, 12-27> [1-32]
S S-adenosyl-L-methionine + succinamide <20> (Reversibility: ? <20> [32]) [32]
P S-adenosyl-L-homocysteine + ?
S S-adenosyl-L-methionine + synapsin I <5, 13> (Reversibility: ? <5, 13> [7]) [7]
P S-adenosyl-L-homocysteine + ?

S S-adenosyl-L-methionine + t-BOC-G-isoD-G <13> (Reversibility: ? <13> [12]) [12]
P S-adenosyl-L-homocysteine + ?
S S-adenosyl-L-methionine + t-BOC-isoD-G <13> (Reversibility: ? <13> [12]) [12]
P S-adenosyl-L-homocysteine + ?
S S-adenosyl-L-methionine + t-BOC-isoD-GG <13> (Reversibility: ? <13> [12]) [12]
P S-adenosyl-L-homocysteine + ?
S Additional information <13> (<13> protein-l-aspartate is no substrate [12]; native E. coli histidine phosphocarrier protein is no substrate [14]; <20> O-succinyl-L-homoserine, succinyl choline chloride, succinyl acetone, succinic acid and succinamic acid are no substrates [32]) [12, 14, 32]
P ?

Inhibitors

5'-deoxy-5'-methylthioadenosine <13, 20> (<13> partial inhibition [5]) [5, 32]
S-adenosylhomocysteine <13, 20, 21> (<13> complete inhibition [5]; <20> potent inhibitor at pH 4 and 7.5 [32]) [5, 9, 32]
adenosine <20> [32]

Cofactors/prosthetic groups

S-adenosyl-L-homocysteine <13, 20> (<13,20> tightly bound cofactor [30,32]) [30, 32]
S-adenosyl-L-methionine <20> (<20> active methyl-donating cofactor [26]) [26, 32]
adenosine <20> (<20> tightly bound cofactor [32]) [32]

Specific activity (U/mg)

0.0000057 <8> [28]
0.0000069 <10> [8]
0.0000127 <2> (<2> VYP-L-isoAsp-GHG as substrate [25]) [25]
0.0000175 <27> [28]
0.0000215 <19> [28]
0.00013 <2> (<2> γ-globulin as substrate [25]) [25]
0.000229 <2> (<2> VYP-L-isoAsp-RR as substrate [25]) [25]
0.000233 <13> [5]
0.000433 <2> (<2> ovalbumin as substrate [25]) [25]
0.000439 <2> [25]
0.001441 <2> (<2> VYP-L-isoAsp-CA as substrate [25]) [25]
0.0022 <23> (<23> 37°C [20]) [20]
0.004179 <2> (<2> KASA-L-isoAsp-LAKY as substrate [25]) [25]
0.00427 <13> [4]
0.005028 <2> (<2> VYP-L-isoAsp-HA as substrate [25]) [25]
0.012-0.016 <5> [1]
0.014 <21> [16]

0.0152 <21> [22]
0.016-0.02 <21> (<21> γ-globulin as methyl acceptor [16]) [16]
0.0168 <21> [16]
0.156 <23> [20]

K_m-Value (mM)
0.00026 <20> (KASA (isoAsp) LAKY, <20> 68°C, pH 7.5 [32]) [32]
0.00029 <13> (VYP (isoAsp) HA) [12, 13, 20, 25]
0.00029 <13> (VYP-L-isoAsp-HA, <13> 37°C, pH 7.5 [32]) [32]
0.0004 <13> (KASA (isoD) LAKY) [7]
0.00052 <13> (KASA (isoAsp) LAKY) [13, 32]
0.00052 <13> (KASA (isoD) LAKY) [12, 20, 25]
0.00053 <20> (VYP-L-isoAsp-HA, <20> 68°C, pH 7.5 [32]) [32]
0.00055 <20> (KASA (isoAsp) LAKY) [32]
0.00055 <5, 13> (KQVV-isoD-SAYEVIK) [7, 12]
0.00059 <20> (VYP-L-isoAsp-HA, <20> 37°C, pH 7.5 [32]) [32]
0.00088 <5> (KASA (isoD) LAKY) [7]
0.00089 <21> (S-adenosyl-L-methionine, <21> native brain enzyme [16]) [16]
0.0011 <21> (S-adenosyl-L-methionine, <21> recombinant protein [16]) [16]
0.0012 <5> (synapsin I) [7]
0.0013 <13> (A-isoD-QLTEEQIAEFK) [12]
0.00167 <5> (VYP-isoD-AA) [7]
0.0019 <23> (S-adenosyl-L-methionine, <23> 37°C [20]) [20, 25]
0.002 <13> (S-adenosyl-L-methionine) [20]
0.0022 <13> (S-adenosyl-L-methionine) [25, 32]
0.0022 <13> (ovalbumin) [12]
0.0024 <13> (TS-isoD-YSKY) [12]
0.0027 <13> (KASA (D-Asp) LAKY) [13]
0.0028 <23> (KASA (isoD) LAKY, <23> 37°C [20]) [20, 25]
0.0028 <13> (VYP-isoD-CA) [7]
0.00295 <13> (VYP-isoD-DA) [7]
0.003 <13> (VYP-D-Asp-HA) [13]
0.00304 <5> (VYP-isoD-SA) [7]
0.0031 <6> (S-adenosyl-L-methionine) [20, 25]
0.0031 <13> (VYP-isoD-CA) [12]
0.0033 <13> (VYP-isoD-SA) [12]
0.00341 <20> (S-adenosyl-L-methionine, <20> 68°C, pH 7.5 [32]) [32]
0.0035 <5, 13> (GFDL-isoD-GGGVG) [7, 12]
0.0035 <13> (VYR-isoD-RR) [12]
0.00358 <13> (histidine phosphocarrier protein-1 (isoAsp-38)) [14]
0.0036 <13> (LSH-isoD-GYCL) [7]
0.0039 <13> (LSH-isoD-GYCL) [12]
0.00397 <13> (VYP-isoD-PA) [12]
0.004 <21> (S-adenosyl-L-methionine, <21> brush-border membrane isozyme [9]) [9]
0.0041 <23> (VYP (isoD) HA, <23> 37°C [20]) [20, 25]

0.0042 <13> (WM-D-Asp-F-amide) [13]
0.00425 <5> (VYP-isoD-GA) [7]
0.0046 <13> (YVS-isoD-GCG) [7]
0.00463 <20> (ovalbumin, <20> 68°C, pH 7.5 [32]) [32]
0.0048 <13> (VYP-D-Asp-GA) [13]
0.0049 <13> (isoD-G) [12]
0.005 <13> (WM-D-Asp-F-amide) [12]
0.005 <13> (WM-isoD-F-NH_2) [7]
0.005 <13> (protein L-isoaspartate) [13]
0.0051 <13> (YVS-isoD-GCG) [12]
0.0051 <13> (isoD-11 EGF, <13> proteolyzed [12]) [12]
0.0053 <13> (WAGG-isoD-ASGE) [12]
0.0056 <23> (S-adenosyl-L-methionine, <23> 85°C [20]) [20]
0.00574 <13> (VYG-L-isoAsp-PA) [12, 13]
0.006 <2> (S-adenosyl-L-methionine) [20, 25]
0.00617 <5, 13> (KVTCK-isoD-GQTNCYQSK) [7, 12]
0.0062 <13> (VYP-isoD-GA) [7]
0.0063 <13> (t-BOC-G-isoD-G) [12]
0.0068 <13> (VYP-L-isoAsp-GA) [12-14]
0.0068 <13> (VYP-isoD-GA) [12]
0.008 <20> (VYP-L-isoAsp-HA, <20> 68°C, pH 4 [32]) [32]
0.00841 <13> (acetyl-isoD-G) [12]
0.00912 <6> (KASA (isoAsp) LAKY) [20, 25]
0.011 <13> (LM-isoD-T) [12]
0.0118 <10> (VYP (isoAsp) HA) [20, 25]
0.0127 <25> (KASA (isoAsp) LAKY) [20, 25]
0.013 <5> (isoACTH) [1]
0.0144 <13> (acetyl-G-isoD-G) [12]
0.0155 <13> (YVS-isoAsp-GHG) [14]
0.0157 <13> (VYP-isoD-LA) [12]
0.0164 <21> (isoAspDSIP) [16]
0.017 <13> (LSH-isoD-GY-Cya-L) [7]
0.0178 <13> (G-isoD-G) [12]
0.0194 <6> (VYP (isoD) HA) [20, 25]
0.0194 <13> (acetyl-Asp) [12]
0.021 <20> (KASA (isoAsp) LAKY, <20> 68°C, pH 4.0 [32]) [32]
0.0217 <25> (VYP (isoD) HA) [20]
0.0224 <13> (KMK-isoD-TDSEEEIR) [12]
0.0229 <13> (T1-isoAsp-12) [14]
0.023 <20> (KASA (D-Asp) LAKY) [32]
0.0235 <13> (acetyl-Asn) [12]
0.025 <23> (KASA (isoD) LAKY, <23> 85°C [20]) [20]
0.0272 <13> (AD) [12]
0.0273 <13> (LSH-isoD-GY(3-sulfo)AL) [12]
0.0274 <13> (V2-isoAsp-12) [14]
0.03 <21> (S-adenosyl-L-methionine, <21> cytosolic isozyme [9]) [9]
0.03 <13> (ovalbumin) [25]

0.0336 <23> (VYP (isoD) HA, <23> 85°C [20]) [20]
0.034 <13> (t-BOC-isoD-GG) [12]
0.035 <13> (ovalbumin) [20, 32]
0.0368 <5> (ovalbumin) [7]
0.0377 <13> (VYP-isoD-DA) [12]
0.043 <27> (VYP-L-isoAsp-HA) [25]
0.0506 <10> (KASA (isoAsp) LAKY) [20, 25]
0.0517 <25> (VYP-L-isoAsp-HA) [25]
0.0546 <13> (AP-isoD-amide) [12]
0.0633 <13> (AA-isoD-F-amide) [12]
0.065 <13> (AA-isoD-amide) [12]
0.08 <2> (KASA-L-isoAsp-LAKY) [25]
0.0827 <13> (KMKDT-isoD-SEEEIR) [12]
0.092 <27> (KASA-L-isoAsp-LAKY) [25]
0.0975 <13> (V4-isoAsp-38) [14]
0.104 <13> (YVS-isoD-G-Cya-G) [7]
0.126 <13> (β-101-lysozyme) [12]
0.13 <27> (VYP-L-isoAsp-HA, <27> enzyme from leaves [28]) [28]
0.13 <13> (YVS-isoD-G(3-sulfo)AG) [12]
0.166 <20> (ovalbumin) [32]
0.175 <13> (VYP-isoD-Cya-G) [7]
0.183 <13> (β-101-lysozyme) [7]
0.195 <13> (VYP-isoD-(3sulfo)AA) [12]
0.197 <13> (HPr-2 (isoAsp-12,-38)) [14]
0.207 <13> (isoD-56 calbindin D9k) [12]
0.24 <8> (VYP-L-isoAsp-HA) [28]
0.29 <27> (VYP-L-isoAsp-HA, <27> enzyme from seeds [28]) [28]
0.31 <2> (VYP-L-isoAsp-HA) [25]
0.326 <23> (ovalbumin, <23> 37°C [20]) [20, 25]
0.33 <19> (VYP-L-isoAsp-HA) [28]
0.339 <13> (T4-isoAsp-38) [14]
0.375 <20> (N-succinyl-AAVAp-nitroanilide, <20> 68°C, pH 7.5 [32]) [32]
0.38 <25> (VYP-L-isoAsp-HA) [28]
0.452 <13> (A-L-isoAsp-T) [13]
0.452 <13> (A-isoD-T) [12]
0.469 <13> (YVS-isoD-GDG) [7]
0.489 <13> (YVS-isoD-GDG) [12]
0.598 <13> (t-BOC-isoD-G) [12]
0.69 <13> (N-succinyl-AAVAp-nitroanilide) [32]
0.727 <10> (ovalbumin) [20, 25]
1.03 <19> (ovalbumin) [28]
1.081 <13> (native lysozyme) [7]
1.2 <2> (ovalbumin, <2> native enzyme [25]) [25]
1.5 <13> (calbindin D9k) [12]
1.6 <13> (isoD-11 EGF, <13> folded [12]) [12]
2 <13> (calbindin D9k, <13> + Ca^{2+} [12]) [12]
2.32 <25> (ovalbumin) [28]

2.53 <8> (ovalbumin) [28]
2.66 <27> (ovalbumin) [25]
2.7 <13> (KASA (D-Asp) LAKY) [32]
2.77 <27> (ovalbumin, <27> enzyme from seeds [28]) [28]
3 <13> (protein D-aspartate) [13]
4.95 <2> (ovalbumin) [25]
5 <13> (WM-L-isoAsp-F-amide) [13]
5 <2> (ovalbumin) [25]
11 <20> (N-succinyl-AAVAp-nitroanilide, <20> 68°C, pH 4.0 [32]) [32]
12.7 <27> (ovalbumin, <27> enzyme from leaves [28]) [28]
17.8 <13> (G-isoD-G) [12]

pH-Optimum
4 <20> [32]
5.5 <10> [20]
5.6 <23> [20]
6-7 <21> [20]
7-8 <2> (<2> recombinant enzyme, expressed in E. coli [25]) [25]
7.5 <21, 25> [9, 20]

pH-Range
3-10 <2, 20> [25, 32]
5-8.5 <23> [20]

Temperature optimum (°C)
45 <2, 10> (<2> recombinant enzyme, expressed in E. coli [25]) [25, 28]
45-55 <2, 8, 19, 20, 25, 27> [28, 32]
55 <13> (<13> recombinant enzyme, expressed in E. coli [28]) [28]
85 <20, 23> [20, 32]

Temperature range (°C)
4-95 <20> [32]
25-65 <6, 10, 13> [28]
37-93 <23> [20]

4 Enzyme Structure

Molecular weight
21000 <21> (<21> brush-border membrane isozyme, SDS-PAGE [9]) [9]
23000 <21> (<21> cytosolic isozyme, SDS-PAGE [9]) [9]
24000 <10> (<10> gel filtration [8]) [8]
24500 <5, 13> (<5,13> amino acid sequence [15]) [15]
24580 <13> (<13> amino acid sequence [6]) [6]
24600 <2> (<2> amino acid sequence [25]) [25]
24630 <21> (<21> amino acid sequence [10]) [10]
25000 <13, 21> (<21> SDS-PAGE [16]; <13> gel filtration [5,6]) [5, 6, 10, 16]
25400-26000 <11> [5]
26500 <5, 13> (<5,13> SDS-PAGE [4,6]) [4, 6]

27000 <5, 13, 21> (<5,13,21> SDS-PAGE [15]; <5> brain enzyme [5]) [5, 15]
28000 <2, 13> (<2,13> SDS-PAGE [5,25]) [5, 25]
29000 <24> [5]
30000 <24> (<24> SDS-PAGE [15]) [15]
34000 <26> (<26> SDS-PAGE [15]) [15]
35000 <5, 10> (<5> thymus enzyme [5]; <10> SDS-PAGE [20]) [5, 20]
36400 <23> (<23> calculated from translated amino acid sequence [20]) [20]

Subunits

monomer <2, 5, 10, 13, 23> (<13> 1 * 25000 [10]; <10> 1 * 24000, SDS-PAGE [8]; <5,13> 1 * 26500, isoenzyme I and II, SDS-PAGE [4]; <13> 1 * 26500, SDS-PAGE [6]; <23> 1 * 34000, equilibrium centrifugation [20]; <2> 1 * 28000, SDS-PAGE [25]) [4, 6, 10, 20, 25]

5 Isolation/Preparation/Mutation/Application

Source/tissue

blood <12, 13, 21> [4, 5, 13, 15, 19, 23, 25]
brain <5, 7, 13, 18, 21> [1-8, 10, 15, 16, 21-25, 27]
electric organ <24> [15]
embryo <2> [17, 25]
erythrocyte <5, 12, 13, 18, 21> [1-8, 10, 12, 13, 15, 20, 23, 25, 27]
flower <2> [17]
germ <25> [11, 15, 20, 25]
heart <7, 18> [5, 10, 15, 27]
kidney <5, 13, 18, 21> [9, 15, 27]
larva <6, 9> [22, 24]
leaf <2, 8, 15, 25, 27> [17, 25, 28]
lens <5> [7]
liver <7, 18, 21> [5, 10, 25, 27]
muscle <7> [5, 10]
oocyte <26> [2, 5, 15, 25, 27]
pituitary gland <21> [5]
placenta <13> [10]
renal cortex <21> [9, 15]
root <2> [17]
seed <2, 8, 15, 19, 25, 27> [17, 25, 28]
seedling <2> [17, 28]
testis <18, 21> [5, 10, 15, 16, 27]
thymus <5> [5]

Localization

brush border <21> [9, 15]
cytoplasm <13, 26> [15, 25]
cytosol <2, 6, 10, 13, 18, 21, 23> [3, 5, 6, 8, 9, 13, 15, 17, 20, 23-25, 27]
membrane <21> [9]

Purification

<2> (recombinant protein [25,28]) [25, 28]
<5> (2 isozymes [4]) [1, 4, 15]
<6> (recombinant protein [25]) [25]
<10> [8, 15, 20]
<12> [18]
<13> (isozyme I, partially [13]; 2 isozymes [4]; recombinant protein [25]) [3-6, 10, 12, 13, 15, 20, 25]
<20> (recombinant protein [32]) [32]
<21> (recombinant protein [16,22,25]) [16, 22, 25]
<23> (recombinant protein [25]) [20, 25]
<25> [15, 20, 25]

Crystallization

<13> [30, 31]
<20> [26, 32]

Cloning

<2> (cDNA isolation, pcm gene cloned and expressed in Escherichia coli BL21(DE3) [25]) [17, 25, 28]
<6> (disruption of the pcm gene [20,24]) [20, 24, 25]
<10> (pcm gene sequenced [8]) [8]
<13> (southern blot analysis of PCR amplified genomic DNA, human chromosome contains a single gene encoding the enzyme, PCMT1, mapped to the q24-25 region of chromosome 6 [10,23]) [10, 14, 23, 25]
<18> [17]
<20> (cloned and expressed in Escherichia coli TOP10 [32]) [32]
<21> (sequence analysis of a cDNA clone of a rat brain enzyme [10]; PCR amplified cDNA cloned into vector pdeltablue and expressed in Escherichia coli [16]) [10, 16, 22, 25]
<23> (pcm gene PCR amplified, overexpressed in Escherichia coli as a fusion protein [20]) [20, 25]
<25> (cDNA isolation [25]) [25, 28]

Application

biotechnology <23> (<23> useful in biotechnical applications [20]) [20]
medicine <13, 21> (<21> a cloned source of this enzyme should be useful in the identification of cellular substrates and for quantitation and localization of isoaspartyl sites in purified proteins, including recombinant proteins used as pharmaceuticals [16]; <13> aging research, accumulation of damaged proteins have detrimental effects over the human lifespan [23]) [16, 23]
nutrition <12> (<12> enzymatic detection of L-isoaspartyl residues in food proteins [18]) [18]

6 Stability

Temperature stability

35-50 <27> (<27> enzyme from leaves [28]) [28]
35-65 <27> (<27> enzyme from seeds [28]) [28]
37-50 <13> [23]
55 <2> (<2> loses activity rapidly at [25]) [25]
85-95 <20> (<20> little or no loss of activity [32]) [32]
100 <23> (<23> extremely heat stable, no loss of activity after 60 min [20]) [20]

General stability information

<21>, no stability problems during purification [16]

Storage stability

<21>, -70°C, can be stored in 10% glycerol [16]
<23>, -80°C, stored in liquid nitrogen, no apparent loss of activity [20]

References

[1] Johnson, B.A.; Aswad, D.W.: Enzymatic protein carboxyl methylation at physiological pH: cyclic imide formation explains rapid methyl turnover. Biochemistry, **24**, 2581-2586 (1985)
[2] Aswad, D.W.; Johnson, B.A.: The unusual substrate specificity of eukaryotic protein carboxyl methyltranferases. Trends Biochem. Sci., **12**, 155-158 (1987)
[3] Ota, I.M.; Ding, L.; Clarke, S.: Methylation at specific altered aspartyl and asparaginyl residues in glucagon by the erythrocyte protein carboxyl methyltransferase. J. Biol. Chem., **262**, 8522-8531 (1987)
[4] Gilbert, J.M.; Fowler, A.; Bleibaum, J.; Clarke, S.: Purification of homologous protein carboxyl methyltransferase isozymes from human and bovine erythrocytes. Biochemistry, **27**, 5227-5233 (1988)
[5] Ota, I.M.; Gilbert, J.M.; Clarke, S.: Two major isozymes of the protein D-aspartyl/L-isoaspartyl methyltransferase from human erythrocytes. Biochem. Biophys. Res. Commun., **151**, 1136-1143 (1988)
[6] Ingrosso, D.; Fowler, A.V.; Bleibaum, J.; Clarke, S.: Sequence of the D-aspartyl/L-isoaspartyl protein methyltransferase from human erythrocytes. Common sequence motifs for protein, DNA, RNA, and small molecule S-adenosylmethionine-dependent methyltransferases. J. Biol. Chem., **264**, 20131-20139 (1989)
[7] Lowenson, J.D.; Clarke, S.: Identification of isoaspartyl-containing sequences in peptides and proteins that are usually poor substrates for the class II protein carboxyl methyltransferase. J. Biol. Chem., **265**, 3106-3110 (1990)
[8] Fu, J.C.; Ding, L.; Clarke, S.: Purification, gene cloning, and sequence analysis of an L-isoaspartyl protein carboxyl methyltransferase from Escherichia coli [published erratum appears in J. Biol. Chem. 1992 Jun 5;267(16):11660]. J. Biol. Chem., **266**, 14562-14572 (1991)

[9] Gingras, D.; Menard, P.; Beliveau, R.: Protein carboxyl methylation in kidney brush-border membranes. Biochim. Biophys. Acta, **1066**, 261-267 (1991)

[10] Ingrosso, D.; Kagan, R.M.; Clarke, S.: Distinct C-terminal sequences of isozymes I and II of the human erythrocyte L-isoaspartyl/D-aspartyl protein methyltransferase [published erratum appears in Biochem. Biophys. Res. Commun 1991 Apr 15;176(1):549]. Biochem. Biophys. Res. Commun., **175**, 351-358 (1991)

[11] Johnson, B.A.; Ngo, S.Q.; Aswad, D.W.: Widespread phylogenetic distribution of a protein methyltransferase that modifies L-isoaspartyl residues. Biochem. Int., **24**, 841-847 (1991)

[12] Lowenson, J.D.; Clarke, S.: Structural elements affecting the recognition of L-isoaspartyl residues by the L-isoaspartyl/D-aspartyl protein methyltransferase. Implications for the repair hypothesis. J. Biol. Chem., **266**, 19396-19406 (1991)

[13] Lowenson, J.D.; Clarke, S.: Recognition of D-aspartyl residues in polypeptides by the erythrocyte L-isoaspartyl/D-aspartyl protein methyltransferase. Implications for the repair hypothesis. J. Biol. Chem., **267**, 5985-5995 (1992)

[14] Brennan, T.V.; Anderson, J.W.; Jia, Z.; Waygood, E.B.; Clarke, S.: Repair of spontaneously deamidated HPr phosphocarrier protein catalyzed by the L-isoaspartate-(D-aspartate) O-methyltransferase. J. Biol. Chem., **269**, 24586-24595 (1994)

[15] Boivin, D.; Bilodeau, D.; Beliveau, R.: Immunochemical characterization of L-isoaspartyl-protein carboxyl methyltransferase from mammalian tissues. Biochem. J., **309**, 993-998 (1995)

[16] David, C.L.; Aswad, D.W.: Cloning, expression, and purification of rat brain protein L-isoaspartyl methyltransferase. Protein Expr. Purif., **6**, 312-318 (1995)

[17] Mudgett, M.B.; Clarke, S.: A distinctly regulated protein repair L-isoaspartylmethyltransferase from Arabidopsis thaliana. Plant Mol. Biol., **30**, 723-737 (1996)

[18] Ingrosso, D.; Perna, A.F.; D'Angelo, S.; Buro, M.D.; Malanga, P.; Galletti, P.; De Rosa, M.; Zappia, V.: Enzymic detection of L-isoaspartyl residues in food proteins and the protective properties of trehalose. J. Nutr. Biochem., **8**, 535-540 (1997)

[19] Weber, D.J.; McFadden, P.N.: Detection and characterization of a protein isoaspartyl methyltransferase which becomes trapped in the extracellular space during blood vessel injury. J. Protein Chem., **16**, 257-267 (1997)

[20] Ichikawa, J.K.; Clarke, S.: A highly active protein repair enzyme from an extreme thermophile: the L-isoaspartyl methyltransferase from Thermotoga maritima. Arch. Biochem. Biophys., **358**, 222-231 (1998)

[21] O'Connor, M.B.; O'Connor, C.M.: Complex interactions of the protein L-isoaspartyl methyltransferase and calmodulin revealed with the yeast two-hybrid system. J. Biol. Chem., **273**, 12909-12913 (1998)

[22] David, C.L.; Pierce, V.A.; Aswad, D.W.; Gibbs, A.G.: The effect of urea exposure on isoaspartyl content and protein L-isoaspartate methyltransferase activity in Drosophila melanogaster. Comp. Biochem. Physiol. B, **124B**, 423-427 (1999)

[23] DeVry, C.G.; Clarke, S.: Polymorphic forms of the protein L-isoaspartate (D-aspartate) O-methyltransferase involved in the repair of age-damaged proteins. J. Hum. Genet., **44**, 275-288 (1999)
[24] Niewmierzycka, A.; Clarke, S.: Do damaged proteins accumulate in Caenorhabditis elegans L-isoaspartate methyltransferase (pcm-1) deletion mutants?. Arch. Biochem. Biophys., **364**, 209-218 (1999)
[25] Thapar, N.; Clarke, S.: Expression, Purification, and Characterization of the Protein Repair L-Isoaspartyl Methyltransferase from Arabidopsis thaliana. Protein Expr. Purif., **20**, 237-251 (2000)
[26] Griffith, S.C.; Sawaya, M.R.; Boutz, D.R.; Thapar, N.; Katz, J.E.; Clarke, S.; Yeates, T.O.: Crystal Structure of a Protein Repair Methyltransferase from Pyrococcus furiosus with its L-Isoaspartyl Peptide Substrate. J. Mol. Biol., **313**, 1103-1116 (2001)
[27] Lowenson, J.D.; Kim, E.; Young, S.G.; Clarke, S.: Limited accumulation of damaged proteins in L-isoaspartyl (D-aspartyl) O-methyltransferase-deficient mice. J. Biol. Chem., **276**, 20695-20702 (2001)
[28] Thapar, N.; Kim, A.K.; Clarke, S.: Distinct patterns of expression but similar biochemical properties of protein L-isoaspartyl methyltransferase in higher plants. Plant Physiol., **125**, 1023-1035 (2001)
[29] Young, A.L.; Carter, W.G.; Doyle, H.A.; Mamula, M.J.; Aswad, D.W.: Structural integrity of histone H2B in vivo requires the activity of protein L-isoaspartate O-methyltransferase, a putative protein repair enzyme. J. Biol. Chem., **276**, 37161-37165 (2001)
[30] Ryttersgaard, C.; Griffith, S.C.; Sawaya, M.R.; MacLaren, D.C.; Clarke, S.; Yeates, T.O.: Crystal structure of human L-isoaspartyl methyltransferase. J. Biol. Chem., **277**, 10642-10646 (2002)
[31] Smith, C.D.; Carson, M.; Friedman, A.M.; Skinner, M.M.; Delucas, L.; Chantalat, L.; Weise, L.; Shirasawa, T.; Chattopadhyay, D.: Crystal structure of human L-isoaspartyl-O-methyltransferase with S-adenosyl homocysteine at 1.6 ANG. resolution and modeling of an isoaspartyl-containing peptide at the active site. Protein Sci., **11**, 625-635 (2002)
[32] Thapar, N.; Griffith, S.C.; Yeates, T.O.; Clarke, S.: Protein repair methyltransferase from the hyperthermophilic archaeon Pyrococcus furiosus: unusual methyl-accepting affinity for D-aspartyl and N-succinyl-containing peptides. J. Biol. Chem., **277**, 1058-1065 (2002)

Isoorientin 3'-O-methyltransferase 2.1.1.78

1 Nomenclature

EC number
2.1.1.78

Systematic name
S-adenosyl-L-methionine:isoorientin 3'-O-methyltransferase

Recommended name
isoorientin 3'-O-methyltransferase

Synonyms
isoorientin 3'-methyltransferase

CAS registry number
83061-51-2

2 Source Organism

<1> *Silene pratensis* [1]

3 Reaction and Specificity

Catalyzed reaction
S-adenosyl-L-methionine + isoorientin = S-adenosyl-L-homocysteine + isoscoparin (also acts on isoorientin 2''-O-rhamnoside. Involved in the biosynthesis of flavones)

Reaction type
methyl group transfer

Natural substrates and products
- **S** S-adenosyl-L-methionine + isoorientin <1> (<1> involved in biosynthesis of flavones [1]) (Reversibility: ? <1> [1]) [1]
- **P** S-adenosyl-L-homocysteine + isoscoparin

Substrates and products
- **S** S-adenosyl-L-methionine + isoorientin <1> (Reversibility: ? <1> [1]) [1]
- **P** S-adenosyl-L-homocysteine + isoscoparin
- **S** S-adenosyl-L-methionine + isoorientin 2''-O-rhamnoside <1> (Reversibility: ? <1> [1]) [1]
- **P** S-adenosyl-L-homocysteine + isoscoparin 2''-O-rhamnoside

Inhibitors
EDTA <1> [1]
Hg^{2+} <1> [1]
Zn^{2+} <1> [1]

Metals, ions
Co^{2+} <1> (<1> slightly stimulatory [1]) [1]
Mn^{2+} <1> (<1> slightly stimulatory [1]) [1]

K_m-Value (mM)
0.0012 <1> (S-adenosyl-L-methionine, <1> second substrate caffeic acid [1]) [1]
0.004 <1> (S-adenosyl-L-methionine, <1> second substrate isoorientin [1]) [1]
0.005 <1> (S-adenosyl-L-methionine, <1> second substrate isoorientin 2"-O-rhamnoside [1]) [1]
0.007 <1> (isoorientin 2"-O-rhamnoside, <1> second substrate S-adenosyl-L-methionine [1]) [1]
0.32 <1> (isoorientin) [1]

pH-Optimum
8 <1> (<1> isoorientin as substrate [1]) [1]

pH-Range
7-9 <1> [1]

Temperature optimum (°C)
45 <1> [1]

Temperature range (°C)
30-55 <1> (<1> 35% of maximum activity at 30°C and 55°C [1]) [1]

5 Isolation/Preparation/Mutation/Application

Source/tissue
leaf <1> [1]

Purification
<1> (partial [1]) [1]

6 Stability

Temperature stability
40 <1> (<1> after 20 min: 55% activity remains, after 40 min 20% activity remains [1]) [1]

References

[1] Van Brederode, J.; Kamps-Heinsbroek, R.; Mastenbroek, O.: Biochemical and ontogenetic evidence that the ferulic acid and isocoparin formation in silene are catalyzed by different enzymes. Z. Pflanzenphysiol., **106**, 43-53 (1982)

Cyclopropane-fatty-acyl-phospholipid synthase 2.1.1.79

1 Nomenclature

EC number
2.1.1.79

Systematic name
S-adenosyl-L-methionine:unsaturated-phospholipid methyltransferase (cyclizing)

Recommended name
cyclopropane-fatty-acyl-phospholipid synthase

Synonyms
CFA synthase
CFAS
SfCPA-FAS <6> [12]
cyclopropane fatty acid synthase
cyclopropane fatty acid synthetase
cyclopropane synthase
synthetase, cyclopropane fatty acid
unsaturated-phospholipid methyltransferase

CAS registry number
51845-48-8

2 Source Organism

<1> *Serratia marcescens* [1]
<2> *Clostridium butyricum* [1, 2]
<3> *Escherichia coli* (B [3]; BL21 transformed with plasmid pAYW58 [4]) [3, 4, 5, 8, 9]
<4> *Lactobacillus plantarum* [6, 7]
<5> *Eubacterium HX* (moderately-halophilic [10]) [10]
<6> *Sterculia foetida* [11, 12]

3 Reaction and Specificity

Catalyzed reaction
S-adenosyl-L-methionine + phospholipid olefinic fatty acid = S-adenosyl-L-homocysteine + phospholipid cyclopropane fatty acid

Reaction type

methyl group transfer

Natural substrates and products

- **S** S-adenosyl-L-methionine + phospholipid olefinic fatty acid <3, 4> (<3>, methylation of unsaturated fatty acid moieties of phospholipids in the phospholipid bilayer [8]; <4>, S-adenosylmethionine levels in conjugation with cyclopropane fatty acid synthase activities regulates cyclopropane fatty acid synthesis in Lactobacillus plantarum [7]; <3>, increased level of cyclopropane fatty acid synthase activity as bacterial cultures enter stationary phase is transient, activity quickly declines to the basal level, the loss of activity is due to proteolytic degradation dependent on expression of the heat shock regulon [9]) (Reversibility: ? <3, 4> [7, 8, 9]) [7, 8, 9]
- **P** S-adenosyl-L-homocysteine + phospholipid cyclopropane fatty acid

Substrates and products

- **S** S-adenosyl-L-homocysteine + phosphatidylcholine <6> (<6>, addition of the methylene group to oleic acid occurs at the sn-1 position [12]; <2>, phosphatidylcholine is not a substrate [2]) [12]
- **P** S-adenosyl-L-homocysteine + ?
- **S** S-adenosyl-L-methionine + cardiolipin <3> (Reversibility: ? <3> [8]) [8]
- **P** S-adenosyl-L-homocysteine + ?
- **S** S-adenosyl-L-methionine + oleate <6> (Reversibility: ? <6> [11, 12]) [11, 12]
- **P** S-adenosyl-L-homocysteine + dihydrosterculate: <6> [11, 12]
- **S** S-adenosyl-L-methionine + phosphatidylethanolamine <3> (Reversibility: ? <3> [8]) [8]
- **P** S-adenosyl-L-homocysteine + ?
- **S** S-adenosyl-L-methionine + phosphatidylglycerol <3> (Reversibility: ? <3> [8]) [8]
- **P** S-adenosyl-L-homocysteine + ?
- **S** S-adenosyl-L-methionine + phospholipid olefinic fatty acid <1, 2, 3, 4, 5, 6> (<1>, Serratia marcescens contains endogenous lipid substrates which cannot be removed or replaced by simple manipulations [1]; <3>, reaction is not affected by the order-disorder state of the lipid substrate [8]; <6>, addition of a methylene group from S-adenosylmethionine to the cis-double bond of monoenoic phospholipid-bound fatty acids [11]) (Reversibility: ? <1, 2, 3, 4, 5, 6> [1, 2, 3, 4, 5, 6, 7, 8, 9, 10, 11, 12]) [1-12]
- **P** S-adenosyl-L-homocysteine + phospholipid cyclopropane fatty acid <1-4> (<1,2>, phosphatidylethanolamine which contains cyclopropane fatty acids [1]) [1-8]

Inhibitors

5,5'-dithiobis(2-nitrobenzoic acid) <3> (<3>, reversed by addition of dithiothreitol [8]) [3, 8]
A9145C <3, 4> (<4>, linear competitive [6]) [4, 6]
Cutsum <2> [2]
N-butylmaleimide <3> [4]

N-heptylmaleimide <3> [4]
N-hexylmaleimide <3> [4]
N-pentylmaleimide <3> [4]
NEM <3> [3, 4, 8]
NaCl <5> (<5>, 1 M, 97% inhibition [10]) [10]
S-adenosylhomocysteine <3> (<3>, product inhibition [4]) [3, 4, 8]
Tween 80 <2> [2]
hexadecyltrimethylammonium bromide <2> [2]
oleoyl-CoA <6> (<6>, above 0.05 M [11]) [11]
p-hydroxymercuribenzoate <1, 3> [1, 3, 8]
phosphatidylcholine <2> [2]
sinefungin <3, 4> (<4>, linear competitiove [6]) [4, 6]
sorbitol monolaurate ester <3> [3]
sorbitol monooleate ester <3> [3]

Activating compounds

glycinebetaine <5> (<5>, activity is extremely low in vitro in 100 mM buffers, 100fold stimulation by exogenous addition of 2-3 M glycinebetaine [10]) [10]
oleoyl-CoA <6> (<6>, 0.02-0.05 mM, enhances activity about 2fold [11]) [11]
sodium lauryl sulfate <2> (<2>, stimulates [2]) [2]

Specific activity (U/mg)

0.00000113 <3> [8]
Additional information <2, 3> [2, 3, 4]

K_m-Value (mM)

0.02 <6> (S-adenosylmethionine, <6>, in cell-free extract [11]) [11]
0.057 <2> (S-adenosyl-L-methionine) [2]
0.09 <3> (S-adenosyl-L-methionine) [3, 8]
0.53 <2> (phosphatidylethanolamine) [2]

K_i-Value (mM)

0.000011 <4> (A9145C) [6]
0.00022 <4> (sinefungin) [6]
0.001 <3> (A9145C) [4]
0.22 <3> (S-adenosylhomocysteine) [3, 8]

pH-Optimum

7 <1, 6> [1, 11]

pH-Range

5.5-8.5 <1> (<1>, about 50% of activity maximum at pH 5.5 and 8.5 [1]) [1]

4 Enzyme Structure

Molecular weight

90000 <3> (<3>, equilibrium sedimentation, gel filtration [8]) [8]

Subunits

? <3, 6> (<3>, x * 43913, calculation from nucleotide sequence [4]; <3>, x * 80000-100000, SDS-PAGE [3]; <6>, x * 98000, calculation from nucleotide sequence [12]) [3, 4, 12]
monomer <3> (1 * 90000, SDS-PAGE [8]) [8]

5 Isolation/Preparation/Mutation/Application

Source/tissue

seed <6> (<6>, developing [12]) [11, 12]
Additional information <6> (<6>, no activity in leaves [12]) [12]

Localization

cytoplasm <3> [5]
membrane <3, 6> (<3>, loosely associated with the inner membrane [8]) [8, 12]
microsome <6> [12]

Purification

<2> (partial [2]) [2]
<3> (partial [8]) [3, 4, 8]

Cloning

<3> (overproduction of the enzyme via multicopy cfa plasmids [5]) [5]
<6> (fusion protein of an FAD-containing oxidase at the N-terminus and a methyltransferase at the C-terminus, expression in tobacco cells [11]) [11]

6 Stability

Temperature stability

37 <3> (<3>, 30 min, complete loss of activity in absence of lipid [3]) [3]

General stability information

<3>, sorbitol monolaurate ester stabilizes [3]
<3>, the enzyme is a short-lived protein in vivo and its degradation is dependent on expression of the heat shock regulon [9]
<3, 3>, sorbitol monooleate ester stabilizes [3]

Storage stability

<3>, -20°C, in presence of phospholipid, stable for 2 months [4]

References

[1] Zalkin, H.; Law, J.H.; Goldfine, H.: Enzymic synthesis of cyclopropane fatty acids catalyzed by bacterial extracts. J. Biol. Chem., **238**, 1242-1248 (1963)

[2] Chung, A.E.; Law, J.H.: Cyclopropane fatty acid synthetase: partial purification and properties. Biochemistry, **3**, 967-974 (1964)
[3] Taylor, F.R.; Grogan, D.W.; Cronan, J.E.: Cyclopropane fatty acid synthase from Escherichia coli. Methods Enzymol., **71**, 133-139 (1981)
[4] Wang, A.Y.; Grogan, D.W.; Cronan, J.E.: Cyclopropane fatty acid synthase of Escherichia coli: deduced amino acid sequence, purification, and studies of the enzyme active site. Biochemistry, **31**, 11020-11028 (1992)
[5] Grogan, D.W.; Cronan, J.E.: Cloning and manipulation of the Escherichia coli cyclopropane fatty acid synthase gene: physiological aspects of enzyme overproduction. J. Bacteriol., **158**, 286-295 (1984)
[6] Smith, D.D.; Norton, S.J.: Inhibition of cyclopropane fatty acid synthase by sinefungin and A9145C. Biochem. Biophys. Res. Commun., **94**, 1458-1462 (1980)
[7] Smith, D.D.; Norton, S.J.: S-Adenosylmethionine, cyclopropane fatty acid synthase, and the production of lactobacillic acid in Lactobacillus plantarum. Arch. Biochem. Biophys., **205**, 564-570 (1980)
[8] Taylor, F.R.; Cronan, J.E.: Cyclopropane fatty acid synthase of Escherichia coli. Stabilization, purification, and interaction with phospholipid vesicles. Biochemistry, **18**, 3292-3300 (1979)
[9] Chang, Y.Y.; Eichel, J.; Cronan, J.E., Jr.: Metabolic instability of Escherichia coli cyclopropane fatty acid synthase is due to RpoH-dependent proteolysis. J. Bacteriol., **182**, 4288-4294 (2000)
[10] Kuchta, T.; Russell, N.: Glycinebetaine stimulates, but NaCl inhibits, fatty acid biosynthesis in the moderately halophilic Eubacterium HX. Arch. Microbiol., **161**, 234-238 (1994)
[11] Bao, X.; Katz, S.; Pollard, M.; Ohlrogge, J.: Carbocyclic fatty acids in plants: biochemical and molecular genetic characterization of cyclopropane fatty acid synthesis of Sterculia foetida. Proc. Natl. Acad. Sci. USA, **99**, 7172-7177 (2002)
[12] Bao, X.; Thelen, J.J.; Bonaventure, G.; Ohlrogge, J.B.: Characterization of cyclopropane fatty acid synthase from Sterculia foetida. J. Biol. Chem., **31**, 1-36 (2003)

Protein-glutamate O-methyltransferase 2.1.1.80

1 Nomenclature

EC number
2.1.1.80

Systematic name
S-adenosyl-L-methionine:protein-L-glutamate O-methyltransferase

Recommended name
protein-glutamate O-methyltransferase

Synonyms
EC 2.1.1.24 (formerly)
MCP methyltransferase I
MCP methyltransferase II
S-adenosylmethionine-glutamyl methyltransferase
S-adenosylmethionine:protein-carboxyl O-methyltransferase
methyl-accepting chemotaxis protein O-methyltransferase
methyl-accepting chemotaxis protein methyltransferase II
methyltransferase, protein O-
protein O-methyltransferase
protein carboxyl-O-methyltransferase
protein carboxyl-methylase
protein carboxylmethyltransferase II
protein carboxymethylase
protein carboxymethyltransferase
protein methylase II
protein methyltransferase II
protein(aspartate)methyltransferase
protein(carboxyl)methyltransferase

CAS registry number
9055-09-8

2 Source Organism

<1> *Salmonella typhimurium* [1, 2, 5, 6, 8, 10]
<2> *Bacillus subtilis* (MCP methyltransferase I and II [6]) [3, 6, 8, 9]
<3> *Escherichia coli* [4-8]
<4> *Bos taurus* [6]
<5> *Homo sapiens* [6]

<6> *mammalia* (occurence of the enzyme in mammalia e.g. bovine and horse is uncertain [8]) [6, 8]
<7> *Crithidia luciliae* [11]

3 Reaction and Specificity

Catalyzed reaction

S-adenosyl-L-methionine + protein L-glutamate = S-adenosyl-L-homocysteine + protein L-glutamate methyl ester (<1> random mechanism with no abortive complexes being formed, sequential random bi bi mechanism [10])

Reaction type

methyl group transfer

Natural substrates and products

S S-adenosyl-L-methionine + protein L-glutamate <1> (<1> membrane chemoreceptor proteins [1]; <1> enzyme is essential for maintaining the appropriate rate constants and levels of the regulator of the chemotactic response [2]) (Reversibility: ? <1> [1]) [1, 2]
P S-adenosyl-L-homocysteine + protein L-glutamate methyl ester <1> [1, 2]

Substrates and products

S S-adenosyl-L-methionine + protein L-glutamate <1-3> (<1> membrane chemoreceptor protein [1]; <1-3> membrane protein involved in chemotaxis [4,8,9]; <1> specific for proteins in Salmonella typhimurium and E. coli membranes [2]; <2> membrane-bound methyl-accepting chemotaxis proteins [3]; <1,3> specific for the methyl-accepting chemotaxis proteins [5,7]; <2> methyltransferase I is unable to methylate E. coli membranes [9]; <1> enzyme does not catalyze the methylation of a variety of soluble proteins, including ovalbumin, ribonuclease and lysozyme [8]; <1-3> highly specific for S-adenosylmethionine as methyl donor [6]; <1-3> S-adenosyl-L-ethionine shows 2.4% donor efficiency compared to S-adenosyl-L-methionine [6]) (Reversibility: ? <1-3> [1-10]) [1-10]
P S-adenosyl-L-homocysteine + protein L-glutamate methyl ester <1-3> [1-10]

Inhibitors

A9145C <3, 4> (<3,4> S-adenosylmethionine analog, antifungal antibiotic [6]; <3> 0.000001 mM, 50% inhibition [7]) [6, 7]
Ca^{2+} <1-3> (<1-3> MCP methyltransferase I [6,9]) [6, 9]
S-adenosyl-L-homocysteine <1-5> (<1-3> D-isomer is inactive [6]; <3> 0.01 mM, 50% inhibition [7]; <1> competitive vs. S-adenosyl-L-methionine [10]) [3, 6, 7, 10]
methylated receptor protein <1> [10]
p-chloromercuribenzoate <4> (<4> 0.4 mM, 40% inhibition, can be reversed by 12 mM 2-mercaptoethanol [6]) [6]
sinefungin <3, 4> (<3,4> S-adenosylmethionine analog, antifungal antibiotic [6]; <3> 0.00001 mM, 50% inhibition [7]) [6, 7]

Metals, ions

Ca^{2+} <1-3> (<2> activates, Ca^{2+} is somewhat more effective than Mg^{2+}, optimum: 5 mM Ca^{2+} [3]; <2> Mg^{2+} and Ca^{2+} activate MCP methyltransferase II [6]; <1-3> Ca^{2+}, Mn^{2+}, Cu^{2+}, Zn^{2+}, Co^{2+}, Fe^{2+}, Fe^{3+} and Mg^{2+} at 2 mM have no effect on enzyme activity of methyltransferase I [6]) [3, 6]
Mg^{2+} <1-3> (<2> activates, Ca^{2+} is somewhat more effective than Mg^{2+}, optimum: 5 mM Ca^{2+} [3]; <2> Mg^{2+} and Ca^{2+} activate MCP methyltransferase II [6]; <1-3> Ca^{2+}, Mn^{2+}, Cu^{2+}, Zn^{2+}, Co^{2+}, Fe^{2+}, Fe^{3+} and Mg^{2+} at 2 mM have no effect on enzyme activity of methyltransferase I [6]) [3, 6]

Turnover number (min^{-1})

10 <1> (protein L-glutamate) [1]

Specific activity (U/mg)

0.0000036 <7> (<7> methylation of histone II-A in purine starved cells [11]) [11]
0.000076 <2> [9]
0.0024 <2> [3]
0.0043 <4> (<4> thymus enzyme [6]) [6]
0.027 <5> (<5> erythrocyte enzyme [6]) [6]
0.03 <4> (<4> brain enzyme [6]) [6]
0.398 <1> [1]

K_m-Value (mM)

0.00087 <4> (S-adenosyl-L-methionine, <4> thymus enzyme [6]) [6]
0.002 <2> (S-adenosyl-L-methionine, <2> MCP methyltransferase I [6]) [6]
0.005 <2> (S-adenosyl-L-methionine, <2> MCP methyltransferase II [6]) [3, 6]
0.0077 <4> (follicle-stimulating hormone) [6]
0.01 <1> (S-adenosyl-L-methionine) [2, 6]
0.017 <1> (S-adenosyl-L-methionine) [1]
0.046 <4> (γ-globulin) [6]
0.1 <4> (histone) [6]
0.34 <4> (pancreatic ribonuclease) [6]

K_i-Value (mM)

0.000065 <2> (Ca^{2+}) [9]
0.0002 <2> (S-adenosyl-L-homocysteine) [3, 6, 9]
0.0004 <4> (A1945C) [6]
0.0009 <4, 5> (S-adenosyl-L-homocysteine) [6]
0.0026 <4> (sinefungin) [6]

pH-Optimum

6.5-8.5 <1, 3> [5]
6.9 <2> [3]

pH-Range

6-8 <2> (<2> approx. 30% of maximal activity at pH 6.0, approx. 65% of maximal activity at pH 8.0 [3]) [3]

Temperature optimum (°C)

20-25 <2> [3]

4 Enzyme Structure

Molecular weight

25000 <6> (<6> gel filtration [6]) [6]
30000 <2> (<2> gel filtration, SDS-PAGE [3,6]) [3, 6]
32900 <1> (<1> deduced from nucleotide sequence [1]) [1]
33000 <1> (<1> gel filtration, SDS-PAGE [1]) [1]
38000 <1> (<1> gel filtration [2]) [2]
44000 <2> (<2> MCP methyltransferase I [6,9]; <2> gel filtration [9]) [6, 9]

Subunits

monomer <1, 2> (<2> 1 * 30000, SDS-PAGE [3]; <1> 1 * 33000, SDS-PAGE [1]; <2> 1 * 44000, MCP methyltransferase I, SDS-PAGE [9]) [1, 3, 9]

5 Isolation/Preparation/Mutation/Application

Source/tissue

erythrocyte <5> [6]
thymus <4> [6]

Localization

cytoplasm <2> [3]
cytosol <1-3> [8]

Purification

<1> (partial [5]; ammonium sulfate, DEAE-cellulose, Phenyl-Sepharose, Biogel P-60, hydroxylapatite [1]) [1, 2, 5]
<2> (DEAE-Biogel, CM-Biogel, ammonium sulfate, S-adenosylhomocystein affinity column [3]) [3, 9]
<4> (brain enzyme, pH 5.1 treatment, ammonium sulfate, SAH-Sepharose, Sephadex G-100 [6]) [6]
<5> (ammonium sulfate, Sephadex G-75, SAH-Sepharose, Sephadex G-75 [6]) [6]

Cloning

<1> (expression in Escherichia coli [5]) [5]

6 Stability

Storage stability

<2>, 0°C, 0.33 mM dithiothreitol, 0.33 mg/ml bovine serum albumin, 1 week, no loss of activity [3]
<5>, -20°C, 20-50% glycerol, 5 mM sodium borate, 5 mM EDTA, 2.4 mM 2-mercaptoethanol, pH 9.3, 1 year, no loss of activity [6]

References

[1] Simms, S.A.; Stock, A.M.; Stock, J.B.: Purification and characterization of the S-adenosylmethionine:glutamyl methyltransferase that modifies membrane chemoreceptor proteins in bacteria. J. Biol. Chem., **262**, 8537-8543 (1987)

[2] Springer, W.R.; Koshland, D.E.: Identification of a protein methyltransferase as the cheR gene product in the bacterial sensing system. Proc. Natl. Acad. Sci. USA, **74**, 533-537 (1977)

[3] Burgess-Cassler, A.; Ullah, A.H.J.; Ordal, G.W.: Purification and characterization of Bacillus subtilis methyl-accepting chemotaxis protein methyltransferase II. J. Biol. Chem., **257**, 8412-8417 (1982)

[4] Kleene, S.J.; Toews, M.L.; Adler, J.: Isolation of glutamic acid methyl ester from an Escherichia coli membrane protein involved in chemotaxis. J. Biol. Chem., **252**, 3214-3218 (1977)

[5] Stock, J.B.; Clarke, S.; Koshland, D.E.: The protein carboxylmethyltransferase involved in Escherichia coli and Salmonella typhimurium chemotaxis. Methods Enzymol., **106**, 310-321 (1984)

[6] Kim, S.: S-Adenosylmethionine: protein-carboxyl O-methyltransferase (protein methylase II). Methods Enzymol., **106**, 295-309 (1984)

[7] Rollins, C.M.; Dahlquist, F.W.: Methylation of chemotaxis-specific proteins in Escherichia coli cells permeable to S-adenosylmethionine. Biochemistry, **19**, 4627-4632 (1980)

[8] Clarke, S.: Protein carboxyl methyltransferases: two distinct classes of enzymes. Annu. Rev. Biochem., **54**, 479-506 (1985)

[9] Ullah, A.H.J.; Ordal, G.W.: Purification and characterization of methyl-accepting chemotaxis protein methyltransferase I in Bacillus subtilis. Biochem. J., **199**, 795-805 (1981)

[10] Simms, S.A.; Subbaramaiah, K.: The kinetic mechanism of S-adenosyl-L-methionine: glutamylmethyltransferase from Salmonella typhimurium. J. Biol. Chem., **266**, 12741-12746 (1991)

[11] Alleman, M.M.; Mann, V.H.; Bacchi, C.J.; Yarlett, N.; Gottlieb, M.; Dwyer, D.M.: Crithidia luciliae: effect of purine starvation on S-adenosyl-L-methionine uptake and protein methylation. Exp. Parasitol., **81**, 519-528 (1995)

Nicotine N-methyltransferase 2.1.1.81

1 Nomenclature

EC number
2.1.1.81 (deleted, included in EC 2.1.1.49)

Recommended name
nicotine N-methyltransferase

3-Methylquercitin 7-O-methyltransferase 2.1.1.82

1 Nomenclature

EC number
2.1.1.82

Systematic name
S-adenosyl-L-methionine:3',4',5,7-tetrahydroxy-3-methoxyflavone 7-O-methyltransferase

Recommended name
3-methylquercitin 7-O-methyltransferase

Synonyms
7-OMT
flavonol 7-O-methyltransferase
methyltransferase, flavonol 7-

CAS registry number
96477-62-2

2 Source Organism

<1> *Chrysosplenium americanum* [1, 2]

3 Reaction and Specificity

Catalyzed reaction
S-adenosyl-L-methionine + 3',4',5,7-tetrahydroxy-3-methoxyflavone = S-adenosyl-L-homocysteine + 3',4',5-trihydroxy-3,7-dimethoxyflavone (<1> ordered bi bi mechanism, S-adenosyl-L-methionine binds first, S-adenosyl-L-homocysteine is the last product released [2])

Reaction type
methyl group transfer

Natural substrates and products
- **S** S-adenosyl-L-methionine + 3',4',5,7-tetrahydroxy-3-methoxyflavone <1> (<1> involved with EC 2.1.1.76 and 2.1.1.83 in the methylation of quercetin [1]) (Reversibility: ? <1> [1]) [1]
- **P** S-adenosyl-L-homocysteine + 3',4',5-trihydroxy-3,7-dimethoxyflavone <1> (<1> trivial name 3,7-dimethylquercetin [1]) [1]

Substrates and products

S S-adenosyl-L-methionine + 3',4',5,7-tetrahydroxy-3-methoxyflavone <1> (<1> trivial name 3-methylquercetin, specific for 3-methylquercetin [1,2]) (Reversibility: ? <1> [1]) [1, 2]

P S-adenosyl-L-homocysteine + 3',4',5-trihydroxy-3,7-dimethoxyflavone <1> (<1> trivial name 3,7-dimethylquercetin [1]) [1]

Inhibitors

Ca^{2+} <1> (<1> 1 mM, 20-50% inhibition [2]) [2]
Co^{2+} <1> (<1> 1 mM, 20-50% inhibition [2]) [2]
Cu^{2+} <1> (<1> 1 mM, 20-50% inhibition [2]) [2]
N-ethylmaleimide <1> [1, 2]
Zn^{2+} <1> (<1> 1 mM, 20-50% inhibition [2]) [2]
p-chloromercuribenzoate <1> [1, 2]

Specific activity (U/mg)

0.011 <1> [2]

K_m-Value (mM)

0.007 <1> (3',4',5,7-tetrahydroxy-3-methoxyflavone) [2]
0.065 <1> (S-adenosyl-L-methionine) [2]

pH-Optimum

7.8-8.2 <1> [1]

pH-Range

7.2-9.5 <1> (<1> approx. 50% of maximal activity at pH 7.2 and 9.5 [1]) [1]

4 Enzyme Structure

Molecular weight

57000 <1> (<1> gel filtration [1]) [1]

5 Isolation/Preparation/Mutation/Application

Source/tissue

leaf <1> (<1> shoot tip [2]) [1, 2]

Purification

<1> (loss of activity during chromatofocusing possibly due to instability of the enzyme in the purified form or to the loss of some factors required for activity [1]; ammonium sulfate, gel filtration, chromatofocusing, ion exchange [2]) [1, 2]

6 Stability

General stability information

<1>, loss of activity during chromatofocusing [1]

Storage stability

<1>, -15°C, 200 mM potassium phosphate buffer, pH 8, 20% v/v glycerol, 5 mM dithiothreitol, several months, no loss of activity [1]
<1>, -20°C, 200 mM Tris-HCl, pH 8.0, 10 mM dithiothreitol, 2 months, no loss of activity [2]
<1>, 4°C, 4-6 weeks, 80% loss of activity [1]

References

[1] De Luca, V.; Ibrahim, R.K.: Enzymatic synthesis of polymethylated flavonols in Chrysosplenium americanum. I. Partial purification and some properties of S-adenosyl-L-methionine:flavonol 3-, 6-, 7-, and 4-O-methyltransferases. Arch. Biochem. Biophys., **238**, 596-605 (1985)
[2] Khouri, H.E.; De Luca, V.; Ibrahim, R.K.: Enzymic synthesis of polymethylated flavonols in Chrysosplenium americanum. III. Purification and kinetic analysis of S-adenosyl-L-methionine:3-methylquercetin 7-O-methyltransferase. Arch. Biochem. Biophys., **265**, 1-7 (1988)

3,7-Dimethylquercitin 4'-O-methyltransferase 2.1.1.83

1 Nomenclature

EC number
2.1.1.83

Systematic name
S-adenosyl-L-methionine:3',4',5-trihydroxy-3,7-dimethoxyflavone 4'-O-methyltransferase

Recommended name
3,7-dimethylquercitin 4'-O-methyltransferase

Synonyms
4'-OMT
flavonol 4'-O-methyltransferase
methyltransferase, flavonol 4'-

CAS registry number
96477-60-0

2 Source Organism

<1> *Chrysosplenium americanum* [1, 2, 3]

3 Reaction and Specificity

Catalyzed reaction
S-adenosyl-L-methionine + 3',4',5-trihydroxy-3,7-dimethoxyflavone = S-adenosyl-L-homocysteine + 3',5-dihydroxy-3,4',7-trimethoxyflavone (mechanism, <1> [2])

Reaction type
methyl group transfer

Natural substrates and products
S S-adenosyl-L-methionine + 3',4',5-trihydroxy-3,7-dimethoxyflavone <1> (<1> trivial name 3,7-dimethylquercetin, enzyme does not act on flavones, dihydroflavonols or their glucosides, involved with EC 2.1.1.76 and 2.1.1.82 in the methylation of quercitin to 3,7,4'-trimethylquercitin [1]) (Reversibility: ? <1> [1]) [1]

P S-adenosyl-L-homocysteine + 3',5-dihydroxy-3,4',7-trimethoxyflavone <1> (<1> trivial name 3,7,4'-trimethylquercetin [1]) [1]

Substrates and products

S S-adenosyl-L-methionine + 3',4',5-trihydroxy-3,7-dimethoxyflavone <1> (<1> trivial name 3,7-dimethylquercetin, enzyme does not act on flavones, dihydroflavonols or their glucosides [1]) (Reversibility: ? <1> [1]) [1]

P S-adenosyl-L-homocysteine + 3',5-dihydroxy-3,4',7-trimethoxyflavone <1> (<1> trivial name 3,7,4'-trimethylquercetin [1]) [1]

S S-adenosyl-L-methionine + 3,7-dimethylquercetagetin <1> [1]

P S-adenosyl-L-homocysteine + 3,7,4'-trimethylquercetagetin <1> [1]

Inhibitors

3,5-dihydroxy-3,4,7'-trimethoxyflavone <1> [2]
N-ethylmaleimide <1> [1]
S-adenosyl-L-homocysteine <1> (<1> product inhibition [2]) [2]
p-chloromercuribenzoate <1> [1]

Specific activity (U/mg)

7.2 <1> [1]

K_m-Value (mM)

0.015 <1> (3,7-dimethylquercetin) [2]
0.13 <1> (S-adenosyl-L-methionine) [2]

pH-Optimum

8.8-9 <1> [1]

pH-Range

7.2-9.5 <1> (<1> approx. 50% of maximal activity at pH 7.2 and 9.5 [1]) [1]

4 Enzyme Structure

Molecular weight

57000 <1> (<1> gel filtration [1]) [1]

5 Isolation/Preparation/Mutation/Application

Source/tissue

leaf <1> [1, 2]

Purification

<1> (chromatofocusing on Superose 12 and Mono P [3]) [1]

6 Stability

Storage stability

<1>, -15°C, 200 mM potassium phosphate buffer, pH 8, 20% glycerol, 5 mM dithiothreitol, several months, no loss of activity [1]
<1>, 4°C, 4-6 weeks, 80% loss of activity [1]

References

[1] De Luca, V.; Ibrahim, R.K.: Enzymatic synthesis of polymethylated flavonols in Chrysosplenium americanum. I. Partial purification and some properties of S-adenosyl-L-methionine:flavonol 3-, 6-, 7-, and 4-O-methyltransferases. Arch. Biochem. Biophys., **238**, 596-605 (1985)
[2] De Luca, V.; Ibrahim, R.K.: Enzymatic synthesis of polymethylated flavonols in Chrysosplenium americanum. II. Substrate interaction and product inhibition studies of flavonol 3-, 6-, and 4-O-methyltransferases. Arch. Biochem. Biophys., **238**, 606-618 (1985)
[3] Khouri, H.E.; Ibrahim, R.K.: Resolution of five position-specific flavonoid O-methyltransferases by fast protein liquid chromatofocusing. J. Chromatogr., **407**, 291-297 (1987)

Methylquercetagetin 6-O-methyltransferase 2.1.1.84

1 Nomenclature

EC number
2.1.1.84

Systematic name
S-adenosyl-L-methionine:3',4',5,6-tetrahydroxy-3,7-dimethoxyflavone 6-O-methyltransferase

Recommended name
methylquercetagetin 6-O-methyltransferase

Synonyms
6-OMT
flavonol 6-O-methyltransferase
methyltransferase, flavonol 6-

CAS registry number
96477-61-1

2 Source Organism

<1> *Chrysosplenium americanum* [1-3]

3 Reaction and Specificity

Catalyzed reaction
S-adenosyl-L-methionine + 3',4',5,6-tetrahydroxy-3,7-dimethoxyflavone = S-adenosyl-L-homocysteine + 3',4',5-trihydroxy-3,6,7-trimethoxyflavone (<1> mechanism [2])

Reaction type
methyl group transfer

Natural substrates and products
S S-adenosyl-L-methionine + 3',4',5,6-tetrahydroxy-3,7-dimethoxyflavone <1> (Reversibility: ? <1> [1]) [1]
P S-adenosyl-L-homocysteine + 3',4',5-trihydroxy-3,6,7-trimethoxyflavone <1> [1]

Substrates and products

- **S** 3,7,3'-trimethylquercetagetin + S-adenosyl-L-methionine <1> (Reversibility: ? <1> [1]) [1]
- **P** 3,6,7',3'-tetrametylquercetagetin + S-adenosyl-L-homocysteine <1> [1]
- **S** 8-hydroxykaempferol + S-adenosyl-L-methionine <1> (<1> 25% of activity with 3',4',5,6-tetrahydroxy-3,7-dimethoxyflavone [1]) (Reversibility: ? <1> [1]) [1]
- **P** ?
- **S** S-adenosyl-L-methionine + 3',4',5,6-tetrahydroxy-3,7-dimethoxyflavone <1> (Reversibility: ? <1> [1]) [1]
- **P** S-adenosyl-L-homocysteine + 3',4',5-trihydroxy-3,6,7-trimethoxyflavone <1> [1]

Inhibitors

3,6,7,3'-tetramethylquercetagetin <1> (<1> noncompetitive product inhibition [2]) [2]
EDTA <1> [1]
N-ethylmaleimide <1> [1]
S-adenosyl-L-homocysteine <1> (<1> competitive product inhibition [2]) [2]
p-chloromercuribenzoate <1> [1]

Metals, ions

Mg^{2+} <1> (<1> activates [1]) [1]

Specific activity (U/mg)

2.64 <1> [1]

K_m-Value (mM)

0.018 <1> (3,7,3'-trimethylquercetagetin) [2]
0.051 <1> (S-adenosyl-L-methionine) [2]

pH-Optimum

8.8-9 <1> [1]

4 Enzyme Structure

Molecular weight

57000 <1> (<1> gel filtration [1]) [1]

5 Isolation/Preparation/Mutation/Application

Source/tissue

leaf <1> [1-3]

Purification

<1> (ammonium sulfate, Sephacryl S-200, hydroxylapatite, chromatofocusing [1]; chromatofocusing on Superose 12 and Mono P [3]) [1-3]

6 Stability

Storage stability

<1>, -15°C, 0.2 M potassium phosphate, pH 8, 20% glycerol, 5 mM dithiothreitol, for several months, no loss of activity [1]
<1>, 4°C, 4-6 weeks, 80% loss of activity [1]

References

[1] De Luca, V.; Ibrahim, R.K.: Enzymatic synthesis of polymethylated flavonols in Chrysosplenium americanum. I. Partial purification and some properties of S-adenosyl-L-methionine:flavonol 3-, 6-, 7-, and 4-O-methyltransferases. Arch. Biochem. Biophys., **238**, 596-605 (1985)
[2] De Luca, V.; Ibrahim, R.K.: Enzymatic synthesis of polymethylated flavonols in Chrysosplenium americanum. II. Substrate interaction and product inhibition studies of flavonol 3-, 6-, and 4-O-methyltransferases. Arch. Biochem. Biophys., **238**, 606-618 (1985)
[3] Khouri, H.E.; Ibrahim, R.K.: Resolution of five position-specific flavonoid O-methyltransferases by fast protein liquid chromatofocusing. J. Chromatogr., **407**, 291-297 (1987)

Protein-histidine N-methyltransferase 2.1.1.85

1 Nomenclature

EC number
2.1.1.85

Systematic name
S-adenosyl-L-methionine:protein-L-histidine N-tele-methyltransferase

Recommended name
protein-histidine N-methyltransferase

Synonyms
S-adenosyl methionine:protein-histidine N-methyltransferase
actin-specific histidine methyltransferase
methyltransferase, protein (histidine)
protein methylase IV

CAS registry number
108022-17-9

2 Source Organism

<1> *Oryctolagus cuniculus* (rabbit [1]) [1, 2]

3 Reaction and Specificity

Catalyzed reaction
S-adenosyl-L-methionine + protein L-histidine = S-adenosyl-L-homocysteine + protein N^{τ}-methyl-L-histidine

Reaction type
methyl group transfer

Natural substrates and products
S S-adenosyl-L-methionine + protein L-histidine <1> (Reversibility: ? <1> [1, 2]) [1, 2]
P S-adenosyl-L-homocysteine + protein N^{τ}-methyl-L-histidine
S S-adenosyl-L-methionine + synthetic peptide corresponding to residue 69-77 of actin <1> (Reversibility: ? <1> [2]) [2]
P S-adenosyl-L-homocysteine + peptide N^{τ}-methyl-L-histidine

Substrates and products

S S-adenosyl-L-methionine + protein L-histidine <1> (<1> chicken β-actin expressed in E. coli as a fusion protein with 80 amino acids of an influenza protein, NS1 [2]; <1> specifically methylates histidine residues of actin [1,2]; <1> some activity with gluten [1]; <1> negligible activity with: bovine serum albumin, fibrinogen, casein, histone type IIA, gelatin, human γ-globulin [1]; <1> naturally occuring actins are poor substrates, due to preexistent methylation at His 73 [2]) (Reversibility: ? <1> [1, 2]) [1, 2]
P S-adenosyl-L-homocysteine + protein N^{τ}-methyl-L-histidine <1> [1, 2]
S S-adenosyl-L-methionine + synthetic peptide corresponding to residue 69-77 of actin <1> (<1> Tyr-Pro-Ile-Glu-His-Gly-Ile-Ile-Thr [2]) (Reversibility: ? <1> [2]) [2]
P S-adenosyl-L-homocysteine + peptide N^{τ}-methyl-L-histidine

Inhibitors

ATP <1> (<1> 47% inhibition at 0.4 mM [1]) [1]
Ca^{2+} <1> (<1> 30% inhibition at 0.4 mM [1]) [1]
EDTA <1> (<1> slight inhibition [1]) [1]
K^{+} <1> (<1> 72% inhibition at 200 mM [1]) [1]
Mg^{2+} <1> (<1> 60% inhibition at 8 mM [1]) [1]
N-ethylmaleimide <1> (<1> 70% inhibition at 4 mM [1]) [1]
Na^{+} <1> (<1> 95% inhibition at 200 mM [1]) [1]
borate <1> (<1> slight inhibition [1]) [1]

Activating compounds

Triton X-100 <1> (<1> stimulation at 1% [1]) [1]
Additional information <1> (<1> reducing agents e.g. mercaptoethanol, required for activity, 4-fold stimulation [1]) [1]

Metals, ions

Additional information <1> (<1> no requirement for any metal ion [1]) [1]

Specific activity (U/mg)

Additional information <1> (<1> different fractions after ammonium sulfate precipitation [1]) [1]

pH-Optimum

8 <1> [1]

pH-Range

6.8-9 <1> (<1> about 50% of activity maximum at pH 6.8 and 9.0 [1]) [1]

Temperature optimum (°C)

37 <1> (<1> assay at [1]) [1]

4 Enzyme Structure

Molecular weight

70000 <1> (<1> gel filtration [2]) [2]

5 Isolation/Preparation/Mutation/Application

Source/tissue
skeletal muscle <1> [1, 2]

Localization
myofibril <1> [1, 2]

Purification
<1> (partial [1]) [1]

6 Stability

Temperature stability
40 <1> (<1> 10 min, 75% loss of activity [1]) [1]
50 <1> (<1> 10 min, complete loss of activity [1]) [1]

General stability information
<1>, 2-mercaptoethanol or glycerol does not stabilize [1]

Storage stability
<1>, -20°C [1]
<1>, 2°C, 4 days, 75% loss of activity, no stabilization by mercaptoethanol or glycerol [1]

References

[1] Vijayasarathy, C.; Rao, B.S.N.: Partial purification and characterisation of S-adenosylmethionine:protein-histidine N-methyltransferase from rabbit skeletal muscle. Biochim. Biophys. Acta, **923**, 156-165 (1987)
[2] Raghavan, M.; Lindberg, U.; Schutt, C.: The use of alternative substrates in the characterization of actin-methylating and carnosine-methylating enzymes. Eur. J. Biochem., **210**, 311-318 (1992)

Tetrahydromethanopterin S-methyltransferase 2.1.1.86

1 Nomenclature

EC number
2.1.1.86

Systematic name
5-methyl-5,6,7,8-tetrahydromethanopterin:2-mercaptoethanesulfonate 2-methyltransferase

Recommended name
tetrahydromethanopterin S-methyltransferase

Synonyms
methyltransferase, tetrahydromethanopterin
tetrahydromethanopterin methyltransferase

CAS registry number
103406-60-6

2 Source Organism

<1> *Methanobacterium thermoautotrophicum* [1]

3 Reaction and Specificity

Catalyzed reaction
5-methyl-5,6,7,8-tetrahydromethanopterin + 2-mercaptoethanesulfonate = 5,6,7,8-tetrahydromethanopterin + 2-(methylthio)ethanesulfonate

Reaction type
methyl group transfer

Natural substrates and products
S 5-methyl-5,6,7,8-tetrahydromethanopterin + 2-mercaptoethanesulfonate <1> (<1> involved in the formation of methane from CO [1]) (Reversibility: r <1> [1]) [1]
P 5,6,7,8-tetrahydromethanopterin + 2-(methylthio)ethanesulfonate <1> [1]

Substrates and products
S 5-methyl-5,6,7,8-tetrahydromethanopterin + 2-mercaptoethanesulfonate <1> (Reversibility: r <1> [1]) [1]
P 5,6,7,8-tetrahydromethanopterin + 2-(methylthio)ethanesulfonate <1> [1]

K_m-Value (mM)

0.88 <1> (5,6,7,8-tetrahydromethanopterin) [1]
3.7 <1> ((2-methylthio)ethanesulfonate) [1]

pH-Optimum

6.7 <1> [1]

pH-Range

6-7 <1> (<1> sharp decrease in activity below pH 6.0 and above pH 7.0 [1]) [1]

Temperature optimum (°C)

60 <1> (<1> assay at [1]) [1]

6 Stability

Temperature stability

100 <1> (<1> 5 min, inactivation [1]) [1]

Oxidation stability

<1>, oxygen-sensitive [1]

References

[1] Sauer, F.D.: Tetrahydromethanopterin methyltransferase, a component of the methane synthesizing complex of Methanobacterium thermoautotrophicum. Biochem. Biophys. Res. Commun., **136**, 542-547 (1986)

Pyridine N-methyltransferase 2.1.1.87

1 Nomenclature

EC number
2.1.1.87

Systematic name
S-adenosyl-L-methionine:pyridine N-methyltransferase

Recommended name
pyridine N-methyltransferase

Synonyms
methyltransferase, pyridine

CAS registry number
104327-10-8

2 Source Organism

<1> *Oryctolagus cuniculus* (rabbit, strain New Zealand White [1]) [1]

3 Reaction and Specificity

Catalyzed reaction
S-adenosyl-L-methionine + pyridine = S-adenosyl-L-homocysteine + N-methylpyridinium

Reaction type
methyl group transfer

Natural substrates and products
S S-adenosyl-L-methionine + pyridine <1> (Reversibility: ? <1> [1]) [1]
P S-adenosyl-L-homocysteine + N-methylpyridinium <1> [1]

Substrates and products
S S-adenosyl-L-methionine + pyridine <1> (<1> N^5-methyltetrahydrofolic acid cannot replace S-adenosyl-L-methionine [1]) (Reversibility: ? <1> [1]) [1]
P S-adenosyl-L-homocysteine + N-methylpyridinium <1> [1]

Inhibitors

S-adenosyl-L-homocysteine <1> [1]
Additional information <1> (<1> endogenous low-molecular-weight substance, enzyme activity increases after dialysis [1]) [1]

pH-Optimum

7.9 <1> (<1> assay at [1]) [1]

Temperature optimum (°C)

37 <1> (<1> assay at [1]) [1]

5 Isolation/Preparation/Mutation/Application

Source/tissue

kidney <1> [1]
liver <1> [1]
lung <1> [1]
Additional information <1> (<1> no activity in brain [1]) [1]

Localization

cytosol <1> (<1> no activity in mitochondria or microsomes [1]) [1]

6 Stability

Storage stability

<1>, -20°C, up to a month [1]

References

[1] Damani, L.A.; Shaker, M.S.; Crooks, P.A.; Godin, C.S.; Nwosu, C.: N-methylation and quaternization of pyridine in vitro by rabbit lung, liver and kidney N-methyltransferases: an S-adenosyl-L-methionine-dependent reaction. Xenobiotica, **16**, 645-650 (1986)

8-Hydroxyquercitin 8-O-methyltransferase 2.1.1.88

1 Nomenclature

EC number
2.1.1.88

Systematic name
S-adenosyl-L-methionine:3,3',4',5,7,8-hexahydroxyflavone 8-O-methyltransferase

Recommended name
8-hydroxyquercitin 8-O-methyltransferase

Synonyms
flavonol 8-O-methyltransferase
methyltransferase, flavonol 8-

CAS registry number
99775-17-4

2 Source Organism

<1> *Lotus corniculatus* [1]

3 Reaction and Specificity

Catalyzed reaction
S-adenosyl-L-methionine + 3,3',4',5,7,8-hexahydroxyflavone = S-adenosyl-L-homocysteine + 3,3',4',5,7-pentahydroxy-8-methoxy-flavone (<1> mechanism [1])

Reaction type
methyl group transfer

Natural substrates and products
S S-adenosyl-L-methionine + 3,3',4',5,7,8-hexahydroxyflavone <1> (Reversibility: ? <1> [1]) [1]
P S-adenosyl-L-homocysteine + 3,3',4',5,7-pentahydroxy-8-methoxyflavone <1> [1]
S S-adenosyl-L-methionine + 8-hydroxykaempferol <1> (Reversibility: ? <1> [1]) [1]
P S-adenosyl-L-homocysteine + 8-methoxykaempferol <1> [1]

Substrates and products

- **S** S-adenosyl-L-methionine + 3,3',4',5,7,8-hexahydroxyflavone <1> (<1> i.e. 8-hydroxyquercetin [1]) (Reversibility: ? <1> [1]) [1]
- **P** S-adenosyl-L-homocysteine + 3,3',4',5,7-pentahydroxy-8-methoxyflavone <1> (<1> i.e. 8-methoxyquercetin [1]) [1]
- **S** S-adenosyl-L-methionine + 8-hydroxykaempferol <1> (Reversibility: ? <1> [1]) [1]
- **P** S-adenosyl-L-homocysteine + 8-methoxykaempferol <1> [1]
- **S** Additional information <1> (<1> not: glycosides of 8-hydroxyflavonols [1]) [1]
- **P** ?

Inhibitors

8-methoxykaempferol <1> (<1> product inhibition, competitive with respect to 8-hydroxykaempferol [1]) [1]
EDTA <1> (<1> in absence of Mg^{2+} [1]) [1]
S-adenosyl-L-homocysteine <1> (<1> product inhibition, non-competitive with respect to S-adenosyl-L-methionine [1]) [1]

Metals, ions

Mg^{2+} <1> (<1> required [1]) [1]

Specific activity (U/mg)

Additional information <1> [1]

K_m-Value (mM)

0.0013 <1> (8-hydroxykaempferol) [1]
0.053 <1> (S-adenosyl-L-methionine) [1]

pH-Optimum

7.9 <1> [1]

4 Enzyme Structure

Molecular weight

55000 <1> (<1> gel filtration [1]) [1]

5 Isolation/Preparation/Mutation/Application

Source/tissue

bud <1> [1]

Purification

<1> (1230fold [1]) [1]

References

[1] Jay, M.; De Luca, V.; Ibrahim, R.K.: Purification, properties and kinetic mechanism of flavonol 8-O-methyltransferase from Lotus corniculatus L. Eur. J. Biochem., **153**, 321-325 (1985)

Tetrahydrocolumbamine 2-O-methyltransferase 2.1.1.89

1 Nomenclature

EC number
2.1.1.89

Systematic name
S-adenosyl-L-methionine:5,8,13,13a-tetrahydrocolumbamine 2-O-methyltransferase

Recommended name
tetrahydrocolumbamine 2-O-methyltransferase

Synonyms
methyltransferase, tetrahydrocolumbamine
tetrahydrocolumbamine methyltransferase

CAS registry number
93792-09-7

2 Source Organism

<1> *Berberis aggregata* [1]

3 Reaction and Specificity

Catalyzed reaction
S-adenosyl-L-methionine + 5,8,13,13a-tetrahydrocolumbamine = S-adenosyl-L-homocysteine + tetrahydropalmatine

Reaction type
methyl group transfer

Natural substrates and products
- **S** S-adenosyl-L-methionine + tetrahydrocolumbamine <1> (<1> enzyme is involved in the biosynthesis of berberine alkaloids [1]) (Reversibility: ? <1> [1]) [1]
- **P** S-adenosyl-L-homocysteine + tetrahydropalmatine <1> [1]

Substrates and products

S S-adenosyl-L-methionine + tetrahydrocolumbamine <1> (<1> no activity with tetrahydrojatrorhizine, tetrahydropalmatrubine, columbamine, jatrorrhizine and palmatrubine [1]) (Reversibility: ? <1> [1]) [1]

P S-adenosyl-L-homocysteine + tetrahydropalmatine <1> [1]

Temperature optimum (°C)

30 <1> (<1> assay at [1]) [1]

5 Isolation/Preparation/Mutation/Application

Source/tissue

callus <1> [1]

References

[1] Beecher, C.W.W.; Kelleher, W.J.: enzymatic study of the late stages of protoberberine alkaloid biosynthesis. Tetrahedron Lett., 25, 4595-4598 (1984)

Methanol-5-hydroxybenzimidazolylcobamide Co-methyltransferase 2.1.1.90

1 Nomenclature

EC number
2.1.1.90

Systematic name
methanol:5-hydroxybenzimidazolylcobamide Co-methyltransferase

Recommended name
methanol-5-hydroxybenzimidazolylcobamide Co-methyltransferase

Synonyms
MT 1
methanol cobalamin methyltransferase
methanol:5-hydroxybenzimidazolylcobamide methyltransferase
methyltransferase, methanol-cobalamin

CAS registry number
86611-98-5

2 Source Organism

<1> *Methanosarcina barkeri* [1-3, 6, 7]
<2> *Methanobacterium bryantii* [5]
<3> *Methanobacterium thermoautotrophicum* [4]
<4> *Eubacterium limosum* [4]

3 Reaction and Specificity

Catalyzed reaction
methanol + 5-hydroxybenzimidazolylcobamide = Co-methyl-Co-5-hydroxybenzimidazolylcob(I)amide + H_2O

Reaction type
methyl group transfer

Natural substrates and products
S methanol + 5-hydroxybenzimidazolylcobamide <1> (<1> one of the enzymes responsible for the transmethylation from methanol to coenzyme M [1]) (Reversibility: ? <1> [1]) [1]
P Co-methyl-Co-5-hydroxybenzimidazolylcobamide + H_2O <1> [1]

Substrates and products

S methanol + 5-hydroxybenzimidazolylcobamide <1> (Reversibility: ? <1> [1]) [1, 7]
P Co-methyl-Co-5-hydroxybenzimidazolylcobamide + H_2O <1> [1, 7]
S methanol + cob(I)alamin <1> (<1> catalyzed by β subunit MtaB alone [6]) (Reversibility: r <1> [6]) [6, 7]
P methyl-cob(III)alamin + H_2O <1> [6, 7]

Inhibitors

$HgCl_2$ <1> (<1> 1 mM, 90% inhibition [1]) [1]
NH_4Cl <1> (<1> 400 mM, 93% inhibition [1]) [1]
Na_2SO_3 <1> (<1> 1 mM, 50% inhibition [1]) [1]
$NaNO_2$ <1> (<1> 0.1 mM, 90% inhibition [1]) [1]
O_2 <1> (<1> inactivation by O_2 and other oxidizing agents [1]) [1]
pyridoxal 5'-phosphate <1> (<1> 1 mM, 74% inhibition [1]) [1]

Cofactors/prosthetic groups

5-hydroxybenzimidazolylcobamide <1> (<1> 3-4 molecules bound, acting as methyl acceptor [1]; <1> corrinoid protein [7]) [1, 3, 7]

Activating compounds

ATP <1> (<1> half-maximal activation at 0.16 mM ATP [1]) [1]
Additional information <1> (<1> reductive activation by a mixture of H_2, ferredoxin needed in the reductive activation process [1,2]; <1> reductive activation by a mixture of H_2, ferredoxin and hydrogenase but also by CO, ATP, GTP or CTP needed in the reductive activation process [1]) [1, 2]

Metals, ions

Ba^{2+} <1> (<1> metal-free enzyme preparation has no activity, addition of Ba^{2+} restores 27% of the original activity [1]) [1]
Ca^{2+} <1> (<1> metal-free enzyme preparation has no activity, addition of Ca^{2+} restores 35% of the original activity [1]) [1]
Co^{2+} <1> (<1> metal-free enzyme preparation has no activity, addition of Co^{2+} restores 46% of the original activity [1]) [1]
Mg^{2+} <1> (<1> metal-free enzyme preparation has no activity, addition of Mg^{2+} restores 66% of the original activity [1]) [1]
Mn^{2+} <1> (<1> metal-free enzyme preparation has no activity, addition of Mn^{2+} restores 27% of the original activity [1]) [1]
Ni^{2+} <1> (<1> metal-free enzyme preparation has no activity, addition of Ni^{2+} restores 21% of the original activity [1]) [1]
Sr^{2+} <1> (<1> metal-free enzyme preparation has no activity, addition of Sr^{2+} restores 20% of the original activity [1]) [1]
Zn^{2+} <1> (<1> β subunit MtaB contains 1 mol Zn^{2+}/mol subunit, Zn^{2+} can be substituted by Co^{2+} [6]) [6]

Specific activity (U/mg)

750 <1> [7]

K_m-Value (mM)

40 <1> (methanol, <1> β subunit MtaB [6]) [6]

pH-Optimum

7.2 <1> (<1> assay at [1]) [1]

Temperature optimum (°C)

37 <1> (<1> assay at [1]) [1]

4 Enzyme Structure

Molecular weight

122000 <1> (<1> PAGE with different gel concentrations [1]) [1]

Subunits

trimer <1> (<1> α_2,β, 2 * 34000 + 1 * 53000, SDS-PAGE in presence of 2-mercaptoethanol [1]; <1> α_2,β, 2 * 49000 + 1 * 24000, β subunit MtaB harbours the corrinoid prosthetic group [7]) [1, 7]

5 Isolation/Preparation/Mutation/Application

Purification

<1> [1, 2, 7]

Cloning

<1> (expression of β subunit MtaB in Escherichia coli [6]) [6, 7]

6 Stability

Oxidation stability

<1>, oxygen-sensitive [1, 3]

General stability information

<1>, presence of divalent cations e.g. Mg^{2+}, Mn^{2+}, Sr^{2+}, Ca^{2+} or Ba^{2+} are required for stability [1]

References

[1] Van der Meijden, P.; te Brömmelstroet, B.W.; Poirot, C.M.; van der Drift, C.; Vogels, G.D.: Purification and properties of methanol:5-hydroxybenzimidazolylcobamide methyltransferase from Methanosarcina barkeri. J. Bacteriol., **160**, 629-635 (1984)

[2] Van der Meijden, P.; van der Lest, C.; van der Drift, C.; Vogels, G.D.: Reductive activation of methanol: 5-hydroxybenzimidazolylcobamide methyltransferase of Methanosarcina barkeri. Biochem. Biophys. Res. Commun., **118**, 760-766 (1984)

[3] Van der Meijden, P.; Heythuysen, H.J.; Pouwels, A.; Houwen, F.; van der Drift, C.; Vogels, G.D.: Methyltransferases involved in methanol conversion by Methanosarcina barkeri. Arch. Microbiol., **134**, 238-242 (1983)
[4] Van der Meijden, P.; van der Drift, C.; Vogels, G.D.: Methanol conversion in Eubacterium limosum. Arch. Microbiol., **138**, 360-364 (1984)
[5] Taylor, C.D.; Wolfe, R.S.: A simplified assay for coenzyme M ($HSCH_2CH_2SO_3$). Resolution of methylcobalamin-coenzyme M methyltransferase and use of sodium borohydride. J. Biol. Chem., **249**, 4886-4890 (1974)
[6] Sauer, K.; Thauer, R.K.: Methanol:coenzyme M methyltransferase from Methanosarcina barkeri. Zinc dependence and thermodynamics of the methanol:cob(I)alamin methyltransferase reaction. Eur. J. Biochem., **249**, 280-285 (1997)
[7] Sauer, K.; Harms, U.; Thauer, R.K.: Methanol:coenzyme M methyltransferase from Methanosarcina barkeri. Purification, properties, and encoding genes of the corrinoid protein MT1. Eur. J. Biochem., **243**, 670-677 (1997)

Isobutyraldoxime O-methyltransferase 2.1.1.91

1 Nomenclature

EC number
2.1.1.91

Systematic name
S-adenosyl-L-methionine:2-methylpropanal-oxime O-methyltransferase

Recommended name
isobutyraldoxime O-methyltransferase

Synonyms
S-adenosylmethionine:aldoxime O-methyltransferase
aldoxime methyltransferase
methyltransferase, aldoxime O-

CAS registry number
95471-32-2

2 Source Organism

<1> *Pseudomonas sp.* (N.C.I.B. 11652 [1]) [1]

3 Reaction and Specificity

Catalyzed reaction
S-adenosyl-L-methionine + 2-methylpropanal oxime = S-adenosyl-L-homocysteine + 2-methylpropanal O-methyloxime

Reaction type
methyl group transfer

Substrates and products
S 2-methylbutyraldoxime + S-adenosyl-L-methionine <1> (<1> highest activity [1]) (Reversibility: ? <1> [1]) [1]
P 2-methylbutyraldoxime methyl ether + S-adenosyl-L-homocysteine <1> [1]
S 3-methylbutyraldoxime + S-adenosyl-L-methionine <1> (Reversibility: ? <1> [1]) [1]
P 3-methylbutyraldoxime methyl ether + S-adenosyl-L-homocysteine <1> [1]

S isobutyraldoxime + S-adenosyl-L-methionine <1> (Reversibility: ? <1> [1]) [1]
P isobutyraldoxime methyl ether + S-adenosyl-L-homocysteine <1> [1]
S methacrylaldoxime + S-adenosyl-L-methionine <1> (Reversibility: ? <1> [1]) [1]
P methacrylaldoxime methyl ether + S-adenosyl-L-homocysteine <1> [1]
S n-butyraldoxime + S-adenosyl-L-methionine <1> (Reversibility: ? <1> [1]) [1]
P n-butyraldoxime methyl ether + S-adenosyl-L-homocysteine <1> [1]
S n-pentanaldoxime + S-adenosyl-L-methionine <1> (Reversibility: ? <1> [1]) [1]
P n-pentanaldoxime methyl ether + S-adenosyl-L-homocysteine <1> [1]
S n-propionaldoxime + S-adenosyl-L-methionine <1> (Reversibility: ? <1> [1]) [1]
P n-propionaldoxime methyl ether + S-adenosyl-L-homocysteine <1> [1]
S Additional information <1> (<1> no reaction with n-hexanaldoxime, benzaldoxime, 2-butoxime [1]) [1]
P ?

Inhibitors

Ag^{2+} <1> (<1> 100% inhibition at 1 mM [1]) [1]
Hg^{2+} <1> (<1> 100% inhibition at 0.05 mM [1]) [1]
N-ethylmaleimide <1> (<1> 56% inhibition at 0.05 mM [1]) [1]
S-adenosyl-L-homocysteine <1> (<1> competitive inhibition [1]) [1]
iodoacetamide <1> (<1> 55% inhibition at 0.5 mM [1]) [1]
phenylmercuriacetate <1> (<1> 95% inhibition at 0.025 mM [1]) [1]
Additional information <1> (<1> EDTA and Mg^{2+} are not inhibitory [1]) [1]

Activating compounds

Additional information <1> (<1> requires thiol-protecting agents for activity [1]) [1]

Metals, ions

Additional information <1> (<1> Mg^{2+} has no significant effect [1]) [1]

Specific activity (U/mg)

247.4 <1> [1]

K_m-Value (mM)

0.14 <1> (2-methylbutyraldoxime) [1]
0.15 <1> (S-adenosyl-L-methionine) [1]
0.16 <1> (3-methylbutyraldoxime) [1]
0.24-0.27 <1> (isobutyraldoxime) [1]
0.67 <1> (n-pentanaldoxime) [1]
0.98 <1> (n-butyraldoxime) [1]
1.11 <1> (methacrylaldoxime) [1]
1.54 <1> (n-propionaldoxime) [1]

K_i-Value (mM)

0.027 <1> (S-adenosyl-L-homocysteine) [1]

pH-Optimum
10.25 <1> [1]

pH-Range
9-10.4 <1> (<1> less than 50% of maximal activity below pH 9.0, almost no activity above pH 10.4 [1]) [1]

Temperature optimum (°C)
25 <1> (<1> assay at [1]) [1]

4 Enzyme Structure

Molecular weight
84000 <1> (<1> gel filtration [1]) [1]

Subunits
dimer <1> (<1> 2 * 37500, SDS-PAGE [1]) [1]

5 Isolation/Preparation/Mutation/Application

Purification
<1> (220fold [1]) [1]

6 Stability

Temperature stability
35 <1> (<1> rapid inactivation above [1]) [1]
Additional information <1> (<1> temperature sensitive: 0-3°C are necessary during purification [1]) [1]

Oxidation stability
<1>, labile to O_2, thiol-protecting agents are necessary [1]

Storage stability
<1>, -15°C, 100 mM potassium phosphate buffer, pH 7.5, 5 mM DTT, several months [1]
<1>, 0°C, $(NH_4)_2SO_4$-paste has a half life of 10 weeks [1]

References

[1] Harper, D.B.; Kennedy, J.T.: Purification and properties of S-adenosylmethionine: aldoxime O-methyltransferase from Pseudomonas sp. N.C.I.B. 11652. Biochem. J., **226**, 147-153 (1985)

Bergaptol O-methyltransferase 2.1.1.92

1 Nomenclature

EC number
2.1.1.92

Systematic name
S-adenosyl-L-methionine:bergaptol O-methyltransferase

Recommended name
bergaptol O-methyltransferase

Synonyms
BMT
methyltransferase, bergaptol

CAS registry number
101637-31-4

2 Source Organism

<1> *Petroselinum crispum* [1-6]

3 Reaction and Specificity

Catalyzed reaction
S-adenosyl-L-methionine + bergaptol = S-adenosyl-L-homocysteine + O-methylbergaptol

Reaction type
methyl group transfer

Natural substrates and products

S S-adenosyl-L-methionine + bergaptol <1> (<1>, final methylation in the biosynthesis of the furanocoumarin bergapten [1]; <1>, part of plant/pathogen interaction, enzyme of the furanocoumarin pathway [3]; <1>, late enzyme of the furanocoumarin pathway [4]) (Reversibility: ? <1> [1, 3, 4]) [1, 3, 4]

P ?

Substrates and products

S S-adenosyl-L-methionine + 5,7-dihydroxycoumarin <1> (<1>, 6% of the activity with bergaptol [1]) (Reversibility: ? <1> [1]) [1]

P S-adenosyl-L-homocysteine + citropten <1> [1]
S S-adenosyl-L-methionine + 5,8-dihydroxypsoralen <1> (<1>, 2.6% of the activity with bergaptol [1]) (Reversibility: ? <1> [1]) [1]
P S-adenosyl-L-homocysteine + ?
S S-adenosyl-L-methionine + 5-hydroxyxanthotoxin <1> (<1>, best substrate [1]) (Reversibility: ? <1> [1]) [1]
P S-adenosyl-L-homocysteine + isopimpinellin <1> [1]
S S-adenosyl-L-methionine + bergaptol <1> (Reversibility: ? <1> [1-6]) [1-6]
P S-adenosyl-L-homocysteine + bergapten <1> [1-6]

Specific activity (U/mg)
0.041 <1> [1]

K_m-Value (mM)
0.001 <1> (5-hydroxyxanthotoxin) [1]
0.0031 <1> (S-adenosyl-L-methionine, <1>, reaction with bergaptol or 5-hydroxyxanthotoxin [1]) [1]
0.004 <1> (bergaptol) [1]

pH-Optimum
8-8.5 <1> [1]

pH-Range
6.5-9.3 <1> (<1>, pH 6.5: about 30% of maximal activity, pH 9.3: about 45% of maximal activity [1]) [1]

4 Enzyme Structure

Molecular weight
73000 <1> (<1>, gel filtration [1]) [1]

Subunits
dimer <1> (<1>, 2 * 36000, SDS-PAGE [1]) [1]

5 Isolation/Preparation/Mutation/Application

Source/tissue
anther <1> (<1>, tapetum and pollen [6]) [6]
cell suspension culture <1> (<1>, from leaf petiols, cells activated by an elicitor from Phytophtora megasperma, no activity in untreated cells [1]; <1>, uninfected cells from cotyledons [3]) [1, 2, 3]
cotyledon <1> (<1>, activity decreases from high activity in young cotyledons, where it is assumed to participate in flavonoid biosynthesis, to low activity at late stages of cotyledon development [3]) [3]
disc <1> (<1>, epidermis, subepidermal cells and oil duct epithelial cells [6]) [6]

leaf <1> (<1>, in young leaf the enzyme is active exclusively in the epithelial cells of oil ducts [4,5]; <1>, vascular bundle, oil duct epithelial cells and palisade parenchyma of young and old leaves [6]) [4, 5, 6]
mesophyll <1> (<1>, very young leaf buds [5]) [5]
ovary <1> (<1>, oil duct epithelial cells, ovule and funiculus [6]) [6]
stem <1> (<1>, vascular bundle, oil duct epithelial cells and collenchymatic parenchyma [6]) [6]

Localization

protoplast <1> (<1>, activated by UV-light or fungal elicitor [2]) [2]

Purification

<1> [1]

6 Stability

pH-Stability

5 <1> (<1>, below: inactivation within 30 min [1]) [1]

General stability information

<1>, 2-mercaptoethanol stabilizes during purification and storage [1]

Storage stability

<1>, -20 C, stable for 20 days in 50% v/v glycerol [1]

References

[1] Hauffe, K.D.; Hahlbrock, K.; Scheel, D.: Elicitor-stimulated furanocoumarin biosynthesis in cultured parsley cells: S-adenosyl-L-methionine:bergaptol and S-adenosyl-L-methionine:xanthotoxol O-methyltransferase. Z. Naturforsch. C, **41**, 228-239 (1986)
[2] Dangl, J.L.; Hauffe, K.D.; Lipphardt, S.; Hahlbrock, K.; Scheel, D.: Parsley protoplasts retain differential responsiveness to u.v. light and fungal elicitor. EMBO J., **6**, 2551-2556 (1987)
[3] Knogge, W.; Kombrink, E.; Schmelzer, E.; Hahlbrock, K.: Occurence of phytoalexins and other putative defense-related substances in uninfected parsley plants. Planta, **171**, 279-287 (1987)
[4] Wu, S.C.; Hahlbrock, K.: In situ localization of phenylpropanoid-related gene expression in different tissues of light- and dark-grown parsley seedlings. Z. Naturforsch. C, **47**, 591-600 (1992)
[5] Jahnen, W.; Hahlbrock, K.: Differential regulation and tissue specific distribution of enzymes of phenylpropanoid pathways in developing parsley seedlings. Planta, **173**, 453-458 (1988)
[6] Reinold, S.; Hahlbrock, K.: In situ localization of phenylpropanoid biosynthetic mRNAs and proteins in parsley (Petroselinum crispum). Bot. Acta, **110**, 431-443 (1997)

Xanthotoxol O-methyltransferase 2.1.1.93

1 Nomenclature

EC number
2.1.1.93

Systematic name
S-adenosyl-L-methionine:xanthotoxol O-methyltransferase

Recommended name
xanthotoxol O-methyltransferase

Synonyms
XMT
methyltransferase, xanthotoxol

CAS registry number
101637-32-5

2 Source Organism

<1> *Petroselinum crispum* (cv Hamburger Schnitt [1]) [1, 2, 3]

3 Reaction and Specificity

Catalyzed reaction
S-adenosyl-L-methionine + xanthotoxol = S-adenosyl-L-homocysteine + O-methylxanthotoxol

Reaction type
methyl group transfer

Natural substrates and products
S S-adenosyl-L-methionine + xanthotoxol <1> (<1>, final methylation in the biosynthesis of the furanocoumarin xanthotoxin [1,3]) (Reversibility: ? <1> [1, 3]) [1, 3]
P S-adenosyl-L-homocysteine + xanthotoxin

Substrates and products
S S-adenosyl-L-methionine + 5-hydroxyxanthotoxin <1> (<1>, 0.7% of the activity with xanthotoxol [1]) (Reversibility: ? <1> [1]) [1]
P S-adenosyl-L-homocysteine + isopimpinellin <1> [1]

S S-adenosyl-L-methionine + 8-hydroxybergapten <1> (<1>, 2% of the activity with xanthotoxol [1]) (Reversibility: ? <1> [1]) [1]
P S-adenosyl-L-homocysteine + isopimpinellin <1> [1]
S S-adenosyl-L-methionine + bergaptol <1> (<1>, 0.5% of the activity with xanthotoxol [1]) (Reversibility: ? <1> [1]) [1]
P S-adenosyl-L-homocysteine + bergapten <1> [1]
S S-adenosyl-L-methionine + daphnetin <1> (<1>, 0.6% of the activity with xanthotoxol [1]) (Reversibility: ? <1> [1]) [1]
P S-adenosyl-L-homocysteine + hydrangetin <1> [1]
S S-adenosyl-L-methionine + xanthotoxol <1> (Reversibility: ? <1> [1-3]) [1-3]
P S-adenosyl-L-homocysteine + xanthotoxin <1> [1]

Inhibitors
phosphate <1> [1]

Specific activity (U/mg)
0.155 <1> [1]

K_m-Value (mM)
0.0044 <1> (S-adenosyl-L-methionine) [1]
0.0098 <1> (xanthotoxol) [1]

Temperature optimum (°C)
7.5-8 <1> [1]

4 Enzyme Structure

Molecular weight
67000 <1> (<1>, gel filtration [1]) [1]

Subunits
dimer <1> (<1>, 2 * 34000, SDS-PAGE [1]) [1]

5 Isolation/Preparation/Mutation/Application

Source/tissue
cell suspension culture <1> (<1>, from leaf petiole, cells treated with an elicitor from Phytophtora megasperma, no activity in untreated cells [1]) [1]
cotyledon <1> (<1>, activity decreases from high activity in young cotyledons to low activity in late stages of cotyledon development [3]) [3]
petiole <1> (<1> cell suspension culture [1]) [1]

Localization
protoplast <1> (<1>, enzyme activity is stimulated by elicitor treatment of protoplasts, increase in activity up to 30 h post treatment, no activity is detected in untreated or UV-irradiated protoplasts [2]) [2]

Purification

<1> [1]

6 Stability

pH-Stability

5 <1> (<1>, 30 min, inactivation below [1]) [1]

General stability information

<1>, 2-mercaptoethanol stabilizes during purification and storage [1]

Storage stability

<1>, -20°C, stable for 20 days in 50% v/v glycerol [1]

References

[1] Hauffe, K.D.; Hahlbrock, K.; Scheel, D.: Elicitor-stimulated furanocoumarin biosynthesis in cultured parsley cells: S-adenosyl-L-methionine: bergaptol and S-adenosyl-L-methionine:xanthotoxol O-methyltransferase. Z. Naturforsch. C, **41**, 228-239 (1986)

[2] Dangl, J.L.; Hauffe, K.D.; Lippardt, S.; Hahlbrock, K.; Scheel, D.: Parsley protoplasts retain differential responsiveness to U.V. light and fungal elicitor. EMBO J., **6**, 2551-2556 (1987)

[3] Knogge, W.; Kombrink, E.; Schmelzer, E.; Hahlbrock, K.: Occurence of phytoalexins and other putative defense-related substances in uninfected parsley plants. Planta, **171**, 279-287 (1987)

11-O-Demethyl-17-O-deacetylvindoline O-methyltransferase 2.1.1.94

1 Nomenclature

EC number
2.1.1.94

Systematic name
S-adenosyl-L-methionine:11-O-demethyl-17-O-deacetylvindoline 11-O-methyltransferase

Recommended name
11-O-demethyl-17-O-deacetylvindoline O-methyltransferase

Synonyms
11-O-demethyl-17-O-deacetylvindoline 11-O-methyltransferase
11-demethyl-17-deacetylvindoline 11-methyltransferase
methyltransferase, 11-demethyl-17-deacetylvindoline 11-

CAS registry number
100984-95-0

2 Source Organism

<1> *Catharanthus roseus* [1]

3 Reaction and Specificity

Catalyzed reaction
S-adenosyl-L-methionine + 11-O-demethyl-17-O-deacetylvindoline = S-adenosyl-L-homocysteine + 17-O-deacetylvindoline

Reaction type
methyl group transfer

Natural substrates and products
S S-adenosyl-L-methionine + 11-O-demethyl-17-O-deacetylvindoline <1> (<1> involved in biosynthesis of the alkaloid vindoline [1]) (Reversibility: ? <1> [1]) [1]
P S-adenosyl-L-homocysteine + 17-O-deacetylvindoline

Substrates and products

S S-adenosyl-L-methionine + 11-O-demethyl-17-O-deacetylvindoline <1> (<1> 11-O-demethylvindoline is not accepted as substrate [1]) (Reversibility: ? <1> [1]) [1]

P S-adenosyl-L-homocysteine + 17-O-deacetylvindoline <1> [1]

Temperature optimum (°C)

30 <1> (<1> assay at [1]) [1]

5 Isolation/Preparation/Mutation/Application

Source/tissue

cell suspension culture <1> [1]

References

[1] Fahn, W.; Laussermair, E.; Deus-Neumann, B.; Stöckigt, J.: Late enzymes of vindoline biosynthesis. S-adenosyl-L-methionine:11-O-demethyl-17-O-deacetylvindoline 11-O-methyltransferase and unspecific acetylesterase. Plant Cell Rep., **4**, 337-340 (1985)

Tocopherol O-methyltransferase 2.1.1.95

1 Nomenclature

EC number
2.1.1.95

Systematic name
S-adenosyl-L-methionine:γ-tocopherol 5-O-methyltransferase

Recommended name
tocopherol O-methyltransferase

Synonyms
γ-tocopherol methyltransferase

CAS registry number
84788-82-9

2 Source Organism

<1> *Capsicum annuum* [1-4, 7, 8]
<2> *Euglena gracilis* [5]
<3> *Spinacia oleracea* [6]
<4> *Arabidopsis thaliana* [8]

3 Reaction and Specificity

Catalyzed reaction
S-adenosyl-L-methionine + γ-tocopherol = S-adenosyl-L-homocysteine + α-tocopherol

Reaction type
methyl group transfer

Natural substrates and products
S S-adenosyl-L-methionine + γ-tocopherol <1-4> (<1-4> biosynthesis of α-tocopherol [1-8]) (Reversibility: ? <1-4> [1-8]) [1-8]
P S-adenosyl-L-homocysteine + α-tocopherol <1-4> [1-8]

Substrates and products
S S-adenosyl-L-methionine + N-(4-azido-2-nitrophenyl)-β-alanyl-γ-tocopherol <1> (Reversibility: ? <1> [4]) [4]
P ?

S S-adenosyl-L-methionine + β-tocopherol <2, 3> (<2> 70% as effective as γ-tocopherol [5]; <3> 28% as effective as γ-tocopherol [6]) (Reversibility: ? <2, 3> [5, 6]) [5, 6]
P ?
S S-adenosyl-L-methionine + δ-tocopherol <1-4> (<2> 35% as effective as γ-tocopherol [5]; <3> 16% as effective as γ-tocopherol [6]) (Reversibility: ? <1-4> [5-8]) [5-8]
P S-adenosyl-L-homocysteine + β-tocopherol <1, 3, 4> [6-8]
S S-adenosyl-L-methionine + γ-tocopherol <1-4> (<4> iso-ordered bi-bi type reaction [8]) (Reversibility: ? <1-4> [1-8]) [1-8]
P S-adenosyl-L-homocysteine + α-tocopherol <1-4> [1-8]
S Additional information <1, 4> (<1,4> β-tocopherol not accepted as substrate [7,8]) [7, 8]
P ?

Inhibitors
Cu^{2+} <2> [5]
Fe^{2+} <2> [5]
N-(4-azido-2-nitrophenyl)-β-alanyl-γ-tocopherol <1> (<1> in presence of ultraviolet light [4]) [4]
S-adenosyl-L-homocysteine <4> (<4> noncompetitive [8]) [8]
Triton X-100 <1> (<1> no activity when used to solubilize enzyme [3]) [3]
α-tocopherol <4> (<4> noncompetitive [8]) [8]

Activating compounds
dithiothreitol <2> (<2> slight stimulation [5]) [5]

Metals, ions
Additional information <2> (<2> no effect with Mg^{2+}, Ca^{2+}, K^+ and Na^+ [5]) [5]

Specific activity (U/mg)
0.000003 <1> [3]
0.00007 <3> [6]
0.004 <2> (<2> purified enzyme [5]) [5]

K_m-Value (mM)
0.002 <1> (S-adenosyl-L-methionine) [8]
0.0025 <1> (S-adenosyl-L-methionine) [4]
0.003 <1> (δ-tocopherol) [8]
0.003 <4> (δ-tocopherol) [8]
0.003 <1> (γ-tocopherol) [8]
0.004 <1> (S-adenosyl-L-methionine) [7]
0.004 <1> (δ-tocopherol) [7]
0.005 <1> (S-adenosyl-L-methionine) [3]
0.005 <4> (S-adenosyl-L-methionine) [8]
0.005 <1> (γ-tocopherol) [7]
0.005 <4> (γ-tocopherol) [8]
0.009 <3> (S-adenosyl-L-methionine) [6]

0.014 <1> (γ-tocopherol) [4]
0.05 <2> (S-adenosyl-L-methionine) [5]

pH-Optimum
7.5 <1-3> [5, 6, 8]
8.2 <1> [4]
8.5 <4> [8]

pH-Range
5.5-10 <1, 4> [4, 8]

Temperature optimum (°C)
34 <1, 4> [8]
35 <2, 3> [5, 6]

4 Enzyme Structure

Molecular weight
33000 <1> (<1> SDS-PAGE [4]) [4]
36000 <1> (<1> SDS-PAGE [8]) [8]
150000 <2> (<2> gel filtration [5]) [5]
430000 <1> (<1> gel filtration [4]) [4]
600000 <1> (<1> gel filtration [8]) [8]

5 Isolation/Preparation/Mutation/Application

Source/tissue
fruit <1> (semi-ripe [1,3]; higher activity in red then in yellow fruits [7]) [1-4, 7, 8]
leaf <3> [6]

Localization
chloroplast <2, 3> [5, 6]
chromoplast <1> [1-4, 7]
membrane <1, 2, 3> (<2> loosely bound to outer chloroplast membrane [5]) [1-6, 8]

Purification
<1> [4, 8]
<2> (partial [5]) [5]
<4> (recombinant enzyme from E. coli using His-tag [8]) [8]

Cloning
<4> (in Escherichia coli [8]) [8]

6 Stability

pH-Stability

7-8 <2> [5]

Temperature stability

90 <3> (<3> complete inactivation within 5 min [6]) [6]

General stability information

<1>, stable in presence of 0.5-1% Tween 80 [4]

<2>, very labile in absence of $MnCl_2$, complete inactivation within 36 h at 4°C, 50% loss of activity after 7 h at 4°C, stabilized by addition of 1 mM $MnCl_2$, only 15% loss of activity after 90 h at 4°C [5]

<3>, complete inactivation within 72 h at 4°C, half time for inactivation is 30 h [6]

References

[1] Camara, B.; D'Harlingue, A.: Demonstration and solubilization of S-adenosylmethionine:γ-tocopherol methyltransferase from Capsicum chromoplasts. Plant Cell Rep., **4**, 31-32 (1985)

[2] Camara, B.; Bardat, F.; Seye, A.; D'Harlingue, A.; Moneger, R.: Terpenoid metabolism in plastids. Localization of α-tocopherol synthesis in Capsicum chromoplasts. Plant Physiol., **70**, 1562-1563 (1982)

[3] Camara, B.: S-Adenosylmethionine:γ-tocopherol methyltransferase (Capsicum chromoplasts). Methods Enzymol., **111**, 544-547 (1985)

[4] D'Harlingue, A.; Camara, B.: Plastid enzymes of terpenoid biosynthesis. Purification and characterization of γ-tocopherol methyltransferase from Capsicum chromoplasts. J. Biol. Chem., **260**, 15200-15203 (1985)

[5] Shigeoka, S.; Ishiko, H.; Nakano, Y.; Mitsunaga, T.: Isolation and properties of γ-tocopherol methyltransferase in Euglena gracilis. Biochim. Biophys. Acta, **1128**, 220-226 (1992)

[6] Ishiko, H.; Shigeoka, S.; Nakano, Y.; Mitsunaga, T.: Some properties of γ-tocopherol methyltransferase solubilized from spinach chloroplasts. Phytochemistry, **31**, 1499-1500 (1992)

[7] Koch, M.; Arango, Y.; Mock, H.P.; Heise, K.P.: Factors influencing a-tocopherol synthesis in pepper fruits. J. Plant Physiol., **159**, 1015-1019 (2002)

[8] Koch, M.; Lemke, R.; Heise, K.P.; Mock, H.P.: Characterization of γ-tocopherol methyltransferases from Capsicum annuum L. and Arabidopsis thaliana. Eur. J. Biochem., **270**, 84-92 (2003)

Thioether S-methyltransferase 2.1.1.96

1 Nomenclature

EC number
2.1.1.96

Systematic name
S-adenosyl-L-methionine:dimethyl-sulfide S-methyltransferase

Recommended name
thioether S-methyltransferase

Synonyms
S-adenosyl-L-methionine:thioether S-methyltransferase
thioether methyltransferase

CAS registry number
114797-02-3

2 Source Organism

<1> *Mus musculus* [1, 3]
<2> *Mus musculus* [2]

3 Reaction and Specificity

Catalyzed reaction
S-adenosyl-L-methionine + dimethyl sulfide = S-adenosyl-L-homocysteine + trimethylsulfonium

Reaction type
methyl group transfer

Substrates and products
- **S** S-adenosyl-L-methionine + 1,4-dithiane <1> (Reversibility: ? <1> [1]) [1]
- **P** S-adenosyl-L-homocysteine + methyl-1,4-dithiane <1> [1]
- **S** S-adenosyl-L-methionine + 2-methylthioethanol <1> (Reversibility: ? <1> [1]) [1]
- **P** S-adenosyl-L-homocysteine + 2,2-dimethylthioethanol <1> [1]
- **S** S-adenosyl-L-methionine + 2-methylthioethylamine <1> (Reversibility: ? <1> [1]) [1]
- **P** S-adenosyl-L-homocysteine + 2,2-dimethylthioethylamine <1> [1]

S S-adenosyl-L-methionine + 3-methylthiopropionaldehyde <1> (Reversibility: ? <1> [1]) [1]
P S-adenosyl-L-homocysteine + 3,3-dimethylthiopropionaldehyde <1> [1]
S S-adenosyl-L-methionine + benzyl methyl sulfide <1> (Reversibility: ? <1> [1]) [1]
P S-adenosyl-L-homocysteine + benzyl dimethyl sulfonium <1> [1]
S S-adenosyl-L-methionine + diethyl sulfide <1> (Reversibility: ? <1> [1]) [1]
P S-adenosyl-L-homocysteine + diethylmethylsulfonium <1> [1]
S S-adenosyl-L-methionine + dimethyl selenide <1> (Reversibility: ? <1> [1]) [1]
P S-adenosyl-L-homocysteine + trimethylselenonium <1> [1]
S S-adenosyl-L-methionine + dimethyl sulfide <1> (Reversibility: ? <1> [1]) [1]
P S-adenosyl-L-homocysteine + trimethylsulfonium <1> [1]
S S-adenosyl-L-methionine + dimethyl telluride <1> (Reversibility: ? <1> [1]) [1]
P S-adenosyl-L-homocysteine + trimethyltelluronium <1> [1]
S S-adenosyl-L-methionine + ethyl vinyl sulfide <1> (<1> suicide substrate leading to inactivation [3]) (Reversibility: ? <1> [1]) [1, 3]
P S-adenosyl-L-homocysteine + methyl ethyl vinyl sulfonium <1> [1, 3]
S S-adenosyl-L-methionine + methyl n-propyl sulfide <1> (Reversibility: ? <1> [1]) [1]
P S-adenosyl-L-homocysteine + dimethyl n-propyl sulfonium <1> [1]
S S-adenosyl-L-methionine + pentamethylene sulfide <1> (Reversibility: ? <1> [1]) [1]
P S-adenosyl-L-homocysteine + pentamethylenemethylsulfonium <1> [1]
S S-adenosyl-L-methionine + tetrahydrothiophene <1> (Reversibility: ? <1> [1]) [1]
P S-adenosyl-L-homocysteine + methyltetrahydrothiophene <1> [1]
S S-adenosyl-L-methionine + thiomorpholine <1> (Reversibility: ? <1> [1]) [1]
P S-adenosyl-L-homocysteine + methylthiomorpholine <1> [1]

Inhibitors

S-adenosyl-L-homocysteine <1> (<1> 0.04 mM, 50% inhibition [1]) [1]
ethyl vinyl sulfide <1> (<1> suicide substrate, inactivation rate constant: 0.05/min, dimethyl sulfide protects [3]) [3]
methyl ethyl vinyl sulfonium <1> (<1> product of ethyl vinyl sulfide methylation inactivates [3]) [3]
sinefungin <1> (<1> 0.025 mM, 50% inhibition [1]) [1]

K_m-Value (mM)

0.0004 <1> (dimethyl selenide) [1]
0.001 <1> (S-adenosylmethionine) [1]
0.001 <1> (dimethyl sulfide) [1]

K_i-Value (mM)
0.275 <1> (ethyl vinyl sulfide, <1> suicide substrate [3]) [3]

pH-Optimum
6.3 <1> (<1> sharp drop in activity below [1]) [1]

4 Enzyme Structure

Molecular weight
28000 <1> (<1> gel filtration [1]) [1]

Subunits
monomer <1> (<1> 1 * 28000, SDS-PAGE [1]; <2> 1 * 29460, deduced from nucleotide sequence [2]) [1, 2]

5 Isolation/Preparation/Mutation/Application

Source/tissue
liver <1> [1]
lung <1> [1]

Localization
cytosol <1> [1]

Purification
<1> (DEAE, gel filtration, chromatofocusing [1]) [1, 2]

Cloning
<2> (cloning of cDNA [2]) [2]

References

[1] Mozier, N.M.; McConnel, K.P.; Hoffman, J.L.: S-Adenosyl-L-methionine:thioether S-methyltransferase, a new enzyme in sulfur and selenium metabolism. J. Biol. Chem., **263**, 4527-4531 (1988)
[2] Warner, D.R.; Mozier, N.M.; Pearson, J.D.; Hoffman, J.L.: Cloning and base sequence analysis of a cDNA encoding mouse lung thioether S-methyltransferase. Biochim. Biophys. Acta, **1246**, 160-166 (1995)
[3] Warner, D.R.; Hoffman, J.L.: Suicide inactivation of thioether S-methyltransferase by ethyl vinyl sulfide. Biochemistry, **35**, 4480-4484 (1996)

3-Hydroxyanthranilate 4-C-methyltransferase 2.1.1.97

1 Nomenclature

EC number
2.1.1.97

Systematic name
S-adenosyl-L-methionine:3-hydroxyanthranilate 4-C-methyltransferase

Recommended name
3-hydroxyanthranilate 4-C-methyltransferase

Synonyms
3-hydroxyanthranilate 4-methyltransferase

CAS registry number
112445-22-4

2 Source Organism

<1> *Streptomyces antibioticus* [1, 2]
<2> *Streptomyces sp.* [1]
<3> *Streptomyces chrysomallus* [1]

3 Reaction and Specificity

Catalyzed reaction
S-adenosyl-L-methionine + 3-hydroxyanthranilate = S-adenosyl-L-homocysteine + 3-hydroxy-4-methylanthranilate (Involved in the biosynthesis of the antibiotic actinomycin in Streptomyces antibioticus)

Reaction type
methyl group transfer

Natural substrates and products
- **S** S-adenosyl-L-methionine + 3-hydroxyanthranilate <1-3> (<1> involved in the biosynthesis of the antibiotic actinomycin [1,2]) (Reversibility: ? <1-3> [1]) [1, 2]
- **P** S-adenosyl-L-homocysteine + 3-hydroxy-4-methylanthranilate <1> [1, 2]

Substrates and products
- **S** S-adenosyl-L-methionine + 3-hydroxyanthranilic acid <1-3> (Reversibility: ? <1-3> [1]) [1, 2]

P S-adenosyl-L-homocysteine + 4-methyl-3-hydroxyanthranilic acid <1> [1, 2]
S Additional information <1> (<1> does not catalyze the methylation of 3-hydroxy-DL-kynurenine [1]) [1]
P ?

Inhibitors
HEPES buffer <1> [1]

Activating compounds
2-mercaptoethanol <1> [2]
Na_2EDTA <1> [2]

Specific activity (U/mg)
3.052 <1> [2]

K_m-Value (mM)
0.18 <1> (S-adenosyl-L-methionine) [2]
0.64 <1> (3-hydroxyanthranilate) [2]

pH-Optimum
6-7 <1> (<1> sodium phosphate buffer [1]) [1]
7-8 <1> (<1> Tris buffer [1]) [1]

pH-Range
6-9 <1> [1]

4 Enzyme Structure

Molecular weight
37000 <1> (<1> gel filtration, chromatography in 0.5 M NaCl [1,2]) [1, 2]

Subunits
monomer <1> (<1> 1 * 36000, SDS-PAGE [1,2]) [1, 2]

5 Isolation/Preparation/Mutation/Application

Source/tissue
mycelium <1> [1, 2]

Localization
membrane <1> (<1> intracellular membrane associated [1]) [1]

Purification
<1> [1, 2]

Cloning
<1> (gene for the 88000 Mr PHS subunit cloned [1]) [1]

6 Stability

Storage stability

<1>, -70°C, 5% glycerol, 50 mM sodium phosphate, pH 7.0, 10 mM Na_2EDTA, 5 mM 2-mercaptoethanol, 1 mM phenylmethylsulfonyl fluoride, quite stable, can be frozen and thawed several times without appreciable loss of activity [1, 2]

References

[1] Jones, G.H.: Actinomycin synthesis in Streptomyces antibioticus: Enzymatic conversion of 3-hydroxyanthranilic acid to 4-methyl-3-hydroxyanthranilic acid. J. Bacteriol., **169**, 5575-5578 (1987)
[2] Fawaz, F.; Jones, G.H.: Actinomycin synthesis in Streptomyces antibioticus. J. Biol. Chem., **263**, 4602-4606 (1988)

Diphthine synthase 2.1.1.98

1 Nomenclature

EC number
2.1.1.98

Systematic name
S-adenosyl-L-methionine:2-(3-carboxy-3-aminopropyl)-L-histidine methyl-transferase

Recommended name
diphthine synthase

Synonyms
S-adenosyl-L-methionine:elongation factor 2 methyltransferase
diphthine methyltransferase
methyltransferase, diphthine

CAS registry number
114514-25-9

2 Source Organism

<1> *Cricetulus griseus* (CHO-K1, resistant to Diphtheria toxin [1]) [1]
<2> *Saccharomyces cerevisiae* (strain dpH5, resistant to Diphtheria toxin [2]) [1, 2]
<3> *Mesocricetus auratus* (Syrian hamster [1]) [1]
<4> *Mus musculus* (mouse [1]) [1]
<5> *Rattus norvegicus* [1]
<6> *Oryctolaus cuniculus* (rabbit [1]) [1]
<7> *Homo sapiens* [1]
<8> *Cercopithecus aethiops* (monkey [1]) [1]
<9> *Bos taurus* (bovine [1]) [1]
<10> *Muntiacus muntjak* (Indian muntjac [1]) [1]
<11> *Gallus gallus* (chicken [1]) [1]
<12> *Notopthalmus viridescens* (newt [1]) [1]
<13> *Portheria despar* (gypsy moth [1]) [1]
<14> *Triticum vulgare* (wheat [1]) [1]
<15> *Schizophyllum commune* (basidiomycete [1]) [1]
<16> *Thermoplasma acidophilum* (archaebacterium [1]) [1]
<17> *Streptococcus sanguis* [1]

3 Reaction and Specificity

Catalyzed reaction

S-adenosyl-L-methionine + 2-(3-carboxy-3-aminopropyl)-L-histidine = S-adenosyl-L-homocysteine + 2-[3-carboxy-3-(methylammonio)propyl]-L-histidine

Reaction type

methyl group transfer

Natural substrates and products

S S-adenosyl-L-methionine + 2-(3-carboxy-3-aminopropyl)-L-histidine <1, 2> (Reversibility: ? <1, 2> [1, 2]) [1, 2]

P S-adenosyl-L-homocysteine + 2-[3-carboxy-3-(methylammonio)propyl]-L-histidine

Substrates and products

S S-adenosyl-L-methionine + 2-(3-carboxy-3-aminopropyl)-histidine <1, 2> (Reversibility: ? <1, 2> [1, 2]) [1, 2]

P S-adenosyl-L-homocysteine + 2-[3-carboxy-3-(methylammonio)propyl]-histidine <1, 2> [1, 2]

S S-adenosyl-L-methionine + 2-[3-carboxy-3-(dimethylammonio)propyl]-L-histidine <1-17> (Reversibility: ? <1-17> [1, 2]) [1, 2]

P S-adenosyl-L-homocysteine + 2-[3-carboxy-3-(trimethylammonio)propyl]histidine <1-17> (<1,2> i.e. diphthine [1,2]) [1, 2]

S S-adenosyl-L-methionine + 2-[3-carboxy-3-(methylammonio)propyl]-L-histidine <1, 2> (Reversibility: ? <1, 2> [1, 2]) [1, 2]

P S-adenosyl-L-homocysteine + 2-[3-carboxy-3-(dimethylammonio)propyl]histidine <1, 2> [1, 2]

Temperature optimum (°C)

30 <1, 2> (<1,2> assay at [1,2]) [1, 2]

5 Isolation/Preparation/Mutation/Application

Source/tissue

ovary <1> (<1> CHO-K1 cells [1]) [1]

Localization

cytosol <2> [2]

Purification

<1> [1]

<2> (partial [2]) [2]

6 Stability

Storage stability

<2>, -20°C, 20 mM Tris-HCl, pH 7.5, 1 mM EDTA, 1 mM EGTA, 5 mM 2-mercaptoethanol, 50% glycerol [2]

References

[1] Moehring, J.M.; Moehring, T.J.: The post-translational trimethylation of diphthamide studied in vitro. J. Biol. Chem., **263**, 3840-3844 (1988)
[2] Chen, J.Y. C.; Bodley, J.W.: Biosynthesis of diphthamide in Saccharomyces cerevisiae. Partial purification and characterization of a specific S-adenosylmethionine:elongation factor 2 methyltransferase. J. Biol. Chem., **263**, 11692-11696 (1988)

16-Methoxy-2,3-dihydro-3-hydroxytabersonine N-methyltransferase 2.1.1.99

1 Nomenclature

EC number
2.1.1.99

Systematic name
S-adenosyl-L-methionine:16-methoxy-2,3-dihydro-3-hydroxy-tabersonine N-methyltransferase

Recommended name
16-methoxy-2,3-dihydro-3-hydroxytabersonine N-methyltransferase

Synonyms
NMT
methyltransferase, 16-methoxy-2,3-dihydro-3-hydroxytabersonine

CAS registry number
113478-40-3

2 Source Organism

<1> *Catharanthus roseus* [1-3]

3 Reaction and Specificity

Catalyzed reaction
S-adenosyl-L-methionine + 16-methoxy-2,3-dihydro-3-hydroxytabersonine = S-adenosyl-L-homocysteine + deacetoxyvindoline (<1> summerization of biosynthetic pathway of vindoline [2,3])

Reaction type
methyl group transfer

Natural substrates and products
S S-adenosyl-L-methionine + 16-methoxy-2,3-dihydro-3-hydroxytabersonine <1> (<1> enzyme is involved in the biosynthesis of vindoline from tabersonine [1-3]) [1-3]
P S-adenosyl-L-homocysteine + deacetoxyvindoline <1> [1, 2]

Substrates and products

S S-adenosyl-L-methionine + 16-methoxy-2,3-dihydro-3-hydroxytabersonine <1> (<1> high degree of specifity for the N-indole of 2,3-dihydro-3-hydroxytabersonine [3]) (Reversibility: ? <1> [1-3]) [1-3]

P S-adenosyl-L-homocysteine + deacetoxyvindoline <1> (<1> i.e. N(1)-methyl-16-methoxy-2,3-dihydro-3-hydroxytabersonine [1,2]) [1, 2]

S Additional information <1> (<1> hydrogenation of 2,3 double bond of tabersonine instead of hydration results in a 75% decrease in N-methylation, double bound present: no N-methylation, the presence of 6,7 double bound is a necessary structural requirement for N-methylation with membrane-bound enzyme, whereas the detergent-solubilized enzyme is also able to N-methylate the 6,7 anhydro analogue [3]) [3]

P ?

Inhibitors

lysolecitin <1> (<1> completely elimination of activity [3]) [3]

Activating compounds

CHAPS <1> (<1> highest activity with 3 mg/ml [3]) [3]

Specific activity (U/mg)

Additional information <1> [3]

4 Enzyme Structure

Molecular weight

60000 <1> (<1> gel chromatography [3]) [3]

5 Isolation/Preparation/Mutation/Application

Source/tissue

leaf <1> [1]
seed <1> [2]

Localization

chloroplast <1> (<1> associated with thylakoids [1,3]) [1, 3]

Purification

<1> (partially [3]) [3]

6 Stability

Oxidation stability

<1>, labile to O_2, DTT or 2-mercaptoethanol stabilizes [1]

Organic solvent stability

Triton X-100 <1> (<1> 0.1% Triton X-100 90% of activity [3]) [3]
Additional information <1> (<1> optimal pH for solubilization: 7.5 [1]; <1> octyl-β-D-glucoside: 0.15% solution 100% of activity [3]; <1> CHAPS: 232% activity is solubilized [3]) [1, 3]

General stability information

<1>, reducing agent, e.g. DTT, necessary for the retention of activity after breakage of thylakoid membranes of chloroplasts [1]

Storage stability

<1>, -20°C, in absence of CHAPS, 4 days, 50% loss of activity [3]
<1>, -20°C, in absence of detergent by adding 25% glycerol, one week, 9% loss of activity [3]
<1>, -20°C, in presence of 0.6% CHAPS, 7 days, 50% loss of activity [3]

References

[1] De Luca, V.; Cutler, A.J.: Subcellular localization of enzymes involved in indole alkaloid biosynthesis in Catharanthus roseus. Plant Physiol., **85**, 1099-1102 (1987)
[2] De Luca, V.; Balsevich, J.; Tyler, R.T.; Eilert, U.; Panchuk, B.D.; Kurz, W.G.W.: Biosynthesis of indole alkaloids: developmental regulation of the biosynthethic pathway from tabersonine to vindoline in Catheranthus roseus. J. Plant Physiol., **125**, 147-156 (1986)
[3] Dethier, M.; De Luca, V.: Partial purification of an N-methyltransferase involved in vindoline biosynthesis in Catharanthus roseus. Phytochemistry, **32**, 673-678 (1993)

Protein-S-isoprenylcysteine O-methyltransferase 2.1.1.100

1 Nomenclature

EC number
2.1.1.100

Systematic name
S-adenosyl-L-methionine:protein-C-terminal-S-farnesyl-L-cysteine O-methyltransferase

Recommended name
protein-S-isoprenylcysteine O-methyltransferase

Synonyms
S-farnesylcysteine methyltransferase
farnesyl cysteine C-terminal methyltransferase
farnesyl-protein carboxymethyltransferase
farnesylated protein C-terminal O-methyltransferase
isoprenylated protein methyltransferase
methyltransferase, protein C-terminal farnesylcysteine O-
prenylated protein methyltransferase
protein S-farnesylcysteine C-terminal methyltransferase

CAS registry number
130731-20-3

2 Source Organism

<1> *Rattus norvegicus* (<1> activity increases during development [16]) [1, 2, 5, 9, 10, 11, 16, 17, 18, 19]
<2> *Saccharomyces cerevisiae* [3, 4, 10]
<3> *Bos taurus* [6, 7, 14, 16, 21, 24]
<4> *Oryctolagus cuniculus* [8]
<5> *Homo sapiens* [11, 12, 13, 15, 22, 25, 27]
<6> *Nicotiana tabacum* [20]
<7> *Leishmania donovani* [23]
<8> *Trypanosoma brucei* [26]
<9> *Mus musculus* [26]

3 Reaction and Specificity

Catalyzed reaction

S-adenosyl-L-methionine + protein C-terminal S-farnesyl-L-cysteine = S-adenosyl-L-homocysteine + protein C-terminal S-farnesyl-L-cysteine methyl ester (<3> kinetic mechanism [6]; <5> ordered bi bi mechanism, S-adenosyl-L-methionine binds first, S-adenosyl-L-homocysteine departs last [15])

Reaction type

methyl group transfer

Natural substrates and products

S S-adenosyl-L-methionine + protein C-terminal S-farnesyl-L-cysteine <1, 2> (<1> reversible methylation of ras-proteins may play an important role in the modulation of their signaling properties [1]; <1> carboxyl methylation may play a role in the regulation of receptor mediated signal transduction processes in eukaryotic cells [5]; <2> a-factor mating pheromone, Ras1 and Ras2 are putative substrates of the enzyme [3]) (Reversibility: ? <1, 2> [1, 3]) [1, 3, 5]

P S-adenosyl-L-homocysteine + protein C-terminal S-farnesyl-L-cysteine methyl ester <1, 2> [1, 3, 5]

Substrates and products

S S-adenosyl-L-methionine + protein C-terminal S-farnesyl-L-cysteine <1-9> (<2> a-factor, Ras1 and Ras2 [3]; <1,2> ras-proteins [1,3,5]; <1> γ-subunit of heterotrimeric G-proteins [5]; <4> fusion protein of heat-inducible ubiquitin and the 11 C-terminal amino acid sequence of Ha-ras protein [8]; <1> enzyme can methylate both peptides and proteins [2]; <1> methyl esterification of proteins containing a C-terminal-CXXX sequence requires not only the removal of the 3 terminal amino acids but the isoprenylation of the sulfhydryl group as well [2]; <5> 21000-24000 Da GTP-binding proteins [12]; <5> methylation of a 23000 Da small G protein is indicated [22]) (Reversibility: ? <1-9> [1-3, 5, 6, 8, 12, 20, 23, 26]) [1-3, 5, 6, 8, 12, 20, 22, 23, 26]

P S-adenosyl-L-homocysteine + protein C-terminal S-farnesyl-L-cysteine methyl ester <1-9> [1-3, 5, 6, 8, 12, 20, 22, 23, 26]

S S-adenosylmethionine + N-((benzoylglycyl)glycyl)-S-farnesyl-L-cysteine <5> (Reversibility: ? <5> [15]) [15]

P S-adenosylhomocysteine + N-((benzoylglycyl)glycyl)-S-farnesyl-L-cysteine methyl ester <5> [15]

S S-adenosylmethionine + N-(benzoylglycyl)-S-farnesyl-L-cysteine <5> (Reversibility: ? <5> [15]) [15]

P S-adenosylhomocysteine + N-(benzoylglycyl)-S-farnesyl-L-cysteine methyl ester <5> [15]

S S-adenosylmethionine + N-acetyl-S-(E)-geranyl-L-cysteine <1> (Reversibility: ? <1> [5]) [5]

P S-adenosylhomocysteine + N-acetyl-S-(E)-geranyl-L-cysteine methyl ester <1> [5]

S S-adenosylmethionine + N-acetyl-S-(E,E)-farnesyl-D-cysteine <1> (Reversibility: ? <1> [5]) [5]
P S-adenosylhomocysteine + N-acetyl-S-(E,E)-farnesyl-D-cysteine methyl ester <1> [5]
S S-adenosylmethionine + N-acetyl-S-(E,E)-farnesyl-L-cysteine <1, 3, 5, 6, 7, 8, 9> (<8> recombinant enzyme: 50fold higher activity than with N-acetyl-S-geranylgeranyl-L-cysteine, native enzyme: 40fold higher activity than with N-acetyl-S-geranylgeranyl-L-cysteine [26]) (Reversibility: ? <1, 3, 5, 6, 7, 8, 9> [5, 7, 13, 20, 23, 26]) [5, 7, 13, 14, 20, 23, 26]
P S-adenosylhomocysteine + N-acetyl-S-(E,E)-farnesyl-L-cysteine methyl ester <1, 3, 5, 6, 7, 8> [5, 7, 13, 14, 20, 23, 26]
S S-adenosylmethionine + N-acetyl-S-3,3-dimethylallyl-L-cysteine <1> (Reversibility: ? <1> [5]) [5]
P S-adenosylhomocysteine + N-acetyl-S-3,3-dimethylallyl-L-cysteine methyl ester <1> [5]
S S-adenosylmethionine + N-acetyl-S-geranylgeranyl-L-cysteine <1, 3, 5, 6, 7, 8, 9> (<6> no activity with N-acetyl-S-trans-geranyl-L-cysteine [20]) (Reversibility: ? <1, 3, 5, 6, 7, 8, 9> [7, 13, 17, 20, 23, 26]) [7, 13, 17, 20, 23, 26]
P S-adenosylhomocysteine + N-acetyl-S-geranylgeranyl-L-cysteine methyl ester <1, 3, 5, 6, 7, 8> [7, 13, 17, 20, 23, 26]
S S-adenosylmethionine + N-acetyl-S-geranylgeranyl-L-cysteinyl-L-alanyl-S-geranylgeranyl-L-cysteine <3> (<3> no activity with + N-acetyl-S-geranylgeranyl-L-cysteinyl-S-geranylgeranyl-L-cysteine [14]) (Reversibility: ? <3> [14]) [14]
P S-adenosylhomocysteine + N-acetyl-S-geranylgeranyl-L-cysteinyl-L-alanyl-S-geranylgeranyl-L-cysteine methyl ester <3> [14]
S S-adenosylmethionine + N-isobutyryl-S-farnesyl-L-cysteine <5> (Reversibility: ? <5> [15]) [15]
P S-adenosylhomocysteine + N-isobutyryl-S-farnesyl-L-cysteine methyl ester <5> [15]
S S-adenosylmethionine + N-isovaleryl-S-farnesyl-L-cysteine <5> (Reversibility: ? <5> [15]) [15]
P S-adenosylhomocysteine + N-isovaleryl-S-farnesyl-L-cysteine methyl ester <5> [15]
S S-adenosylmethionine + S-decyl-Leu-Ala-Arg-Tyr-Lys-Cys <1> (Reversibility: ? <1> [2]) [2]
P S-adenosylhomocysteine + S-decyl-Leu-Ala-Arg-Tyr-Lys-Cys methyl ester <1> [2]
S S-adenosylmethionine + S-farnesyl-Leu-Ala-Arg-Tyr-Lys-Cys <1-3> (Reversibility: ? <1-3> [2-6]) [2-6]
P S-adenosylhomocysteine + S-farnesyl-Leu-Ala-Arg-Tyr-Lys-Cys methyl ester <1-3> [2-6]
S S-adenosylmethionine + S-geranyl-Leu-Ala-Arg-Tyr-Lys-Cys <1> (Reversibility: ? <1> [2]) [2]
P S-adenosylhomocysteine + S-geranyl-Leu-Ala-Arg-Tyr-Lys-Cys methyl ester <1> [2]

S S-adenosylmethionine + S-geranylgeranyl-Leu-Ala-Arg-Tyr-Lys-Cys <1> (Reversibility: ? <1> [2]) [2]

P S-adenosylhomocysteine + S-geranylgeranyl-Leu-Ala-Arg-Tyr-Lys-Cys methyl ester <1> [2]

S S-adenosylmethionine + S-octyl-Leu-Ala-Arg-Tyr-Lys-Cys <1> (Reversibility: ? <1> [2]) [2]

P S-adenosylhomocysteine + S-octyl-Leu-Ala-Arg-Tyr-Lys-Cys methyl ester <1> [2]

S S-adenosylmethionine + S-pentadecyl-Leu-Ala-Arg-Tyr-Lys-Cys <1> (Reversibility: ? <1> [2]) [2]

P S-adenosylhomocysteine + S-pentadecyl-Leu-Ala-Arg-Tyr-Lys-Cys methyl ester <1> [2]

S S-adenosylmethionine + S-tridecyl-Leu-Ala-Arg-Tyr-Lys-Cys <1> (Reversibility: ? <1> [2]) [2]

P S-adenosylhomocysteine + S-tridecyl-Leu-Ala-Arg-Tyr-Lys-Cys methyl ester <1> [2]

Inhibitors

1-ethyl-3-(3-dimethylaminopropyl)carbodiimide <1> (<1> 22% inhibition [18]) [18]

2584 Da liver peptide <1> (<1> 0.000071 mM, 50% inhibition [11]) [11]

4-hydroxyphenylglyoxal <1> (<1> 48% inhibition [18]) [18]

5'-deoxymethylthioadenosine <5> [13]

cetyltrimethylammonium <3> (<3> 0.1%, more than 90% inhibition [24]) [24]

Ca^{2+} <3> (<3> moderate inhibition [24]) [24]

Cu^{2+} <3> (<3> complete inhibition [24]) [24]

Hg^{2+} <3> (<3> complete inhibition [24]) [24]

Mg^{2+} <3> (<3> moderate inhibition [24]) [24]

Mg^{2+} <5> (<5> inhibition of 23000 Da protein methylation [22]) [22]

Mn^{2+} <3> (<3> moderate inhibition [24]) [24]

N,N'-dicyclohexylcarbodiimide <1> (<1> 76% inhibition [18]) [18]

N-acetyl-S-(E,E)-farnesyl-L-cysteine methyl ester <3> (<3> mixed-type vs. N-acetyl-S-(E,E)-farnesyl-L-cysteine [6]) [6]

N-acetyl-S-geranylgeranyl-L-cysteine <3> (<3> inhibition of N-acetyl-S-farnesyl-L-cysteine methylation [7]) [7]

N-bromosuccinimide <1> (<1> 56% inhibition in the presence of S-adenosyl-L-homocysteine, 87% inhibition in the absence of S-adenosyl-L-homocysteine [18]) [18]

N-ethylmaleimide <1, 3> (<1> 2 mM, 94% inhibition after preincubation for 30 min, 45% inhibition in the presence of S-adenosyl-L-homocysteine [18]; <3> moderate inhibition [24]) [18, 24]

Nonidet P40 <3> (<3> 0.1%, 73% inhibition [24]) [24]

S-(geranylgeranyl-2-thio)acetic acid <3> [7]

S-adenosyl-L-homocysteine <1, 3, 5, 7> (<3> competitive vs. S-adenosyl-L-methionine, mixed-type vs. N-acetyl-S-(E,E)-farnesyl-L-cysteine [6]; <1> 0.02 mM, 50% inhibition [17]; <7> 0.025 mM; 50% inhibition [23]; <3>

0.005 mM, 50% inhibition [24]; <5> 0.2 mM, 67% inhibition of the endogenous methylation of 21000-25000 Da proteins [25]) [2, 6, 12, 13, 14, 17, 23, 24, 25]
S-adenosylethionine <3> [14]
S-farnesyl thiosalicylic acid <1, 7> (<7> 0.05 mM, 50% inhibition [23]) [19, 23]
S-farnesyl-4-chlorothiosalicylic acid <1> [19]
S-farnesyl-4-fluorothiosalicylic acid <1> [19]
S-farnesyl-5-chlorothiosalicylic acid <1> [19]
S-farnesyl-5-fluorothiosalicylic acid <1> [19]
S-farnesylthioacetic acid <3, 5> (<3> competitive vs. N-acetyl-S-(E,E)-farnesyl-L-cysteine [6]; <3> 0.048 mM, 50% inhibition [24]) [6, 7, 12, 24]
Triton X-100 <3> (<3> 0.1%, 63% inhibition [24]) [24]
Tween 80 <3> (<3> 0.1%, 33% inhibition [24]) [24]
Zn^{2+} <3> (<3> complete inhibition [24]) [24]
deoxycholate <3> (<3> 0.1%, 51% inhibition [24]) [24]
diethylpyrocarbonate <1> (<1> 96% inhibition [18]) [18]
dimethylsulfoxide <3> (<3> more than 4% cause inhibition [24]) [24]
iodoacetic acid <1> (<1> 16% inhibition [18]) [18]
N-acetyl-S-farnesyl-L-cysteine methyl ester <3> [6]
methylthioadenosine <7> (<7> 0.9 mM, 50% inhibition [23]) [23]
monoiodoacetate <3> (<3> moderate inhibition [24]) [24]
n-octyl β-D-glucoside <9> (<9> 0.2%, 55-75% inhibition [26]) [26]
phenylglyoxal <1> (<1> 74% inhibition, inhibition increases with the alkanity of the preincubation medium, S-adenosyl-L-methionine protects against inhibition [18]) [18]
retinoic acid <5> (<5> 20% decrease of N-acetyl-S-(E,E)-farnesyl-L-cysteine methylation in cells treated with 0.01 mM retinoic acid [25]) [25]
sinefungin <1, 3, 5, 7> (<1> 0.0009 mM, 50% inhibition [17]; <7> 0.00002 mM; 50% inhibition, competitive vs. S-adenosyl-L-methionine [23]; <3> 0.0035 mM, 50% inhibition [24]) [13, 14, 17, 23, 24]

Activating compounds

formyl-methionine-leucine-phenylalanine <5> (<5> stimulation of carboxyl methylation of GTP-binding proteins through a substrate-dependent mechanism [13]) [13]
guanosine 5'-[γ-thio]-triphosphate <3, 5, 7> (<3> 2 mM, 40fold increase in activity [21]; <5> stimulation of 23000 Da protein methylation [22]; <7> stimulation of carboxymethylation of endogenous substrates [23]) [21, 22, 23]
Additional information <1, 7> (<1> no increase in methyltransferase activity in cell-free preparations by guanosine 5'-[γ-thio]-triphosphate [17]; <7> no increase in N-acetyl-S-(E,E)-farnesyl-L-cysteine methylation [23]) [17, 23]

Metals, ions

Li^+ <3> (<3> slight stimulation [24]) [24]

Specific activity (U/mg)

0.000143 <3> [21]

K_m-Value (mM)

0.00013 <7> (S-adenosyl-L-methionine) [23]
0.00017 <8> (N-acetyl-S-(E,E)-farnesyl-L-cysteine, <8> recombinant enzyme [26]) [26]
0.0013 <5> (S-adenosyl-L-methionine) [13]
0.0014 <8> (N-acetyl-S-(E,E)-farnesyl-L-cysteine, <8> native enzyme [26]) [26]
0.0014 <5> (N-acetyl-S-geranylgeranyl-L-cysteine) [13]
0.0021 <1> (S-adenosyl-L-methionine) [2]
0.0022 <1> (S-farnesyl-Leu-Ala-Arg-Tyr-Lys-Cys) [2]
0.0023 <8> (N-acetyl-S-geranylgeranyl-L-cysteine, <8> recombinant enzyme [26]) [26]
0.00255 <5> (N-(benzoylglycyl)-S-farnesyl-L-cysteine) [15]
0.00255 <5> (N-isovaleryl-S-farnesyl-L-cysteine) [15]
0.00295 <5> (N-isobutyryl-S-farnesyl-L-cysteine) [15]
0.0036 <3> (S-adenosyl-L-methionine) [14]
0.0038 <1> (S-adenosyl-L-methionine) [10]
0.004 <8> (N-acetyl-S-geranylgeranyl-L-cysteine, <8> native enzyme [26]) [26]
0.004 <3> (S-adenosyl-L-methionine) [24]
0.0105 <5> (N-((benzoylglycyl)glycyl)-S-farnesyl-L-cysteine) [15]
0.0109 <1> (S-geranylgeranyl-Leu-Ala-Arg-Tyr-Lys-Cys) [2]
0.0115 <9> (N-acetyl-S-(E,E)-farnesyl-L-cysteine, <9> native enzyme [26]) [26]
0.0116 <5> (N-acetyl-S-(E,E)-farnesyl-L-cysteine) [13]
0.0116 <9> (N-acetyl-S-geranylgeranyl-L-cysteine, <9> native enzyme [26]) [26]
0.018 <5> (N-acetyl-S-(E,E)-farnesyl-L-cysteine, <5> enzyme from endometrial carcinoma [12]) [12]
0.0192 <3> (N-acetyl-S-geranylgeranyl-L-cysteine) [7]
0.02 <1> (N-acetyl-S-(E,E)-farnesyl-L-cysteine) [11]
0.02 <1, 3> (N-acetyl-S-(E,E)-farnesyl-L-cysteine) [5, 6]
0.02 <5> (N-acetyl-S-(E,E)-farnesyl-L-cysteine, <5> enzyme from endometrium [12]) [12]
0.021 <6> (N-acetyl-S-geranylgeranyl-L-cysteine) [20]
0.0223 <3> (N-acetyl-S-farnesyl-L-cysteine) [7]
0.03 <1> (N-acetyl-S-(E,E)-farnesyl-L-cysteine) [10]
0.03 <1> (N-acetyl-S-(E,E)-farnesyl-L-cysteine) [17]
0.04 <1> (N-acetyl-S-(E,E)-farnesyl-D-cysteine) [5]
0.052 <3> (N-acetyl-S-geranylgeranyl-L-cysteine) [24]
0.0565 <3> (N-acetyl-S-(E,E)-farnesyl-L-cysteine) [14]
0.06 <7> (N-acetyl-S-(E,E)-farnesyl-L-cysteine) [23]
0.073 <6> (N-acetyl-S-(E,E)-farnesyl-L-cysteine) [20]
0.09 <3> (N-acetyl-S-(E,E)-farnesyl-L-cysteine) [24]
0.166 <7> (N-acetyl-S-geranylgeranyl-L-cysteine) [23]
0.389 <1> (S-geranyl-Leu-Ala-Arg-Tyr-Lys-Cys) [2]
0.48 <1> (S-decyl-Leu-Ala-Arg-Tyr-Lys-Cys) [2]

0.594 <3> (N-acetyl-S-geranylgeranyl-L-cysteinyl-L-alanyl-S-geranylgeranyl-L-cysteine) [14]
0.64 <1> (S-tridecyl-Leu-Ala-Arg-Tyr-Lys-Cys) [2]
1 <1> (N-acetyl-S-(E)-geranyl-L-cysteine) [5]
1.12 <1> (S-pentadecyl-Leu-Ala-Arg-Tyr-Lys-Cys) [2]
1.76 <1> (S-octyl-Leu-Ala-Arg-Tyr-Lys-Cys) [2]
3.8 <3> (S-adenosyl-L-methionine) [16]
5 <1> (N-acetyl-S-3,3-dimethylalkyl-L-cysteine) [5]
11.1 <3> (N-acetyl-S-(E,E)-farnesyl-L-cysteine) [16]

K_i-Value (mM)

0.00007 <5> (sinefungin) [13]
0.000296 <3> (sinefungin, <3> inhibition of N-acetyl-S-geranylgeranyl-L-cysteinyl-L-alanyl-S-geranylgeranyl-L-cysteine [14]) [14]
0.000318 <3> (sinefungin, <3> inhibition of N-acetyl-S-(E,E)-farnesyl-L-cysteine methylation [14]) [14]
0.0012 <3> (S-farnesylthioacetic acid) [6]
0.00157 <5> (S-adenosyl-L-homocysteine) [13]
0.0019 <3> (S-adenosyl-L-homocysteine, <3> vs. S-adenosyl-L-methionine [6]) [6]
0.002 <5> (S-farnesylthioacetic acid) [12]
0.00207 <5> (N-acetyl-S-farnesyl-L-cysteine methyl ester, <5> mixed-type vs. N-acetyl-S-farnesyl-L-cysteine [15]) [15]
0.0023 <3> (N-acetyl-S-(E,E)-geranylgeranyl-L-cysteine, <3> competitive vs. N-acetyl-S-(E,E)-farnesyl-L-cysteine [7]) [7]
0.0026 <3> (N-acetyl-S-(E,E)-farnesyl-L-cysteine, <3> competitive vs. N-acetyl-S-(E,E)-geranylgeranyl-L-cysteine [7]) [7]
0.0028 <1> (S-farnesyl thiosalicylic acid) [19]
0.0033 <3> (S-adenosyl-L-homocysteine, <3> vs. N-acetyl-S-(E,E)-farnesyl-L-cysteine [6]) [6]
0.0039 <3> (S-(geranylgeranyl-2-thio)acetic acid, <3> competitive vs. N-acetyl-S-(E,E)-farnesyl-L-cysteine [7]) [7]
0.00413 <5> (S-adenosyl-L-homocysteine, <5> competitive vs. S-adenosyl-L-methionine [15]) [15]
0.0044 <5> (S-adenosyl-L-homocysteine, <5> mixed-type vs. N-acetyl-S-farnesyl-L-cysteine [15]) [15]
0.0046 <3> (S-farnesylthioacetic acid, <3> competitive vs. N-acetyl-S-(E,E)-farnesyl-L-cysteine [7]) [7]
0.005 <5> (N-acetyl-S-farnesyl-L-cysteine methyl ester, <5> competitive vs. S-adenosyl-L-methionine [15]) [15]
0.0063 <1> (S-farnesyl-5-fluorothiosalicylic acid) [19]
0.00645 <3> (S-adenosylethionine, <3> inhibition of N-acetyl-S-(E,E)-farnesyl-L-cysteine methylation [14]) [14]
0.00658 <3> (S-adenosyl-L-homocysteine, <3> inhibition of N-acetyl-S-(E,E)-farnesyl-L-cysteine methylation [14]) [14]
0.0085 <1> (S-farnesyl-5-chlorothiosalicylic acid) [19]
0.0092 <1> (S-adenosyl-L-homocysteine) [2]

0.0112 <3> (S-farnesylthioacetic acid, <3> inhibition of N-acetyl-S-(E,E)-farnesyl-L-cysteine methylation [14]) [14]
0.0115 <3> (S-adenosyl-L-homocysteine, <3> inhibition of N-acetyl-S-geranylgeranyl-L-cysteinyl-L-alanyl-S-geranylgeranyl-L-cysteine [14]) [14]
0.0137 <5> (S-farnesylthioacetic acid, <5> competitive vs. N-acetyl-S-farnesyl-L-cysteine [15]) [15]
0.0218 <3> (S-adenosylethionine, <3> inhibition of N-acetyl-S-geranylgeranyl-L-cysteinyl-L-alanyl-S-geranylgeranyl-L-cysteine [14]) [14]
0.023 <1> (S-farnesyl-4-fluorothiosalicylic acid) [19]
0.028 <1> (S-farnesyl-4-chlorothiosalicylic acid) [19]
0.041 <3> (N-acetyl-S-(E,E)-farnesyl-L-cysteine methyl ester) [6]
1.8 <3> (S-adenosyl-L-homocysteine) [16]
3.95 <5> (5'-deoxymethylthioadenosine) [13]

pH-Optimum
6.5-8 <7> [23]
7 <1, 6> [2, 20]

pH-Range
5.6-8.6 <1> [2]
7.4-8 <3> [24]

4 Enzyme Structure

Molecular weight
24000 <2> (<2> radiation inactivation [10]) [10]
30000 <3> (<3> gel filtration, photoaffinity labelling [21]) [21]
86000 <3> (<3> PAGE, gel filtration [16]; <3> non denaturing PAGE [21]) [16, 21]
98000 <1> (<1> radiation inactivation [10]) [10]
180000 <4> (<4> gel filtration [8]) [8]

Subunits
? <3> (<3> x * 30000, SDS-PAGE [21]; <8> x * 28301, deduced from nucleotide sequence [26]) [21, 26]
monomer <2> (<2> 1 * 24000, deduced from nucleotide sequence [10]) [10]
Additional information <1> (<1> enzyme may be a homotetramer [10]) [10]

5 Isolation/Preparation/Mutation/Application

Source/tissue
SH-SY5Y cell <5> [25]
adrenal gland <3> (<3> adrenal chromaffin cells [24]) [24]
blood <5> [13, 15]
blood platelet <5> [15]
brain <1, 3> [11, 14, 16, 21]

cell culture <1> (<1> insulin-secreting [17]) [17]
embryo <1> [1, 5]
endometrioid carcinoma cell <5> [12]
fibroblast <1> [1, 5]
heart <3> [14]
intestine <3> [14]
kidney <1, 3> (<1> cortex [9,18]) [9, 10, 14, 18]
liver <1, 3, 9> [2, 14, 26]
lung <3> [14]
myeloid cell <5> [27]
neutrophil <5> (<5> light membrane fraction [13]) [13]
pancreas <1> [17]
reticulocyte <4> (<4> lysate [8]) [8]
retina <3> [6, 7, 14]
retinal rod <3> (<3> outer segment [14]) [14]
suspension culture <6> [20]
testis <3> [14]

Localization

basolateral plasma membrane <1> [10]
membrane <2, 3, 5, 6, 9> [4, 9, 10, 14, 15, 16, 17, 20, 24, 26]
microsome <1, 5> [2, 5, 12, 17]
mitochondrial membrane <3, 5> (<3> 22% of the activity [16]) [12, 16]
nuclear membrane <3, 5> (<3> 48% of the activity [16]) [12, 16]

Purification

<3> (acid treatment, DEAE-cellulose, Superose 6, Superdex 75, partially purified [21]) [21]
<4> (Q-Sepharose, partially purified [8]) [8]

Cloning

<2> (expression of TrpE-STE14 methyl transferase fusion protein in Escherichia coli [3]) [3, 4]
<5> (overepression in tissue culture cells [27]) [27]
<8> (expression in Sf9 cells [26]) [26]

6 Stability

pH-Stability

5.5-9.5 <1> (<1> stable [18]) [18]

General stability information

<5>, complete loss of activity after exposure to detergent e.g. CHAPS, octyl-β-glucoside and sodium deoxycholate, activity can be reconstituted by removal of the detergent in the presence of phospholipid [13]

Storage stability

<1>, -70°C, 6 months [2]

References

[1] Clarke, S.; Vogel, J.P.; Deschenes, R.J.; Stock, J.: Posttranslational modification of the Ha-ras oncogene protein: evidence for a third class of protein carboxyl methyltransferases [published erratum appears in Proc Natl Acad Sci U S A 1988 Oct;85(20):7556]. Proc. Natl. Acad. Sci. USA, **85**, 4643-4647 (1988)

[2] Stephenson, R.C.; Clarke, S.: Identification of a C-terminal protein carboxyl methyltransferase in rat liver membranes utilizing a synthetic farnesyl cysteine-containing peptide substrate. J. Biol. Chem., **265**, 16248-16254 (1990)

[3] Hrycyna, C.A.; Sapperstein, S.K.; Clarke, S.; Michaelis, S.: The Saccharomyces cerevisiae STE14 gene encodes a methyltransferase that mediates C-terminal methylation of a-factor and RAS proteins. EMBO J., **10**, 1699-1709 (1991)

[4] Hrycyna, C.A.; Clarke, S.: Farnesyl cysteine C-terminal methyltransferase activity is dependent upon the STE14 gene product in Saccharomyces cerevisiae. Mol. Cell. Biol., **10**, 5071-5076 (1990)

[5] Volker, C.; Miller, R.A.; McCleary, W.R.; Rao, A.; Poenie, M.; Backer, J.M.; Stock, J.B.: Effects of farnesylcysteine analogs on protein carboxyl methylation and signal transduction. J. Biol. Chem., **266**, 21515-21522 (1991)

[6] Shi, Y.Q.; Rando, R.R.: Kinetic mechanism of isoprenylated protein methyltransferase. J. Biol. Chem., **267**, 9547-9551 (1992)

[7] Perez-Sala, D.; Gilbert, B.A.; Tan, E.W.; Rando, R.R.: Prenylated protein methyltransferases do not distinguish between farnesylated and geranylgeranylated substrates. Biochem. J., **284**, 835-840 (1992)

[8] Yoo, Y.; Watts, S.; Rechsteiner, M.: Ubiquitin-RAS peptide extensions as substrates for farnesyl-protein transferase and carboxymethyltransferase. Biochem. J., **285**, 55-60 (1992)

[9] Gingras, D.; Beliveau, R.: Cytosolic proteins of 21-23 kDa are methylated by kidney cortex membrane-associated C-terminal carboxyl methyltransferases. Biochim. Biophys. Acta, **1204**, 149-156 (1994)

[10] Boivin, D.; Potier, M.; Beliveau, R.: Functional size of C-terminal protein carboxyl methyltransferase from kidney basolateral plasma membranes. Biochim. Biophys. Acta, **1207**, 114-119 (1994)

[11] Park, S.H.; Lee, H.W.: Inhibition of C-terminal O-methyltransferase by a rat liver cytosolic peptide. Arch. Pharmacol. Res., **17**, 354-359 (1994)

[12] Klein, Z.; Ben-Baruch, G.; Marciano, D.; Solomon, R.; Altaras, M.; Kloog, Y.: Characterization of the prenylated protein methyltransferase in human endometrial carcinoma. Biochim. Biophys. Acta, **1226**, 330-336 (1994)

[13] Pillinger, M.H.; Volker, C.; Stock, J.B.; Weissmann, G.; Philips, M.R.: Characterization of a plasma membrane-associated prenylcysteine-directed a carboxyl methyltransferase in human neutrophils. J. Biol. Chem., **269**, 1486-1492 (1994)

[14] Giner, J.L.; Rando, R.R.: Novel methyltransferase activity modifying the carboxy terminal bis(geranylgeranyl)-Cys-Ala-Cys structure of small GTP-binding proteins. Biochemistry, **33**, 15116-15123 (1994)

[15] Ma, Y.T.; Shi, Y.Q.; Lim, Y.H.; McGrail, S.H.; Ware, J.A.; Rando, R.R.: Mechanistic studies on human platelet isoprenylated protein methyltransferase: farnesylcysteine analogs block platelet aggregation without inhibiting the methyltransferase. Biochemistry, **33**, 5414-5420 (1994)
[16] Park, K.S.; Choi, S.Y.; Park, G.H.; Kim, S.: Studies on mammalian brain protein farnesyl-cysteine carboxyl-methyltransferase. Korean J. Biochem., **26**, 47-57 (1994)
[17] Li, G.; Kowluru, A.; Metz, S.A.: Characterization of prenylcysteine methyltransferase in insulin-secreting cells. Biochem. J., **316**, 345-351 (1996)
[18] Boivin, D.; Lin, W.; Beliveau, R.: Essential arginine residues in isoprenylcysteine protein carboxyl methyltransferase. Biochem. Cell Biol., **75**, 63-69 (1997)
[19] Marciano, D.; Aharonson, Z.; Varsano, T.; Haklai, R.; Kloog, Y.: Novel inhibitors of the prenylated protein methyltransferase reveal distinctive structural requirements. Bioorg. Med. Chem. Lett., **7**, 1709-1714 (1997)
[20] Crowell, D.N.; Sen, S.E.; Randall, S.K.: Prenylcysteine a-carboxyl methyltransferase in suspension-cultured tobacco cells. Plant Physiol., **118**, 115-123 (1998)
[21] Yoo, B.C.; Kang, M.S.; Kim, S.; Lee, Y.S.; Choi, S.Y.; Ryu, C.K.; Park, G.H.; Han, J.S.: Partial purification of protein farnesyl cysteine carboxyl methyltransferase from bovine brain. Exp. Mol. Med., **30**, 227-234 (1998)
[22] Hong, K.M.; Choi, Y.B.; Hong, J.H.; Chang, H.S.; Rhee, K.I.; Park, H.; Paik, M.K.: Cysteine carboxyl O-methylation of human placental 23 kDa protein. Exp. Mol. Med., **31**, 30-35 (1999)
[23] Hasne, M.P.; Lawrence, F.: Characterization of prenylated protein methyltransferase in Leishmania. Biochem. J., **342**, 513-518 (1999)
[24] De Busser, H.M.; Van Dessel, G.A.F.; Lagrou, A.R.: Identification of prenylcysteine carboxymethyltransferase in bovine adrenal chromaffin cells. Int. J. Biochem. Cell Biol., **32**, 1007-1016 (2000)
[25] Van Dessel, G.A.F.; De Busser, H.M.; Lagrou, A.R.: Prenylcysteine carboxymethyltransferase type III activity is decreased in retinoic acid-treated SH-SY5Y neuroblastoma cells. Int. J. Biochem. Cell Biol., **34**, 477-486 (2002)
[26] Buckner, F.S.; Kateete, D.P.; Lubega, G.W.; Van Voorhis, W.C.; Yokoyama, K.: Trypanosoma brucei prenylated-protein carboxyl methyltransferase prefers farnesylated substrates. Biochem. J., **367**, 809-816 (2002)
[27] Choy, E.; Philips, M.: Expression and activity of human prenylcysteine-directed carboxyl methyltransferase. Methods Enzymol., **325**, 101-114 (2000)

Macrocin O-methyltransferase 2.1.1.101

1 Nomenclature

EC number
2.1.1.101

Systematic name
S-adenosyl-L-methionine:macrocin 3'''-O-methyltransferase

Recommended name
macrocin O-methyltransferase

Synonyms
S-adenosyl-L-methionine-macrocin O-methyltransferase
macrocin methyltransferase

CAS registry number
79468-52-3

2 Source Organism

<1> *Streptomyces fradiae* [1-4]

3 Reaction and Specificity

Catalyzed reaction
S-adenosyl-L-methionine + macrocin = S-adenosyl-L-homocysteine + tylosin (the 3-hydroxyl group of a 2-O-methyl-6-deoxy-D-allose residue in the macrolide antibiotic macrosin acts as methyl acceptor; also converts lactenosin into desmycocin. Not identical with EC 2.1.1.102 demethylmacrocin O-methyltransferase)

Reaction type
methyl group transfer

Natural substrates and products
S S-adenosyl-L-methionine + macrocin <1> (<1> terminal and main rate-limiting step in tylosin biosynthesis [1-3]; <1> enzyme catalyzes 3'''-O-methylation of bound 2'''-O-methylated 6-deoxy-D-allose [4]) (Reversibility: ? <1> [1-4]) [1-4]
P S-adenosylhomocysteine + tylosin

Substrates and products

S S-adenosyl-L-methionine + 20-dihydrolactenocin <1> (Reversibility: ? <1> [2]) [2]
P S-adenosyl-L-homocysteine + ?
S S-adenosyl-L-methionine + dihydromacrocin <1> (Reversibility: ? <1> [2]) [2]
P S-adenosyl-L-homocysteine + ?
S S-adenosyl-L-methionine + lactenocin <1> (Reversibility: ? <1> [2]) [2]
P S-adenosyl-L-homocysteine + desmycosin
S S-adenosyl-L-methionine + macrocin <1> (Reversibility: ? <1> [1-4]) [1-4]
P S-adenosyl-L-homocysteine + tylosin
S Additional information <1> (<1> demethylmacrocin, demethyllactenocin, N^5-methyltetrahydrofolate, betaine, choline, methylthioadenosine and methionine are not substrates [2]) [2]
P ?

Inhibitors

A9145C <1> [4]
S-adenosyl-L-homocysteine <1> (<1> not D-enantiomer [4]) [2, 4]
demethyllactenocin <1> [2, 4]
demethylmacrocin <1> [2, 4]
desmycosin <1> (<1> weak inhibition [2,4]) [2, 4]
macrocin <1> [1]
relomycin <1> (<1> weak inhibition [2,4]) [2-4]
sinefungin <1> (<1> competitive inhibition, potent inhibitor [2,4]) [2, 4]
tylosin <1> (<1> weak inhibition [4]) [1, 4]

Metals, ions

Co^{2+} <1> (<1> can substitute Mg^{2+} [2,4]; <1> stimulated by cobalt chloride [3]) [2-4]
Fe^{2+} <1> (<1> stimulated by iron (II) sulfate [3]) [3]
Mg^{2+} <1> (<1> required for maximal activity [1,2,4]; <1> stimulated by magnesium sulfate [3]) [1-4]
Mn^{2+} <1> (<1> can substitute Mg^{2+} [2,4]; <1> stimulated by manganese sulfate [3]) [2-4]
Zn^{2+} <1> (<1> stimulated by zinc sulfate [3]) [3]

Specific activity (U/mg)

0.061 <1> [2]

K_m-Value (mM)

0.005 <1> (macrocin) [4]
0.007 <1> (lactenocin) [4]
0.022 <1> (S-adenosyl-L-methionine, <1> lactenocin as substrate [4]) [4]
0.023 <1> (S-adenosyl-L-methionine, <1> macrocin as substrate [4]) [4]
0.029 <1> (S-adenosyl-L-methionine) [3]
0.033 <1> (macrocin) [3]

K_i-Value (mM)

0.00006 <1> (sinefungin) [2, 4]
0.00025 <1> (sinefungin) [2]
0.005 <1> (demethylmacrocin) [2]
0.006 <1> (demethylactenocin) [2]
0.008 <1> (S-adenosyl-L-homocysteine, <1> fixed substrate macrocin [2]) [2]
0.023 <1> (desmycosin, <1> fixed substrate S-adenosyl-L-methionine [2]) [2]
0.025 <1> (S-adenosyl-L-homocysteine, <1> fixed substrate S-adenosyl-L-methionine [2]) [2]
0.026 <1> (desmycosin, <1> fixed substrate macrocin [2]) [2]
0.079 <1> (tylosin, <1> fixed substrate S-adenosyl-L-methionine [2]) [2]
0.217 <1> (tylosin, <1> fixed substrate lactenocin [2]) [2]

pH-Optimum

7.5-8 <1> [2]
7.5-8.2 <1> [1]
7.8 <1> [3]
7.8-8 <1> [4]

Temperature optimum (°C)

31 <1> [1, 2, 4]

4 Enzyme Structure

Molecular weight

65000 <1> (<1> gel filtration [2]) [2, 4]

Subunits

dimer <1> (<1> 2 * 32000, SDS-PAGE [2,4]) [2, 4]

5 Isolation/Preparation/Mutation/Application

Purification

<1> [2, 4]

Application

agriculture <1> (<1> tylosin is used in the swine industry as a growth promotant [2]) [2]
pharmacology <1> (<1> tylosin is used in veterinary medicine for treatment of infections caused by gram-positive bacteria and mycoplasma [2]; <1> tylosin fermentation, antibiotic biosynthesis, enzyme catalyzes conversion of macrocin to tylosin in vivo [4]) [2, 4]

6 Stability

Storage stability

<1>, 4°C, enzyme is unstable, about 30% activity remains after 3 h [3]
<1>, 4°C, quite stable in presence of 0.2 mM S-adenosyl-L-methionine and 10% ethanol, pH 7.0, stable over 6 months [2, 4]
<1>, 4°C, very stable for 3 h in presence of 10% ethanol [3]

References

[1] Seno, E.T.; Baltz, R.H.: Properties of S-adenosyl-L-methionine:macrocin O-methyltransferase in extracts of Streptomyces fradiae strains which produce normal or elevated levels of tylosin and in mutants blocked in specific O-methylations. Antimicrob. Agents Chemother., **20**, 370-377 (1981)
[2] Bauer, N.J.; Kreuzman, A.J.; Dotzlaf, J.E.; Yeh, W.K.: Purification, characterization, and kinetic mechanism of S-adenosyl-L-methionine:macrocin O-methyltransferase from Streptomyces fradiae. J. Biol. Chem., **263**, 15619-15625 (1988)
[3] Yeh, W.K.; Bauer, N.J.; Dotzlaf, J.E.: High-performance liquid chromatographic assay for S-adenosyl-L-methionine: macrocin O-methyltransferase. J. Chromatogr., **288**, 157-165 (1984)
[4] Kreuzman, A.J.; Turner, J.R.; Yeh, W.K.: Two distinctive O-methyltransferases catalyzing penultimate and terminal reactions of macrolide antibiotic (tylosin) biosynthesis. Substrate specificity, enzyme inhibition, and kinetic mechanism. J. Biol. Chem., **263**, 15626-15633 (1988)

Demethylmacrocin O-methyltransferase 2.1.1.102

1 Nomenclature

EC number
2.1.1.102

Systematic name
S-adenosyl-L-methionine:demethylmacrocin 2'''-O-methyltransferase

Recommended name
demethylmacrocin O-methyltransferase

Synonyms
demethylmacrocin methyltransferase

CAS registry number
120313-64-6

2 Source Organism

<1> *Streptomyces fradiae* [1]

3 Reaction and Specificity

Catalyzed reaction
S-adenosyl-L-methionine + demethylmacrocin = S-adenosyl-L-homocysteine + macrocin (the 2-hydroxyl group of a 6-deoxy-D-allose residue in demethylmacrocin acts as a methyl acceptor. Not identical with EC 2.1.1.101 macrocin O-methyltransferase)

Reaction type
methyl group transfer

Natural substrates and products
S S-adenosyl-L-methionine + demethylmacrocin <1> (<1> penultimate step in tylosin biosynthesis [1]; <1> enzyme catalyzes 2'''-O-methylation of bound 6-deoxy-D-allose [1]) (Reversibility: ? <1> [1]) [1]
P S-adenosyl-L-homocysteine + macrocin

Substrates and products
S S-adenosyl-L-methionine + demethyllactenocin <1> (Reversibility: ? <1> [1]) [1]
P S-adenosyl-L-homocysteine + lactenocin <1> [1]

S S-adenosyl-L-methionine + demethylmacrocin <1> (Reversibility: ? <1> [1]) [1]
P S-adenosyl-L-homocysteine + macrocin

Inhibitors
20-dihydro-23-deoxymycaminosyltylonolide <1> [1]
20-dihydrolactenocin <1> (<1> inhibition of macrocin formation [1]) [1]
20-dihydromacrocin <1> [1]
23-demycinosyloxytylosin <1> [1]
23-deoxy-O-mycaminosyltylonolide <1> [1]
A9145C <1> [1]
Fe^{2+} <1> [1]
O-mycaminosyltylonolide <1> [1]
S-adenosyl-L-homocysteine <1> (<1> not D-enantiomer [1]) [1]
demycinosyltylosin <1> [1]
lactenocin <1> (<1> inhibition of macrocin formation [1]) [1]
macrocin <1> (<1> inhibition of lactenocin formation [1]) [1]
mycaminosyltylactone <1> [1]
sinefungin <1> (<1> competitive inhibition, poor inhibitor [1]) [1]

Metals, ions
Mg^{2+} <1> (<1> required for maximal activity [1]) [1]

Specific activity (U/mg)
0.25 <1> [1]

K_m-Value (mM)
0.006 <1> (demethylmacrocin) [1]
0.01 <1> (demethyllactenocin) [1]
0.111 <1> (S-adenosylmethionine) [1]

K_i-Value (mM)
0.022 <1> (lactenocin, <1> fixed substrate demethylmacrocin [1]) [1]
0.024 <1> (lactenocin, <1> fixed substrate S-adenosylmethionine [1]) [1]
0.037 <1> (macrocin, <1> fixed substrate S-adenosylmethionine [1]) [1]
0.057 <1> (A914C, <1> fixed substrate demethylmacrocin [1]) [1]
0.065 <1> (mycaminosyltylonolide) [1]
0.067 <1> (macrocin, <1> fixed substrate demethyllactenocin [1]) [1]
0.073 <1> (23-deoxy-O-mycaminosyltylonolide) [1]
0.077 <1> (sinefungin) [1]
0.108 <1> (A914C, <1> fixed substrate S-adenosylmethionine [1]) [1]
0.11 <1> (S-adenosyl-L-homocysteine, <1> fixed substrate demethylmacrocin [1]) [1]
0.186 <1> (23-demycinosyloxytylosin) [1]
0.226 <1> (S-adenosyl-L-homocysteine, <1> fixed substrate S-adenosylmethionine [1]) [1]

pH-Optimum
7.8-8.5 <1> [1]

Temperature optimum (°C)

42 <1> [1]

4 Enzyme Structure

Molecular weight

122000-126000 <1> (<1> gel filtration [1]) [1]

Subunits

trimer <1> (<1> 3 * 42000, SDS-PAGE [1]) [1]

5 Isolation/Preparation/Mutation/Application

Purification

<1> [1]

Application

pharmacology <1> (<1> tylosin fermentation, antibiotic biosynthesis [1]) [1]

6 Stability

Storage stability

<1>, 4°C, quite stable in presence of S-adenosylmethionine and 10% ethanol, pH 7.0-8.0, stable for a minimum of 2 months [1]

References

[1] Kreuzman, A.J.; Turner, J.R.; Yeh, W.K.: Two distinctive O-methyltransferases catalyzing penultimate and terminal reactions of macrolide antibiotic (tylosin) biosynthesis. Substrate specificity, enzyme inhibition, and kinetic mechanism. J. Biol. Chem., **263**, 15626-15633 (1988)

Phosphoethanolamine N-methyltransferase 2.1.1.103

1 Nomenclature

EC number
2.1.1.103

Systematic name
S-adenosyl-L-methionine:ethanolamine-phosphate N-methyltransferase

Recommended name
phosphoethanolamine N-methyltransferase

Synonyms
phosphodimethylethanolamine methyltransferase

CAS registry number
145539-90-8
171040-79-2

2 Source Organism

<1> *Lemna paucicostata* (duck weed) [1]
<2> *Glycine max* (soybean) [1, 7]
<3> *Daucus carota* (carrot) [1]
<4> *Rattus norvegicus* (Sprague-Dawley) [2, 3]
<5> *Spinacia oleracea* (highest activity after prolonged light period [4,7,11]) [4, 7, 11]
<6> *Spinacia oleracea* [5]
<7> *Arabidopsis sp.* [6]
<8> *Pisum sativum* [7]
<9> *Gossypium hirsutum* [7]
<10> *Lamonium perezii* (highest activity after prolonged light period [7]) [7]
<11> *Beta vulgaris* (highest activity after prolonged light period [7]) [7]
<12> *Amaranthus caudatus* (highest activity after prolonged light period [7]) [7]
<13> *Nicotiana tobacum* [8]
<14> *Arabidopsis thaliana* [9]
<15> *Triticum aestivum* [10]

3 Reaction and Specificity

Catalyzed reaction

S-adenosyl-L-methionine + ethanolamine phosphate = S-adenosyl-L-homocysteine + N-methylethanolamine phosphate

Reaction type

methyl group transfer

Natural substrates and products

S S-adenosyl-L-methionine + ethanolamine phosphate <1-15> (<1-15> involved in synthesis of phosphocholine and phosphatidylcholine [1-11]; <4,6,14> involved in production of free choline [2,5,9]) (Reversibility: ? <1-15> [1-11]) [1-11]

P S-adenosyl-L-homocysteine + N-methylethanolamine phosphate <1-15> [1-11]

Substrates and products

S S-adenosyl-L-methionine + ethanolamine phosphate <1-15> (Reversibility: ? <1-15> [1-11]) [1-11]

P S-adenosyl-L-homocysteine + N-methylethanolamine phosphate <1-15> [1-11]

S S-adenosyl-L-methionine + phosphodimethylethanolamine <1, 3-7, 9-13, 15> (<4> reaction not catalyzed by phosphoethanolamine N-methyltransferase but by another enzyme [3]; <5> poor substrate [11]) (Reversibility: ? <1, 3-7, 9-13, 15> [1, 2, 5-8, 10, 11]) [1, 2, 3, 5-8, 10, 11]

P S-adenosyl-L-homocysteine + phosphocholine <1, 3-7, 9-13, 15> [1, 2, 5-8, 10, 11]

S S-adenosyl-L-methionine + phosphomethylethanolamine <1, 3, 4-7, 9-13, 15> (Reversibility: ? <1, 3, 4-7, 9-13, 15> [1-3, 5-8, 10, 11]) [1-3, 5-8, 10, 11]

P S-adenosyl-L-homocysteine + phosphodimethylethanolamine <1, 3, 4-7, 9-13, 15> [1-3, 5-8, 10, 11]

S Additional information <1, 2, 8> (<1> no activity with ethanolamine, methylethanolamine and dimethylethanolamine [1]; <2,8> no activity with phosphomethylethanolamine and phosphodimethylethanolamine [1,7]; <4> no activity with phosphodimethylethanolamine [3]) [1, 3, 7]

P ?

Inhibitors

Co^{2+} <5> [11]
Mn^{2+} <5> [11]
S-adenosyl-L-homocysteine <5> [11]
phosphate <5> [11]
phosphocholine <5, 6> [5, 11]

Activating compounds

EDTA <1> (<1> removal of EDTA leads to 60% loss of activity [1]) [1]
glutathione <1> (<1> removal of glutathione leads to almost complete loss of activity [1]) [1]

Specific activity (U/mg)

0.0085 <7> (<7> with phosphodimethylethanolamine as substrate [6]) [6]
0.013 <7> (<7> with phosphoethanolamine as substrate [6]) [6]
0.189 <5> (<5> purified enzyme [11]) [11]

K_m-Value (mM)

0.056 <15> (S-adenosyl-L-methionine) [10]
0.065 <15> (ethanolamine phosphate) [10]
0.096 <6> (ethanolamine phosphate) [5]
0.132 <6> (S-adenosyl-L-methionine) [5]

pH-Optimum

7.8 <1, 5> [1, 11]
7.8-8.5 <6> [5]

pH-Range

6.8-8.3 <1> [1]

4 Enzyme Structure

Molecular weight

54000 <5> (<5> SDS-PAGE [11]) [11]
56100 <14> (<14> calculated from DNA sequence [9]) [9]
57000 <6, 15> (<6> gel filtration [5]; <15> calculated from DNA sequence [10]) [5, 10]
65000 <15> (<15> SDS-PAGE, enzyme from E. coli, contains of 57 kD enzyme protein + 8 kD His-tag [10]) [10]
77000 <5> (<5> gel filtration [11]) [11]

Subunits

monomer <6> (<6> gel filtration [5]) [5]

5 Isolation/Preparation/Mutation/Application

Source/tissue

brain <4> [2, 3]
leaf <5, 6> [4, 5, 11]
liver <4> (<4> less than 20% activity compared to brain extracts [2]) [2]

Localization

cytosol <4, 5> [2, 3, 4]

Purification

<5> [11]
<15> (from E. coli, His-tag used for affinity chromatography [10]) [10]

Cloning

<6> (in Schizosaccharomyces pombe [5]) [5]
<7> (in Saccharomyces cerevisiae opi3 [6]) [6]
<13> (tobacco transformed with spinach cDNA [8]) [8]
<14> (generation of transgenic tobacco plants [9]) [9]
<15> (in Escherichia coli [10]) [10]

Engineering

Additional information <6, 14> (<6> deleting the C-terminal domain does not abolish ethanolamine phosphate reactivity, C-terminal domain catalyzes reactions with methylated ethanolamine phosphate [5]; <14> Arabidopsis mutant strain t365 with silenced gene for EC 2.1.1.103 enzyme shows pale green leaves, early senescence, temperature-sensitive male sterility and hypersensitivity to salt stress [9]; <15> several point mutations and N- and C-terminal domains investigated, N-terminal domain catalyzes all three activities, regulatory role for C-terminal domain suggested [10]) [5, 9, 10]

References

[1] Datko, A.H.; Mudd, H.S.: Enzymes of phosphatidylcholine synthesis in lemna, soybean, and carrot. Plant Physiol., **88**, 1338-1348 (1988)

[2] Andriamampandry, C.; Massarelli, R.; Freysz, L.; Kanfer, J.N.: A rat brain cytosolic N-methyltransferase(s) activity converting phosphorylethanolamine into phosphorylcholine. Biochem. Biophys. Res. Commun., **171**, 758-763 (1990)

[3] Andriamampandry, C.; Massarelli, R.; Kanfer, J.N.: Properties of a partially purified phosphodimethylethanolamine methyltransferase from rat brain cytosol. Biochem. J., **288**, 267-272 (1992)

[4] Weretilnyk, E.A.; Smith, D.D.; Wilch, G.A.; Summers, P.S.: Enzymes of choline synthesis in spinach. Response of phospho-base N-methyltransferase activities to light and salinity. Plant Physiol., **109**, 1085-1091 (1995)

[5] Nuccio, M.L.; Ziemak, M.J.; Henry, S.A.; Weretilnyk, E.A.; Hanson, A.D.: cDNA cloning of phosphoethanolamine N-methyltransferase from spinach by complementation in Schizosaccharomyces pombe and characterization of the recombinant enzyme. J. Biol. Chem., **275**, 14095-14101 (2000)

[6] Bolognese, C.P.; McGraw, P.: The isolation and characterization in yeast of a gene for Arabidopsis S-adenosylmethionine:phospho-ethanolamine N-methyltransferase. Plant Physiol., **124**, 1800-1813 (2000)

[7] Lorenzin, D.; Webb, C.; Summers, P.S.; Weretilnyk, E.A.: Enzymes of choline synthesis in diverse plants: Variation in phospho-base N-methyltransferase activities. Can. J. Bot., **79**, 897-904 (2001)

[8] McNeil, S.D.; Nuccio, M.L.; Ziemak, M.J.; Hanson, A.D.: Enhanced synthesis of choline and glycine betaine in transgenic tobacco plants that overexpress

phosphoethanolamine N-methyltransferase. Proc. Natl. Acad. Sci. USA, **98**, 10001-10005 (2001)
[9] Mou, Z.; Wang, X.; Fu, Z.; Dai, Y.; Han, C.; Ouyang, J.; Bao, F.; Hu, Y.; Li, J.: Silencing of phosphoethanolamine N-methyltransferase results in temperature-sensitive male sterility and salt hypersensitivity in Arabidopsis. Plant Cell, **14**, 2031-2043 (2002)
[10] Charron, J.B.F.; Breton, G.; Danyluk, J.; Muzac, I.; Ibrahim, R.K.; Sarhan, F.: Molecular and biochemical characterization of a cold-regulated phosphoethanolamine N-methyltransferase from wheat. Plant Physiol., **129**, 363-373 (2002)
[11] Smith, D.D.; Summers, P.S.; Weretilnyk, E.A.: Phosphocholine synthesis in spinach: characterization of phosphoethanolamine N-methyltransferase. Physiol. Plant., **108**, 286-294 (2000)

Caffeoyl-CoA O-methyltransferase 2.1.1.104

1 Nomenclature

EC number
2.1.1.104

Systematic name
S-adenosyl-L-methionine:caffeoyl-CoA 3-O-methyltransferase

Recommended name
caffeoyl-CoA O-methyltransferase

Synonyms
CCoAOMT <6-13> [1-13]
caffeoyl coenzyme A methyltransferase
caffeoyl-CoA 3-O-methyltransferase
trans-caffeoyl-CoA 3-O-methyltransferase

CAS registry number
120433-42-3

2 Source Organism

<1> *Petroselinum crispum* (parsley [9,15,16,17,18]) [9, 15, 16, 17, 18]
<2> *Dianthus caryophyllus* (cv. Mei Ling, carnation [17]) [17]
<3> *Carthamus tinctorius* (cv. US10, safflower [17]) [17]
<4> *Daucus carota* (ssp. sativus, carrot [17]) [17]
<5> *Ammi majus* (bishop's weed [17]) [17]
<6> *Medicago sativa* (alfalfa [1,5]; cv Apollo for non-transgenic and cv Regen SY for transgenic line with downregulated CCoAOMT [5]; yeast elicitor-treated cv Apollo [14]) [1, 5, 12, 14]
<7> *Zinnia sp.* [2]
<8> *Linum usitatissimum* (cv Barbara, flax [3]) [3]
<9> *Nicotiana tabacum* (tobacco [4,8]; cv Samsun NN, 4 different clones were isolated from infected leaves [8,10]; cv Havana SR1 [9]) [4, 8, 9, 10]
<10> *Populus tremula x Populus alba* (woody poplar [6]; transgenic poplar [7]; under stress conditions [7]) [6, 7]
<11> *Populus trichocarpa* (poplar [7]; under stress conditions [7]) [7]
<12> *Populus tremuloides* (aspen [11]) [11]
<13> *Vitis vinifera* (cv Gamay Fréaux, fungal elicitation [13]) [13]

3 Reaction and Specificity

Catalyzed reaction

S-adenosyl-L-methionine + caffeoyl-CoA = S-adenosyl-L-homocysteine + feruloyl-CoA

Reaction type

methyl group transfer

Natural substrates and products

S S-adenosyl-L-methionine + caffeoyl-CoA <1, 2, 3, 4, 5, 6, 7> (<1> reaction of the overall disease resistance response of plants [16]; <1> inducible [18]; <1,6> bi-functional enzyme, involved in lignin biosynthesis [1,9]; <7> essential role in the synthesis of guaiacyl lignin units as well as in the supply of substrates for the synthesis of syringyl lignin units [2]; <1> enzyme plays a pivotal role in cell wall reinforcement during the induced disease resistance response [9]) (Reversibility: ? <1, 2, 3, 4, 5, 6, 7> [1, 2, 9, 16, 17, 18]) [1, 2, 9, 16, 17, 18]

P S-adenosyl-L-homocysteine + feruloyl-CoA

Substrates and products

S S-adenosyl-L-methionine + 5-hydroxyconiferaldehyde <6> (Reversibility: ? <6> [5]) [5]

P S-adenosyl-L-homocysteine + 5-hydroxy-3-methoxyconiferyl aldehyde

S S-adenosyl-L-methionine + 5-hydroxyferulic acid <12> (Reversibility: ? <12> [11]) [11]

P S-adenosyl-L-homocysteine + sinapic acid

S S-adenosyl-L-methionine + 5-hydroxyferuoyl-CoA <6, 9, 10, 12, 13> (Reversibility: ? <6, 9, 10, 12, 13> [5, 6, 9, 11, 12, 13]) [5, 6, 9, 11, 12, 13]

P S-adenosyl-L-homocysteine + sinapoyl-CoA

S S-adenosyl-L-methionine + caffeoyl-CoA <1, 2, 3, 4, 5, 6, 7, 8, 9, 10, 11, 12, 13> (<1> trans-caffeoyl-CoA preferred [18]) (Reversibility: ? <1, 2, 3, 4, 5, 6, 7, 8, 9, 10, 11, 12, 13> [1, 2, 3, 4, 5, 6, 7, 8, 9, 10, 11, 12, 13, 15, 16, 17, 18]) [1, 2, 3, 4, 5, 6, 7, 8, 9, 10, 11, 12, 13, 15, 16, 17, 18]

P S-adenosyl-L-homocysteine + feruloyl-CoA <1, 2, 3, 4, 5, 6, 8, 12> [1, 3, 5, 11, 12, 15, 16, 17, 18]

S S-adenosyl-L-methionine + trans-caffeic acid esters <1> (<1> such as methyl caffeate, chlorogenic acid, trans-5-O-caffeoylshikimate, rosmarinic acid, poor substrates [18]) (Reversibility: ? <1> [18]) [18]

P S-adenosyl-L-homocysteine + ?

S Additional information <1, 9, 13> (<1> no substrates are caffeic acid, bergaptol, xanthotoxol, luteolin, esculetin [18]; <9,13> caffeoyl-CoA preferred to 5-hydroxyferuoyl-CoA [8,13]) [8, 13, 18]

P ?

Inhibitors

Ca^{2+} <1> [18]
EDTA <1> [18]
Mn^{2+} <1> [18]

S-adenosyl-L-homocysteine <1> (<1> product inhibition, non-competitive to both substrates [16]) [16]
feruloyl-CoA <1> (<1> competitive to caffeoyl-CoA, non-competitive to S-adenosyl-L-methionine [16]) [16]

Activating compounds
Mg^{2+} <1> (<1> activation, 0.2 mM [18]) [16, 18]
NaCl <1> (<1> activation, 0.4 M [18]) [18]
ethanol <1> (<1> activation, 12% v/v [18]) [18]

Metals, ions
Mg^{2+} <1> (<1> activation, 0.2 mM [18]) [16, 18]
NaCl <1> (<1> activation, 0.4 M [18]) [18]

Specific activity (U/mg)
0.002 <12> (<12> with 5-hydroxyferulic acid [11]) [11]
0.008 <12> (<12> with 5-hydroxyferuloyl-CoA [11]) [11]
0.021 <12> (<12> with caffeoyl-CoA [11]) [11]
0.0313 <1> (<1> unreliable, great loss during purification [15]) [15]
0.059 <1> [18]

K_m-Value (mM)
0.00024 <1> (caffeoyl-CoA) [9]
0.0007 <1> (5-hydroxyferuoyl-CoA) [9]
0.0043 <6> (caffeoyl-CoA, <6> recombinant enzyme [5]) [5]
0.0062 <6> (5-hydroxyferuoyl-CoA, <6> recombinant enzyme [5]) [5]
0.0275 <9> (caffeoyl-CoA, <9> wild-type [4]) [4]
0.29 <9> (caffeoyl-CoA, <9> Q61S-mutant [4]) [4]
0.64 <9> (caffeoyl-CoA, <9> D58A-mutant [4]) [4]

pH-Optimum
7.5 <1, 6, 12> [11, 12, 18]

pH-Range
6.9-8.6 <1> (<1> about half-maximal activity at pH 6.9 and 8.6 [18]) [18]

Temperature optimum (°C)
30 <1> (<1> assay at [15,18]) [15, 18]

4 Enzyme Structure

Molecular weight
27000 <1, 9> (<9> SDS-PAGE, class 1 recombinant protein [8]; <1> gel filtration, recombinant protein [9]) [8, 9]
27230 <13> (<13> calculated mass for the translated polypeptide [13]) [13]
28000 <10, 11> (<10,11> SDS-PAGE [6,7]) [6, 7]
30000 <1, 6> (<6> SDS-PAGE [5]; <1> gel filtration, crude extract [9]) [5, 9]
33000 <8> (<8> SDS-PAGE [3]) [3]
48000 <1> (<1> gel filtration [18]) [18]
50000 <1> (<1> PAGE [15]) [15]

Subunits

dimer <1> (<1> 2 * 24000, SDS-PAGE [15]) [15]

5 Isolation/Preparation/Mutation/Application

Source/tissue

cell suspension culture <1, 2, 3, 4, 5, 13> (<1> treated with crude cell wall elicitor from Phytophthora megasperma f.sp. glycinea [15,18]) [15, 17, 18, 13]
leaf <1, 9, 10, 11> (<10,11> sprayed with spores of Melampsora pinitorca [7]; <9> infected with tobacco mosaic virus [8,10]) [7, 8, 9, 10, 17]
root <1> [17]
stem <6, 8, 10> (<8> maximal activity in the flowering phase, at the basal part of the stem [3]; <6> first to the tenth stem internodes [5]; <10> expression in all lignifying cells including vessel elements and fibers as well as in xylem ray parenchyma cells [6]) [3, 5, 6]
xylem <12> (<12> secondary xylem during seasonal development [11]) [11]

Purification

<1> (affinity chromatography on Blue Sepharose CL-6B, isoelectric focusing [15]) [15, 18]
<9> [4]
<13> [13]

Cloning

<1> (expression in Escherichia coli JM109 DE3 [9]) [9, 17]
<6> (expression in Escherichia coli DH5α [12]) [12]
<9> (expressed in Escherichia coli strain TG2 [4]; 4 clones expressed in Escherichia coli [8,10]; transformed with parsley gene [9]; expression in Escherichia coli DH5α [10]) [4, 8, 9, 10]
<12> (expression in Escherichia coli BL21 [11]) [11]
<13> (expression in Escherichia coli BL21 [13]) [13]

Engineering

D58A <9> (<9> decrease of catalytic activity [4]) [4]
Q61S <9> (<9> decrease of catalytic activity [4]) [4]
R220T <9> (<9> total loss of activity, R220 is involved in the electrostatic interaction with CoA-moiety of the substrate [4]) [4]

Application

agriculture <10> (<10> antisense repression of the enzyme is an efficient means for genetic engineering of trees with low lignin content [6]) [6]

6 Stability

pH-Stability

7 <1> (<1> unstable below [15]) [15]

References

[1] Marita, J.M.; Ralph, J.; Hatfield, R.D.; Guo, D.; Chen, F.; Dixon, R.A.: Structural and compositional modifications in lignin of transgenic alfalfa down-regulated in caffeic acid 3-O-methyltransferase and caffeoyl coenzyme A 3-O-methyltransferase. Phytochemistry, **62**, 53-65 (2003)

[2] Ye, Z.H.; Zhong, R.; Morrison, W.H.; Himmelsbach, D.S.: Caffeoyl coenzyme A O-methyltransferase and lignin biosynthesis. Phytochemistry, **57**, 1177-1185 (2001)

[3] Day, A.; Dehorter, B.; Neutelings, G.; Czeszak, X.; Chabbert, B.; Belingheri, L.; David, H.: Caffeoyl-coenzyme A 3-O-methyltransferase enzyme activity, protein and transcript accumulation in flax (Linum usitatissimum) stem during development. Physiol. Plant., **113**, 275-284 (2001)

[4] Hoffmann, L.; Maury, S.; Bergdoll, M.; Thion, L.; Erard, M.; Legrand, M.: Identification of the enzymatic active site of tobacco caffeoyl-coenzyme A O-methyltransferase by site-directed mutagenesis. J. Biol. Chem., **276**, 36831-36838 (2001)

[5] Parvathi, K.; Chen, F.; Guo, D.; Blount, J.W.; Dixon, R.A.: Substrate preferences of O-methyltransferases in alfalfa suggest new pathways for 3-O-methylation of monolignols. Plant J., **25**, 193-202 (2001)

[6] Zhong, R.; Morrison, W.H.; Himmelsbach, D.S.; Poole, F.L.; Ye, Z.H.: Essential role of caffeoyl coenzyme A O-methyltransferase in lignin biosynthesis in woody poplar plants. Plant Physiol., **124**, 563-578 (2000)

[7] Chen, C.; Meyermans, H.; Burggraeve, B.; De Rycke, R.M.; Inoue, K.; De Vleesschauwer, V.; Steenackers, M.; Van Montagu, M.C.; Engler, G.J.; Boerjan, W.A.: Cell-specific and conditional expression of caffeoyl-coenzyme A-3-O-methyltransferase in poplar. Plant Physiol., **123**, 853-867 (2000)

[8] Maury, S.; Geoffroy, P.; Legrand, M.: Tobacco O-methyltransferases involved in phenylpropanoid metabolism. The different caffeoyl-coenzyme A/5-hydroxyferuloyl-coenzyme A 3/5-O-methyltransferase and caffeic acid/5-hydroxyferulic acid 3/5-O-methyltransferase classes have distinct substrate specificities and expression patterns. Plant Physiol., **121**, 215-224 (1999)

[9] Grimmig, B.; Kneusel, R.E.; Junghanns, K.T.; Matern, U.: Expression of bifunctional caffeoyl-CoA 3-O-methyltransferase in stress compensation and lignification. Plant Biol., **1**, 299-310 (1999)

[10] Martz, F.; Maury, S.; Pincon, G.; Legrand, M.: cDNA cloning, substrate specificity and expression study of tobacco caffeoyl-CoA 3-O-methyltransferase, a lignin biosynthetic enzyme. Plant Mol. Biol., **36**, 427-437 (1998)

[11] Meng, H.; Campbell, W.H.: Substrate profiles and expression of caffeoyl coenzyme A and caffeic acid O-methyltransferases in secondary xylem of aspen during seasonal development. Plant Mol. Biol., **38**, 513-520 (1998)

[12] Inoue, K.; Sewalt, V.J.H.; Ballance, G.M.; Ni, W.; Sturzer, C.; Dixon, R.A.: Developmental expression and substrate specificities of alfalfa caffeic acid 3-O-methyltransferase and caffeoyl coenzyme A 3-O-methyltransferase in relation to lignification. Plant Physiol., **117**, 761-770 (1998)

[13] Busam, G.; Junghanns, K.T.; Kneusel, R.E.; Kassemeyer, H.H.; Matern, U.: Characterization and expression of caffeoyl-coenzyme A 3-O-methyltransferase proposed for the induced resistance response of Vitis vinifera L. Plant Physiol., **115**, 1039-1048 (1997)

[14] Ni, W.; Sewalt, V.J.H.; Korth, K.L.; Blount, J.W.; Ballance, G.M.; Dixon, R.A.: Stress responses in alfalfa. XXI. Activation of caffeic acid 3-O-methyltransferase and caffeoyl coenzyme A 3-O-methyltransferase genes does not contribute to changes in metabolite accumulation in elicitor-treated cell-suspension cultures. Plant Physiol., **112**, 717-726 (1996)

[15] Pakusch, A.E.; Matern, U.; Schiltz, E.: Elicitor-inducible caffeoyl-coenzyme A 3-O-methyltransferase from Petroselinum crispum cell suspensions. Plant Physiol., **95**, 137-143 (1991)

[16] Pakusch, A.E.; Matern, U.: Kinetic characterization of caffeoyl-coenzyme A-specific 3-O-methyltransferase from elicited parsley cell suspensions. Plant Physiol., **96**, 327-330 (1991)

[17] Schmidt, D.; Pakusch, A.E.; Matern, U.: Molecular cloning, induction and taxonomic distribution of caffeoyl-CoA 3-O-methyltransferase, an enzyme involved in disease resistance. J. Biol. Chem., **266**, 17416-17423 (1991)

[18] Pakusch, A.E.; Kneusel, R.E.; Matern, U.: S-Adenosyl-L-methionine:trans-caffeoyl-coenzyme A 3-O-methyltransferase from elicitor-treated parsley cell suspension cultures. Arch. Biochem. Biophys., **271**, 488-494 (1989)

N-Benzoyl-4-hydroxyanthranilate 4-O-methyltransferase 2.1.1.105

1 Nomenclature

EC number
2.1.1.105

Systematic name
S-adenosyl-L-methionine:N-benzoyl-4-O-hydroxyanthranilate 4-O-methyltransferase

Recommended name
N-benzoyl-4-hydroxyanthranilate 4-O-methyltransferase

Synonyms
N-benzoyl-4-hydroxyanthranilate 4-methyltransferase
benzoyl-CoA:anthranilate N-benzoyltransferase

CAS registry number
125498-68-2

2 Source Organism

<1> *Dianthus caryophyllus* (enzyme activity is induced by treatment of the cells with elicitor [1]) [1]

3 Reaction and Specificity

Catalyzed reaction
S-adenosyl-L-methionine + N-benzoyl-4-hydroxyanthranilate = S-adenosyl-L-homocysteine + N-benzoyl-4-methoxyanthranilate

Reaction type
methyl group transfer

Substrates and products
S S-adenosyl-L-methionine + N-benzoyl-4-hydroxyanthranilate <1> (Reversibility: ? <1> [1]) [1]
P S-adenosyl-L-homocysteine + N-benzoyl-4-methoxyanthranilate <1> [1]

5 Isolation/Preparation/Mutation/Application

Source/tissue

cell culture <1> [1]

References

[1] Reinhard, K.; Matern, U.: The biosynthesis of phytoalexins in Dianthus caryophyllus L. cell cultures: induction of benzoyl-CoA:anthranilate N-benzoyltransferase activity. Arch. Biochem. Biophys., **275**, 295-301 (1989)

Tryptophan 2-C-methyltransferase 2.1.1.106

1 Nomenclature

EC number
2.1.1.106

Systematic name
S-adenosyl-L-methionine:L-tryptophan 2-C-methyltransferase

Recommended name
tryptophan 2-C-methyltransferase

Synonyms
S-adenosylmethionine:tryptophan 2-methyltransferase
tryptophan 2-methyltransferase

CAS registry number
126626-83-3

2 Source Organism

<1> *Streptomyces laurentii* [1]

3 Reaction and Specificity

Catalyzed reaction
S-adenosyl-L-methionine + L-tryptophan = S-adenosyl-L-homocysteine + L-2-methyltryptophan

Reaction type
methyl group transfer

Substrates and products
S S-adenosyl-L-methionine + D,L-tryptophan <1> (<1> significant slower reaction rate with D-tryptophan [1]) (Reversibility: ? <1> [1]) [1]
P S-adenosyl-L-homocysteine + D,L-2-methyltryptophan <1> [1]

Inhibitors
1-methyltryptophan <1> (<1> product inhibition [1]) [1]
S-adenosyl-L-homocysteine <1> [1]

K_m-Value (mM)
0.12 <1> (S-adenosyl-L-methionine, <1> cell free extract [1]) [1]

K_i-Value (mM)
0.48 <1> (S-adenosyl-L-homocysteine) [1]

pH-Optimum
7.8 <1> (<1> cell free extract [1]) [1]

pH-Range
6.8-9 <1> (<1> half-maximal activity at pH 6.8 and pH 9.0 [1]) [1]

Temperature optimum (°C)
45 <1> [1]

5 Isolation/Preparation/Mutation/Application

Localization
membrane <1> (<1> possibly loosely membrane-bound [1]) [1]

6 Stability

Temperature stability
30 <1> (<1> half-life: 2 h [1]) [1]

Storage stability
<1>, 4°C, 1 d, loss of 80% activity [1]

References

[1] Frenzel, T.; Zhou, P.; Floss, H.G.: Formation of 2-methyltryptophan in the biosynthesis of thiostrepton: isolation of S-adenosyl-methionine:tryptophan 2-methyltransferase. Arch. Biochem. Biophys., **278**, 35-40 (1990)

Uroporphyrin-III C-methyltransferase 2.1.1.107

1 Nomenclature

EC number
2.1.1.107

Systematic name
S-adenosyl-L-methionine:uroporphyrin-III C-methyltransferase

Recommended name
uroporphyrin-III C-methyltransferase

Synonyms
S-adenosyl-L-methionine dependent uroporphyrinogen III methylase
adenosylmethionine-uroporphyrinogen III methyltransferase
sirohaem synthase (<1> CysG protein complex, catalyzes four different reactions one of which is EC 2.1.1.107 activity [9])
uroporphyrinogen III methylase
uroporphyrinogen methyltransferase,

CAS registry number
73665-99-3

2 Source Organism

<1> *Escherichia coli* [1, 2, 9, 10]
<2> *Pseudomonas denitrificans* [3, 10]
<3> *Pseudomonas denitrificans* [4]
<4> *Propionibacterium freudenreichii* [5]
<5> *Geobacillus stearothermophilus* [6]
<6> *Zea mays* [7]
<7> *Arabidopsis thaliana* (genomic DNA, cDNA has accession number L47479 [8]) [8]
<8> *Saccharomyces cerevisiae* [10]
<9> *Salmonella thyphimurium* [10]

3 Reaction and Specificity

Catalyzed reaction
2 S-adenosyl-L-methionine + uroporphyrin III = 2 S-adenosyl-L-homocysteine + sirohydrochlorin

Reaction type
methyl group transfer

Natural substrates and products

S S-adenosyl-L-methionine + uroporphyrinogen III <1-9> (<1,4> involved in biosynthesis of sirohydrochlorin [1, 2, 5]; <1-4> involved in biosynthesis of cobalamin [3, 4, 5, 9]; <6-8> involved in biosynthesis of siroheme [7, 8, 10]; <7> no synthesis of cobalamin [8]) (Reversibility: ? <1-9> [1-10]) [1-10]

P S-adenosyl-L-homocysteine + dihydrosirohydrochlorin <1-9> (<1-9> product is precorrin-2 [1-10]) [1-10]

Substrates and products

S S-adenosyl-L-methionine + heptacarboxyporphyrinogen III <2> (Reversibility: ? <2> [3]) [3]

P ?

S S-adenosyl-L-methionine + hexacarboxyporphyrinogen III <2> (Reversibility: ? <2> [3]) [3]

P ?

S S-adenosyl-L-methionine + pentacarboxyporphyrinogen III <2> (Reversibility: ? <2> [3]) [3]

P ?

S S-adenosyl-L-methionine + uroporphyrinogen I <1, 2, 4> (<1,4> methylates positions 2 and 7 [1,5]) (Reversibility: ? <1, 2, 4> [1, 3, 5]) [1, 3, 5]

P ?

S S-adenosyl-L-methionine + uroporphyrinogen III <1-9> (<1> methylates positions 2 and 7 [1-10]) (Reversibility: ? <1-9> [1-10]) [1-10]

P S-adenosyl-L-homocysteine + dihydrosirohydrochlorin <1-9> (<1-9> product is precorrin-2 [1-10]; <2> produces equal amounts of precorrin-1 and precorrin-2 [3]; <4> after prolonged incubation (15 h) also tri- and tetra-methylated compounds formed, third methyl group at position 12 [5]) [1-10]

S Additional information <2, 7> (<2> uroporphyrin III is not a substrate of the enzyme [3]; <7> no formation of sirohydrochlorin in presence of NAD^+ or $NADP^+$ [8]) [3, 8]

P ?

Inhibitors
S-adenosyl-L-homocysteine <2> (<2> competitive [3]) [3]
uroporphyrinogen III <2> (<2> substrate inhibition above 0.002 mM [3]) [3]
Additional information <2> (<2> no inhibition by cobalamine or intermediates of synthesis [3]) [3]

Metals, ions
Co^{2+} <2> (<2> complete inhibition by 0.1 mM, mechanism unclear [3]) [3]

Turnover number (min^{-1})
2280 <2> (uroporphyrinogen III) [3]

Specific activity (U/mg)
0.0062 <2> (<2> purified enzyme [3]) [3]
0.0091 <7> (<7> purified enzyme [8]) [8]

K_m-Value (mM)
0.001 <2> (uroporphyrinogen III) [3]
0.0034 <2> (pentacarboxyporphyrinogen III) [3]
0.0042 <2> (heptacarboxyporphyrinogen III) [3]
0.0047 <2> (hexacarboxyporphyrinogen III) [3]
0.0063 <2> (S-adenosyl-L-methionine) [3]

K_i-Value (mM)
0.00032 <2> (S-adenosyl-L-homocysteine) [3]

pH-Optimum
7.7-7.8 <2> [3]

4 Enzyme Structure

Molecular weight
27000 <4> (<4> SDS-PAGE [5]) [5]
27610 <5> (<5> calculated from DNA sequence [6]) [6]
28000 <5> (<5> SDS-PAGE [6]) [6]
29200 <3> (<3> calculated from DNA sequence [4]) [4]
30000 <2> (<2> SDS-PAGE [3]) [3]
39900 <7> (<7> calculated from DNA sequence [8]) [8]
41000 <7> (<7> SDS-PAGE [8]) [8]
49930 <1> (<1> calculated from DNA-sequence [2]) [2]
50000 <5> (<5> gel filtration [6]) [6]
52000 <1> (<1> SDS-PAGE [2]) [2]
56000 <2> (<2> gel filtration [3]) [3]
66000 <1> (<1> gel filtration [2]) [2]
67000 <7> (<7> SDS-PAGE, uncleaved fusion protein [8]) [8]

Subunits
dimer <2, 5> (<2> homodimer, 2 * 30000, SDS-PAGE, gel filtration [3]; <5> homodimer, 2 * 27000, SDS-PAGE, gel filtration [6]) [3, 6]
monomer <1> (<1> SDS-PAGE, gel filtration [2]) [2]

5 Isolation/Preparation/Mutation/Application

Localization
chloroplast <7> [8]
soluble <4> [5]

Purification
<1> [2, 9]
<5> (from recombinant E. coli [6]) [6]

<7> (from recombinant E. coli [8]) [8]
<8> (from recombinant E. coli using His-tag [10]) [10]

Cloning

<1> (overexpression of cysG gene [1,2]; cysG containing plasmid transformed into Escherichia coli CBK103 [2]; complementation experiments with cysteine autotroph strain [9]) [1, 2, 9]
<2> (50fold increase of enzyme activity when overexpressed [3]) [3]
<3> (expression of 5 open reading frames, codon 1120-1959 (cobA) encodes EC 2.1.1.107 activity [4]) [4]
<4> (overexpression of cobA gene in Escherichia coli TB1 [5]) [5]
<5> (in Escherichia coli JM83 and MV1184 [6]) [6]
<6> (in Escherichia coli CBK103 [7]) [7]
<7> (in Escherichia coli CBK103, amino acids 202-457 responsible for EC 2.1.1.107 activity [8]) [8]
<8> (EC 2.1.1.107 deficient mutants complemented with cysG from Salmonella typhimurium and Escherichia coli and cobA from Pseudomonas denitrificans, overexpression of yeast gene product in Escherichia coli [10]) [10]

Engineering

D227A <1> (<1> full activity [9]) [9]
D248A <1> (<1> no transmethylase activity [9]) [9]
D303A <1> (<1> full activity [9]) [9]
G224A <1> (<1> unable to bind S-adenosyl-L-methionine [9]) [9]
K270I <1> (<1> full activity [9]) [9]
R298L <1> (<1> unable to bind S-adenosyl-L-methionine [9]) [9]
R309L <1> (<1> no transmethylase activity [9]) [9]

6 Stability

Temperature stability

65 <5> (<5> thermostable, 17% of activity remaining after 10 min incubation [6]) [6]

General stability information

<7>, complete loss of activity when treated with Triton X-100 [8]

Storage stability

<2>, 4°C, 10 mM Tris, pH 7.7, stable for several weeks [3]

References

[1] Warren, M.J.; Gonzalez, M.D.; Williams, H.J.; Stolowich, N.J.; Scott, A.I.: Uroporpyrinogen III methylase catalyzes the enzymatic synthesis of sirohydrochlorines II and IV by a clockwise mechanism. J. Am. Chem. Soc., **112**, 5343-5345 (1990)

[2] Warren, M.J.; Roessner, C.A.; Santander, P.J.; Scott, A.I.: The Escherichia coli cysG gene encodes S-adenosylmethionine-dependent uroporphyrinogen III methylase. Biochem. J., **265**, 725-729 (1990)
[3] Blanche, F.; Debussche, L.; Thibaut, D.; Crouzet, J.; Cameron, B.: Purification and characterization of S-adenosyl-L-methionine: uroporphyrinogen III methyltransferase from Pseudomonas denitrificans. J. Bacteriol., **171**, 4222-4231 (1989)
[4] Crouzet, J.; Cauchois, L.; Blanche, F.; Debussche, L.; Thibaut, D.; Rouyez, M.C.; Rigault, S.; Mayaux, J.F.; Cameron, B.: Nucleotide sequence of a Pseudomonas denitrificans 5.4-kilobase DNA fragment containing five cob genes and identification of structural genes encoding S-adenosyl-L-methionine: uroporphyrinogen III methyltransferase and cobyrinic acid a,c-diamide synthase. J. Bacteriol., **172**, 5968-5979 (1990)
[5] Sattler, I.; Roessner, C.A.; Stolowich, N.J.; Hardin, S.H.; Harris-Haller, L.W.; Yokubaitis, N.T.; Murooka, Y.; Hashimoto, Y.; Scott, A.I.: Cloning, sequencing, and expression of the uroporphyrinogen III methyltransferease cobA gene of Propionibacterium freudenreichii (shermanii). J. Bacteriol., **177**, 1564-1569 (1995)
[6] Chou, P.L.; Ohtsuka, M.; Minowa, T.; Yamasato, K.; Sakano, Y.; Matsuzawa, H.; Ohta, T.; Sakai, H.: Reddish Escherichia coli cells caused by overproduction of Bacillus stearothermophilus uroporphyrinogen III methylase: cloning, sequencing, and expression of the gene. Biosci. Biotechnol. Biochem., **59**, 1817-1824 (1995)
[7] Sakakibara, H.; Takei, K.; Sugiyama, T.: Isolation and characterization of a cDNA that encodes maize uroporphyrinogen III methyltransferase, an enzyme involved in the synthesis of siroheme, which is a prosthetic group of nitrite reductase. Plant J., **10**, 883-892 (1996)
[8] Leustek, T.; Smith, M.; Murillo, M.; Singh, D.P.; Smith, A.G.; Woodcock, S.C.; Awan, S.J.; Warren, M.J.: Siroheme biosynthesis in higher plants. Analysis of an S-adenosyl-L-methionine-dependent uroporphyrinogen III methyltransferase from Arabidopsis thaliana. J. Biol. Chem., **272**, 2744-2752 (1997)
[9] Woodcock, S.C.; Raux, E.; Levillayer, F.; Thermes, C.; Rambach, A.; Warren, M.J.: Effect of mutations in the transmethylase and dehydrogenase/chelatase domains of siroheme synthase (CysG) on siroheme and cobalamin biosynthesis. Biochem. J., **330**, 121-129 (1998)
[10] Raux, E.; McVeigh, T.; Peters, S.E.; Leustek, T.; Warren, M.J.: The role of Saccharomyces cerevisiae MET1p and MET8p in sirohaem and cobalamin biosynthesis. Biochem. J., **338**, 701-708 (1999)

6-Hydroxymellein O-methyltransferase 2.1.1.108

1 Nomenclature

EC number
2.1.1.108

Systematic name
S-adenosyl-L-methionine:6-hydroxymellein 6-O-methyltransferase

Recommended name
6-hydroxymellein O-methyltransferase

Synonyms
6-hydroxymethyllein-methyltransferase
S-adenosyl-L-methionine:6-hydroxymellein O-methyltransferase
o-methyltransferase, 6-hydroxymellein

CAS registry number
124149-02-6

2 Source Organism

<1> *Daucus carota* (treated with fungal elicitors [1]; treated with 2-chloroethylphosphonic acid [1,2]) [1-4]

3 Reaction and Specificity

Catalyzed reaction
S-adenosyl-L-methionine + 6-hydroxymellein = S-adenosyl-L-homocysteine + 6-methoxymellein (<1> mechanism [3])

Reaction type
methyl group transfer

Natural substrates and products

S S-adenosyl-L-methionine + 6-hydroxymellein <1> (<1> enzyme is involved in the biosynthesis of the phytoalexin 6-methoxymellein in carrot tissue [1-3]; <1> biosynthetic pathway [2,3]) (Reversibility: ? <1> [1-4]) [1-4]

P S-adenosyl-L-homocysteine + 6-methoxymellein <1> [1]

Substrates and products

- **S** S-adenosyl-L-methionine + 3,4-dehydro-6-hydroxymellein <1> (Reversibility: ? <1> [1, 2]) [1, 2]
- **P** S-adenosyl-L-homocysteine + 3,4-dehydro-6-methoxymellein <1> [1]
- **S** S-adenosyl-L-methionine + 6-hydroxymellein <1> (Reversibility: ? <1> [1-4]) [1-4]
- **P** S-adenosyl-L-homocysteine + 6-methoxymellein <1> [1-4]
- **S** S-adenosyl-L-methionine + 6-hydroxymellein + 3,4-dehydro-6-hydroxymellein <1> (Reversibility: ? <1> [1]) [1]
- **P** S-adenosyl-L-homocysteine + 6-methoxymellein + 3,4-dehydro-6-methoxymellein <1> [1]
- **S** Additional information <1> (<1> no substrates: 3,4-dehydro-6-methoxymellein, 6-methoxymellein, esculetin [1]; <1> no substrates: umbelliferone, scopoletin [1,4]; <1> no substrates: naringenin, apigenein [4]) [1, 4]
- **P** ?

Inhibitors

6-methoxymellein <1> [3]
Co^{2+} <1> [3]
Cu^{2+} <1> [3]
Fe^{2+} <1> [3]
Hg^{2+} <1> [3]
Mn^{2+} <1> [3]
S-adenosyl-L-homocysteine <1> [2, 3]
iodoacetoamide <1> [3]
p-chloromercuribenzoate <1> [3]
sulfhydryl reagents <1> [3]

Activating compounds

2-chloroethylphosphonic acid <1> (<1> elicitor, the enzyme activity increases after a 2 day lag period and attains a maximum 3 days after the addition, then decreases gradually [1,2]) [1, 2]
Fusarium moniliforme endo-polygalacturonsae <1> (<1> elicitor, the enzyme activity reaches a maximum 40 h after the addition, then decreased graduelly [4]) [4]
Phythophthora megasperma cell wall fraction <1> (<1> elicitor, the enzyme activity reaches a maximum 50 h after the addition [4]) [4]
uronide <1> (<1> elicitor [1]) [1]

Specific activity (U/mg)

Additional information <1> [3]

K_i-Value (mM)

0.026 <1> (S-adenosyl-L-homocysteine) [2, 3]
0.047 <1> (6-methoxymellein) [3]

pH-Optimum

7.5-8 <1> (<1> pI: 5.7 [3]) [3]

Temperature optimum (°C)
37 <1> (<1> assay at [1]) [1]

4 Enzyme Structure

Molecular weight
76000 <1> (<1> gel filtration [3]) [3]

5 Isolation/Preparation/Mutation/Application

Source/tissue
root <1> [1, 2]

Purification
<1> (partial [3]) [3]

6 Stability

Storage stability
<1>, 4°C, 7 d, completely loss of activity [3]
<1>, frozen overnight, completely loss of activity [3]

References

[1] Kurosaki, F.; Nishi, A.: A methyltransferase forsynthesis of the phytoalexin 6-methoxymellein in carrot cells. FEBS Lett., **227**, 183-186 (1988)
[2] Kurosaki, F.; Kizawa, Y.; Nishi, A.: Biosynthesis of dihydroisocoumarin by extracts of elicitor treated carrot root. Phytochemistry, **28**, 1843-1845 (1989)
[3] Kurosaki, F.: Partial purification and characterization of 6-hydroxymellein-O-methyltransferase from elicitor-treated carrot cells. Phytochemistry, **41**, 1023-1027 (1996)
[4] Marinelli, F.; Ronchi, V.N.; Salvadori, P.: Elicitor induction of enzyme activities and 6-methoxymellein production in carrot cell suspension culture. Phytochemistry, **35**, 1457-1460 (1994)

Demethylsterigmatocystin 6-O-methyltransferase

2.1.1.109

1 Nomenclature

EC number
2.1.1.109

Systematic name
S-adenosyl-L-methionine:6-demethylsterigmatocystin 6-O-methyltransferase

Recommended name
demethylsterigmatocystin 6-O-methyltransferase

Synonyms
O-methyltransferase I
methyltransferase, demethylsterigmatocystin

CAS registry number
123516-47-2

2 Source Organism

<1> *Aspergillus parasiticus* (NIAH-26 [1,3]) [1, 3]
<2> *Aspergillus nidulans* [2]
<3> *Aspergillus parasiticus* [4]

3 Reaction and Specificity

Catalyzed reaction
S-adenosyl-L-methionine + 6-demethylsterigmatocystin = S-adenosyl-L-homocysteine + sterigmatocystin

Reaction type
methyl group transfer

Natural substrates and products
S S-adenosyl-L-methionine + 6-demethylsterigmatocystin <1> (<1> involved in biosynthesis of aflatoxins in fungi [1]) (Reversibility: ? <1> [1]) [1]
P S-adenosyl-L-homocysteine + sterigmatocystin <1> [1]

Substrates and products
S S-adenosyl-L-methionine + 6-demethylsterigmatocystin <1, 2> (Reversibility: ? <1, 2> [1, 2]) [1, 2, 3]

P S-adenosyl-L-homocysteine + sterigmatocystin <1, 2> [1, 2, 3]
S S-adenosyl-L-methionine + dihydrodemethylsterigmatocystin <1> (Reversibility: ? <1> [1]) [1, 3]
P S-adenosyl-L-homocysteine + dihydrosterigmatocystin <1> [1, 3]

Inhibitors
6-demethylsterigmatocystin <1> (<1> at higher concentrations [3]) [3]
N-ethylmaleimide <1> [1]
dihydrodemethylsterigmatocystin <1> (<1> at higher concentrations [3]) [3]

Specific activity (U/mg)
0.043 <1> (<1> with dihydrodemethylsterigmatocystin [3]) [3]
0.061 <1> (<1> with 6-demethylsterigmatocystin [3]) [3]

K_m-Value (mM)
0.00094 <1> (6-demethylsterigmatocystin) [3]
0.0025 <1> (dihydrodemethylsterigmatocystin) [3]

K_i-Value (mM)
0.108 <1> (6-demethylsterigmatocystin) [3]
0.171 <1> (dihydrodemethylsterigmatocystin) [3]

pH-Optimum
6.6-9 <1> [3]

4 Enzyme Structure

Molecular weight
15000 <1> (<1> gel filtration [3]) [3]

Subunits
? <1> (<1> x * 43000, SDS-PAGE [3]) [3]

5 Isolation/Preparation/Mutation/Application

Purification
<1> (phenyl-Sepharose, Sephacryl S-300, Matrex gel Green A [3]) [3]

Cloning
<3> (cloning of cDNA [4]) [4]

References

[1] Yabe, K.; Ando, Y.; Hashimoto, J.; Hamasaki, T.: Two distinct O-methyltransferases in aflatoxin biosynthesis. Appl. Environ. Microbiol., **55**, 2172-2177 (1989)

[2] Kelkar, H.S.; Keller, N.P.; Adams, T.H.: Aspergillus nidulans stcP encodes an O-methyltransferase that is required for sterigmatocystin biosynthesis. Appl. Environ. Microbiol., **62**, 4296-4298 (1996)
[3] Yabe, K.; Matsushima, K.I.; Koyama, T.; Hamasaki, T.: Purification and characterization of O-methyltransferase I involved in conversion of demethylsterigmatocystin to sterigmatocystin and of dihydrodemethylsterigmatocystin to dihydrosterigmatocystin during aflatoxin biosynthesis. Appl. Environ. Microbiol., **64**, 166-171 (1998)
[4] Motomura, M.; Chihaya, N.; Shinozawa, T.; Hamasaki, T.; Yabe, K.: Cloning and characterization of the O-methyltransferase I gene (dmtA) from Aspergillus parasiticus associated with the conversions of demethylsterigmatocystin to sterigmatocystin and dihydrodemethylsterigmatocystin to dihydrosterigmatocystin in aflatoxin biosynthesis. Appl. Environ. Microbiol., **65**, 4987-4994 (1999)

Sterigmatocystin 7-O-methyltransferase 2.1.1.110

1 Nomenclature

EC number
2.1.1.110

Systematic name
S-adenosyl-L-methionine:sterigmatocystin 7-O-methyltransferase

Recommended name
sterigmatocystin 7-O-methyltransferase

Synonyms
O-methyltransferase II
methyltransferase, sterigmatocystin
sterigmatocystin methyltransferase

CAS registry number
116958-29-3

2 Source Organism

<1> *Aspergillus parasiticus* (NIAH-26 [1]; SRCC 163 [2,4,5]; CP461, SRCC 2043 [3,6]) [1-6]
<2> *Aspergillus flavus* (strain 3357 [6]) [4, 6]

3 Reaction and Specificity

Catalyzed reaction
S-adenosyl-L-methionine + sterigmatocystin = S-adenosyl-L-homocysteine + 7-O-methylsterigmatocystin

Reaction type
methyl group transfer

Natural substrates and products
S S-adenosyl-L-methionine + sterigmatocystin <1, 2> (<1> involved in the biosynthesis of aflatoxins in fungi [2,3]; <1,2> penultimate step in aflatoxin biosynthetic pathway [4]) (Reversibility: ? <1, 2> [2-4]) [2-4]
P S-adenosyl-L-homocysteine + 7-O-methylsterigmatocystin <1, 2> [2-4]

Substrates and products

S S-adenosyl-L-methionine + dihydrosterigmatocystin <1> (Reversibility: ? <1> [1]) [1]

P S-adenosyl-L-homocysteine + dihydro-7-O-methylsterigmatocystin <1> [1]

S S-adenosyl-L-methionine + sterigmatocystin <1, 2> (Reversibility: ? <1, 2> [1-4]) [1-4, 6]

P S-adenosyl-L-homocysteine + 7-O-methylsterigmatocystin <1, 2> [1-4, 6]

Inhibitors

Cu^{2+} <1> (<1> 0.1 mM, 58% inhibition [6]) [6]
Fe^{2+} <1> (<1> 0.1 mM, 87% inhibition [6]) [6]
Hg^{2+} <1> (<1> 0.1 mM, complete inhibition [6]) [6]
S-adenosyl-L-homocysteine <1> [6]
phenylmercuriacetate <1> (<1> 0.1 mM, 96% inhibition [6]) [6]

Turnover number (min^{-1})

1.32 <1> (sterigmatocystin) [2]
84 <1> (sterigmatocystin) [4]
132 <1> (sterigmatocystin) [5]

Specific activity (U/mg)

0.0195 <1> [6]
0.021 <1> [4]

K_m-Value (mM)

0.0018 <1> (sterigmatocystin) [2, 5]
0.002 <1> (sterigmatocystin) [4]
0.0096 <1> (S-adenosyl-L-methionine) [4]
0.042 <1> (S-adenosyl-L-methionine) [2, 5]

K_i-Value (mM)

0.0012 <1> (S-adenosyl-L-homocysteine) [6]

pH-Optimum

7-9.4 <1> [4]
7.5 <1> [2]
7.5-8 <1> [5]

Temperature optimum (°C)

25-35 <1> [2, 5]
40-45 <1> [4]

Temperature range (°C)

25-60 <1> (<1> approx. 35% of maximal activity at 25°C, approx. 30% of maximal activity at 60°C [4]) [4]

4 Enzyme Structure

Molecular weight

160000 <1> (<1> gel filtration [2]) [2]
168000 <1> [5]

Subunits

? <1> (<1> x * 40000, SRRC 163, SDS-PAGE, the 40000 Da protein is distinct from the 168000 Da methyl transferase purified from the same fungal strain [4]) [4]
dimer <1> (<1> 1 * 110000 + 1 * 58000, SRCC 163, SDS-PAGE [2,5]) [2, 5]

5 Isolation/Preparation/Mutation/Application

Source/tissue

mycelium <2> [2, 4]

Localization

soluble <1> [4]

Purification

<1> (NIAH-26 [2]; strain SRRC 163, anion exchange, hydroxylapatite, DEAE-Sperodex, octyl avidgel, anion-exchange [4]; strain SRRC 2043, ultracentrifugation, ammonium sulfate, DEAE-Sephadex, immunoaffinity column [6]) [2, 4]

References

[1] Yabe, K.; Ando, Y.; Hashimoto, J.; Hamasaki, T.: Two distinct O-methyltransferases in aflatoxin biosynthesis. Appl. Environ. Microbiol., **55**, 2172-2177 (1989)
[2] Bhatnagar, D.; Ullah, A.H.J.; Cleveland, T.E.: Purification and characterization of a methyltransferase from Aspergillus parasiticus SRRC 163 involved in aflatoxin biosynthetic pathway. Prep. Biochem., **18**, 321-349 (1988)
[3] Bhatnagar, D.; McCormick, S.P.; Lee, L.S.; Hill, R.A.: Identification of O-methylsterigmatocystin as an aflatoxin B1 and G1 precursor in Aspergillus parasiticus. Appl. Environ. Microbiol., **53**, 1028-1033 (1987)
[4] Keller, N.P.; Dischinger, H.C.; Bhatnagar, D.; Cleveland, T.E.; Ullah, A.H.J.: Purification of a 40-kilodalton methyltransferase active in the aflatoxin biosynthetic pathway. Appl. Environ. Microbiol., **59**, 479-484 (1993)
[5] Bhatnagar, D.; Ullah, A.H.J.; Cleveland, T.E.: Purification and characterization of a methyltransferase from Aspergillus parasiticus SRRC 163 involved in aflatoxin biosynthetic pathway. Prep. Biochem., **18**, 321-349 (1988)
[6] Liu, B.H.; Bhatnagar, D.; Chu, F.S.: Purification and characterization of 40-kDa sterigmatocystin O-methyltransferase involved in aflatoxin biosynthesis. Nat. Toxins, **7**, 63-69 (1999)

Anthranilate N-methyltransferase 2.1.1.111

1 Nomenclature

EC number
2.1.1.111

Systematic name
S-adenosyl-L-methionine:anthranilate N-methyltransferase

Recommended name
anthranilate N-methyltransferase

Synonyms
anthranilic acid N-methyltransferase

CAS registry number
123779-15-7

2 Source Organism

<1> *Ruta graveolens* [1-4]
<2> *Choisya ternata* [2]

3 Reaction and Specificity

Catalyzed reaction
S-adenosyl-L-methionine + anthranilate = S-adenosyl-L-homocysteine + N-methylanthranilate

Reaction type
methyl group transfer

Natural substrates and products
S S-adenosyl-L-methionine + anthranilate <1, 2> (<1> first specific step in acridone alkaloid biosynthesis [1,3]; <1,2> first specific step in quinoline alkaloid biosynthesis [2]) (Reversibility: ? <1, 2> [1-3]) [1-3]
P S-adenosyl-L-homocysteine + N-methylanthranilate

Substrates and products
S S-adenosyl-L-methionine + anthranilate <1, 2> (Reversibility: ? <1, 2> [1-3]) [1-3]
P S-adenosyl-L-homocysteine + N-methylanthranilate

pH-Optimum

7.8 <1> [4]
8.2 <1, 2> (<1,2> assay at [2]) [2]

Temperature optimum (°C)

32 <1, 2> (<1,2> assay at [1,2]) [1, 2]

4 Enzyme Structure

Molecular weight

70000 <1> (<1> gel filtration [4]) [4]

Subunits

monomer <1> (<1> 1 * 62000, SDS-PAGE [4]) [4]

5 Isolation/Preparation/Mutation/Application

Source/tissue

root <1> [1]

Purification

<1> [1, 4]

References

[1] Eilert, U.; Wolters B.: Elicitor induction of S-adenosyl-L-methionine: anthranilic acid N-methyltransferase activity in cell suspension and organ cultures of Ruta graveolens L.. Plant Cell Tissue Organ Cult., **18**, 1-18 (1989)
[2] Baumert, A.; Creche, J.; Rideau, M.; Chenieux, J.C.; Groeger, D.: Anthranilic acid-specific enzymes in alkaloid-producing tissue cultures of Choisya ternata and Ruta graveolens. Plant Physiol. Biochem., **28**, 587-592 (1990)
[3] Baumert, A.; Maier, W.; Schumann, B.; Gröger, D.: Increased accumulation of Acridone alkaloids by cell suspension cultures of Ruta graveolens in response to elicitors. J. Plant Physiol., **139**, 224-228 (1991)
[4] Maier, W; Baumert, A.; Groeger, D.: Partial purification and characterization of S-adenosyl-L-methionine:anthranilic acid N-methyltransferase from Ruta cell suspension cultures. J. Plant Physiol., **145**, 1-6 (1995)

Glucuronoxylan 4-O-methyltransferase 2.1.1.112

1 Nomenclature

EC number
2.1.1.112

Systematic name
S-adenosyl-L-methionine:glucuronoxylan-D-glucuronate 4-O-methyltransferase

Recommended name
glucuronoxylan 4-O-methyltransferase

CAS registry number
123644-79-1

2 Source Organism

<1> *Pisum sativum* [1, 2]

3 Reaction and Specificity

Catalyzed reaction
S-adenosyl-L-methionine + glucuronoxylan D-glucuronate = S-adenosyl-L-homocysteine + glucuronoxylan 4-O-methyl-D-glucuronate

Substrates and products
S S-adenosyl-L-methionine + glucuronoxylan D-glucuronate <1> (Reversibility: ? <1> [1]) [1]
P S-adenosyl-L-homocysteine + glucuronoxylan 4-O-methyl-D-glucuronate

Inhibitors
acetyl-CoA <1> [1]

Activating compounds
UDP-D-glucuronic acid <1> [1]
UDP-D-xylose <1> [1]

Metals, ions
Co^{2+} <1> (<1> 1 mM [1]) [1]

pH-Optimum
8 <1> [1]

5 Isolation/Preparation/Mutation/Application

Source/tissue
epicotyl <1> [1, 2]

Localization
Golgi membrane <1> (<1> low and medium membrane [2]) [2]

Purification
<1> [1, 2]

References

[1] Baydoun, E.A.H.; Usta, J.A.R.; Waldron, K.W.; Brett, C.T.: A methyltransferase involved in the biosynthesis of 4-O-methylglucuronoxylan in ethiolated pea epicotyls. J. Plant Physiol., **135**, 81-85 (1989)
[2] Baydoun, E.A.H.; Rizk, S.E, Brett, C.T.: Localisation of methyltransferases involved in glucuronoxylan and pectin methylation in the Golgi apparatus in etiolated pea epicotyls. J. Plant Physiol., **155**, 240-244 (1999)

Site-specific DNA-methyltransferase (cytosine-N^4-specific) 2.1.1.113

1 Nomenclature

EC number
2.1.1.113

Recommended name
site-specific DNA-methyltransferase (cytosine-N^4-specific)

Synonyms
BamHI [cytosine-N^4] MTase <3> [3, 5, 6]
DNA-(N^4-cytosine)-methyltransferase
DNA[cytosine-N^4]methyltransferase
M.AvaI <5> [8]
M.BalI <6> [8]
M.BcnIA <28> [10]
M.BcnIB <28> [10]
M.BglII <30> [8]
M.BsoBI <6> [8]
M.BsoBI <8> [8]
M.CsyAIP <9> [8]
M.CsyBIP <31> [8]
M.CsyBIP <9> [8]
M.Hpy99ORF244P <10> [8]
M.Hpy99ORF244P <33> [8]
M.Hpy99ORF248P <10> [8]
M.Hpy99ORF248P <35> [8]
M.Hpy99ORF629P <10> [8]
M.Hpy99ORF629P <34> [8]
M.HpyAIIP <10> [8]
M.HpyAIIP <32> [8]
M.HpyAXIIBP <10> [8]
M.MjaIP <11> [8]
M.MjaV <11> [8]
M.MjaV <36> [8]
M.MthZI <12> [8]
M.MwoI <13> [8]
M.NcoI <14> [8]
M.NgoMXV <15> [8]
M.Pac25I <16> [8]
M.PhiGIP <18> [8]
M.PhiHII <19> [8]

M.PhoIIIP <17> [8]
M.PspGI <20> [8]
M.PvuII <4> [4, 7, 9, 11]
M.PvuII <42> [8]
M.SapIA <22> [8]
M.SapIB <22> [8]
M.ScaI <23> [8]
M.SfiI <24> [8]
M.SmaI <25> [8]
M.StyCIP <26> [8]
M.XcyI <27> [8]
M.XmaI <21> [8]
N^{4}mC MTase
cytosine-N^5-methyltransferase
m^{4}C-forming Mtase
modification methylase
restriction-modification system
Additional information <29> (<29>, this is a large group of enzymes most of which, with enzymes of similar specificity listed as EC 3.1.21.3, 4 or 5, form so-called "restriction-modification systems". A complete listing of all these enzymes has been produced by R.J. Roberts, this list is updated annually [2]) [2]

2 Source Organism

<1> *Micrococcus varians* (strain RFL19, enzyme MvaI [1]) [1]
<2> *Citrobacter freundii* (strain RFL9, enzyme M.Cfr9I [1]) [1]
<3> *Bacillus amyloliquefaciens* [3, 5, 6, 12]
<4> *Proteus vulgaris* (enzyme M.PvuII [4,7,9]) [4, 7, 9, 11]
<5> *Anabaena variabilis* (enzyme M.AvaI [8]) [8]
<6> *Brevibacterium albidum* (enzyme M.BalI [8]) [8]
<7> *Bacillus globigii* (enzyme M.BglI [8]) [8]
<8> *Geobacillus stearothermophilus* (strain JN2091, enzyme M.BsoBI [8]) [8]
<9> *Cenarchaeum symbiosum* (strain A with enzyme M.CsyAIP [8]) [8]
<10> *Helicobacter pylori* (strain 26695 with the enzyme M.HpyAXIIBP [8]) [8]
<11> *Methanococcus jannaschii* (enzyme M.MjaIP [8]) [8]
<12> *Methanobacterium thermoformicicum* (strain Z-250, enzyme M.MthZI [8]) [8]
<13> *Methanobacterium wolfei* (enzyme M.MwoI [8]) [8]
<14> *Nocardia corallina* (enzyme M.NcoI [8]) [8]
<15> *Neisseria gonorrhoeae* (strain ME11, enzyme M.NgoMXV [8]) [8]
<16> *Pseudomonas alcaligenes* (strain NCIB 986, enzyme M.Pac25I [8]) [8]
<17> *Pyrococcus horikoshii* (strain OT3, enzyme M.PhoIIIP [8]) [8]
<18> *Lactobacillus phage phi gle* (enzyme M.PhiGIP [8]) [8]
<19> *Halobacterium salinarium phage phi-H* (enzyme M.PhiHII [8]) [8]
<20> *Pyrococcus sp.* (strain GI-H, enzyme M.PspGI [8]) [8]

<21> *Xanthomonas malvacearum* (enzyme M.XmaI [8]) [8]
<22> *Saccharopolyspora sp.* (enzymes M.SapIA and M.SapIB [8]) [8]
<23> *Streptomyces caespitosus* (enzyme M. ScaI [8]) [8]
<24> *Streptomyces fimbriatus* (enzyme M.SfiI [8]) [8]
<25> *Serratia marcescens* (enzyme M.SmaI [8]) [8]
<26> *Salmonella typhi* (strain CT18, enzyme M.StyCIP [8]) [8]
<27> *Xanthomonas cyanopsidis* (enzyme M.XcyI [8]) [8]
<28> *Bacillus centrosporus* (M.BcnIA and M.BcnIB [10]) [10]
<29> *prokaryote* [2]
<30> *Bacillus globigii* (enzyme M.BglII from plasmidpTsp45s enzyme [8]) [8]
<31> *Cenarchaeum symbiosum* (strain B with enzyme M.CsyBIP [8]) [8]
<32> *Helicobacter pylori* (strain 26695 with the enzyme M.HpyAIIP [8]) [8]
<33> *Helicobacter pylori* (strain J99 with enzyme M.Hpy99ORF244P [8]) [8]
<34> *Helicobacter pylori* (strain J99 with enzyme M.Hpy99ORF629P [8]) [8]
<35> *Helicobacter pylori* (strain J99 with enzyme M.Hpy99ORF248P [8]) [8]
<36> *Methanococcus jannaschii* (enzyme M.MjaV [8]) [8]
<37> *Bacillus amyloliquefaciens* (strain H [8]) [8]
<38> *Bacillus amyloliquefaciens H prophage H2* [8]
<39> *Citrobacter freundii* (strain RFL9 [8]) [8]
<40> *Citrobacter freundii* (strain 4111, plasmid pZE8 [8]) [8]
<41> *Micrococcus varians* (strain RFL19, enzyme MvaI [8]) [8]
<42> *Proteus vulgaris* (enzyme M.PvuII [8]) [8]

3 Reaction and Specificity

Catalyzed reaction

S-adenosyl-L-methionine + DNA cytosine = S-adenosyl-L-homocysteine + DNA N^4-methylcytosine (<3>, the product inhibition pattern is consistent with a steady-state random bi-bi mechanism in which the dominant order of substrate binding and product release: S-adenosyl-L-methionine, methylated DNA, DNAMe, S-adenosyl-L-homocysteine [3,5])

Reaction type

methyl group transfer

Substrates and products

S S-adenosyl-L-methionine + DNA cytosine <1-42> (<3>, 20-mer oligodeoxynucleotide duplex containing the palindromic recognition site GGATCC. The enzyme catalyzes methyl group transfer to the internal cytosine residue in the palindromic recognition site GGATCC. The enzyme transfers the methyl group to the exocyclic NH_2-group without formation of a transient covalent bond [3]; <3>, 5'-GTGAAAT*GGATCC*TAAACTG-3'/3'-CACTTTA*CCTAGG*ATTTGAC-5'. The target base is the internal cytosine in the recognition sequence [6]; <4>, the enzyme generates N^4-methylcytosine in duplex 5'-CAGCTG-3' DNA. PvuII methyltransferase is catalytically competent with one and with two bound molecules of S-adenosyl-L-methionine [7]; <5-27,30-42>, target sequences [8]; <4>, the enzyme catalyzes the transfer of the methyl group from S-adenosyl-L-methionine to the exocyclic aminoN^4 nitrogen of the central cytosine in its recognition sequence 5'-CAGCTG-3' [9]; <28>, M.BcnIA and M.BcnIB act on 5'-CC(C/G)GG-3' sites in double-stranded DNA. M.BcnIA can also, with a comparable efficiency, modify the specific targets in single-stranded DNA [10]; <4>, the enzyme modifies the second cytosine in CAGCTG sequences, also methylates adenine residues in CAGATG/CAGCTG substrates in which the target cytosine is replaced by adenine in one strand of the recognition sequence [11]; <3>, 20-mer duplex 5'-CAGTTTAGGATCCATTTCAC-3'/3'-GTCAAATCCTAGGTAAAGAG-5' containing a palindromic recognition site GGATCC [12]) (Reversibility: ? <1-42> [1-12]) [1-12]

P S-adenosyl-L-homocysteine + DNA N^4-methylcytosine <1-42> [1-12]

Inhibitors

S-adenosyl-L-homocysteine <3> (<3>, competitive with regard to S-adenosylmethionine and non-competitive with respect to unmethylated 20-mer duplex [3]) [3]

Additional information <3> (<3>, no apparent substrate inhibition at concentrations 100fold higher than the K_m for DNA and 20fold higher than K_m for S-adenosyl-L-methionine [3]; <3>, no substrate inhibition even at concentrations many times higher than the K_m-values [5]) [3, 5]

Turnover number (min^{-1})

Additional information <3> [6, 12]

K_m-Value (mM)

0.000196 <3> (5'-GTGAAAT*GGATCC*TAAACTG-3'/3'-CACTTTA*CCTAGG*ATTTGAC-5') [6]

0.00035 <3> (5'-CAGTTTAGGATCCATTTCAC-3'/3'-GTCAAATCCTAGGTAAAGAG-5') [12]

0.00122 <3> (S-adenosyl-L-methionine) [6]

0.0016 <3> (S-adenosyl-L-methionine) [12]

4 Enzyme Structure

Subunits

? <4> (<4>, x * 36847, MALDI-TOF mass spectrometry [4]) [4]

Additional information <3> (<3>, enzyme in a free state exists as a dimer. Introduction of substoichiometric amounts of DNA into the reaction mixture results in pronounced multimerization of the enzyme. However, addition of SAM in saturating concentration at an excess of the oligonucleotide duplex over BamHI Mtase converts most of the enzyme into a monomeric state [12]) [12]

5 Isolation/Preparation/Mutation/Application

Purification

<4> (native and selenomethionyl M.PvuII [4]) [4]

<28> (purified from E. coli cells expressing the genes for M.BcnIA and M.BcnIB [10]) [10]

Crystallization

<4> (native and selenomethionyl M.PvuII crystallized separately as binary complexs of the methyl donor S-adenosyl-L-methionine in the monoclinic space group P2. Hanging-drop vapor-diffusion technique [4]; determination of the structure of PvuII methyltransferase complexed with S-adenosyl-L-methionine by multiwavelength anomalous diffraction using a crystal of the selenomethionine substituted protein [9]) [4, 9]

Cloning

<4> (overexpression in Escherichia coli DH10B [4]) [4]

References

[1] Klimasauskas, S.; Timinskas, A.; Menkevicius, S.; Butkiene, D.; Butkus, V.; Janulaitis, A.: Sequence motifs characteristic of DNA[cytosine-N^4]methyltransferases: similarity to adenine and cytosine-C^5 DNA-methylases. Nucleic Acids Res., **17**, 9823-9832 (1989)

[2] Roberts, R.J.; Macelis, D.: Restriction enzymes and their isoschizomers. Nucleic Acids Res., **20**, 2167-2180 (1992)

[3] Malygin, E.G.; Zinoviev, V.V.; Evdokimov, A.A.; Lindstrom, W.M., Jr.; Reich, N.O.; Hattman, S.: DNA-(cytosine-N^4)- and (adenine-N^6)-methyltransferases have different kinetic mechanisms but the same reaction route: a comparison of M.BamHI and T4 Dam. J. Biol. Chem., **20**, 1-31 (2003)

[4] O'Gara, M.; Adams, G.M.; Gong, W.; Kobayashi, R.; Blumenthal, R.M.; Cheng, X.: Expression, purification, mass spectrometry, crystallization and multiwavelength anomalous diffraction of selenomethionyl PvuII DNA methyltransferase (cytosine-N^4-specific). Eur. J. Biochem., **247**, 1009-1018 (1997)

[5] Zinoviev, V.V.; Evdokimov, A.A.; Malygin, E.G.: DNA-(N^4-Cytosine)-methyltransferase from Bacillus amyloliquefaciens: mechanism of action derived from steady-state kinetics. Mol. Biol. (Mosk.), **37**, 116-124 (2003)

[6] Lindstrom, W.M., Jr.; Malygin, E.G.; Ovechkina, L.G.; Zinoviev, V.V.; Reich, N.O.: Functional analysis of BamHI DNA cytosine-N^4 methyltransferase. J. Mol. Biol., **325**, 711-720 (2003)

[7] Adams, G.M.; Blumenthal, R.M.: The PvuII DNA (cytosine-N^4)-methyltransferase comprises two trypsin-defined domains, each of which binds a molecule of S-adenosyl-L-methionine. Biochemistry, **36**, 8284-8292 (1997)

[8] Bujnicki, J.M.; Radlinska, M.: Molecular evolution of DNA-(cytosine-N^4) methyltransferases: evidence for their polyphyletic origin. Nucleic Acids Res., **27**, 4501-4509 (1999)
[9] Gong, W.; O'Gara, M.; Blumenthal, R.M.; Cheng, X.: Structure of pvu II DNA-(cytosine N^4) methyltransferase, an example of domain permutation and protein fold assignment. Nucleic Acids Res., **25**, 2702-2715 (1997)
[10] Merkiene, E.; Vilkaitis, G.; Klimasauskas, S.: A pair of single-strand and double-strand DNA cytosine-N^4 methyltransferases from Bacillus centrosporus. Biol. Chem., **379**, 569-571 (1998)
[11] Jeltsch, A.: The cytosine N^4-methyltransferase M.PvuII also modifies adenine residues. Biol. Chem., **382**, 707-710 (2001)
[12] Malygin, E.G.; Ovechkina, L.G.; Zinoviev, V.V.; Lindstrem, U.M.; Reich, N.O.: DNA-(N^4-cytosine)-methyltransferase from Bacillus amyloliquefaciens: kinetic and substrate-binding properties. Mol. Biol. (Mosk.), **35**, 35-44 (2001)

Hexaprenyldihydroxybenzoate methyltransferase 2.1.1.114

1 Nomenclature

EC number
2.1.1.114

Systematic name
S-adenosyl-L-methionine:3-hexaprenyl-4,5-dihydroxylate O-methyltransferase

Recommended name
hexaprenyldihydroxybenzoate methyltransferase

Synonyms
3,4-dihydroxy-5-hexaprenylbenzoate methyltransferase
dihydroxyhexaprenylbenzoate methyltransferase

CAS registry number
139569-30-5

2 Source Organism

<1> *Saccharomyces cerevisiae* [1]
<2> *Arabidopsis thaliana* [2]
<3> *Rattus norvegicus* [3]
<4> *Escherichia coli* [3]
<5> *Saccharomyces cerevisiae* [3, 4]
<6> *Homo sapiens* (enzyme may be involved in coenzyme Q defiency diseases [4]) [4]

3 Reaction and Specificity

Catalyzed reaction
S-adenosyl-L-methionine + 3-hexaprenyl-4,5-dihydroxybenzoate = S-adenosyl-L-homocysteine + 3-hexaprenyl-4-hydroxy-5-methoxybenzoate

Reaction type
methyl group transfer

Natural substrates and products
S S-adenosyl-L-methionine + 2-polyprenyl-3-methyl-5-hydroxy-6-methoxy-1,4-benzoquinol <3, 4, 5, 6> (<5> 30fold higher activity than with 3,4-

dihydroxy-5-hexaprenylbenzoate [4]) (Reversibility: ? <3, 4, 5, 6> [3, 4]) [3, 4]

P S-adenosyl-L-homocysteine + ubiquinol-n <3, 4, 5, 6> (<6> ubiqinol-10 [4]; <5> ubiquinol-6 [4]; <3> ubiquinol-8 [3]) [3, 4]

S S-adenosyl-L-methionine + 2-polyprenyl-6-hydroxyphenol <4> (Reversibility: ? <4> [3]) [3]

P S-adenosyl-L-homocysteine + 2-polyprenyl-6-methoxyphenol <4> [3]

S S-adenosyl-L-methionine + 3,4-dihydroxy-5-hexaprenylbenzoate <1, 3, 5> (<1> second step in the biosynthetic pathway of coenzyme Q [1]) (Reversibility: ? <1, 3, 5> [1, 3]) [1, 3]

P S-adenosyl-L-homocysteine + 3-methoxy-4-hydroxy-5-hexaprenylbenzoate <1, 3, 5> [1, 3]

Substrates and products

S S-adenosyl-L-methionine + 2-polyprenyl-3-methyl-5-hydroxy-6-methoxy-1,4-benzoquinol <3, 4, 5, 6> (Reversibility: ? <3, 4, 5, 6> [3, 4]) [3, 4]

P S-adenosyl-L-homocysteine + ubiquinol-n <3, 4, 5, 6> (<6> ubiqinol-10 [4]; <5> ubiquinol-6 [4]; <3> ubiquinol-8 [3]) [3, 4]

S S-adenosyl-L-methionine + 2-polyprenyl-6-hydroxyphenol <3, 4, 5, 6> (<5,6> low activity [4]) (Reversibility: ? <3, 4, 5, 6> [3, 4]) [3, 4]

P S-adenosyl-L-homocysteine + 2-polyprenyl-6-methoxyphenol <3, 4, 5, 6> [3, 4]

S S-adenosyl-L-methionine + 3,4-dihydroxy-5-hexaprenylbenzoate <1, 3, 4, 5, 6> (Reversibility: ? <1, 3, 4, 5, 6> [1, 3, 4]) [1, 3, 4]

P S-adenosyl-L-homocysteine + 3-methoxy-4-hydroxy-5-hexaprenylbenzoate <1, 3, 4, 5, 6> [1, 3, 4]

Cofactors/prosthetic groups

NADH <3, 5> (<3,5> methylation of 2-polyprenyl-3-methyl-5-hydroxy-6-methoxy-1,4-benzoquinol, it is likely that NADH provides the reducing equivalents for the generation of the hydroquinone [3]) [3]

4 Enzyme Structure

Subunits

? <2> (<2> x * 35360, deduced from nucleotide sequence [2]) [2]

5 Isolation/Preparation/Mutation/Application

Source/tissue

leaf <2> [2]

Localization

mitochondrial membrane <2, 5> (<5> peripherally associated inner membrane protein [3]) [2, 3]

Purification

<3> (recombinant His-tagged enzyme [3]) [3]

Cloning

<1> (cloning of COQ3 gene [1]) [1]
<2> (cloning of cDNA [2]) [2]
<6> (cloning of cDNA [4]) [4]

References

[1] Clarke, C.F.; Williams, W.; Teruya, J.H.: Ubiquinone biosynthesis in Saccharomyces cerevisae. Isolation and sequence of COQ3 the 3,4-dihydroxy-5-hexaprenylbenzoate methyltransferase gene. J. Biol. Chem., **266**, 16636-16644 (1991)
[2] Avelange-Macheral, M.H.; Joyard, J.: Cloning and functional expression of AtCOQ3, the Arabidopsis homologue of the yeast COQ3 gene, encoding a methyltransferase from the plant mitochondria involved in ubiquinone biosynthesis. Plant J., **14**, 203-213 (1998)
[3] Poon, W.W.; Barkovich, R.J.; Hsu, A.Y.; Frankel, A.; Lee, P.T.; Sheperd, J.N.; Myles, D.C.; Clarke, C.F.: Yeast and rat Coq3 and Escherichia coli UbiG polypeptides catalyze both O-methyltransferase steps in coenzyme Q biosynthesis. J. Biol. Chem., **274**, 21665-21672 (1999)
[4] Jonassen, T.; Clarke, C.F.: Isolation and functional expression of human COQ3, a gene encoding a methyltransferase required for ubiquinone biosynthesis. J. Biol. Chem., **275**, 12381-12387 (2000)

(RS)-1-Benzyl-1,2,3,4-tetrahydroisoquinoline N-methyltransferase 2.1.1.115

1 Nomenclature

EC number
2.1.1.115

Systematic name
S-adenosyl-L-methionine:(RS)-1-benzyl-1,2,3,4-tetrahydroisoquinoline N-methyltransferase

Recommended name
(RS)-1-benzyl-1,2,3,4-tetrahydroisoquinoline N-methyltransferase

Synonyms
(R,S)-tetrahydrobenzylisoquinoline-N-methyltransferase
NMT
methyltransferase, tetrahydrobenzylisoquinoline
norreticuline N-methyltransferase

CAS registry number
132084-82-3

2 Source Organism

<1> *Berberis koetineana* (3 isoenzymes: NMT-I, NMT-II, NMT-III [1]) [1]
<2> *Berberis henryana* [1]
<3> *Berberis angulosa* [1]
<4> *Berberis beaniana* [1]
<5> *Berberis taliensis* [1]
<6> *Berberis stolonifera* (V 29) [1]
<7> *Berberis julianae* [1]
<8> *Berberis brettschneideri* [1]
<9> *Berberis wilsoniae* [1]
<10> *Berberis dictophylla* [1]
<11> *Mahonia nervosa* [1]
<12> *Berberis gagnepainii* [1]
<13> *Berberis laxifolia* [1]
<14> *Berberis crataegina* [1]
<15> *Berberis carminea* [1]
<16> *Podophyllum hexandrum* [1]
<17> *Berberis atropurpurea* [1]
<18> *Berberis turcomanica* [1]

<19> *Podophyllum peltatum* [1]
<20> *Berberis stenophylla* [1]
<21> *Berberis vulgaris* [1]
<22> *Thalictrum dipterocarpum* [1]
<23> *Coptis japonica* [1]
<24> *Glaucium corniculatum* [1]
<25> *Glaucium flavum* [1]
<26> *Eschscholtzia californica* [1]
<27> *Bocconia cordata* [1]
<28> *Glaucium elegans* [1]
<29> *Argemone platyceras* [1]
<30> *Corydalis vaginans* [1]
<31> *Tinospora caffra* [1]
<32> *Stephania delavyi* [1]

3 Reaction and Specificity

Catalyzed reaction

S-adenosyl-L-methionine + (RS)-1-benzyl-1,2,3,4-tetrahydroisoquinoline = S-adenosyl-L-homocysteine + N-methyl-(RS)-1-benzyl-1,2,3,4-tetrahydroisoquinoline

Reaction type

methyl group transfer

Natural substrates and products

S S-adenosyl-L-methionine + coclaurine <1> [1]
P ?

Substrates and products

S S-adenosyl-L-methionine + (R)-6-O-methylnorlaudanosoline <1> [1]
P ?
S S-adenosyl-L-methionine + (R)-norcoclaurine <1> [1]
P ?
S S-adenosyl-L-methionine + (R)-norprotosinomenine <1> (<1>, isoenzyme NMT-I: 6% of maximal activity with (R)-tetrahydropapaverine [1]) [1]
P ?
S S-adenosyl-L-methionine + (R,S)-4'-O-methylnorlaudanosoline <1> (<1>, isoenzyme NMT-I: 3% of maximal activity with (R)-tetrahydropapaverine [1]) [1]
P ?
S S-adenosyl-L-methionine + (R,S)-7-O-methylnorlaudanosoline <1> (<1>, isoenzyme NMT-I: 12% of maximal activity with (R)-tetrahydropapaverine [1]) [1]
P ?
S S-adenosyl-L-methionine + (R,S)-isococlaurine <1> [1]

P ?
S S-adenosyl-L-methionine + (R,S)-norarmepavine <1> [1]
P ?
S S-adenosyl-L-methionine + (R,S)-norcodamine <1> [1]
P ?
S S-adenosyl-L-methionine + (R,S)-nororientaline <1> [1]
P ?
S S-adenosyl-L-methionine + (RS)-1-benzyl-1,2,3,4-tetrahydroisoquinoline <1> [1]
P ?
S S-adenosyl-L-methionine + (S)-6-O-methylnorlaudanosoline <1> [1]
P ?
S S-adenosyl-L-methionine + (S)-norcoclaurine <1> [1]
P ?
S S-adenosylmethionine + (R)-coclaurine <1> (<1>, isoenzyme NMT-II and NMT-III show maximal activity with (R)-coclaurine [1]) [1]
P S-adenosyl-L-homocysteine + N-methyl-(RS)-1-benzyl-1,2,3,4-tetrahydro-isoquinoline <1> [1]
S S-adenosylmethionine + (R)-norreticuline <1> [1]
P ?
S S-adenosylmethionine + (R)-tetrahydropapaverine <1> (<1>, isoenzyme NMT-I shows maximal activity with (R)-tetrahydropapaverine [1]) [1]
P ?
S S-adenosylmethionine + (S)-coclaurine <1> [1]
P S-adenosylhomocysteine + N-methylcoclaurine <1> [1]
S S-adenosylmethionine + (S)-norreticuline <1> [1]
P S-adenosylhomocysteine + (S)-scoulerine <1> [1]
S S-adenosylmethionine + (S)-tetrahydropapaverine <1> [1]
P ?

K_m-Value (mM)

0.0015 <1> ((S)-6-O-methylnorlaudanosoline, isoenzyme NMT-III, <1> [1]) [1]
0.003 <1> ((S)-6-O-methylnorlaudanosoline, isoenzyme NMT-II, <1> [1]) [1]
0.004 <1> ((R)-6-O-methylnorlaudanosoline, isoenzyme NMT-II, <1> [1]) [1]
0.005 <1> ((S)-norreticuline, isoenzyme NMT-III, <1> [1]; (R)-6-O-methyl-norlaudanosoline, isoenzyme NMT-I, <1> [1]) [1]
0.009 <1> ((S)-norreticuline, isoenzyme NMT-II, <1>) [1]
0.012 <1> ((R)-norreticuline, isoenzyme NMT-III, <1> [1]; (S)-6-O-methyl-norlaudanosoline, isoenzyme NMT-I, <1> [1]) [1]
0.015 <1> ((R)-norreticuline, isoenzyme NMT-II, <1>) [1]
0.017 <1> ((S)-norreticuline, isoenzyme NMT-I, <1>) [1]
0.018 <1> ((R)-norreticuline, isoenzyme NMT-I, <1>) [1]
0.05 <1> ((R)-6-O-methylnorlaudanosoline, isoenzyme NMT-III, <1> [1]) [1]

pH-Optimum

6.8 <1> (<1>, isoenzyme NMT-III [1]) [1]
7.4 <1> (<1>, isoenzyme NMT-I and NMT-II [1]) [1]

Temperature optimum (°C)

35-40 <1> (<1>, isoenzyme NMT-I, NMT-II and NMT-III [1]) [1]

4 Enzyme Structure

Molecular weight

60000 <1> (<1>, isoenzyme NMT-I and NMT-III, gel filtration [1]) [1]
68000 <21> (<21>, gel filtration) [1]
78000 <1> (<1>, isoenzyme NMT-II, gel filtration [1]) [1]

5 Isolation/Preparation/Mutation/Application

Source/tissue

cell culture <1> [1]

Purification

<1> [1]

Application

synthesis <1> (<1>, the enzyme immobilized on CH-Sepharose or CPG-10 glass beads is a useful tool for the preparative synthesis of isotopically labelled N-methylated benzylisoquinoline alkaloids) [1]

6 Stability

Temperature stability

6 <1> (<1>, in presence of 20% glycerol, half-life of immobilized enzyme: 20 days) [1]
20 <1> (<1>, in presence of 20% glycerol, half-life of immobilized enzyme: 7 days) [1]

General stability information

<1>, -20°C, in presence of 20% glycerol, half-life of immobilized enzyme: 40 days [1]
<1>, 6°C, in presence of 20% glycerol, half-life of immobilized enzyme: 20 days [1]

Storage stability

<1>, -20°C, in presence of 20% glycerol, half-life of immobilized enzyme: 40 days [1]

References

[1] Frenzel, T.; Zenk, M.H.: Purification and characterization of three isoforms of S-adenosyl-L-methionine: (R,S)-tetrahydrobenzylisoquinoline-N-methyltransferase from Berberis koetineana cell cultures. Phytochemistry, **29**, 3491-3497 (1990)

3'-Hydroxy-N-methyl-(S)-coclaurine 4'-O-methyltransferase 2.1.1.116

1 Nomenclature

EC number
2.1.1.116

Systematic name
S-adenosyl-L-methionine:3'-hydroxy-N-methyl-(S)-coclaurine 4'-O-methyltransferase

Recommended name
3'-hydroxy-N-methyl-(S)-coclaurine 4'-O-methyltransferase

Synonyms
4'-OMT
S-adenosyl-L-methionine: 3'-hydroxy-N-methyl-(S)-coclaurine-4'-O-methyl transferase
S-adenosyl-L-methionine:3'-hydroxy-N-methylcoclaurine 4'-O-methyltransferase
SAM:3'-hydroxy-N-methyl-(S)-coclaurine-4'-O-methyltransferase
methyltransferase, (S)-3'-hydroxy-N-methylcoclaurine 4'-

CAS registry number
132084-81-2

2 Source Organism

<1> *Berberis koetineana* [1]

3 Reaction and Specificity

Catalyzed reaction
S-adenosyl-L-methionine + 3'-hydroxy-N-methyl-(S)-coclaurine = S-adenosyl-L-homocysteine + (S)-reticuline

Reaction type
methyl group transfer

Natural substrates and products
S S-adenosyl-L-methionine + 3'-hydroxy-N-methyl-(S)-coclaurine <1> (<1>, the enzyme catalyzes the last step in (S)-reticuline biosynthesis) [1]
P ?

Substrates and products

S S-adenosyl-L-methionine + (R,S)-7-O-methyllaudanosoline <1> (<1>, 93% of the activity relative to 3'-hydroxy-N-methyl-(S)-coclaurine) [1]
P S-adenosyl-L-homocysteine + norprotosinomenine <1> [1]
S S-adenosyl-L-methionine + (R,S)-laudanosoline <1> (<1>, 84% of the activity relative to 3'-hydroxy-N-methyl-(S)-coclaurine) [1]
P S-adenosyl-L-homocysteine + 4'-O-methyllaudanosoline <1> [1]
S S-adenosyl-L-methionine + (S)-3'-hydroxy-coclaurine <1> (<1>, 10% of the activity relative to 3'-hydroxy-N-methyl-(S)-coclaurine) [1]
P S-adenosyl-L-homocysteine + (S)-norreticuline <1> [1]
S S-adenosyl-L-methionine + 3'-hydroxy-N-methyl-(S)-coclaurine <1> [1]
P S-adenosyl-L-homocysteine + (S)-reticuline <1> [1]

Inhibitors

(R)-3'-hydroxy-coclaurine <1> [1]
(S)-3'-hydroxy-coclaurine <1> [1]
(S)-norreticuline <1> [1]
(S)-reticuline <1> [1]
3'-hydroxy-N-methyl-(R)-coclaurine <1> [1]
berberine <1> [1]
jatrorrhizine <1> [1]

Specific activity (U/mg)

Additional information <1> [1]

K_m-Value (mM)

0.0045 <1> (3'-hydroxy-N-methyl-(S)-coclaurine, <1>) [1]
0.03 <1> (S-adenosyl-L-methionine, <1>) [1]

pH-Optimum

8.3-8.5 <1> [1]

pH-Range

7-9 <1> (<1>, about 50% of maximal activity at pH 7 and at pH 9) [1]

Temperature optimum (°C)

35-40 <1> [1]

4 Enzyme Structure

Molecular weight

51000 <1> (<1>, gel filtration) [1]

Subunits

monomer <1> (<1>, 1 * 40000, SDS-PAGE) [1]

5 Isolation/Preparation/Mutation/Application

Source/tissue
cell culture <1> [1]

Purification
<1> [1]

6 Stability

Temperature stability
6 <1> (<1>, half-life: 15 days) [1]
20 <1> (<1>, half-life: 3 days) [1]

Storage stability
<1>, -20°C, 20% glycerol, stable for more than 1 year [1]

References

[1] Frenzel, T.; Zenk, M.H.: S-Adenosyl-L-methionine: 3'-hydroxy-N-methyl-(S)-coclaurine-4'-O-methyl transferase, a regio- and stereoselective enzyme of the (S)-reticuline pathway. Phytochemistry, **29**, 3505-3511 (1990)

(S)-Scoulerine 9-O-methyltransferase 2.1.1.117

1 Nomenclature

EC number
2.1.1.117

Systematic name
S-adenosyl-L-methionine:(S)-scoulerine 9-O-methyltransferase

Recommended name
(S)-scoulerine 9-O-methyltransferase

Synonyms
(S)-adenosyl-L-methionine:(S)-scoularine 9-O-methyltransferase
SMT
methyltransferase, (S)-scoularine 9-
methyltransferase, (S)-scoularine 9- (Coptis japonica clone pCJSMT)

CAS registry number
164205-21-4 (methyltransferase, (S)-scoularine 9- (Coptis japonica clone pCJSMT))
96380-65-3

2 Source Organism

<1> *Coptis japonica* [1-4]
<2> *Berberis stolonifera* [4]
<3> *Berberis wilsoniae* (var. subcauliata) [5]

3 Reaction and Specificity

Catalyzed reaction
S-adenosyl-L-methionine + (S)-scoulerine = S-adenosyl-L-homocysteine + (S)-tetrahydrocolumbamine

Reaction type
methyl group transfer

Natural substrates and products
S S-adenosyl-L-methionine + (S)-scoulerine <1-3> (<1>, the enzyme is responsible for a late step in the biosynthesis of berberine [1,3,4]; <3>, central role in orotoberberine biosynthesis [5]) [1, 3-5]
P ?

Substrates and products

S S-adenosyl-L-methionine + (R,S)-2,3,9,10-tetrahydroxytetrahydroprotoberberine <3> (<3>, methylated at 50% of the rate of (S)-scoulerine [5]) [5]

P ?

S S-adenosyl-L-methionine + (S)-3,9-dihydroxy-2,10-dimethoxytetrahydroprotoberberine <3> (<3>, methylated at 4% of the rate of (S)-scoulerine [5]) [5]

P ?

S S-adenosyl-L-methionine + (S)-scoulerine <1-3> [1-5]

P S-adenosyl-L-homocysteine + (S)-tetrahydrocolumbamine <1, 3> [1, 5]

S S-adenosyl-L-methionine + 6-O-methylnorlaudanosoline <1> (<1>, 8% of the activity relative to (S)-scoulerine) [2]

P ?

S S-adenosyl-L-methionine + norlaudanosoline <1> (<1>, 2% of the activity relative to (S)-scoulerine) [2]

P ?

S S-adenosyl-L-methionine + norreticuline <1> (<1>, 9% of the activity relative to (S)-scoulerine) [2]

P ?

Inhibitors

Ca^{2+} <1> (<1>, 1 mM, 64% inhibition) [2]
Cu^{2+} <1, 3> (<1>, 1 mM, 72% inhibition [2]; <3>, 5 mM, 97% inhibition [5]) [2, 5]
Mn^{2+} <1> (<1>, 1 mM, 59% inhibition) [2]
PCMB <1> (<1>, 0.5 mM, 51% inhibition) [2]
S-adenosylhomocysteine <1, 3> [2, 5]
berberine <1> (<1>, 5 mM, 70% inhibition) [2]
palmatine <1> (<1>, 5 mM, 70% inhibition) [2]

Specific activity (U/mg)

Additional information <3> [5]

K_m-Value (mM)

0.042 <3> (S-adenosyl-L-methionine, <3>) [5]
0.1 <1> (scoulerine, <1>) [2]
0.17 <1> (S-adenosyl-L-methionine, <1>) [2]
1.6 <3> (scoulerine, <3>) [5]

pH-Optimum

8.9 <3> [5]
9-9.5 <1> [2]

pH-Range

7.5-10 <1> (<1>, 50% of maximal activity at pH 7.5 and at pH 10) [2]

Temperature optimum (°C)

37 <3> [5]

4 Enzyme Structure

Molecular weight

63000 <3> [5]
120000-140000 <1> (<1>, gel filtration) [2]

Subunits

? <1> (<1>, x * 38364, calculation from nucleotide sequence [3]; <1>, x * 41000, SDS-PAGE [2]) [2, 3]

5 Isolation/Preparation/Mutation/Application

Source/tissue

cell culture <1, 3> (<1>, activity of high berberine-producing cells is higher than that of non-selected cells [1]) [1, 5]
cell suspension culture <3> [5]
leaf <1> (<1>, low activity) [1]
root <1> (lateral; <1>, high activity) [1]
root <1> (main root; <1>, low activity) [1]
stem <1> (<1>, high activity) [1]

Purification

<1> [2]
<3> [5]

Cloning

<1> (expression in Escherichia coli) [3]

6 Stability

Storage stability

<3>, -20°C, 20% glycerol, 50% loss of activity after 7 days [5]

References

[1] Fujiwara, H.; Takeshita, N.; Terano, Y.; Fitchen, J.H.; Tsujita, T.; Katagiri, Y.; Sato, F.; Yamada, Y.: Expression of (S)-scoulerine 9-O-methyltransferase in Coptis japonica plants. Phytochemistry, **34**, 949-954 (1993)
[2] Sato, F.; Takeshita, N.; Fitchen, J.H.; Fujiwara, H.; Yamada, Y.: S-Adenosyl-L-methionine:scoulerine-9-O-methyltransferase from cultured Coptis japonica cells. Phytochemistry, **32**, 659-664 (1993)
[3] Takeshita, N.; Fujiwara, H.; Mimura, H.; Fitchen, J.H.; Yamada, Y.; Sato, F.: Molecular cloning and characterization of S-adenosyl-L-methionine:scoulerine 9-O-methyltransferase from cultured cells of Coptis japonica. Plant Cell Physiol., **36**, 29-36 (1995)

[4] Galneder, E.; Rueffer, M.; Wanner, G.; Tabata, M.; Zenk, M.H.: Alternative final steps in berberine biosynthesis in Coptis japonica cell cultures. Plant Cell Rep., 7, 1-4 (1988)
[5] Muemmler, S.; Rueffer, M.; Nagakura, N.; Zenk, M.H.: S-Adenosyl-L-methionine:(S)-scoulerine 9-O-methyltransferase, a highly stereo- and regio-specific enzyme in tetrahydroprotoberberine biosynthesis. Plant Cell Rep., **4**, 36-39 (1985)

Columbamine O-methyltransferase 2.1.1.118

1 Nomenclature

EC number
2.1.1.118

Systematic name
S-adenosyl-L-methionine:columbamine O-methyltransferase

Recommended name
columbamine O-methyltransferase

Synonyms
methyltransferase, columbamine

CAS registry number
105843-76-3

2 Source Organism

<1> *Berberis wilsoniae* (var. subcaulialate [1]) [1]
<2> *Berberis aggregata* [1]
<3> *Coptis japonica* [2]

3 Reaction and Specificity

Catalyzed reaction
S-adenosyl-L-methionine + columbamine = S-adenosyl-L-homocysteine + palmatine

Reaction type
methyl group transfer

Substrates and products
S S-adenosyl-L-methionine + columbamine <1-3> (Reversibility: ? <1-3> [1, 2]) [1, 2]
P S-adenosyl-L-homocysteine + palmatine

Inhibitors
S-adenosyl-L-homocysteine <1> [1]

K_m-Value (mM)
0.0013 <1, 2> (columbamine) [1]
0.0015 <1, 2> (S-adenosyl-L-methionine) [1]

Temperature optimum (°C)
20 <1, 2> [1]

4 Enzyme Structure

Molecular weight
52000 <1, 2> (<1,2> gel filtration [1]) [1]

5 Isolation/Preparation/Mutation/Application

Source/tissue
suspension culture <1, 2> [1]

Localization
vesicular fraction <1, 2> [1]

Purification
<1, 2> (partial [1]) [1]

Cloning
<3> (expression in Escherichia coli [2]) [2]

References

[1] Rueffer, M.; Amann, M.; Zenk, M.H.: S-Adenosyl-L-methionine:columbamine-O-methyl transferase, a compartmentalized enzyme in protoberberine biosynthesis. Plant Cell Rep., **3**, 182-185 (1986)
[2] Morishige, T.; Dubouzet, E.; Choi, K.B.; Yazaki, K.; Sato, F.: Molecular cloning of columbamine O-methyltransferase from cultured Coptis japonica cells. Eur. J. Biochem., **269**, 5659-5667 (2002)

10-Hydroxydihydrosanguinarine 10-O-methyltransferase   2.1.1.119

1 Nomenclature

EC number
2.1.1.119

Systematic name
S-adenosyl-L-methionine:10-hydroxydihydrosanguinarine 10-O-methyltransferase

Recommended name
10-hydroxydihydrosanguinarine 10-O-methyltransferase

Synonyms
10-hydroxydihydrosanguinarine-10-O-methyltransferase
methyltransferase, 10-dihydroxydihydrosanguinarine O-

CAS registry number
144388-39-6

2 Source Organism

<1> *Eschscholtzia californica* [1]

3 Reaction and Specificity

Catalyzed reaction
S-adenosyl-L-methionine + 10-hydroxydihydrosanguinarine = S-adenosyl-L-homocysteine + dihydrochelirubine

Reaction type
methyl group transfer

Natural substrates and products
S S-adenosyl-L-methionine + 10-hydroxydihydrosanguinarine <1> (<1>, the enzyme is involved in benzophenanthridine alkaloid biosynthesis [1]) [1]
P ?

Substrates and products
S S-adenosyl-L-methionine + 10-hydroxydihydrosanguinarine <1> (<1>, high substrate specificity [1]) [1]
P S-adenosyl-L-homocysteine + dihydrochelirubine <1> [1]

Specific activity (U/mg)
Additional information <1> [1]

K_m-Value (mM)
0.003 <1> (S-adenosyl-L-methionine, <1>) [1]
0.01 <1> (10-hydroxydihydrosanguinarine, <1>) [1]

pH-Optimum
8.5 <1> [1]

Temperature optimum (°C)
30-40 <1> [1]

4 Enzyme Structure

Molecular weight
49000 <1> (<1>, gel filtration) [1]

5 Isolation/Preparation/Mutation/Application

Purification
<1> [1]

References

[1] De-Eknamkul, W.; Tanahashi, T.; Zenk, M.H.: Enzymic 10-hydroxylation and 10-O-methylation of dihydrosanguinarine in dihydrochelirubine formation by Eschscholtzia. Phytochemistry, **31**, 2713-2717 (1992)

12-Hydroxydihydrochelirubine 12-O-methyltransferase 2.1.1.120

1 Nomenclature

EC number
2.1.1.120

Systematic name
S-adenosyl-L-methionine:12-hydroxydihydrochelirubine 12-O-methyltransferase

Recommended name
12-hydroxydihydrochelirubine 12-O-methyltransferase

Synonyms
12-hydroxydihydrochelirubine 12-O-methyltransferase
SAM:12-hydroxydihydrochelirubine-12-O-methyltransferase
methyltransferase, 12-hydroxydihydrochelirubine-12-O

CAS registry number
158736-40-4

2 Source Organism

<1> *Thalictrum bulgaricum* [1]

3 Reaction and Specificity

Catalyzed reaction
S-adenosyl-L-methionine + 12-hydroxydihydrochelirubine = S-adenosyl-L-homocysteine + dihydromacarpine

Reaction type
methyl group transfer

Natural substrates and products
S S-adenosyl-L-methionine + 12-hydroxydihydrochelirubine <1> (<1>, involved in biosynthesis of macarpine, the most highly oxidized benzophenanthridine alkaloid, not significantly affected by elicitation [1]) [1]
P ?

Substrates and products
S S-adenosyl-L-methionine + 12-hydroxydihydrochelirubine <1> [1]
P S-adenosyl-L-homocysteine + dihydromacarpine <1> [1]

pH-Optimum

7.5 <1> [1]

Temperature optimum (°C)

30 <1> [1]

5 Isolation/Preparation/Mutation/Application

Source/tissue

cell culture <1> [1]

Purification

<1> [1]

References

[1] Kammerer, L.; De-Eknamkul, W.; Zenk, M.H.: Enzymic 12-hydroxylation and 12-O-methylation of dihydrochelirubine in dihydromacarpine formation by Thalictrum bulgaricum. Phytochemistry, **36**, 1409-1416 (1994)

6-O-Methylnorlaudanosoline 5'-O-methyltransferase 2.1.1.121

1 Nomenclature

EC number
2.1.1.121

Systematic name
S-adenosyl-L-methionine:6-O-methylnorlaudanosoline 5'-O-methyltransferase

Recommended name
6-O-methylnorlaudanosoline 5'-O-methyltransferase

Synonyms
(6-O-methylnorlaudanosoline)-5'-O-methyltransferase
methyltransferase, 6-methylnorlaudanosoline 5'-

CAS registry number
89511-99-9

2 Source Organism

<1> *Argemone platyceras* [1]

3 Reaction and Specificity

Catalyzed reaction
S-adenosyl-L-methionine + 6-O-methylnorlaudanosoline = S-adenosyl-L-homocysteine + nororientaline

Reaction type
methyl group transfer

Natural substrates and products
S S-adenosyl-L-methionine + 6-O-methylnorlaudanosoline <1> (<1>, the product nororientaline is a precursor of the alkaloid papaverine [1]) [1]
P ?

Substrates and products
S S-adenosyl-L-methionine + 6-O-methylnorlaudanosoline <1> [1]
P S-adenosyl-L-homocysteine + nororientaline <1> [1]

Inhibitors
Fe^{3+} <1> (<1>, 5 mM, complete inhibition [1]) [1]
Hg^{2+} <1> (<1>, 5 mM, complete inhibition [1]) [1]
NEM <1> (<1>, 5 mM, 82% inhibition) [1]
PCMB <1> (<1>, 5 mM, 84% inhibition) [1]
S-adenosyl-L-homocysteine <1> [1]
iodobenzoate <1> (<1>, 10 mM, 79% inhibition) [1]

Specific activity (U/mg)
Additional information <1> [1]

K_m-Value (mM)
0.05 <1> (S-adenosyl-L-methionine, <1>) [1]
0.4 <1> (6-O-methylnorlaudanosoline, <1>) [1]

pH-Optimum
7.5 <1> [1]

Temperature optimum (°C)
35 <1> [1]

5 Isolation/Preparation/Mutation/Application

Source/tissue
cell suspension culture <1> [1]

Purification
<1> [1]

6 Stability

Temperature stability
30 <1> (<1>, half-life: 33 h) [1]

General stability information
<1>, complete inactivation by freezing even in presence of 30% glycerol [1]

Storage stability
<1>, 4°C, 0.05% NaN_3, stable for 4 weeks [1]

References

[1] Rueffer, M.; Nagakura, N.; Zenk, M.H.: A highly specific O-methyltransferase for nororientaline synthesis isolated from Argemone platyceras cell cultures. Planta Med., **49**, 196-198 (1983)

(S)-Tetrahydroprotoberberine N-methyltransferase 2.1.1.122

1 Nomenclature

EC number
2.1.1.122

Systematic name
S-adenosyl-L-methionine:(S)-7,8,13,14-tetrahydroprotoberberine cis-N-methyltransferase

Recommended name
(S)-tetrahydroprotoberberine N-methyltransferase

Synonyms
(S)tetrahydroprotoberberine-cis-N-methyltransferase
S-adenosyl-L-methionine: (S)-7,8,13,14-tetrahydroberberine-cis-N-methyltransferase
S-adenosyl-L-methionine:tetrahydroberberine-cis-N-methyltransferase
S-adenosylmethionine:tetrahydroberberine-N-methyltransferase
methyltransferase, tetrahydroberberine N-
tetrahydroberberine cis-N-methyltransferase

CAS registry number
106878-42-6

2 Source Organism

<1> *Sanguinaria canadensis* [1]
<2> *Eschscholtzia californica* [2]
<3> *Corydalis vaginans* [2, 3]
<4> *Berberis stenophylla* [3]
<5> *Dicentra spectabilis* [3]
<6> *Fumaria officinalis* [3]
<7> *Papaver somniferum* [3]

3 Reaction and Specificity

Catalyzed reaction
S-adenosyl-L-methionine + (S)-7,8,13,14-tetrahydroprotoberberine = S-adenosyl-L-homocysteine + cis-N-methyl-(S)-7,8,13,14-tetrahydroprotoberberine

Reaction type

methyl group transfer

Substrates and products

S S-adenosyl-L-methionine + (S)-7,8,13,14-tetrahydroprotoberberine <1-7> [1-3]
P S-adenosyl-L-homocysteine + cis-N-methyl-(S)-7,8,13,14-tetrahydroprotoberberine <1-7> [1-3]
S S-adenosyl-L-methionine + (S)-stylopine <2-7> (<3>, at 77% of the activity with (S)-canadine [3]) [2, 3]
P ?
S S-adenosyl-L-methionine + tetrahydrojatrorrhizine <1> [1]
P ?
S S-adenosyl-L-methionine + tetrahydropalatrubine <1> [1]
P ?

Inhibitors

Ca^{2+} <1> (<1>, in Hepes buffer, not inhibitory in sodium phosphate buffer [1]) [1]
Co^{2+} <1> (<1>, in sodium phosphate buffer and in Hepes buffer [1]) [1]
Mg^{2+} <1> (<1>, slight increase in activity with increasing concentrations of Mg^{2+} in sodium phosphate buffer, inhibitory in Hepes buffer [1]) [1]
NEM <1> [1]
S-adenosyl-L-homocysteine <1, 2> [1, 2]
Zn^{2+} <1> (<1>, in sodium phosphate buffer and in Hepes buffer [1]) [1]
berberine <1> [1]

Metals, ions

Mg^{2+} <1> (<1>, slight increase in activity with increasing concentrations of Mg^{2+} in sodium phosphate buffer, inhibitory in Hepes buffer [1]) [1]

Specific activity (U/mg)

Additional information <1, 2> [1, 2]

K_m-Value (mM)

0.0012 <1> (S-adenosyl-L-methionine, <1>) [1]
0.0017 <3> (S-adenosyl-L-methionine, <3>) [2]
0.0021 <1> ((S)-7,8,13,14-tetrahydroprotoberberine, <1>) [1]
0.0031 <3> ((S)-stylopine, <2>) [2]
0.004 <2> ((S)-stylopine, <3>) [2]
0.0064 <2> ((S)-canadine, <2>) [2]
0.007 <3> ((S)-canadine, <3>) [2]
0.012 <2> (S-adenosyl-L-methionine, <2>) [2]

pH-Optimum

7 <1> [1]
8 <3> [3]
8.9 <2> [2]

pH-Range

6.5-8 <1> (<1>, pH 6.5: about 55% of maximal activity, pH 8.0: about 65% of maximal activity) [1]
6.6-9.1 <2> (<2>, about 50% of maximal activity at pH 6.6 and pH 9.1) [2]

Temperature optimum (°C)

25-30 <1> [1]
37-40 <2> [2]

Temperature range (°C)

15-40 <1> (<1>, 15°C: about 30% of maximal activity, 40°C: about 25% of maximal activity) [1]

4 Enzyme Structure

Molecular weight

70000 <1> (<1>, gel filtration) [1]
72000 <3> (<3>, gel filtration) [2, 3]
78000 <2> (<2>, gel filtration) [2]

Subunits

dimer <1> (<1>, 2 * 39000, SDS-PAGE) [1]

5 Isolation/Preparation/Mutation/Application

Source/tissue

cell suspension culture <1, 2> [1-3]

Localization

Additional information <2> (<2>, the enzyme is not compartmentalized [2]) [2]

Purification

<1> [1]
<2> (partial [2]) [2]
<3> [3]

6 Stability

Temperature stability

20 <2> (<2>, half-life: 5 days) [2]

Storage stability

<2>, -20°C, 20% glycerol, half-life: 5 days [2]
<2>, 4°C, 20% glycerol, half-life: 10 days [2]

References

[1] O'Keefe, B.R.; Beecher, C.W.W.: Isolation and characterization of S-adenosyl-L-methionine:tetrahydroberberibe-cis-N-methyltransferase from suspension cultures of Sanguinaria canadensis L.. Plant Physiol., **105**, 395-403 (1994)

[2] Rueffer, M.; Zumstein, G.; Zenk, M.H.: Partial purification and properties of S-adenosyl-L-methionine:(S)-tetrahydroprotoberberine-cis-N-methyltransferase from suspension-cultured cells of Eschscholtzia and Corydalis. Phytochemistry, **29**, 3727-3733 (1990)

[3] Rueffer, M.; Zenk, M.H.: S-Adenosyl-L-methionine: (S)-7,8,13,14-tetrahydroberberine-cis-N-methyltransferase, a branch point enzyme in the biosynthesis of benzophenanthridine and protopine alkaloids. Tetrahedron Lett., **27**, 5603-5604 (1986)

[Cytochrome c]-methionine S-methyltransferase 2.1.1.123

1 Nomenclature

EC number
2.1.1.123

Systematic name
S-adenosyl-L-methionine:[cytochrome c]-methionine S-methyltransferase

Recommended name
[cytochrome c]-methionine S-methyltransferase

Synonyms
methyltransferase, cytochrome c (methionine)

CAS registry number
93585-98-9

2 Source Organism

<1> *Euglena gracilis* [1]

3 Reaction and Specificity

Catalyzed reaction
S-adenosyl-L-methionine + [cytochrome c]-methionine = S-adenosyl-L-homocysteine + [cytochrome c]-S-methyl-methionine

Reaction type
methyl group transfer

Substrates and products
S S-adenosyl-L-methionine + [cytochrome c]-methionine <1> (<1>, the methylation site is Met65 [1]; <1>, various types of histones are not modified [1]) [1]
P S-adenosyl-L-homocysteine + [cytochrome c]-S-methyl-methionine <1> [1]

Inhibitors
S-adenosyl-D-homocysteine <1> [1]
S-adenosyl-L-homocysteine <1> [1]
sinefungin <1> [1]

Specific activity (U/mg)
0.0000686 <1> [1]

K_m-Value (mM)
0.0166 <1> (S-adenosyl-L-methionine, <1>) [1]

pH-Optimum
7 <1> [1]

4 Enzyme Structure

Molecular weight
28000 <1> (<1>, gel filtration) [1]

5 Isolation/Preparation/Mutation/Application

Localization
cytosol <1> [1]

Purification
<1> [1]

6 Stability

Storage stability
<1>, -70°C, 50% loss of activity after 1 week [1]

References

[1] Farooqui, J.Z.; Tuck, M.; Paik, W.K.: Purification and characterization of enzymes from Euglena gracilis that methylate methionine and arginine residues of cytochrome c. J. Biol. Chem., **260**, 537-545 (1985)

[Cytochrome c]-arginine N-methyltransferase 2.1.1.124

1 Nomenclature

EC number
2.1.1.124

Systematic name
S-adenosyl-L-methionine:[cytochrome c]-arginine N^{ω}-methyltransferase

Recommended name
[cytochrome c]-arginine N-methyltransferase

Synonyms
EC 2.1.1.23 (deleted entry)
S-adenosyl-L-methionine:protein arginine N-methyltransferase
arginine methylase
methyltransferase, protein (arginine)
protein (arginine) N-methyltransferase
protein (arginine) methyltransferase
protein methylase I
protein methyltransferase I
Additional information (this enzyme was previously listed together with EC 2.1.1.25 and EC 2.1.1.26 as a single, now deleted, entry: protein-arginine N-methyltransferase, EC 2.1.1.23)

CAS registry number
9055-07-6

2 Source Organism

<1> *Euglena gracilis* [1]

3 Reaction and Specificity

Catalyzed reaction
S-adenosyl-L-methionine + [cytochrome c]-arginine = S-adenosyl-L-homocysteine + [cytochrome c]-N^{ω}-methyl-arginine

Reaction type
methyl group transfer

Substrates and products

S S-adenosyl-L-methionine + [cytochrome c]-arginine <1> (<1>, highly specific towards cytochrome c, the methylation site is Arg36 [1]) [1]

P S-adenosyl-L-homocysteine + [cytochrome c]-N^g-monomethyl arginine <1> [1]

Inhibitors

S-adenosyl-D-homocysteine <1> [1]
S-adenosyl-L-homocysteine <1> [1]
sinefungin <1> [1]

Specific activity (U/mg)

Additional information <1> [1]

K_m-Value (mM)

0.04 <1> (S-adenosyl-L-methionine, <1>) [1]
0.083 <1> (cytochrome c, <1>) [1]

pH-Optimum

7 <1> [1]

4 Enzyme Structure

Molecular weight

36000 <1> (<1>, gel filtration) [1]

5 Isolation/Preparation/Mutation/Application

Localization

cytosol <1> [1]

Purification

<1> [1]

6 Stability

Storage stability

<1>, -70°C, approximately 50% loss of activity after 1 week [1]

References

[1] Farooqui, J.Z.; Tuck, M.; Paik, W.K.: Purification and characterization of enzymes from Euglena gracilis that methylate methionine and arginine residues of cytochrome c. J. Biol. Chem., **260**, 537-545 (1985)

Histone-arginine N-methyltransferase 2.1.1.125

1 Nomenclature

EC number
2.1.1.125

Systematic name
S-adenosyl-L-methionine:histone-arginine N^{ω}-methyltransferase

Recommended name
histone-arginine N-methyltransferase

Synonyms
AdoMet:histone-arginine N-methyltransferase
EC 2.1.1.23 (formerly)
S-adenosyl-L-methionine:histone-arginine N-methyltransferase
S-adenosyl-L-methionine:protein arginine N-methyltransferase
arginine methylase
histone protein methylase I
methyltransferase, protein (arginine)
nuclear protein (histone) N-methyltransferase
protein (arginine) N-methyltransferase
protein (arginine) methyltransferase
protein methylase I
protein methyltransferase I
Additional information (the enzyme was previously listed together with EC 2.1.1.24 and 2.1.1.26 as a single, now deleted, entry: protein-arginine N-methyltransferase, EC 2.1.1.23)

CAS registry number
9055-07-6

2 Source Organism

<1> *Bos taurus* (calf [1,2,5,11]) [1-3, 5, 9, 11]
<2> *Homo sapiens* [4, 7]
<3> *Triticum aestivum* [6, 8, 10]

3 Reaction and Specificity

Catalyzed reaction

S-adenosyl-L-methionine + (histone)-arginine = S-adenosyl-L-homocysteine + (histone)-N^{ω}-methyl-arginine (<1>, sequential ordered bi bi mechanism with S-adenosyl-L-methionine as the first substrate, histone H4 as the second substrate, methylated histone H4 as the first product, and S-adenosyl-L-homocysteine as the second product released [9])

Reaction type

methyl group transfer

Substrates and products

S S-adenosyl-L-methionine + (protein A1)-arginine <1> (<1>, protein A1 is a heterogeneous ribonucleoprotein particle A1. Protein A1 possesses multiple sites for methylation, one of which is Arg194 [1]) [1]

P S-adenosyl-L-homocysteine + (protein A1)-N^{γ}-monomethylarginine + (protein A1)-N^{γ},N^{γ}-dimethylarginine <1> (<1>, (protein A1)-N^{γ},N'^{γ}-dimethylarginine is not detectable [1]) [1]

S S-adenosyl-L-methionine + ancephalitogenic basic protein-arginine <1> [9]

P S-adenosyl-L-homocysteine + ancephalitogenic basic protein-methylarginine

S histone-arginine <1-3> (<3>, calf thymus histone [8,10]; <1>, studies on the S-adenosyl-L-methionine binding site [5]; <3>, specific to methylation site Arg35 in calf thymus histone H4 [10]) [1, 2, 4-10]

P S-adenosyl-L-homocysteine + histone-N^{γ}-monomethylarginine + histone-N^{γ},N'^{γ}-dimethylarginine <1-3> (<1,2>, + histone-N^{γ},N^{γ}-monomethylarginine [7,9]; <3>, the ratio of histone-N^{γ}-monomethylarginine to histone-N^{γ},N'^{γ}-dimethylarginine is 75:25 [8]; <1>, the ratio of histone-N^{γ}-monomethylarginine to histone-N^{γ},N^{γ}-dimethylarginine to histone-N^{γ},N'^{γ}-dimethylarginine is 55:40:5 [9]; <1>, histone-N^{γ},N^{γ}-monomethylarginine is not detectable [2]; <3>, histone-N^{γ},N^{γ}-monomethylarginine is barely detectable [8]) [2, 7-9]

S Additional information <1> (<1>, no activity with myelin basic protein) [2]

P ?

Inhibitors

3-deazaadenosine <3> (<3>, moderate [8]) [8]
5'-methylthioadenosine <3> (<3>, moderate [8]) [8]
A9145C <3> (<3>, an analog of S-adenosyl-L-homocysteine [6]; <3>, powerful [8]) [6, 8]
Cu^{2+} <2> [7]
PCMB <1> (<1>, 0.5 mM, most of the activity is lost [2]) [2]
S-adenosyl-L-homocysteine <1, 3> (<3>, powerful [8]) [1, 6, 8, 9]
S-inosyl-L-(2-hydroxy-4-methylthio)butyrate <3> (<3>, powerful [8]) [6, 8]
adenosine <3> (<3>, competitive [8]) [6, 8]

guanidine-HCl <1> (<1>, 2 mM, nearly complete inactivation) [2]
myelin basic protein <1> (<1>, 0.326 mM, 50% inhibition [3]; <1>, neither of the peptide fragments, residues 1-116 and residues 117-170 generated by cleavage of myelin basic protein at the Trp residues retains inhibitory activity for histone-specific protein methylase I [3]) [1-3]
n-butylthioadenosine <3> (<3>, moderate [8]) [8]
sinefungin <1, 3> (<3>, powerful [8]) [2, 6, 8]
thioethanoladenosine <3> (<3>, moderate [8]) [8]

Activating compounds
Low-molecular-weight cofactor <3> (<3>, required for activity [6]; <3>, heat-labile, dialyzable, peptide in nature [8]) [6, 8]

Specific activity (U/mg)
Additional information <1, 3> [2, 3, 6]

K_m-Value (mM)
0.00019 <1> (protein A1, <1>) [1]
0.005 <2> (S-adenosyl-L-methionine, <2>) [7]
0.0057 <3> (S-adenosyl-L-methionine, <3>) [6, 8]
0.0076 <1> (S-adenosyl-L-methionine, <1>) [9]
0.008 <1> (S-adenosyl-L-methionine, with histone as cosubstrate, <1>) [2]
0.0111 <1> (S-adenosyl-L-methionine, with histone as cosubstrate, <1> [1]; histone, <1> [2]) [1, 2]
0.0125 <1> (S-adenosyl-L-methionine, with protein A1 as cosubstrate, <1>) [1]
0.021 <1> (histone, <1>) [1]
0.025 <1> (histone H_4, <1>) [9]
0.055 <3> (calf thymus histone, <3>) [8]
0.071 <1> (ancephalitogenic basic protein, <1>) [9]

pH-Optimum
7.2 <1> [9]
9 <3> [6, 8]

4 Enzyme Structure

Molecular weight
275000 <1> (<1>, gel filtration) [2]

Subunits
? <1> (<1>, x * 110000 + x * 75000, SDS-PAGE [2]) [2]

5 Isolation/Preparation/Mutation/Application

Source/tissue
brain <1> [1-3, 5, 9, 11]
cerebrospinal fluid <2> [4]

germ <3> [6, 8]
placenta <2> (<2>, the endogenous enzyme activity during the pregnancy increases significantly, reaching more than 4 times the initial activity at the end of the term [7]) [7]

Localization
cytosol <1> [1]

Purification
<1> [1-3, 9]
<2> [7, 8]
<3> [6]

6 Stability

Temperature stability
45-65 <1> (<1>, progressively inactivated with a linear decay between 45°C and 65°C [2]) [2]
50 <1> (<1>, 5 min, 40% loss of activity) [2]

General stability information
<1>, not inactivated by dialysis [2]

References

[1] Rajpurohit, R.; Lee, S.O.; Park, J.O.; Paik, W.K.; Kim, S.: Enzymatic methylation of recombinant heterogeneous nuclear RNP protein A1. J. Biol. Chem., **269**, 1075-1082 (1994)
[2] Gosh, S.K.; Paik, W.K.; Kim, S.: Purification and molecular identification of two protein methylases I from calf brain. J. Biol. Chem., **263**, 19024-19033 (1988)
[3] Park, G.H.; Chanderkar, L.P.; Paik, W.K.; Kim, S.: Myelin basic protein inhibits histone-specific protein methylase I. Biochim. Biophys. Acta, **874**, 30-36 (1986)
[4] Park, J.; Greenstein, J.I.; Paik, W.K.; Kim, S.: Studies on protein methyltransferase in human cerebrospinal fluid. J. Mol. Neurosci., **1**, 151-157 (1989)
[5] Syed, S.K.; Kim, S.; Paik, W.K.: Comparative studies on S-adenosyl-L-methionine binding sites of protein N-methyltransferases, using 8-azido-S-adenosyl-L-methionine as photoaffinity probe. J. Protein Chem., **12**, 603-612 (1993)
[6] Tuck, M.; Paik, W.K.: S-Adenosylmethionine:protein (arginine) N-methyltransferase (protein methylase I) (wheat germ). Methods Enzymol., **106**, 268-274 (1984)

[7] Paik, W.K.; Lee, K.H.; Hson, S.S.; Park, I.M.; Hong, J.H.; Hwang, B.D.: Human placental protein methylase - I. Purification and characterization. Int. J. Biochem., **23**, 939-945 (1991)
[8] Gupta, A.; Jensen, D.; Kim, S.; Paik, W.K.: Histone-specific protein-arginine methyltransferase from wheat germ. J. Biol. Chem., **257**, 9677-9683 (1982)
[9] Lee, H.W.; Kim, S.; Paik, W.K.: S-Adenosylmethionine:protein-arginine methyltransferase, purification and mechanism of the enzyme. Biochemistry, **16**, 78-85 (1977)
[10] Disa, S.G.; Gupta, A.; Kim, S.; Paik, W.K.: Site specificity of histone H4 methylation by wheat germ protein-arginine N-methyltransferase. Biochemistry, **25**, 2443-2448 (1986)
[11] Gosh, S.K.; Paik, W.K.; Kim, S.: Erratum: Purification and molecular identification of two protein methylases I from calf brain. J. Biol. Chem., **264**, 5313 (1989)

[Myelin basic protein]-arginine N-methyltransferase 2.1.1.126

1 Nomenclature

EC number
2.1.1.126

Systematic name
S-adenosyl-L-methionine:[myelin-basic-protein]-arginine N^{ω}-methyltransferase

Recommended name
[myelin basic protein]-arginine N-methyltransferase

Synonyms
AdoMet:myelin basic protein-arginine N-methyltransferase
EC 2.1.1.23 (formerly)
Sadenosyl-L-methionine:protein arginine N-methyltransferase
arginine methylase
methyltransferase, protein(arginine)
myelin basic protein methylase I
protein (arginine) N-methyltransferase
protein (arginine) methyltransferase
protein methylase I
protein methyltransferase I

CAS registry number
9055-07-6

2 Source Organism

<1> *Bos taurus* (calf [1,3,4]) [1, 3, 4]
<2> *Homo sapiens* [2]

3 Reaction and Specificity

Catalyzed reaction
S-adenosyl-L-methionine + [myelin basic protein]-arginine = S-adenosyl-L-homocysteine + [myelin basic protein]-N^{ω}-methyl-arginine

Reaction type
methyl group transfer

Substrates and products

S S-adenosyl-L-methionine + [myelin basic protein]-arginine <1> [1]
P S-adenosyl-L-homocysteine + [myelin basic protein]-N-methyl-arginine <1> [1]
S S-adenosyl-L-methionine + histone-arginine <1> (<1>, studies on the S-adenosyl-L-methionine binding site [3]) [1, 3]
P S-adenosyl-L-homocysteine + histone-N-methyl-arginine

Inhibitors

PCMB <1> (<1>, 0.5 mM, about 50% loss of activity [1]) [1]
S-adenosyl-L-homocysteine <1, 2> (<2>, 1 mM, 90% inhibition [2]) [1, 2]
guanidine HCl <1> (<1>, 2 mM, about 10% loss of activity [1]) [1]
sinefungin <1> [1]

Specific activity (U/mg)

Additional information <1> [1]

K_m-Value (mM)

0.00023 <1> (myelin basic protein, <1>) [1]
0.0044 <1> (S-adenosyl-L-methionine, <1>) [1]
0.1 <1> (histone, <1>) [1]

pH-Optimum

7.2 <2> [2]

4 Enzyme Structure

Molecular weight

500000 <1> (<1>, gel filtration) [1]

Subunits

? <1> (<1>, x * 72000 + x * 100000, SDS-PAGE [1]) [1]

5 Isolation/Preparation/Mutation/Application

Source/tissue

brain <1> [1, 3]
cerebrospinal fluid <2> [2]

Purification

<1> [1]

6 Stability

Temperature stability

50 <1> (<1>, stable up to, rapidly inactivated at higher temperatures with an inflection point at about 57°C) [1]

General stability information

<1>, easily inactivated by dialysis [1]

References

[1] Gosh, S.K.; Paik, W.K.; Kim, S.: Purification and molecular identification of two protein methylases I from calf brain. J. Biol. Chem., **263**, 19024-19033 (1988)

[2] Park, J.; Greenstein, J.I.; Paik, W.K.; Kim, S.: Studies on protein methyltransferase in human cerebrospinal fluid. J. Mol. Neurosci., **1**, 151-157 (1989)

[3] Syed, S.K.; Kim, S.; Paik, W.K.: Comparative studies on S-adenosyl-L-methionine binding sites of protein N-methyltransferases, using 8-azido-S-adenosyl-L-methionine as photoaffinity probe. J. Protein Chem., **12**, 603-612 (1993)

[4] Gosh, S.K.; Paik, W.K.; Kim, S.: Erratum: Purification and molecular identification of two protein methylases I from calf brain. J. Biol. Chem., **264**, 5313 (1989)

[Ribulose-bisphosphate carboxylase]-lysine N-methyltransferase 2.1.1.127

1 Nomenclature

EC number
2.1.1.127

Systematic name
S-adenosyl-L-methionine:[3-phospho-D-glycerate-carboxy-lyase (dimerizing)]-lysine N^6-methyltransferase

Recommended name
[ribulose-bisphosphate carboxylase]-lysine N-methyltransferase

Synonyms
Rubisco LSMT
Rubisco large subunit $^{\varepsilon}$N-methyltransferase
Rubisco methyltransferase
[Ribulose-bisphosphate-carboxylase]-lysine N-methyltransferase
methyltransferase, ribulose diphosphate carboxylase large subunit(lysine)
rbcMT
ribulose 1,5-bisphosphate carboxylase/oxygenase large subunit N^{ε}-methyltransferase
ribulose bisphosphate carboxylase large subunit methyltransferase
ribulose-1,5-bisphosphate carboxylase/oxygenase large subunit $^{\varepsilon}$N-methyltransferase
ribulose-bisphosphate-carboxylase/oxygenase N-methyltransferase

CAS registry number
139171-98-5

2 Source Organism

<1> *Pisum sativum* [1, 2]
<2> *Nicotiana sp.* (Nicotiana rustica [4]) [3, 4]

3 Reaction and Specificity

Catalyzed reaction
S-adenosyl-L-methionine + [ribulose-1,5-bisphosphate carboxylase]-lysine = S-adenosyl-L-homocysteine + [ribulose-1,5-bisphosphate carboxylase]-N^6-methyl-L-lysine

Reaction type

methyl group transfer

Natural substrates and products

S S-adenosyl-L-methionine + [ribulose-1,5-bisphosphate-carboxylase]-lysine <2> (<2>, the enzyme catalyzes the posttranslational methylation of the ε-amino group of Lys-14 in the large subunit of Rubisco [3]) [3]

P ?

Substrates and products

S S-adenosyl-L-methionine + [ribulose-1,5-bisphosphate-carboxylase]-lysine <1, 2> (<1,2>, catalyzes methylation of the ε-amino group of Lys-14 in the large subunit of ribulose-bisphosphate-carboxylase from Spinacia oleracea [1,4]; <2>, the enzyme does not distinguish between the enzyme form activated by Mg^{2+} and CO_2 and the unactivated enzyme form [4]; <2>, inactive on ribulose-bisphosphate-carboxylases that contain trimethyllysyl residues at position 14 in the large subunit [4]) [1-4]

P S-adenosyl-L-homocysteine + [ribulose-1,5-bisphosphate-carboxylase]-N^6-methyl-L-lysine <1> [1]

Specific activity (U/mg)

0.3 <1> [1]
Additional information <2> [4]

K_m-Value (mM)

0.0027 <1> (ribulose-1,5-bisphosphate-carboxylase, native enzyme, <1>) [1]
0.00827 <1> (S-adenosyl-L-methionine, recombinant enzyme, <1>) [1]
0.0083 <1> (ribulose-1,5-bisphosphate-carboxylase, recombinant enzyme, <1>) [1]
0.01 <2> (S-adenosyl-L-methionine, <2>) [4]
0.0137 <1> (S-adenosyl-L-methionine, native enzyme, <1>) [1]
0.018 <2> (ribulose-bisphosphate-carboxylase from Spinacia sp., <2>) [4]

pH-Optimum

8.5-10 <2> [4]

4 Enzyme Structure

Molecular weight

Additional information <1> (<1>, the full-length enzyme is processed by removing the N-terminal 36 amino acids when expressed in E. coli cells) [1]

Subunits

? <1, 2> (<2>, x * 56000, calculation from nucleotide sequence [3]; <1>, x * 60000, SDS-PAGE [1]) [1, 3]

5 Isolation/Preparation/Mutation/Application

Source/tissue
leaf <1, 2> [1, 4]

Localization
chloroplast <2> [4]

Purification
<1> (large scale affinity purification [1]; native and recombinant enzyme [2]) [1, 2]
<2> (partial [4]) [4]

Cloning
<1> (expression in Escherichia coli) [1]
<2> [3]

6 Stability

Storage stability
<1>, -80°C, in presence of 2 mg/ml β-lactoglobulin, indefinitely stable [1]
<1>, 4°C, stable for up to 30 days [1]
<2>, -80°C, 300 mM KCl, 5 mM DTT, stable for at least 1 month [4]

References

[1] Wang, P.; Royer, M.; Houtz, R.L.: Affinity purification of ribulose-1,5-bisphosphate carboxylase/oxygenase large subunit $^{\varepsilon}$N-methyltransferase. Protein Expr. Purif., **6**, 528-536 (1995)
[2] Zheng, Q.; Simel, E.J.; Klein, P.E.; Royer, M.T.; Houtz, R.L.: Expression, purification, and characterization of recombinant ribulose-1,5-bisphosphate carboxylase/oxygenase large subunit N$^{\varepsilon}$-methyltransferase. Protein Expr. Purif., **14**, 104-112 (1998)
[3] Ying, Z.; Janney, N.; Houtz, R.L.: Organization and characterization of the ribulose-1,5-bisphosphate carboxylase/oxygenase large subunit $^{\varepsilon}$N-methyltransferase gene in tobacco. Plant Mol. Biol., **32**, 663-671 (1996)
[4] Houtz, R.L.; Royer, M.; Salvucci, M.E.: Partial purification and characterization of ribulose-1,5-bisphosphate carboxylase/oxygenase large subunit $^{\varepsilon}$N-methyltransferase. Plant Physiol., **97**, 913-920 (1991)

(RS)-Norcoclaurine 6-O-methyltransferase 2.1.1.128

1 Nomenclature

EC number
2.1.1.128

Systematic name
S-adenosyl-L-methionine:(RS)-norcoclaurine 6-O-methyltransferase

Recommended name
(RS)-norcoclaurine 6-O-methyltransferase

Synonyms
(S)-coclaurine N-methyltransferase
6-OMT
NLS-OMT
S-adenosyl-L-methionine:(S)-coclaurine-N-methyltransferase
S-adenosyl-L-methionine:norcoclaurine 6-O-methyltransferase
S-adenosylmethionine:(R),(S)-norlaudanosoline-6-O-methyltransferase
methyltransferase, (S)-coclaurine N-
methyltransferase, norlaudanosoline
norcoclaurine 6-O-methyltransferase

CAS registry number
167398-06-3
89700-33-4

2 Source Organism

<-2> no activity in *Rauvolfia mannii* [2]
<-1> no activity in *Catharanthus roseus* [2]
<1> *Argemone platyceras* [1]
<2> *Papaver somniferum* [1]
<3> *Tinospora cordifolia* [2]
<4> *Dicentra spectabilis* [2]
<5> *Chondodendron tomentosum* [2]
<6> *Berberis stolonifera* [2, 3]
<7> *Berberis koetineana* [2]
<8> *Berberis canadensis* [2]
<9> *Corydalis vaginans* [2]
<10> *Fumaria officinalis* [2]
<11> *Cocculus trilobus* [2]

<12> *Stephania delavyi* [2]
<13> *Tinospora fragosa* [2]
<14> *Bocconia cordata* [2]
<15> *Eschscholtzia californica* [2, 3]
<16> *Thalictrum dasycarpum* [2]
<17> *Thalictrum flavum* [2]
<18> *Thalictrum glaucum* [2]
<19> *Peumus boldus* [3]
<20> *Coptis japonica* [4]

3 Reaction and Specificity

Catalyzed reaction
S-adenosyl-L-methionine + (RS)-norcoclaurine = S-adenosyl-L-homocysteine + (RS)-coclaurine (<20>, bi-bi ping-pong mechanism [4])

Reaction type
methyl group transfer

Natural substrates and products
- **S** S-adenosyl-L-methionine + (S)-coclaurine <6, 15, 19> (<6,15,19>, pathway for the biosynthesis of the benzyltetrahydroisoquinoline alkaloid reticuline: (S)-norcoclaurine is stereoselectively metabolized to (S)-reticuline via (S)-coclaurine, (S)-N-methylcoclaurine and (S)-3'-hydroxy-N-methylcoclaurine [3]) [3]
- **P** ?

Substrates and products
- **S** S-adenosyl-L-methionine + (R)-coclaurine <5-10, 13-16, 18> [2]
- **P** S-adenosyl-L-homocysteine + ?
- **S** S-adenosyl-L-methionine + (R)-norlaudanosoline <1, 20> (<1>, 92% of the activity with (S)-norlaudanosoline [1]) [1, 4]
- **P** S-adenosyl-L-homocysteine + 6-O-methylnorlaudanosoline + 7-O-methylnorlaudanosoline <1> (<1>, 6-O-methylnorlaudanosoline + 7-O-methylnorlaudanosoline in the ratio 8:2 [1]) [1]
- **S** S-adenosyl-L-methionine + (R,S)-2,3-dihydroxy-9,10-dimethoxytetrahydroprotoberberine <1> (<1>, 5% of the activity with (S)-norlaudanosoline) [1]
- **P** S-adenosyl-L-homocysteine + ?
- **S** S-adenosyl-L-methionine + (R,S)-4'-O-methylnorlaudanosoline <1, 3> (<1>, 34% of the activity with (S)-norlaudanosoline [1]; <3>, 0.7% of the activity with (S)-norcoclaurine [2]) [1, 2]
- **P** S-adenosyl-L-homocysteine + norprotosinomenine <1> [1]
- **S** S-adenosyl-L-methionine + (R,S)-5'-O-methylnorlaudanosoline <1> (<1>, 81% of the activity with (S)-norlaudanosoline) [1]
- **P** S-adenosyl-L-homocysteine + 6-O-methyllaudanosoline <1> [1]

S S-adenosyl-L-methionine + (S)-norlaudanosoline <1, 20> (<20>, activity with (R,S)-norlaudanosoline is 76% of the activity with (S)-norcoclaurine [4]) [1, 4]

P S-adenosyl-L-homocysteine + 6-O-methylnorlaudanosoline + 7-O-methylnorlaudanosoline <1> (<1>, 6-O-methylnorlaudanosoline + 7-O-methylnorlaudanosoline in the ratio 8:2 [1]) [1]

S S-adenosyl-L-methionine + (S)-scoulerine <1> (<1>, 1% of the activity with (S)-norlaudanosoline) [1]

P S-adenosyl-L-homocysteine + jatrorrhizine <1> [1]

S S-adenosyl-L-methionine + 2,3-dihydroxy-9,10-dimethoxyprotoberberine <1> (<1>, 7% of the activity with (S)-norlaudanosoline) [1]

P S-adenosyl-L-homocysteine + tetrahydrojatrorrhizine <1> [1]

S S-adenosyl-L-methionine + laudanosoline <1> (<1>, 79% of the activity with (S)-norlaudanosoline) [1]

P S-adenosyl-L-homocysteine + tetrahydrocolumbamine <1> [1]

S S-adenosyl-L-methionine + norcoclaurine <3, 20> (<3>, (S)-norcoclaurine: 8% of the activity with (S)-coclaurine [2]; <20>, (S)-norcoclaurine and (R)-norcoclaurine [4]) [2, 4]

P S-adenosyl-L-homocysteine + ?

Inhibitors

5,6-dihydro-9,10-dimethoxybenzo[g]-1,3-benzodioxolo[5,6-a]quinolizium <20> (<20>, 10 mM, 50% inhibition [4]) [4]
Co^{2+} <1, 20> (<20>, 5 mM, severe inhibition [4]) [1, 4]
Cu^{2+} <1, 20> (<20>, 5 mM, severe inhibition [4]) [1, 4]
Fe^{2+} <1, 20> (<20>, 5 mM, severe inhibition [4]) [1, 4]
Hg^{2+} <1> [1]
Mn^{2+} <1, 20> (<20>, 5 mM, severe inhibition [4]) [1, 4]
Ni^{2+} <1, 20> (<20>, 5 mM, severe inhibition [4]) [1, 4]
PCMB <1> [1]
S-adenosyl-L-homocysteine <1, 3, 20> [1, 2, 4]
Sn^{2+} <1> [1]
Zn^{2+} <1, 20> (<20>, 5 mM, severe inhibition [4]) [1, 4]
iodobenzoate <1> [1]

Specific activity (U/mg)

Additional information <3, 20> [2, 4]

K_m-Value (mM)

0.036 <3> ((S)-coclaurine, <3>) [2]
0.044 <3> (S-adenosyl-L-methionine, with (S)-coclaurine as cosubstrate, <3>) [2]
0.05 <1> (S-adenosylmethionine, with (S)-norlaudanosoline as cosubstrate, <1>) [1]
0.052 <3> ((S)-norcoclaurine, <3>) [2]
0.2 <1> ((R,S)-norlaudanosoline) [1]
0.3 <1> ((R,S)-laudanosoline, <1>) [1]

1.1 <1> ((R,S)-4'-O-methylnorlaudanosoline, <1>) [1]
2.23 <20> ((R,S)-norlaudanosoline, <20>) [4]
3.95 <20> (S-adenosyl-L-methionine, <20>) [4]

K_i-Value (mM)
10 <20> (5,6-Dihydro-9,10-dimethoxybenzo[g]-1,3-benzodioxolo[5,6-a]quinolizium) [4]

pH-Optimum
7.5 <1> [1]
9 <20> [4]

pH-Range
7-9.5 <20> (<20>, about 50% of maximal activity at pH 7.0 and pH 9.5) [4]

Temperature optimum (°C)
34-37 <3> [2]
35 <1> [1]

4 Enzyme Structure

Molecular weight
47000 <1> (<1>, gel filtration) [1]
95000 <20> (<20>, gel filtration) [4]

Subunits
? <20> (<20>, x * 40000, SDS-PAGE) [4]

5 Isolation/Preparation/Mutation/Application

Source/tissue
cell culture <1-20> [1-4]

Purification
<1> (partial [1]) [1]
<3> [2]
<20> [4]

6 Stability

Temperature stability
30 <1> (<1>, half-life: 8 h [1]) [1]
Additional information <3> (<3>, no significant difference in stability between the immobilized and the soluble enzyme at room temperature or at 4°C, the immobilized enzyme shows higher stability than the soluble form at -20°C up to 100 days [2]) [2]

General stability information

<1>, complete loss of activity after freezing in 30% glycerol solution [1]
<3>, immobilized enzyme shows higher stability than the soluble form at -20°C up to 100 days [2]

Storage stability

<1>, 4°C, 50% loss of activity after 4 weeks [1]

References

[1] Rueffer, M.; Nagakura, N.; Zenk, M.H.: Partial purification and properties of S-adenosylmethionine:(R),(S)-norlaudanosoline-6-O-methyltransferase from Argemone platyceras cell culture. J. Med. Plant Res., **49**, 131-137 (1983)
[2] Loeffler, S.; Deus-Neumann, B.; Zenk, M.H.: S-Adenosyl-L-methionine:(S)-coclaurine-N-methyltransferase from Tinospora cordifolia. Phytochemistry, **38**, 1387-1395 (1995)
[3] Stadler, R.M.; Zenk, M.H.: A revision of the generally accepted pathway for the biosynthesis of the benzyltetrahydroisoquinoline alkaloid reticuline. Liebigs Ann. Chem., **1990**, 555-562 (1990)
[4] Sato, F.; Tsujita, T.; Katagiri, Y.; Yoshida, S.; Yamada, Y.: Purification and characterization of S-adenosyl-L-methionine:norcoclaurine 6-O-methyltransferase from cultured Coptis japonica cells. Eur. J. Biochem., **225**, 125-131 (1994)

Inositol 4-methyltransferase 2.1.1.129

1 Nomenclature

EC number
2.1.1.129

Systematic name
S-adenosyl-L-methionine:1D-myo-inositol 4-methyltransferase

Recommended name
inositol 4-methyltransferase

Synonyms
EC 2.1.1.134 (formerly)
IMT1 protein
S-adenosyl-L-methionine:myo-inositol 6-O-methyltransferase
inositol 6-O-methyltransferase
m^6OMT
methyltransferase, inositol 6-O
methyltransferase, inositol L-1- (Mesembryanthemum crystallinum clone Imt1 reduced)
myo-inositol 4-O-methyltransferase
myo-inositol 6-O-methyltransferase

CAS registry number
146048-86-4
169277-48-9 (formerly EC 2.1.1.134)

2 Source Organism

<1> *Mesembryanthemum crystallinum* [1]
<2> *Vigna umbellata* [2]

3 Reaction and Specificity

Catalyzed reaction
S-adenosyl-L-methionine + myo-inositol = S-adenosyl-L-homocysteine + 1D-4-O-methyl-myo-inositol

Reaction type
methyl group transfer

Natural substrates and products

S S-adenosyl-L-methionine + myo-inositol <1, 2> (<1>, the enzyme catalyzes the first step in the biosynthesis of the cyclic sugar alcohol pinitol. The presence of high levels of sugar alcohols correlates with osmotolerance [1]; <2>, the product 1D-4-O-methyl-myo-inositol is accumulated in response to abiotic stress [2]) [1, 2]

P ?

Substrates and products

S S-adenosyl-L-methionine + 1L-1,3,4/2,5-cyclohexanepentol <2> (<2>, i.e. L-viburnitol) [2]

P ?

S S-adenosyl-L-methionine + 2,4,6/3,5-pentahydroxy-cyclohexanone <2> (<2>, i.e. myo-inosose-2) [2]

P ?

S S-adenosyl-L-methionine + D,L-2,3,4,6/5-pentahydroxy-cyclohexanone <2> (<2>, i.e. myo-inosose-4) [2]

P ?

S S-adenosyl-L-methionine + myo-inositol <1> [1]

P S-adenosyl-L-homocysteine + 4-methyl-myo-inositol (<1>, i.e. ononitol [1]) [1]

S S-adenosyl-L-methionine + myo-inositol <2> [2]

P S-adenosyl-L-homocysteine + 1D-4-O-methyl-myo-inositol <2> (<2>, 1D-4-O-methyl-myo-inositol is accumulated in response to abiotic stresses) [2]

Inhibitors

1L-1,3,4/2,5-cyclohexanepentol <2> (<2>, i.e. L-viburnitol [2]) [2]
2,2'-anhydro-2-C-hydroxymethyl-myo-inositol <2> [2]
2,4,6/3,5-pentahydroxy-cyclohexanone <2> (<2>, i.e. myo-inosose-2 [2]) [2]
Co^{2+} <2> (<2>, 5 mM [2]) [2]
Cu^{2+} <2> (<2>, 5 mM [2]) [2]
D,L-2,3,4,6/5-pentahydroxy-cyclohexanone <2> (<2>, i.e. myo-inosose-4 [2]) [2]
EDTA <2> (<2>, not EGTA) [2]
Fe^{2+} <2> (<2>, 5 mM [2]) [2]
Mn^{2+} <2> (<2>, 5 mM [2]) [2]
Ni^{2+} <2> (<2>, 5 mM [2]) [2]
S-adenosyl-L-homocysteine <2> (<2>, most potent inhibitor [2]) [2]
Zn^{2+} <2> (<2>, 5 mM [2]) [2]
p-chloromercuribenzenesulfonic acid <2> (<2>, 0.02 mM [2]) [2]

Activating compounds

dithiothreitol <2> (<2>, 1-20 mM, slight activation [2]) [2]

Metals, ions

K^+ <2> (<2>, 5 mM, slight stimulation [2]) [2]
Na^+ <2> (<2>, 5 mM, slight stimulation [2]) [2]

Specific activity (U/mg)
Additional information <2> [2]

K_m-Value (mM)
0.063 <2> (S-adenosyl-L-methionine, <2> with myo-inositol as cosubstrate [2]) [2]
0.3 <2> (D,L-2,3,4,6/5-pentahydroxy-cyclohexanone) [2]
0.67 <2> (2,4,6/3,5-pentahydroxy-cyclohexanone) [2]
2.92 <2> (myo-inositol) [2]
4.47 <2> (1-L-1,3,4/2,5-cyclohexanepentol) [2]

pH-Optimum
7.8 <2> [2]

pH-Range
7-8.8 <2> (<2>, about 35% of maximal activity at pH 7.0 and pH 8.8 [2]) [2]

Temperature optimum (°C)
50 <2> [2]

Temperature range (°C)
30-55 <2> (<2>, 30°C: about 40% of maximal activity, 55°C: about 30% of maximal activity [2]) [2]

4 Enzyme Structure

Molecular weight
106200 <2> (<2>, gel filtration [2]) [2]

Subunits
? <1, 2> (<1>, x * 40250, calculation from nucleotide sequence [1]; <1> x * 40000, SDS-PAGE [1]; <2> 2 * or 3 * 40300, SDS-PAGE [2]) [1, 2]

5 Isolation/Preparation/Mutation/Application

Source/tissue
stem <2> [2]

Purification
<2> [2]

Cloning
<1> (expression in Escherichia coli [1]) [1]

6 Stability

pH-Stability

5.5 <2> (<2>, 3 h, at room temperature, about 45% loss of activity compared to maximal stability at pH 7.0-7.6 [2]) [2]
6.5 <2> (<2>, 3 h, at room temperature, about 10% loss of activity compared to maximal stability at pH 7.0-7.6 [2]) [2]
7-7.6 <2> (<2>, 3 h, at room temperature, maximal stability [2]) [2]
8 <2> (<2>, 3 h, at room temperature, about 30% loss of activity compared to maximal stability at pH 7.0-7.6 [2]) [2]
9 <2> (<2>, 3 h, at room temperature, about 70% loss of activity compared to maximal stability at pH 7.0-7.6 [2]) [2]

Temperature stability

45 <2> (<2>, 30 min, stable up to) [2]
50 <2> (<2>, 30 min, about 40% loss of activity) [2]
60 <2> (<2>, 30 min, complete loss of activity) [2]
70 <1> (<1>, 15 min, sensitive to heat) [1]

General stability information

<1>, sensitive to proteinase K treatment [1]
<2>, activity is lost within a week, when the enzyme is frozen or kept at 4°C without glycerol [2]

Storage stability

<2>, -30°C, 25 mM Tris-HCl, pH 7.6, 2 mM dithiothreitol, 2 mM $MgCl_2$, 0.25 mM ammonium sulfate, 50% glycerol, half-life: 300 d [2]

References

[1] Vernon, D.M.; Bohnert, H.J.: A novel methyl transferase induced by osmotic stress in the facultative halophyte Mesembryanthemum crystallinum. EMBO J., **11**, 2077-2085 (1992)
[2] Wanek, W.; Richter, A.: Purification and characterization of myo-inositol 6-O-methyltransferase from Vigna umbellata Ohwi et Ohashi. Planta, **197**, 427-434 (1995)

Precorrin-2 C^{20}-methyltransferase 2.1.1.130

1 Nomenclature

EC number

2.1.1.130

Systematic name

S-adenosyl-L-methionine:precorrin-4 C^{20}-methyltransferase

Recommended name

precorrin-2 C^{20}-methyltransferase

Synonyms

GenBank AE00098-derived protein GI 2622455
GenBank L12006-derived protein, methyltransferase
GenBank U67522-derived protein GI 1499591
S-adenosyl-L-methionine–precorrin-2 methyltransferase
S-adenosylmethionine-precorrin-2-methyltransferase
S-adenosyl-L-methionine:precorrin-2 methyltransferase
SP2MT
methyltransferase, precorrin 2
methyltransferase, precorrin 2 (Methanococcus jannaschii gene MJ0771)
methyltransferase, precorrin 2- (Methanobacterium thermoautotrophicum strain ΔH gene MTH1348)
methyltransferase, precorrin 2- (Pseudomonas denitrificans clone pXL151)
precorrin 2- (Salmonella typhimurium strain LT2 gene cbiL isoenzyme)
precorrin-2-methyltransferase
protein (Salmonella typhimurium clone pJO26 gene cbiL reduced)

CAS registry number

131554-12-6
132053-02-2 (methyltransferase, precorrin 2- (Pseudomonas denitrificans clone pXL151))
150474-70-7 (protein (Salmonella typhimurium clone pJO26 gene cbiL reduced) /genBank L12006-derived protein, methyltransferase, precorrin 2- (Salmonella typhimurium strain LT2 gene cbiL isoenzyme))
181968-95-6 (methyltransferase, precorrin 2 (Methanococcus jannaschii gene MJ0771) /genBank U67522-derived protein GI 1499591)
198502-83-9 (methyltransferase, precorrin 2- (Methanobacterium thermoautotrophicum strain ΔH gene MTH1348) /genBank AE000898-derived protein GI 2622455)

2 Source Organism

<1> *Salmonella typhimurium* [1, 7]
<2> *Pseudomonas denitrificans* [1-6]

3 Reaction and Specificity

Catalyzed reaction
S-adenosyl-L-methionine + precorrin-2 = S-adenosyl-L-homocysteine + precorrin-3A

Reaction type
methyl group transfer

Natural substrates and products
S S-adenosyl-L-methionine + precorrin-2 <1, 2> (<1,2>, one enzyme of the cobalamin biosynthetic pathway [1]; <2>, enzyme is involved in the biosynthesis of vitamin B_{12} [4]; <2>, enzyme is involved in the cobalamin pathway [5]) [1, 4, 5]
P ?

Substrates and products
S S-adenosyl-L-methionine + precorrin-2 <1, 2> [1-7]
P S-adenosyl-L-homocysteine + precorrin-3A <1, 2> [1-7]

Specific activity (U/mg)
Additional information <2> [6]

4 Enzyme Structure

Molecular weight
53000 <2> (<2>, gel filtration) [6]

Subunits
? <2> (<2>, x * 26000, SDS-PAGE) [3]
dimer <2> (<2>, 2 * 25895, calculation from nucleotide sequence [4]; <2>, 2 * 26000, SDS-PAGE [6]) [4, 6]

5 Isolation/Preparation/Mutation/Application

Purification
<1> [7]
<2> [3, 4, 6]

Cloning
<1> (expression in Escherichia coli [7]) [7]
<2> (expression in Escherichia coli [2-4]) [2-4]

Application

synthesis <2> (<2> preparative multi-enzyme synthesis of precorrin-3A) [2, 3]

References

[1] Roth, J.R.; Lawrence, J.G.; Rubenfield, M.; Kieffer-Higgins, S.; Church, G.M.: Characterization of the cobalamin (vitamin B_{12}) biosynthetic genes of Salmonella typhimurium. J. Bacteriol., **175**, 3303-3316 (1993)

[2] Stamford, N.P.J.; Crouzet, J.; Cameron, B.; Alanine, A.I.D.; Pitt, A.R.; Yeliseev, A.A.; Battersby, A.R.: Biosynthesis of vitamin B_{12}: the preparative multi-enzyme synthesis of precorrin-3A and 20-methylsirohydrochlorin (a 2,7,20-trimethylisobacteriochlorin). Biochem. J., **313**, 335-342 (1996)

[3] Warren, M.J.; Roessner, C.A.; Ozaki, S.I.; Stolowich, N.J.; Santander, P.J.; Scott, A.I.: Enzymatic synthesis and structure of precorrin-3, a trimethyldipyrrocorphin intermediate in vitamin B_{12} biosynthesis. Biochemistry, **31**, 603-609 (1992)

[4] Roessner, C.A.; Spencer, J.B.; Ozaki, S.; Min, C.; Atshaves, B.P.; Nayar, P.; Anousis, N.; Stolowich, N.J.; Holderman, M.T.; Scott, A.I.: Overexpression in Escherichia coli of 12 vitamin B_{12} biosynthetic enzymes. Protein Expr. Purif., **6**, 155-163 (1995)

[5] Crouzet, J.; Cameron, B.; Cauchois, L.; Rigault, S.; Rouyez, M.C.; Blanche, F.; Thibaut, D.; Debussche, L.: Genetic and sequence analysis of an 8.7-kilobase Pseudomonas denitrificans fragment carrying eight genes involved in transformation of precorrin-2 to cobyrinic acid. J. Bacteriol., **172**, 5980-5990 (1990)

[6] Thibaut, D.; Couder, M.; Crouzet, J.; Debussche, L.; Cameron, B.; Blanche, F.: Assay and purification of S-adenosyl-L-methionine:precorrin-2 methyltransferase from Pseudomonas denitrificans. J. Bacteriol., **172**, 6245-6251 (1990)

[7] Roessner, C.A.; Warren, M.J.; Santander, P.J.; Atshaves, B.P.; Ozaki, S.i.; Stolowich, N.J.; Iida, K.: Expression of 9 Salmonella typhimurium enzymes for cobinamide synthesis. Identification of the 11-methyl and 20-methyl transferases of corrin biosynthesis. FEBS Lett., **301**, 73-78 (1992)

Precorrin-3B C^{17}-methyltransferase 2.1.1.131

1 Nomenclature

EC number
2.1.1.131

Systematic name
S-adenosyl-L-methionine:precorrin-3B C^{17}-methyltransferase

Recommended name
precorrin-3B C^{17}-methyltransferase

Synonyms
gene cobJ enzyme
precorrin-3 methylase
precorrin-4 synthase
SAM-dependent C-17 methyl transferase
precorrin-3 C-17 methyltransferase
precorrin-3 methyltransferase
precorrin-3B methyltransferase
synthase, precorrin-4

CAS registry number
152787-64-9

2 Source Organism

<1> *Pseudomonas denitrificans* [1-4]

3 Reaction and Specificity

Catalyzed reaction
S-adenosyl-L-methionine + precorrin-3B = S-adenosyl-L-homocysteine + precorrin-4 (<1>, mechanism for ring contraction [1])

Reaction type
methyl group transfer

Natural substrates and products
S S-adenosyl-L-methionine + precorrin-3B <1> (<1>, the enzyme is involved in the cobalamin biosynthetic pathway [1-4]) [1-4]
P ?

Substrates and products

S S-adenosyl-L-methionine + precorrin-3B <1> [1]

P S-adenosyl-L-homocysteine + precorrin-4 <1> [1]

4 Enzyme Structure

Subunits

dimer <1> (<1>, 2 * 27104, calculation from nucleotide sequence [4]; <1>, x * 27000, SDS-PAGE [2]) [2, 4]

5 Isolation/Preparation/Mutation/Application

Cloning

<1> (overexpression in Escherichia coli) [4]

References

[1] Scott, A.I.; Roessner, C.A.; Stolowich, N.J.; Spencer, J.B.; Min, C.; Ozaki, S.I.: Biosynthesis of vitamin B_{12}. Discovery of the enzymes for oxidative ring contraction and insertion of the fourth methyl group. FEBS Lett., **331**, 105-108 (1993)

[2] Debussche, L.; Thibaut, D.; Cameron, B.; Crouzet, J.; Blanche, F.: Biosynthesis of the corrin macrocycle of coenzyme B_{12} in Pseudomonas denitrificans. J. Bacteriol., **175**, 7430-7440 (1993)

[3] Roessner, C.A.; Spencer, J.B.; Ozaki, S.; Min, C.; Atshaves, B.P.; Nayar, P.; Anousis, N.; Stolowich, N.J.; Holderman, M.T.; Scott, A.I.: Overexpression in Escherichia coli of 12 vitamin B_{12} biosynthetic enzymes. Protein Expr. Purif., **6**, 155-163 (1995)

[4] Crouzet, J.; Cameron, B.; Cauchois, L.; Rigault, S.; Rouyez, M.C.; Blanche, F.; Thibaut, D.; Debussche, L.: Genetic and sequence analysis of an 8.7-kilobase Pseudomonas denitrificans fragment carrying eight genes involved in transformation of precorrin-2 to cobyrinic acid. J. Bacteriol., **172**, 5980-5990 (1990)

Precorrin-6Y $C^{5,15}$-methyltransferase (decarboxylating) 2.1.1.132

1 Nomenclature

EC number
2.1.1.132

Systematic name
S-adenosyl-L-methionine:1-precorrin-6Y $C^{5,15}$-methyltransferase (C-12-decarboxylating)

Recommended name
precorrin-6Y $C^{5,15}$-methyltransferase (decarboxylating)

Synonyms
GenBank L12005-derived protein
GenBank U67594-derived protein GI 1500413
Genbank L12006-derived protein
gene cbiE precorrin 6y methylase
gene cbiT precorrin methylase
methyltransferase, precorrin 6y (Methanococcus jannaschii)
precorrhin-6y methyltransferase
methyltransferase precorrin 6y (Salmonella typhimurium strain LT2 gene cbiE)
methyltransferase, precorrin 6y
methyltransferase, precorrin 6y (Salmonella typhimurium strain LT2 gene cbiT)
precorrin-6 methyltransferase
precorrin-6Y methylase
protein (Salmonella typhimurium clone pJO gene cbiE reduced)
protein (Salmonella typhimurium clone pJO26 gene cbiT reduced)

CAS registry number
150474-63-8 (protein (Salmonella typhimurium clone pJO gene cbiE reduced) / GenBank L12005-derived protein /methyltransferase precorrin 6y (Salmonella typhimurium strain LT2 gene cbiE))
150474-64-9 (Genbank L12006-derived protein /protein (Salmonella typhimurium clone pJO26 gene cbiT reduced) /methyltransferase, precorrin 6y (Salmonella typhimurium strain LT2 gene cbiT))
162995-22-4
181889-87-2 (methyltransferase, precorrin 6y (Methanococcus jannaschii) / GenBank U67594-derived protein GI 1500413)

2 Source Organism

<1> *Pseudomonas denitrificans* (strain SC510 Rifr harboring plasmid pXL253 [1]) [1-4]
<2> *Salmonella typhimurium* [2]

3 Reaction and Specificity

Catalyzed reaction
2 S-adenosyl-L-methionine + precorrin-6Y = 2 S-adenosyl-L-homocysteine + precorrin-8X + CO_2

Reaction type
methyl group transfer

Natural substrates and products
S S-adenosyl-L-methionine + precorrin-6Y <1, 2> (<1,2>, enzyme of the cobalamin biosynthetic pathway [2,4]; <1>, enzyme of the vitamin B_{12} biosynthetic pathway [3]) [2-4]
P ?

Substrates and products
S S-adenosyl-L-methionine + precorrin-6Y <1> [1]
P S-adenosyl-L-homocysteine + precorrin-8X + CO_2 <1> [1]

Specific activity (U/mg)
0.0057 <1> [1]

K_m-Value (mM)
0.0022 <1> (precorrin-6Y, <1>) [1]

4 Enzyme Structure

Molecular weight
320000 <1> (<1>, HPLC gel filtration) [1]

Subunits
? <1> (<1>, x * 42900, calculation from nucleotide sequence [1]; <1>, x * 43000, SDS-PAGE [1]) [1]
dimer <1> (<1>, 2 * 42952, calculation from nucleotide sequence) [3]

5 Isolation/Preparation/Mutation/Application

Purification
<1> [1]

Cloning

<1> (overexpression in Escherichia coli) [3]

References

[1] Blanche, F.; Famechon, A.; Thibaut, D.; Debussche, L.; Cameron, B.; Crouzet, J.: Biosynthesis of vitamin B_{12} in Pseudomonas denitrificans: the biosynthetic sequence from precorrin-6y to precorrin-8x is catalyzed by the cobL gene product. J. Bacteriol., **174**, 1050-1052 (1992)
[2] Roth, J.R.; Lawrence, J.G.; Rubenfield, M.; Kieffer-Higgins, S.; Church, G.M.: Characterization of the cobalamin (vitamin B_{12}) biosynthetic genes of Salmonella typhimurium. J. Bacteriol., **175**, 3303-3316 (1993)
[3] Roessner, C.A.; Spencer, J.B.; Ozaki, S.; Min, C.; Atshaves, B.P.; Nayar, P.; Anousis, N.; Stolowich, N.J.; Holderman, M.T.; Scott, A.I.: Overexpression in Escherichia coli of 12 vitamin B_{12} biosynthetic enzymes. Protein Expr. Purif., **6**, 155-163 (1995)
[4] Crouzet, J.; Cameron, B.; Cauchois, L.; Rigault, S.; Rouyez, M.C.; Blanche, F.; Thibaut, D.; Debussche, L.: Genetic and sequence analysis of an 8.7-kilobase Pseudomonas denitrificans fragment carrying eight genes involved in transformation of precorrin-2 to cobyrinic acid. J. Bacteriol., **172**, 5980-5990 (1990)

Precorrin-4 C^{11}-methyltransferase 2.1.1.133

1 Nomenclature

EC number
2.1.1.133

Systematic name
S-adenosyl-L-methionine:precorrin-4 C^{11} methyltransferase

Recommended name
precorrin-4 C^{11}-methyltransferase

Synonyms
S-adenosyl-L-methionine-dependent cobalt-precorrin-4 transmethylase
cobalt-precorrin-4 methyltransferase
methyltransferase (Pseudomonas denitrificans clone pXL151 gene cobF reduced)
methyltransferase, precorrin 2- (Pseudomonas denitrificans clone pXL151)
precorrin-3 methylase
protein (Pseudomonas denitrificans clone pXL151 gene cobF reduced)

CAS registry number
132052-20-1 (protein (Pseudomonas denitrificans clone pXL151 gene cobF reduced))
132053-02-2 (methyltransferase, precorrin 2- (Pseudomonas denitrificans clone pXL151))
9033-25-4

2 Source Organism

<1> *Bacillus megaterium* [1, 2]
<2> *Pseudomonas denitrificans* [3-5]

3 Reaction and Specificity

Catalyzed reaction
S-adenosyl-L-methionine + precorrin-4 = S-adenosyl-L-homocysteine + precorrin-5

Reaction type
methyl group transfer

Natural substrates and products

S S-adenosyl-L-methionine + precorrin-4 <2> (<2>, enzyme in the biosynthesis of vitamin B_{12} [3,5]; <2>, the enzyme is involved in the cobalamin pathway [4]) [3-5]

P ?

Substrates and products

S S-adenosyl-L-methionine + precorrin-4 <1, 2> [1-5]

P S-adenosyl-L-homocysteine + precorrin-5 <1, 2> [1-5]

4 Enzyme Structure

Molecular weight

62000 <1> (<1>, gel filtration) [1]

Subunits

dimer <1, 2> (<2>, 2 * 26862, calculation from nucleotide sequence [3]; <2>, x * 31000, SDS-PAGE [5]; <1>, 2 * 31000, SDS-PAGE [1]) [1, 3, 5]

5 Isolation/Preparation/Mutation/Application

Purification

<1> [1, 2]
<2> [3, 5]

Crystallization

<1> [1, 2]

Cloning

<1> (overexpression in Escherichia coli) [1]
<2> (overexpression in Escherichia coli) [3]

6 Stability

General stability information

<1>, stable to freezing at -20°C [1]

References

[1] Raux, E.; Schubert, H.L.; Woodcock, S.C.; Wilson, K.S.; Warren, M.J.: Cobalamin (vitamin B_{12}) biosynthesis–cloning, expression and crystallisation of the Bacillus megaterium S-adenosyl-L-methionine-dependent cobalt-precorrin-4 transmethylase CbiF. Eur. J. Biochem., **254**, 341-346 (1998)

[2] Schubert, H.L.; Wilson, K.S.; Raux, E.; Woodcock, S.C.; Warren, M.J.: The X-ray structure of a cobalamin biosynthetic enzyme, cobalt-precorrin-4 methyltransferase. Nat. Struct. Biol., **5**, 585-592 (1998)
[3] Roessner, C.A.; Spencer, J.B.; Ozaki, S.; Min, C.; Atshaves, B.P.; Nayar, P.; Anousis, N.; Stolowich, N.J.; Holderman, M.T.; Scott, A.I.: Overexpression in Escherichia coli of 12 vitamin B_{12} biosynthetic enzymes. Protein Expr. Purif., **6**, 155-163 (1995)
[4] Crouzet, J.; Cameron, B.; Cauchois, L.; Rigault, S.; Rouyez, M.C.; Blanche, F.; Thibaut, D.; Debussche, L.: Genetic and sequence analysis of an 8.7-kilobase Pseudomonas denitrificans fragment carrying eight genes involved in transformation of precorrin-2 to cobyrinic acid. J. Bacteriol., **172**, 5980-5990 (1990)
[5] Debussche, L.; Thibaut, D.; Cameron, B.; Crouzet, J.; Blanche, F.: Biosynthesis of the corrin macrocycle of coenzyme B_{12} in Pseudomonas denitrificans. J. Bacteriol., **175**, 7430-7440 (1993)

myo-Inositol 6-O-methyltransferase 2.1.1.134

1 Nomenclature

EC number

2.1.1.134 (transferred to EC 2.1.1.129)

Recommended name

myo-inositol 6-O-methyltransferase

[Methionine synthase]-cobalamin methyltransferase (cob(II)alamin reducing) 2.1.1.135

1 Nomenclature

EC number

2.1.1.135 (transferred to EC 1.16.1.8)

Recommended name

[methionine synthase]-cobalamin methyltransferase (cob(II)alamin reducing)

Chlorophenol O-methyltransferase 2.1.1.136

1 Nomenclature

EC number
2.1.1.136

Systematic name
S-adenosyl-L-methionine:trichlorophenol O-methyltransferase

Recommended name
chlorophenol O-methyltransferase

Synonyms
halogenated phenol O-methyltransferase
trichlorophenol O-methyltransferase

CAS registry number
196414-37-6

2 Source Organism

<1> *Trichoderma virgatum* (enzyme acts also on a range of mono-, di- and trichlorophenols [1]) [1]

3 Reaction and Specificity

Catalyzed reaction
S-adenosyl-L-methionine + trichlorophenol = S-adenosyl-L-homocysteine + trichloroanisole

Reaction type
transfer of methyl group

Natural substrates and products
S trichlorophenol + S-adenosyl-L-methionine <1> (Reversibility: ? <1> [1]) [1]
P trichloroanisole + S-adenosyl-L-homocysteine <1> [1]

Substrates and products
S trichlorophenol + S-adenosyl-L-methionine <1> (Reversibility: ? <1> [1]) [1]
P trichloroanisole + S-adenosyl-L-homocysteine <1> [1]

Cofactors/prosthetic groups

S-adenosyl-L-methionine <1> [1]

pH-Optimum

8.5 <1> [1]

Temperature optimum (°C)

40-50 <1> [1]

References

[1] Kikuchi, T.; Oe, T.: Halogenated phenol O-methyltransferase, its preparation from Trichoderma and methods for wood products deodorization. Jpn. Kokai Tokkyo Koho, 8 (1996)

Arsenite methyltransferase 2.1.1.137

1 Nomenclature

EC number
2.1.1.137

Systematic name
S-adenosyl-L-methionine:arsenite As-methyltransferase

Recommended name
arsenite methyltransferase

Synonyms
EC 2.1.1.138 (formerly)

CAS registry number
167140-41-2

2 Source Organism

<-3> no activity in *Callithrix jacchus* (marmoset monkey [4]) [4]
<-2> no activity in *Saguinus oedipus* (tamarin monkeys [4]) [4]
<-1> no activity in *Cavia porcellus* (guinea pig [3]) [3]
<1> *Mus musculus* (male B6C3F1 mouse [2]) [2]
<2> *Macaca mulatta* (rhesus monkey [4]) [4]
<3> *Oryctolagus cuniculus* [4, 5]
<4> *Mesocricetur auratus* (male golden syrian hamster [1]) [1]

3 Reaction and Specificity

Catalyzed reaction
S-adenosyl-L-methionine + methylarsonite = S-adenosyl-L-homocysteine + dimethylarsinate
S-adenoysl-L-methionine + arsenite = S-adenosyl-L-homocysteine + methylarsonate

Reaction type
transfer of methyl group

Natural substrates and products
S S-adenosyl-L-methionine + arsenite <1, 2, 3, 4> (Reversibility: ? <1, 2, 3, 4> [1, 2, 4, 5]) [1, 2, 4, 5]
P S-adenosyl-L-homocysteine + methylarsonate <1, 2, 3, 4> [1, 2, 4, 5]

Substrates and products

S S-adenosyl-L-methionine + arsenite <1, 2, 3, 4> (Reversibility: ? <1, 2, 3, 4> [1, 2, 4, 5]) [1, 2, 4, 5]

P S-adenosyl-L-homocysteine + methylarsonate <1, 2, 3, 4> [1, 2, 4, 5]

Inhibitors

$HgCl_2$ <3> [5]
selenate <3> (<3> more than 1 mM selenate necessary for 50% inhibition [5]) [5]
selenite <3> (<3> 50% inhibition with 0.0126 mM selinite [5]) [5]

Cofactors/prosthetic groups

S-adenosyl-L-methionine <1, 2, 3, 4> [1, 2, 4, 5]

Activating compounds

L-cysteine <3, 4> (<4> activates stronger than glutathione [1]) [1, 5]
dithiothreitol <3, 4> (<4> weak activation [1]) [1, 5]
glutathione <3, 4> (<4> a 4fold increase in glutathione concentration, 2 to 8 mM, leads to a 2fold increase in enzyme activity, reducing environment is required for enzyme activity [1]) [1, 5]

Specific activity (U/mg)

0.00000000367 <1> (<1> lung [2]) [2]
0.00000000667 <1> (<1> liver [2]) [2]
0.0000000117 <1> (<1> kidney [2]) [2]
0.0000000241 <1> (<1> testis [2]) [2]
0.0000000283 <4> [1]
0.00000128 <2> [4]
0.00000167 <3> [5]

K_m-Value (mM)

0.002 <4> (arsenite) [1]
0.006 <3> (arsenite) [1, 4]
0.006 <2> (arsenite) [4]

pH-Optimum

6.8 <3> (<3> optimum if enzyme activity is measured in crude extracts [5]) [5]
7.6 <4> [1]
8 <1, 2> (<2> little activity at pH 7.0 [4]) [2, 4]
8.2 <3> [5]

4 Enzyme Structure

Molecular weight

46000 <4> (<4> gel filtration [1]) [1]
60000 <3> (<3> gel filtration [5]) [5]

5 Isolation/Preparation/Mutation/Application

Source/tissue

kidney <1> [2]
liver <1, 2, 3, 4> [1, 2, 4]
lung <1> [2]
testis <1> [2]

Localization

cytosol <1, 2, 3, 4> [1, 2, 4, 5]

Purification

<2> (DEAE-cellulose, Sephadex G-200, Sephadex G-100, anion exchange chromatographie [4]) [4]
<3> (DEAE cellulose, ammonium sulfate precipitation, Sephadex G-200, Sephadex G-100, 2000fold purification [5]) [5]
<4> (DEAE anion exchange chromatography, Sephadex G-200, High Q anion exchange chromatographie, 40fold purification [1]) [1]

References

[1] Wildfang, E.; Zakharyan, R.A.; Aposhian, H.V.: Enzymatic methylation of arsenic compounds. VI. Characterization of hamster liver arsenite and methylarsonic acid methyltransferase activities in vitro. Toxicol. Appl. Pharmacol., **152**, 366-375 (1998)

[2] Healey, S.M.; Casarez, E.A.; Ayalo-Fierro, F.; Aposhian, H.V.: Enzymatic methylation of arsenic compoundes. Arsenite methyltransferase activity in tissues of mice. Toxicol. Appl. Pharmacol., **148**, 65-70 (1998)

[3] Healy S.M.; Zakharyan, R.A.; Aposhian, H.V.: Enzymatic methylation of arsenic compounds. VI. In vitro and in vivo deficiency of the methylation of arsenite and monomethylarsonic acid in the guinea pig. Mutat. Res., **386**, 229-239 (1997)

[4] Zakharyan, R.A.; Wildfang, E.; Aposhian, H.V.: Enzymatic methylation of arsenic compoundes. The marmoset and tamarin, but not the rhesus monkeys are deficient in methyltransferases that methylate inorganic compounds. Toxicol. Appl. Pharmacol., **140**, 77-84 (1996)

[5] Zakharyan, R.; Wu, Y.; Bogdan, G.M.; Aposhian, H.V.: Enzymatic methylation of arsenic compounds: assay, partial purification, and properties of arsenite methyltransferase and monomethylarsonic acid methyltransferase of rabbit liver. Chem. Res. Toxicol., **8**, 1029-1038 (1995)

Methylarsonite methyltransferase 2.1.1.138

1 Nomenclature

EC number
2.1.1.138 (deleted, reaction due to EC 2.1.1.137)

Recommended name
methylarsonite methyltransferase

3'-Demethylstaurosporine O-methyltransferase 2.1.1.139

1 Nomenclature

EC number
2.1.1.139

Systematic name
S-adenosyl-L-methionine:3'-demethylstaurosporine O-methyltransferase

Recommended name
3'-demethylstaurosporine O-methyltransferase

Synonyms
3'-demethoxy-3'-hydroxystaurosporine O-methyltransferase
staurosporine synthase

CAS registry number
212906-74-6

2 Source Organism

<1> *Streptomyces staurosporeus* [1]
<2> *Streptomyces longisporoflavus* [2]

3 Reaction and Specificity

Catalyzed reaction
S-adenosyl-L-methionine + 3'-demethylstaurosporine = S-adenosyl-L-homocysteine + staurosporine

Reaction type
transfer of methyl group

Natural substrates and products
S S-adenosyl-L-methionine + 3'-demethylstaurosporine <1, 2> (Reversibility: ? <1, 2> [1, 2]) [1, 2]
P S-adenosyl-L-homocysteine + staurosporine <1, 2> [1, 2]

Substrates and products
S S-adenosyl-L-methionine + 3'-demethylstaurosporine <1, 2> (Reversibility: ? <1, 2> [1, 2]) [1, 2]
P S-adenosyl-L-homocysteine + staurosporine <1, 2> [1, 2]

Inhibitors

3'-demethylstaurosporine <2> (<2> 69% activity at 1.1 mM [2]) [2]
S-(5'-adenosyl)-L-homocysteine <2> (<2> strong inhibition, 8% activity left at 1.1 mM, probably competitive inhibition [2]) [2]
anthranilate <2> (<2> 75% activity left at 1.1 mM [2]) [2]
indole <2> (<2> 62% activity left at 1.1 mM [2]) [2]
indolepyruvate <2> (<2> 44% activity left at 1.1 mM [2]) [2]
sinefungin <1> (<1> antiprotozoal antibiotic from Streptomyces griseolus, efficient inhibition with concentrations higher than 1 mg/100 ml [1]) [1]
tryptophan <2> (<2> 64% activity left at 1.1 mM [2]) [2]

Activating compounds

tryptophan <2> (<2> 2fold increase at 0.2 mM [2]) [2]

5 Isolation/Preparation/Mutation/Application

Purification

<2> (anion exchange chromatography, partial purification [2]) [2]

6 Stability

Storage stability

<2>, 4°C, 6 days, 80% [2]

References

[1] Yang, S.W.; Lin, L.J.; Cordell, G.A.; Wang, P.; Corley, D.G.: O- and N-methylation in the biosynthesis of staurosporine. J. Nat. Prod., **62**, 1551-1553 (1999)
[2] Weidner, S.; Kittelmann, M.; Goeke, K.; Ghisalba, O.; Zähner, H.: 3'-Demethoxy-3'hydroxystaurosporine O-methyltransferase from Streptomyces longisporoflavus catalyzing the last step in the biosynthesis of staurosporine. J. Antibiot., **51**, 697-682 (1998)

(S)-Coclaurine-N-methyltransferase 2.1.1.140

1 Nomenclature

EC number
2.1.1.140

Systematic name
S-adenosyl-L-methionine:(S)-coclaurine-N-methyltransferase

Recommended name
(S)-coclaurine-N-methyltransferase

CAS registry number
167398-06-3

2 Source Organism

<1> *Coptis japonica* [1]
<2> *Tinospora cordifolia* [2]
<3> *Berberis canadensis* [2]
<4> *Berberis koetineana* [2]
<5> *Berberis stolonifera* [2]
<6> *Corydalis vaginans* [2]
<7> *Dicentra spectabilis* [2]
<8> *Fumaria officinalis* [2]
<9> *Cocculus tribolus* [2]
<10> *Stephania delavayi* [2]
<11> *Tinospora fragosoa* [2]
<12> *Bocconia cordata* [2]
<13> *Eschscholtzia californica* [2]
<14> *Thalictrum dasycarpum* [2]
<15> *Thalictrum flavum* [2]
<16> *Thalictrum glaucum* [2]

3 Reaction and Specificity

Catalyzed reaction
S-adenosyl-L-methionine + (S)-coclaurine = S-adenosyl-L-homocysteine + (S)-N-methylcoclaurine

Reaction type

transfer of methyl group

Natural substrates and products

S (S)-coclaurine + S-adenosyl-L-methionine <1-16> (Reversibility: ? <1-16> [1, 2]) [1, 2]

P S-adenosyl-L-homocysteine + (S)-N-methylcoclaurine <1-16> [1, 2]

Substrates and products

S (R)-coclaurine + S-adenosyl-L-methionine <1> (Reversibility: ? <1> [1]) [1]

P S-adenosyl-L-homocysteine + (R)-N-methylcoclaurine <1> [1]

S (R,S)-6-O-methylnorlaudanosoline + S-adenosyl-L-methionine <1> (Reversibility: ? <1> [1]) [1]

P S-adenosyl-L-homocysteine + (R,S)-N-methyl-6-O-methyl-norlaudanosoline <1> [1]

S (R,S)-norlaudanosoline + S-adenosyl-L-methionine <1> (Reversibility: ? <1> [1]) [1]

P S-adenosyl-L-homocysteine + (R,S)-N-methylnorlaudanosoline <1> [1]

S (R,S)-norreticuline + S-adenosyl-L-methionine <1> (Reversibility: ? <1> [1]) [1]

P S-adenosyl-L-homocysteine + (R,S)-N-methylnorreticuline <1> [1]

S (S)-coclaurine + S-adenosyl-L-methionine <1-16> (<2> strictly stereospecific, only (S)-coclaurine is N-methylated [2]) (Reversibility: ? <1-16> [1, 2]) [1, 2]

P S-adenosyl-L-homocysteine + (S)-N-methylcoclaurine <1-16> [1, 2]

S (S)-norcoclaurine + S-adenosyl-L-methionine <2> (<2> strictly stereospecific, only (S)-norcoclaurine is N-methylated [2]) (Reversibility: ? <2> [2]) [2]

P S-adenosyl-L-homocysteine + (R)-N-methylnorcoclaurine <2> [2]

S 1-methyl-6,7-dimethoxy-1,2,3,4-tetrahydroisoquinoline + S-adenosyl-L-methionine <1> (Reversibility: ? <1> [1]) [1]

P N-methyl-1-methyl-6,7-dimethoxy-1,2,3,4-tetrahydroisoquinoline + S-adenosyl-L-methionine <1> [1]

S 6,7-dimethoxy-1,2,3,4-tetrahydroisoquinoline + S-adenosyl-L-methionine <1> (Reversibility: ? <1> [1]) [1]

P N-methyl-6,7-dimethoxy-1,2,3,4-tetrahydroisoquinoline + S-adenosyl-L-methionine <1> [1]

Inhibitors

Co^{2+} <1> (<1> 75% inhibition at 5 mM Co^{2+} [1]) [1]

Cu^{2+} <1> (<1> 47% inhibition at 5 mM Cu^{2+} [1]) [1]

Mn^{2+} <1> (<1> 57% inhibition at 5 mM Mn^{2+} [1]) [1]

S-adenosyl-L-methionine <2> (<2> 50% inhibition at 0.062 mM, substrate (S)-coclaurine [2]) [2]

Additional information <1> (<1> not inhibited by EDTA, iodoacetamide, *p*-chloromercuribenzoate [1]) [1]

Metals, ions

Ca^{2+} <1> (<1> 25% increase in activity at 5 mM Ca^{2+} [1]) [1]

Specific activity (U/mg)

0.000096 <7> (<7> enzyme extract of cell culture, (S)-coclaurine, (R)coclaurine is not methylated [2]) [2]
0.000096 <8> (<8> enzyme extract of cell culture, (S)-coclaurine, (R)coclaurine is not methylated [2]) [2]
0.0001 <10> (<10> enzyme extract of cell culture, (S)-coclaurine, (R)coclaurine is not methylated [2]) [2]
0.0003 <15> (<15> enzyme extract of cell culture, (S)-coclaurine, (R)coclaurine is not methylated [2]) [2]
0.00065 <16> (<16> enzyme extract of cell culture, (S)-coclaurine, (R)coclaurine is not methylated [2]) [2]
0.00066 <14> (<14> enzyme extract of cell culture, (S)-coclaurine, (R)coclaurine is not methylated [2]) [2]
0.0007 <6> (<6> enzyme extract of cell culture, (S)-coclaurine, (R)coclaurine is not methylated [2]) [2]
0.00075 <12> (<12> enzyme extract of cell culture, (S)-coclaurine, (R)coclaurine is not methylated [2]) [2]
0.00077 <9> (<9> enzyme extract of cell culture, (S)-coclaurine, (R)coclaurine is not methylated [2]) [2]
0.0008 <5> (<5> enzyme extract of cell culture, (S)-coclaurine [2]) [2]
0.0013 <13> (<13> enzyme extract of cell culture, (S)-coclaurine, (R)coclaurine is not methylated [2]) [2]
0.002 <3> (<3> enzyme extract of cell culture, (S)-coclaurine [2]) [2]
0.002 <4> (<4> enzyme extract of cell culture, (S)-coclaurine [2]) [2]
0.003 <11> (<11> enzyme extract of cell culture, (S)-coclaurine, (R)coclaurine is not methylated [2]) [2]
0.0036 <2> (<2> enzyme extract of cell culture, (S)-coclaurine [2]) [2]
0.017 <1> [1]
0.4 <2> [2]

K_m-Value (mM)

0.04 <2> ((S)-coclaurine) [2]
0.04 <2> (S-adenosyl-L-methionine) [2]
0.05 <2> ((S)-norcoclaurine, <2> probably allosteric enzyme, Hill coeffizient 2.6 [2]) [2]
0.38 <1> ((R,S)-norreticuline) [1]
0.65 <1> (S-adenosyl-L-methionine) [1]

K_i-Value (mM)

0.062 <2> (S-adenosyl-L-methionine, substrate (S)-coclaurine) [2]

pH-Optimum

7 <1> (<1> 50% activity at pH 6.0 and 9.0 [1]) [1]
8.6 <2> (<2> narrow optimum [2]) [2]

pH-Range

6-9.5 <2> [2]

Temperature optimum (°C)

34-37 <2> (<2> 23fold purified enzyme preparation, pH 9.0 [2]) [2]

4 Enzyme Structure

Molecular weight

83000 <2> (<2> gel filtration HPLC [2]) [2]
87000 <2> (<2> gel filtration [2]) [2]

Subunits

homotetramer <1> (<1> α, 4 * 45000, SDS-PAGE [1]) [1]

5 Isolation/Preparation/Mutation/Application

Purification

<1> (30-50% ammonium sulfate, phenyl Sepharose, Q-Sepharose, MonoQ, MonoP, 343fold purification [1]) [1]
<2> (40% ammonium sulfate, Phenylsepharose, TSK Fraktogel, MonoQ, Affinity gel, disc electrophoresis, 115fold purification [2]) [2]

References

[1] Choi, K.B.; Morishige, T.; Sato, F.: Purification and characterization of coclaurine N-methyltransferase from cultured Coptis japonica cells. Phytochemistry, **56**, 649-655 (2001)
[2] Loeffler, S.; Deus-Neumann, B.; Zenk, M.H.: S-Adenosyl-L-methionine:(S)-coclaurine-N-methyltransferase from Tinospora cordifolia. Phytochemistry, **38**, 1387-1395 (1995)

Jasmonate O-methyltransferase 2.1.1.141

1 Nomenclature

EC number
2.1.1.141

Systematic name
S-adenosyl-L-methionine:jasmonate O-methyltransferase

Recommended name
jasmonate O-methyltransferase

Synonyms
JMT
S-adenosyl-L-methionine:jasmonic acid carboxyl methyltransferase
jasmonic acid carboxyl methyltransferase

CAS registry number
346420-58-4

2 Source Organism

<1> *Arabidopsis thaliana* [1]
<2> *Brassica campestris* (L. perkinensis [2]) [2]

3 Reaction and Specificity

Catalyzed reaction
S-adenosyl-L-methionine + jasmonate = S-adenosyl-L-homocysteine + methyljasmonate

Reaction type
methyl group transfer

Natural substrates and products
S S-adenosyl-L-methionine + jasmonate <1> (<1> expression of the gene is induced both locally and systematically by wounding or methyl jasmonate treatement. The jasmonic acid carboxyl methyltransferase is a key enzyme for jasmonate-regulated plant responses. Activation of JMT expression leads to production of methyl jasmonate that could act as an intracellular regulator, a diffusible intercellular transducer, and an airborne signal

mediating intra- and interplant communications [1]) (Reversibility: ? <1> [1]) [1]

P S-adenosyl-L-homocysteine + methyljasmonate <1> [1]

Substrates and products

S S-adenosyl-L-methionine + 9,10-dihydrojasmonic acid <1> (<1> poor substrate [1]) (Reversibility: ? <1> [1]) [1]

P S-adenosyl-L-homocysteine + ?

S S-adenosyl-L-methionine + jasmonate <1, 2> (Reversibility: ? <1,2> [1,2]) [1, 2]

P S-adenosyl-L-homocysteine + methyljasmonate <1> [1]

S Additional information <1> (<1> the enzyme does not convert 12-oxo-phytodienoic acid, a precursor of jasmonic acid, salicylic acid, benzoic acid, linolenic acid or cinnamic acid into their corresponding methyl esters [1]) [1]

P ?

Inhibitors

Ca^{2+} <1> (<1> 5 mM, 80% loss of activity [1]) [1]
Cu^{2+} <1> (<1> 5 mM, 80% loss of activity [1]) [1]
Fe^{2+} <1> (<1> 5 mM, 80% loss of activity [1]) [1]
Mg^{2+} <1> (<1> 5 mM, 80% loss of activity [1]) [1]
Mn^{2+} <1> (<1> 5 mM, 80% loss of activity [1]) [1]
NH_4^+ <1> (<1> 5 mM, 50-60% inhibition [1]) [1]
Na^+ <1> (<1> 5 mM, 50-60% inhibition [1]) [1]
Zn^{2+} <1> (<1> 5 mM, 80% loss of activity [1]) [1]

Metals, ions

KCl <1, 2> (<1> highest activity in presence of 100 mM KCl [1]; <2> optimal activity in presence of 100-150 mM KCl [2]) [1, 2]

Turnover number (min^{-1})

1500 <1> (jasmonate, <1> pH 7.5, 20°C [1]) [1]
4200 <1> (S-adenosyl-L-methionine, <1> pH 7.5, 20°C [1]) [1]

K_m-Value (mM)

0.0063 <1> (S-adenosyl-L-methionine, <1> pH 7.5, 20°C [1]) [1]
0.0064 <2> (S-adenosyl-L-methionine) [2]
0.0385 <1> (jasmonate, <1> pH 7.5, 20°C [1]) [1]
0.38 <2> (jasmonate) [2]

pH-Optimum

7-8 <1> [1]
7.5 <2> [2]

Temperature optimum (°C)

20 <1, 2> [1, 2]

4 Enzyme Structure

Molecular weight

44000 <2> (<2> as fusion protein with GST [2]) [2]

5 Isolation/Preparation/Mutation/Application

Localization

cytoplasm <1> [1]

Purification

<1> [1]
<2> [2]

Cloning

<1> (expressed in Escherichia coli [1]) [1]
<2> (expression in Escherichia coli [2]) [2]

References

[1] Seo, H.S.; Song, J.T.; Cheong, J.J.; Lee, Y.H.; Lee, Y.W.; Hwang, I.; Lee, J.S.; Choi, Y.D.: Jasmonic acid carboxyl methyltransferase: A key enzyme for jasmonate-regulated plant responses. Proc. Natl. Acad. Sci. USA, **98**, 4788-4793 (2001)

[2] Seo, H.S.; Song, J.T.; Koo, Y.J.; Jung, C.; Yeu, S.Y.; Kim, M.; Song, S.I.; Lee, J.S.; Hwang, I.; Cheong, J.-J.; Choi, Y.D.: Floral nectary-specific gene NTR1 encodes a jasmonic acid carboxyl methyltransferase. Agric. Chem. Biotechnol., **44**, 119-124 (2001)

Cycloartenol 24-C-methyltransferase 2.1.1.142

1 Nomenclature

EC number
2.1.1.142

Systematic name
S-adenosyl-L-methionine:cycloartenol 24-C-methyltransferase

Recommended name
cycloartenol 24-C-methyltransferase

Synonyms
(S)-adenosyl-L-methionine-Δ^{24}-sterol methyltransferase
(S)-adenosyl-L-methionine:$\Delta^{24(25)}$-sterol methyltransferase
S-adenosylmethionine
SAM-Δ^{24} sterol transmethylase
$\Delta^{24(25)}$-sterol methyltransferase
Δ^{24}-sterol methyltransferase
sterol C-methyltransferase
zymosterol-24-methyltransferase

CAS registry number
37257-07-1

2 Source Organism

<1> *Prototheca wickerhamii* [1]

3 Reaction and Specificity

Catalyzed reaction
S-adenosyl-L-methionine + cycloartenol = S-adenosyl-L-homocysteine + (24R)-24-methylcycloart-25-en-3β-ol (<1> mechanism, S-adenosyl-L-methionine mehtylates the Si face of the 24(25)-double bond with elimination of a hydrogen atom from the pro-Z methyl group at C-25 [1])

Reaction type
transfer of methyl group

Natural substrates and products
S S-adenosyl-L-methionine + cycloartenol <1> (Reversibility: ? <1> [1]) [1]
P S-adenosyl-L-homocysteine + (24R)-24-methylcycloart-25-en-3β-ol

Substrates and products

S 14α-methylzymosterol <1> (Reversibility: ? <1> [1]) [1]
P S-adenosyl-L-homocysteine + ?
S 30,31-dinorcycloartenol <1> (Reversibility: ? <1> [1]) [1]
P S-adenosyl-L-homocysteine + ?
S 31-norcycloartenol <1> (Reversibility: ? <1> [1]) [1]
P S-adenosyl-L-homocysteine + ?
S S-adenosyl-L-methionine + cycloartenol <1> (Reversibility: ? <1> [1]) [1]
P S-adenosyl-L-homocysteine + (24R)-24-methylcycloart-25-en-3β-ol
S cholesta-5,20(22)E,24-trienol <1> (Reversibility: ? <1> [1]) [1]
P S-adenosyl-L-homocysteine + ?
S cholesta-5,7,22E,24-tetraenol <1> (Reversibility: ? <1> [1]) [1]
P S-adenosyl-L-homocysteine + ?
S lanosterol <1> (Reversibility: ? <1> [1]) [1]
P S-adenosyl-L-homocysteine + ?
S parkeol <1> (Reversibility: ? <1> [1]) [1]
P S-adenosyl-L-homocysteine + ?
S zymosterol <1> (Reversibility: ? <1> [1]) [1]
P S-adenosyl-L-homocysteine + ?

Inhibitors

Additional information <1> (<1> non-competitive inhibitors are ammonium group substituted for carbon at C_{25}, C_{24}, C_{23} and C_{22} [1]) [1]

K_m-Value (mM)

0.03 <1> (cycloartenol) [1]
0.038 <1> (zymosterol) [1]
0.063 <1> (desmosterol) [1]
0.1 <1> (cholesta-5,20(22)E,24-trienol) [1]
0.112 <1> (cholesta-5,7,22E,24-tetraenol) [1]
0.125 <1> (14α-methylzymosterol) [1]
0.176 <1> (lanosterol) [1]
0.276 <1> (cholesta-5,20(22)E,24-trienol) [1]
0.61 <1> (31-norcycloartenol) [1]
0.8 <1> (parkeol) [1]
0.97 <1> (30,31-dinorcycloartenol) [1]

4 Enzyme Structure

Molecular weight

154000 <1> (<1> gel filtration [1]) [1]

References

[1] Mangla, A.T.; Nes, W.D.: Sterol C-methyl transferase from Prototheca wickerhamii mechanism, sterol specificity and inhibition. Bioorg. Med. Chem., **8**, 925-936 (2000)

24-Methylenesterol C-methyltransferase 2.1.1.143

1 Nomenclature

EC number
2.1.1.143

Systematic name
S-adenosyl-L-methionine:24-methylenelophenol C-methyltransferase

Recommended name
24-methylenesterol C-methyltransferase

Synonyms
(S)-adenosyl-L-methionine-Δ^{24}-sterol methyltransferase
(S)-adenosyl-L-methionine:$\Delta^{24(25)}$-sterol methyltransferase
24-methylenelophenol C-241-methyltransferase
$\Delta^{24(25)}$-sterol methyltransferase
Δ^{24}-methyltransferase
Δ^{24}-sterol methyltransferase
S-adenosylmethionine:Δ^{24}-sterol methyltransferase
SAM-Δ^{24} sterol transmethylase
SMT2
methyltransferase, Δ^{24}-sterol
zymosterol-24-methyltransferase

CAS registry number
37257-07-1

2 Source Organism

<1> *Arabidopsis thaliana* [1, 3]
<2> *Nicotiana tabacum* [1, 2]
<3> *Oryza sativa* [2]

3 Reaction and Specificity

Catalyzed reaction
S-adenosyl-L-methionine + 24-methylenelophenol = S-adenosyl-L-homocysteine + (Z)-24-ethylidenelophenol

Reaction type
transfer of methyl group

Natural substrates and products

S S-adenosyl-L-methionine + 24-methylenelophenol <1, 2> (<1,2>, second methylation step of plant sterol biosynthesis [1,2,3]) (Reversibility: ? <1, 2> [1, 2, 3]) [1, 2, 3]

P S-adenosyl-L-homocysteine + 24-ethylidenelophenol <1> [1]

Substrates and products

S S-adenosyl-L-methionine + 24-methylenelophenol <1> (Reversibility: ? <1> [1]) [1]

P S-adenosyl-L-homocysteine + 24-ethylidenelophenol <1> [1]

S S-adenosyl-L-methionine + cycloartenol <1, 2> (<1>, 6% of the activity with 24-methylenelophenol [1]) (Reversibility: ? <1, 2> [1, 2]) [1, 2]

P S-adenosyl-L-homocysteine + 24-methylene cycloartenol <1> [1]

S S-adenosyl-L-methionine + lanosterol <1> (<1>, 12.4% of the activity with 24-methylenelophenol [1]) (Reversibility: ? <1> [1]) [1]

P S-adenosyl-L-homocysteine + ? <1> [1]

S S-adenosyl-L-methionine + obtusifoliol <1> (<1>, 11.9% of the activity with 24-methylenelophenol [1]) (Reversibility: ? <1> [1]) [1]

P S-adenosyl-L-homocysteine + ?

S S-adenosyl-L-methionine + zymosterol <1> (<1>, 5.2% of the activity with 24-methylenelophenol [1]) (Reversibility: ? <1> [1]) [1]

P S-adenosyl-L-homocysteine + fecosterol + episterol + stigmasta-8,24-dienol + Δ^7-avenasterol <1> [1]

K_m-Value (mM)

0.0049 <1> (24-methylenelophenol) [1]
0.0052 <1> (zymosterol) [1]
0.0114 <1> (cycloartenol) [1]
0.0119 <1> (obtusifoliol) [1]
0.012 <2> (24-methylene lophenol) [2]
0.0124 <1> (lanosterol) [1]
0.042 <2> (cycloartenol) [2]

5 Isolation/Preparation/Mutation/Application

Cloning

<1> (expression in Nicotiana tabacum [3]) [1, 3]
<2> (expression in yeast null mutant erg6 [2]) [2]

References

[1] Bouvier-Nave, P.; Husselstein, T.; Desprez, T.; Benveniste, P.: Identification of cDNAs encoding sterol methyl-transferases involved in the second methylation step of plant sterol biosynthesis. Eur. J. Biochem., **246**, 518-529 (1997)

[2] Bouvier-Nave, P.; Husselstein, T.; Benveniste, P.: Two families of sterol methyltransferases are involved in the first and the second methylation steps of plant sterol biosynthesis. Eur. J. Biochem., **256**, 88-96 (1998)
[3] Schaeffer, A.; Bronner, R.; Benveniste, P.; Schaller, H.: The ratio of campesterol to sitosterol that modulates growth in Arabidopsis is controlled by sterol methyltransferase 2;1. Plant J., **25**, 605-615 (2001)

Trans-aconitate 2-methyltransferase 2.1.1.144

1 Nomenclature

EC number
2.1.1.144

Systematic name
S-adenosyl-L-methionine:(E)-prop-1-ene-1,2,3-tricarboxylate 2'-O-methyltransferase

Recommended name
trans-aconitate 2-methyltransferase

Synonyms
trans-aconitate methyltransferase

CAS registry number
235107-12-7

2 Source Organism

<1> *Escherichia coli* [1-3]

3 Reaction and Specificity

Catalyzed reaction
S-adenosyl-L-methionine + trans-aconitate = S-adenosyl-L-homocysteine + (E)-3-(methoxycarbonyl)pent-2-enedioate

Reaction type
methyl group transfer

Natural substrates and products
- **S** S-adenosyl-L-methionine + trans-aconitate <1> (Reversibility: ? <1> [1-3]) [1-3]
- **P** S-adenosyl-L-homocysteine + (E)-3-(methoxycarbonyl)pent-2-enedioate

Substrates and products
- **S** S-adenosyl-L-methionine + (2R,3S)-isocitrate <1> (Reversibility: ? <1> [1, 2]) [1, 2]
- **P** S-adenosyl-L-homocysteine + ?
- **S** S-adenosyl-L-methionine + DL-isocitrate <1> (Reversibility: ? <1> [1, 2]) [1, 2]

P S-adenosyl-L-homocysteine + ?
S S-adenosyl-L-methionine + cis-aconitate <1> (Reversibility: ? <1> [1, 2]) [1, 2]
P S-adenosyl-L-homocysteine + (E)-3-(methoxycarbonyl)pent-2-enedioate
S S-adenosyl-L-methionine + citrate <1> (Reversibility: ? <1> [1, 2]) [1, 2]
P S-adenosyl-L-homocysteine + ?
S S-adenosyl-L-methionine + trans-aconitate <1> (Reversibility: ? <1> [1-3]) [1-3]
P S-adenosyl-L-homocysteine + (E)-3-(methoxycarbonyl)pent-2-enedioate

Inhibitors
trans-aconitate <1> (<1> 0.5-1 mM, when cis-aconitate is used as substrate [3]) [3]

K_m-Value (mM)
0.32 <1> (trans-aconitate) [1]
0.52 <1> (trans-aconitate) [2]
5 <1> (DL-isocitrate) [1]
7.1 <1> (citrate) [1]
9.1 <1> ((2R,3S)-isocitrate) [1]
10.4 <1> ((2R,3S)-isocitrate) [2]
23.1 <1> (DL-isocitrate) [2]
33 <1> (cis-aconitate) [1]
36 <1> (cis-aconitate) [2]
45 <1> (citrate) [1]

5 Isolation/Preparation/Mutation/Application

Localization
cytosol <1> [1-3]

Purification
<1> [1, 2]

Cloning
<1> (expression in a Escherichia coli BL21 [1]) [1]

References

[1] Cai, H.; Clarke, S.: A novel methyltransferase catalyzes the methyl esterification of trans-aconitate in Escherichia coli. J. Biol. Chem., **274**, 13470-13479 (1999)
[2] Cai, H.; Dumlao, D.; Katz, J.E.; Clarke, S.: Identification of the gene and characterization of the activity of the trans-aconitate methyltransferase in Saccharomyces cerevisiae. Biochemistry, **40**, 13699-13709 (2001)
[3] Cai, H.; Strouse, J.; Dumlao, D.; Jung, M.E.; Clarke, S.: Distinct reactions catalyzed by bacterial and yeast trans-aconitate methyltransferase. Biochemistry, **40**, 2210-2219 (2001)

Trans-aconitate 3-methyltransferase 2.1.1.145

1 Nomenclature

EC number
2.1.1.145

Systematic name
S-adenosyl-L-methionine:(E)-prop-1-ene-1,2,3-tricarboxylate 3'-O-methyltransferase

Recommended name
trans-aconitate 3-methyltransferase

Synonyms
trans-aconitate methyltransferase

CAS registry number
235107-12-7

2 Source Organism

<1> *Saccharomyces cerevisiae* [1, 2]

3 Reaction and Specificity

Catalyzed reaction
S-adenosyl-L-methionine + trans-aconitate = S-adenosyl-L-homocysteine + (E)-2-(methoxycarbonylmethyl)butenedioate

Reaction type
methyl group transfer

Natural substrates and products
S S-adenosyl-L-methionine + trans-aconitate <1> (Reversibility: ? <1> [1, 2]) [1, 2]
P S-adenosyl-L-homocysteine + (E)-2-(methoxycarbonylmethyl)butenedioate

Substrates and products
S S-adenosyl-L-methionine + cis-aconitate <1> (Reversibility: ? <1> [1]) [1]
P S-adenosyl-L-homocysteine + (Z)-2-(methoxycarbonylmethyl)butenedioate

S S-adenosyl-L-methionine + itaconate <1> (Reversibility: ? <1> [1]) [1]
P S-adenosyl-L-homocysteine + ?
S S-adenosyl-L-methionine + trans-aconitate <1> (Reversibility: ? <1> [1, 2]) [1, 2]
P S-adenosyl-L-homocysteine + (E)-2-(methoxycarbonylmethyl)butenedioate

Inhibitors
trans-aconitate <1> (<1> 0.5-1 mM, when cis-aconitate is used as substrate [2]) [2]

K_m-Value (mM)
0.66 <1> (trans-aconitate) [1]
44 <1> (itaconate) [1]
74 <1> (cis-aconitate) [1]

5 Isolation/Preparation/Mutation/Application

Localization
cytosol <1> [1, 2]

References

[1] Cai, H.; Dumlao, D.; Katz, J.E.; Clarke, S.: Identification of the gene and characterization of the activity of the trans-aconitate methyltransferase in Saccharomyces cerevisiae. Biochemistry, **40**, 13699-13709 (2001)
[2] Cai, H.; Strouse, J.; Dumlao, D.; Jung, M.E.; Clarke, S.: Distinct reactions catalyzed by bacterial and yeast trans-aconitate methyltransferase. Biochemistry, **40**, 2210-2219 (2001)

(Iso)eugenol O-methyltransferase 2.1.1.146

1 Nomenclature

EC number
2.1.1.146

Systematic name
S-adenosyl-L-methionine:isoeugenol O-methyltransferase

Recommended name
(iso)eugenol O-methyltransferase

Synonyms
EOMT
IEMT
S-adenosyl-L-methionine:(iso)eugenol O-methyltransferase
allylphenol O-methyltransferase
chavicol O-methyltransferase
eugenol O-methyltransferase
eugenol methyltransferase
isoeugenol methyltransferase
methyltransferase, (iso)eugenol
methyltransferase, allylphenol O-

CAS registry number
191744-33-9

2 Source Organism

<1> *Clarkia breweri* [1, 2]
<2> *Ocimum basilicum* (variety EMX-1 [3]) [3]

3 Reaction and Specificity

Catalyzed reaction
S-adenosyl-L-methionine + isoeugenol = S-adenosyl-L-homocysteine + isomethyleugenol

Reaction type
methyl group transfer

Substrates and products

S S-adenosyl-L-methionine + chavicol <2> (<2>, 24% of the activity with eugenol [3]) (Reversibility: ? <2> [3]) [3]
P S-adenosyl-L-homocysteine + estragol
S S-adenosyl-L-methionine + eugenol <1, 2> (Reversibility: ? <1, 2> [1, 2, 3]) [1, 2, 3]
P S-adenosyl-L-homocysteine + methyleugenol <1> [1, 2]
S S-adenosyl-L-methionine + guaiacol <2> (<2>, 29.9% of the activity with eugenol [3]) (Reversibility: ? <2> [3]) [3]
P S-adenosyl-L-homocysteine + 1,2-dimethoxybenzene
S S-adenosyl-L-methionine + isoeugenol <1, 2> (<2>, 26% of the activity with eugenol [3]) (Reversibility: ? <1, 2> [1, 2, 3]) [1, 2, 3]
P S-adenosyl-L-homocysteine + isomethyleugenol <1, 2> [1, 2, 3]
S S-adenosyl-L-methionine + trans-isoeugenol <2> (Reversibility: ? <2> [3]) [3]
P S-adenosyl-L-homocysteine + (E)-2',3'-dimethoxy-1-propenylbenzene

Inhibitors

Ca^{2+} <1> (<1>, 5 mM, weak inhibition [1]) [1]
Cu^{2+} <1> (<1>, 5 mM, strong inhibition [1]) [1]
Fe^{2+} <1> (<1>, 5 mM, strong inhibition [1]) [1]
Fe^{3+} <1> (<1>, 5 mM, strong inhibition [1]) [1]
K^+ <1> (<1>, 5 mM, weak inhibition [1]) [1]
Mg^{2+} <1> (<1>, 5 mM, 33% inhibition [1]) [1]
Mn^{2+} <1> (<1>, 5 mM, strong inhibition [1]) [1]
NH_4^+ <1> (<1>, 5 mM, weak inhibition [1]) [1]
Na^+ <1> (<1>, 5 mM, weak inhibition [1]) [1]
Zn^{2+} <1> (<1>, 5 mM, strong inhibition [1]) [1]

Metals, ions

Additional information <1> (<1>, methylation of eugenol and isoeugenol [1]) [1]

Turnover number (min^{-1})

0.0522 <2> (chavicol) [3]
0.078 <2> (trans-isoeugenol) [3]
0.186 <2> (eugenol) [3]

Specific activity (U/mg)

Additional information <1> [1]

K_m-Value (mM)

0.003 <2> (eugenol) [3]
0.005 <2> (S-adenosyl-L-methionine) [3]
0.007 <2> (chavicol) [3]
0.007 <1> (eugenol, <1>, plant-purified enzyme [1]) [1]
0.01 <2> (trans-isoeugenol) [3]
0.019 <1> (S-adenosyl-L-methionine, <1>, recombinant enzyme [1]) [1]
0.027 <1> (S-adenosyl-L-methionine, <1>, plant-purified enzyme [1]) [1]

0.03 <1> (eugenol, <1>, recombinant enzyme [1]) [1]
0.058 <1> (isoeugenol, <1>, plant-purified enzyme [1]) [1]
0.074 <1> (isoeugenol, <1>, recombinant enzyme [1]) [1]

pH-Optimum

7-8 <2> [3]
7.5 <1> (<1>, methylation of eugenol and isoeugenol [1]) [1]

4 Enzyme Structure

Molecular weight

81000 <2> (<2>, gel filtration [3]) [3]
88000 <1> (<1>, gel filtration [1]) [1]

Subunits

dimer <1, 2> (<1,2>, 2 * 40000, SDS-PAGE [1,3]) [1, 3]

5 Isolation/Preparation/Mutation/Application

Source/tissue

flower <1> [1]
leaf <2> (<2>, young [3]) [3]
petal <1> [1, 2]
stamen <1> [2]
style <1> [2]
trichome <2> [3]

Purification

<1> [1]
<2> (partial [3]) [3]

Cloning

<1> (expression in Escherichia coli [1,2]) [1, 2]
<2> [3]

6 Stability

Temperature stability

35 <1> (<1>, 30 min, stable up to [1]) [1]
45 <1> (<1>, 30 min, activity is dramatically decreased [1]) [1]

References

[1] Wang, J.; Pichersky, E.: Characterization of S-adenosyl-L-methionine:(iso)eugenol O-methyltransferase involved in floral scent production in Clarkia breweri. Arch. Biochem. Biophys., **349**, 153-160 (1998)

[2] Wang, J.; Dudareva, N.; Bhakta, S.; Raguso, R.A.; Pichersky, E.: Floral scent production in Clarkia breweri (Onagraceae). II. Localization and development modulation of the enzyme S-adenosyl-L-methionine:(iso)eugenol O-methyltransferase and phenylpropanoid emission. Plant Physiol., **114**, 213-221 (1997)

[3] Gang, D.R.; Lavid, N.; Zubieta, C.; Chen, F.; Beuerle, T.; Lewinsohn, E.; Noel, J.P.; Pichersky, E.: Characterization of phenylpropene O-methyltransferases from sweet basil: facile change of substrate specificity and convergent evolution within a plant O-methyltransferase family. Plant Cell, **14**, 505-519 (2002)

Corydaline synthese 2.1.1.147

1 Nomenclature

EC number
2.1.1.147

Systematic name
S-adenosyl-L-methionine:protoberberine 13-C-methyltransferase

Recommended name
corydaline synthase

Synonyms
S-adenosylmethionine:protoberberine-13C-methyltransferase
methyltransferase, protoberberine
protoberberine 13-methyltransferase

CAS registry number
155807-67-3

2 Source Organism

<1> *Corydalis cava* [1]

3 Reaction and Specificity

Catalyzed reaction
S-adenosyl-L-methionine + palmatine + 2 NADPH + H^+ = S-adenosyl-L-homocysteine + corydaline + 2 $NADP^+$

Reaction type
methyl group transfer

Substrates and products
- **S** S-adenosyl-L-methionine + berberine + NADPH + H^+ <1> (Reversibility: ? <1> [1]) [1]
- **P** S-adenosyl-L-homocysteine + corydaline + $NADP^+$ <1> [1]
- **S** S-adenosyl-L-methionine + columbamine + NADPH + H^+ <1> (<1>, 27% of the activity with berberine [1]) (Reversibility: ? <1> [1]) [1]
- **P** S-adenosyl-L-homocysteine + isocorypalmine + $NADP^+$
- **S** S-adenosyl-L-methionine + coptisine + NADPH + H^+ <1> (<1>, 21% of the activity with berberine [1]) (Reversibility: ? <1> [1]) [1]

P S-adenosyl-L-homocysteine + tetrahydrocorysamine + $NADP^+$
S S-adenosyl-L-methionine + dihydroberberine + NADPH + H^+ <1> (<1>, 93% of the activity with berberine [1]) (Reversibility: ? <1> [1]) [1]
P S-adenosyl-L-homocysteine + thalictricavine + $NADP^+$ <1> [1]
S S-adenosyl-L-methionine + jatrorrhizine + NADPH + H^+ <1> (<1>, 67% of the activity with berberine [1]) (Reversibility: ? <1> [1]) [1]
P S-adenosyl-L-homocysteine + 7,8-dehydrothalictricavine + $NADP^+$
S S-adenosyl-L-methionine + palmatine + NADPH + H^+ <1> (<1>, 93% of the activity with berberine [1]) (Reversibility: ? <1> [1]) [1]
P S-adenosyl-L-homocysteine + corydaline + $NADP^+$ <1> [1]

Inhibitors
S-adenosyl-L-homocysteine <1> [1]

Specific activity (U/mg)
Additional information <1> [1]

K_m-Value (mM)
0.0027 <1> (S-adenosyl-L-methionine) [1]
0.005 <1> (berberine) [1]
0.016 <1> (NADPH) [1]

K_i-Value (mM)
0.0026 <1> (S-adenosyl-L-homocysteine) [1]

pH-Optimum
8 <1> [1]

pH-Range
7-9.5 <1> (<1>, pH 7.0: about 45% of maximal activity, pH 9.5: about 30% of maximal activity [1]) [1]

Temperature optimum (°C)
37 <1> [1]

Temperature range (°C)
20-50 <1> (<1>, 20°C: about 35% of maximal activity, 50°C: about 75% of maximal activity [1]) [1]

4 Enzyme Structure

Molecular weight
56000 <1> (<1>, gel filtration [1]) [1]

5 Isolation/Preparation/Mutation/Application

Source/tissue
bulb <1> [1]

Localization

cytosol <1> [1]

Purification

<1> [1]

References

[1] Rueffer, M.; Bauer, W.; Zenk, M.H.: The formation of corydaline and related alkaloids in Corydalis cava in vivo and in vitro. Can. J. Chem., **72**, 170-175 (1994)

Thymidylate synthase (FAD) 2.1.1.148

1 Nomenclature

EC number
2.1.1.148

Systematic name
5,10-methylenetetrahydrofolate,$FADH_2$:dUMP C-methyltransferase

Recommended name
thymidylate synthase (FAD)

Synonyms
TSCP
Thy1
ThyX
thymidylate synthase complementing protein
Additional information (lacks any sequence similarities with the classical thymidylate synthase, ThyA (EC 2.1.1.45))

CAS registry number
9031-61-2

2 Source Organism

<1> *Helicobacter pylori* [1, 4]
<2> *Campylobacter jejuni* [1, 2, 4]
<3> *Rickettsia prowazekii* [1, 4]
<4> *Borrelia burgdorferi* [1, 4]
<5> *Treponema pallidum* [1]
<6> *Chlamydia sp.* (<6> enzyme found in three different species [1]) [1]
<7> *Mycobacterium tuberculosis* [1]
<8> *Pyrococcus abyssi* [1]
<9> *Pyrococcus horikoshii* [1]
<10> *Sulfolobus solfataricus* [1]
<11> *Dictyostelium discoideum* [1, 3, 4]
<12> *Halobacterium salinarum* [2]
<13> *Thermotoga maritima* [3, 4]
<14> *Mycobacterium tuberculosis* (simultaneous occurrence of thyA and thyX [4]) [4]
<15> *Mycobacterium leprae* (simultaneous occurrence of thyA and thyX [4]) [4]
<16> *Treponema pallidum* [4]
<17> *Treponema denticola* [4]

<18> *Chlamydia trachomatis* [4]
<19> *Chlamydia pneumoniae* [4]
<20> *Chlamydia psittaci* [4]
<21> *Mycobacterium bovis* [4]
<22> *Mycobacterium avium* [4]
<23> *Corynebacterium diphtheriae* [4]
<24> *Clostridium botulinum* [4]
<25> *Clostridium difficile* [4]
<26> *Clostridium perfringens* [4]
<27> *Bacillus anthracis* [4]
<28> *Rickettsia connorii* [4]
<29> *Ehrlichia chaffeensis* [4]
<30> *Rhodococcus sp.* [4]

3 Reaction and Specificity

Catalyzed reaction

5,10-methylenetetrahydrofolate + dUMP + $FADH_2$ = dTMP + tetrahydrofolate + FAD (FMN can replace FAD, reaction shown is distinct from that of the classical thymidylate synthase, ThyA (EC 2.1.1.45))

Reaction type

methyl group transfer

Natural substrates and products

S 5,10-methylenetetrahydrofolate + dUMP + $FADH_2$ <1-30> (<1-11> de novo synthesis of thymidylate [1]) [1-4]
P dTMP + tetrahydrofolate + FAD

Substrates and products

S 5,10-methylenetetrahydrofolate + dUMP + $FADH_2$ <1-30> (Reversibility: ? <1-30> [1-4]) [1-4]
P dTMP + tetrahydrofolate + FAD <1-30> [1-4]
S 5,10-methylenetetrahydrofolate + dUMP + $FMNH_2$ <1-11> (Reversibility: ? <1-11> [1]) [1]
P dTMP + tetrahydrofolate + FMN <1-11> [1]

Inhibitors

FAD <13> [4]
HEPES <13> [4]
O_2 <2> [2]

Cofactors/prosthetic groups

FAD <11, 13> (<11, 13> tightly bound, four molecules per tetramer [3, 4]) [3, 4]
$FADH_2$ <1-11> (<1> activity is dependent on reduced flavin nucleotides [1]) [1]
$FMNH_2$ <1-11> (<1> activity is dependent on reduced flavin nucleotides [1]) [1]

Activating compounds

FAD <1> (<1> increases activity [1]) [1]
FMN <1> (<1> increases activity [1]) [1]
NADH <1, 11, 13> (<11,13> required to reduce bound FAD [3,4]) [3, 4]
NADPH <1, 11, 13> (<11,13> required to reduce bound FAD [3,4]) [3, 4]
UMP <1> [1]

4 Enzyme Structure

Molecular weight

23800 <2> (<2> SDS-PAGE [2]) [2]
31000 <1> (<1> SDS-PAGE [1]) [1]
31500 <1> (<1> calculated from DNA sequence [1]) [1]
111000 <1> (<1> gel filtration [1]) [1]

Subunits

tetramer <1, 13> (<1> 4 * 31000, gel filtration, SDS-PAGE [1]; <13> data from crystal structure [4]) [1, 4]

5 Isolation/Preparation/Mutation/Application

Purification

<1> (from transformed E. coli using His-tag [1]) [1]
<2> (from transformed E. coli [2]) [2]
<13> [4]

Crystallization

<13> [4]

Cloning

<1> (complemetation of Escherichia coli, transformants are able to grow in the absence of thymidine [1]) [1]
<2> (complemetation of Escherichia coli thyA deletion mutant, transformants are able to grow in the absence of thymidine under oxygen limited conditions [2]) [2]
<8> (complemetation of Escherichia coli, transformants are unable to grow in the absence of thymidine [1]) [1]
<12> (complemetation of Haloferax volcanii thyA deletion mutant, transformants are able to grow in the absence of thymidine [2]) [2]
<13> [4]

Engineering

S107A <1> (<1> no activity [1]) [1]

Application

medicine <1-4, 16-30> (<1-4,16-30> promising medical target, thyX is present in a number of pathogenic bacteria but absent in human [1,4]) [1, 4]

References

[1] Myllykallio, H.; Lipowski, G.; Leduc, D.; Filee, J.; Forterre, P.; Liebl, U.: An alternative flavin-dependent mechanism for thymidylate synthesis. Science, **297**, 105-107 (2002)
[2] Giladi, M.; Bitan-Banin, G.; Mevarech, M.; Ortenberg, R.: Genetic evidence for a novel thymidylate synthase in the halophilic archaeon Halobacterium salinarum and in Campylobacter jejuni. FEMS Microbiol. Lett., **216**, 105-109 (2002)
[3] Montfort, W.R.: Complementing thymidylate synthase. Structure, **11**, 607-608 (2003)
[4] Mathews, I.I.; Deacon, A.M.; Canaves, J.M.; McMullan, D.; Lesley, S.A.; Agarwalla, S.; Kuhn, P.: Functional analysis of substrate and cofactor complex structures of a thymidylate synthase-complementing protein. Structure, **11**, 677-690 (2003)

Myricetin O-methyltransferase 2.1.1.149

1 Nomenclature

EC number
2.1.1.149

Systematic name
S-adenosyl-L-methionine:myricetin O-methyltransferase

Recommended name
myricetin O-methyltransferase

Synonyms
CrCOMT2
flavonoid 3',5'-O-dimethyltransferase
flavonoid O-methyltransferase
methyltransferase, flavonoid

CAS registry number
118251-36-8

2 Source Organism

<1> *Catharanthus roseus* (Madagascar periwinkle [1]) [1]

3 Reaction and Specificity

Catalyzed reaction
2 S-adenosyl-L-methionine + myricetin = 2 S-adenosyl-L-homocysteine + syringetin

Reaction type
methyl group transfer

Natural substrates and products
S S-adenosyl-L-methionine + myricetin <1> (<1> involvement in biosynthesis of flavonol glycosides and anthocyanins [1]) [1]
P ?

Substrates and products
S S-adenosyl-L-methionine + dihydromyricetin <1> (<1> two sequential methylations at the 3'- and 5'-positions of the B-ring in dihydromyricetin [1]) (Reversibility: ? <1> [1]) [1]

P ?

S S-adenosyl-L-methionine + myricetin <1> (<1> two sequential methylations at the 3'- and 5'-positions of the B-ring in myricetin, activity is strictly confined to flavonols and dihydroflavonols, it requires at least 2 B-ring hydroxyl groups [1]) (Reversibility: ? <1> [1]) [1]

P ?

Specific activity (U/mg)
0.000219 <1> [1]

4 Enzyme Structure

Molecular weight
39000 <1> (<1> SDS-PAGE [1]) [1]

5 Isolation/Preparation/Mutation/Application

Source/tissue
cell suspension culture <1> [1]

Purification
<1> [1]

Cloning
<1> [1]

References

[1] Cacace, S.; Schröder, G.; Wehinger, E.; Strack, D.; Schmidt, J.; Schröder, J.: A flavonol O-methyltransferase from Catharanthus roseus performing two sequential methylations. Phytochemistry, **62**, 127-137 (2003)

Isoflavone 7-O-methyltransferase 2.1.1.150

1 Nomenclature

EC number
2.1.1.150

Systematic name
S-adenosyl-L-methionine:hydroxyisoflavone 7-O-methyltransferase

Recommended name
isoflavone 7-O-methyltransferase

Synonyms
IOMT

CAS registry number
136111-54-1

2 Source Organism

<1> *Medicago sativa* (alfalfa, two isozymes [1]) [1-3, 5, 6]
<2> *Medicago sativa* (IOMT8 [4]) [4]
<3> *Hordeum vulgare* [7]

3 Reaction and Specificity

Catalyzed reaction
S-adenosyl-L-methionine + 7-hydroxyisoflavone = S-adenosyl-L-homocysteine + 7-methoxyisoflavone (<3> sequential bi-bi mechanism [7])

Reaction type
methyl group transfer

Natural substrates and products
S Additional information <1> (<1> involved in biosynthesis of the phytoalexin medicarpin [2]) [2]
P ?

Substrates and products
S 6,7,4'-trihydroxyisoflavone + S-adenosyl-L-methionine <1, 2> (<1> best substrate for IOMT I and IOMT II [1]) (Reversibility: ? <1> [1,3,4]) [1, 3, 4]

P 7-methoxy-6,4'-dihydroxyisoflavone + S-adenosyl-L-homocysteine <1, 2> (<1> i.e. [1]) [1, 3, 4]

S 6α-hydroxymaackiain + S-adenosyl-L-methionine <2> (Reversibility: ? <2> [4]) [4]

P pisatin + S-adenosyl-L-homocysteine <2> [4]

S apigenin + S-adenosyl-L-methionine <3> (<3> i.e. 4',5,7-trihydroxyflavone [7]) (Reversibility: ? <3> [7]) [7]

P genkwanin + S-adenosyl-L-homocysteine <3> (<3> i.e. 4',5-dihydroxy-7-methoxyflavone [7]) [7]

S daidzein + S-adenosyl-L-methionine <1, 2> (<1> i.e. 7,4'-dihydroxyisoflavone [1]) (Reversibility: ? <1,2> [1-5]) [1-5]

P isoformononetin + S-adenosyl-L-homocysteine <1, 2> (<1> i.e. 7-methoxy-4'-hydroxyisoflavone [1]) [1-5]

S genistein + S-adenosyl-L-methionine <1, 2> (<1> i.e. 5,7,4'-trihydroxyisoflavone [1]) (Reversibility: ? <1,2> [1-4]) [1-4]

P prunetin + S-adenosyl-L-homocysteine <1, 2> (<1> i.e. 7-methoxy-5,4'-dihydroxyisoflavone [1]) [1, 2, 4]

S medicarpin + S-adenosyl-L-methionine <1> (Reversibility: ? <1> [1]) [1]

P ? + S-adenosyl-L-homocysteine <1> (<1> A-ring methylation, only IOMT I [1]) [1]

S vestitol + S-adenosyl-L-methionine <1> (<1> only IOMT II [1]) (Reversibility: ? <1> [1]) [1]

P ? + S-adenosyl-L-homocysteine <1> [1]

S Additional information <1> (<1> IOMT shows specificity for 7-hydroxy-position in vitro, but in contrast, modulation of enzyme in transgenic alfalfa has corresponding effects on the amounts of 4'-O-methylated isoflavoniods present [2]; <1> in vivo reaction is 4'-O-methylation, in vitro reaction is 7-O-methylation [3]; <1> colocalization with isoflavone synthase cytochrome P_{450} ensures rapid B-ring methylation of unstable product of isoflavone synthase [5]) [2, 3, 5]

P ?

K_m-Value (mM)

0.0046 <3> (apigenin) [7]
0.011 <3> (S-adenosylmethionine) [7]
0.02 <1> (daidzein, <1> IOMT II [1]) [1]
0.15 <1> (S-adenosyl-L-methionine, <1> IOMT II [1]) [1]

pH-Optimum

8.5 <1> (<1> IOMT II [1]) [1]

4 Enzyme Structure

Molecular weight

41000 <1> (<1> gel filtration [1]) [1]

Subunits

? <1-3> (<1> x * 41000, SDS-PAGE [3]; <2> x * 41000, SDS-PAGE, recombinant enzyme [4]; <3> x * 43000, SDS-PAGE, recombinant enzyme [7]) [3, 4, 7]
monomer <1> (<1> 1 * 41000, SDS-PAGE [1]) [1]

5 Isolation/Preparation/Mutation/Application

Source/tissue

callus <1> [3]
cell suspension culture <1> [1, 3, 5]
leaf <3> [7]

Localization

endoplasmic reticulum <1> (<1> colocalizes with isoflavone synthase cytochrome P_{450} [5]) [5]

Purification

<1> (IOMT II [1]) [1, 3]

Crystallization

<1> (in complex with products S-adenosyl-L-homocysteine and isoformononetin [6]) [6]

Cloning

<2> [4]

6 Stability

General stability information

<1>, EDTA and thiol-containing reagents stabilize [1]

References

[1] Edwards, R.; Dixon, R.A.: Isoflavone O-methyltransferase activities in elicitor-treated cell suspension cultures of Medicago sativa. Phytochemistry, **30**, 2597-2606 (1991)
[2] He, X.Z.; Dixon, R.A.: Genetic manipulation of isoflavone 7-O-methyltransferase enhances biosynthesis of 4'-O-methylated isoflavonoid phytoalexins and disease resistance in alfalfa. Plant Cell, **12**, 1689-1702 (2000)
[3] He, X.-Z.; Dixon, R.A.: Affinity chromatography, substrate/product specificity, and amino acid sequence analysis of an isoflavone O-methyltransferase from alfalfa (Medicago sativa L.). Arch. Biochem. Biophys., **336**, 121-129 (1996)

[4] He, X.Z.; Reddy, J.T.; Dixon, R.A.: Stress responses in alfalfa (Medicago sativa L). XXII. cDNA cloning and characterization of an elicitor-inducible isoflavone 7-O-methyltransferase. Plant Mol. Biol., **36**, 43-54 (1998)

[5] Liu, C.-J.; Dixon, R.A.: Elicitor-induced association of isoflavone O-methyltransferase with endomembranes prevents the formation and 7-O-methylation of daidzein during isoflavonoid phytoalexin biosynthesis. Plant Cell, **13**, 2643-2658 (2001)

[6] Zubieta, C.; He, X.-Z.; Dixon, R.A.; Noel, J.P.: Structures of two natural product methyltransferases reveal the basis for substrate specificity in plant O-methyltransferases. Nat. Struct. Biol., **8**, 271-279 (2001)

[7] Christensen, A.B.; Gregersen, P.L.; Olsen, C.E.; Collinge, D.B.: A flavonoid 7-O-methyltransferase is expressed in barley leaves in response to pathogen attack. Plant Mol. Biol., **36**, 219-227 (1998)

Cobalt-factor II C^{20}-methyltransferase 2.1.1.151

1 Nomenclature

EC number
2.1.1.151

Systematic name
S-adenosyl-L-methionine:cobalt-factor-II C^{20}-methyltransferase

Recommended name
cobalt-factor II C^{20}-methyltransferase

Synonyms
CbiL

2 Source Organism

<1> *Salmonella typhimurium* [1]

3 Reaction and Specificity

Catalyzed reaction
S-adenosyl-L-methionine + cobalt-factor II = S-adenosyl-L-homocysteine + cobalt-factor III (Involved in the anaerobic biosynthesis of vitamin B_{12})

Reaction type
methyl group transfer

Natural substrates and products
S S-adenosyl-L-methionine + cobalt-factor II <1> (Reversibility: ? <1> [1]) [1]
P S-adenosyl-L-homocysteine + cobalt-factor III

Substrates and products
S S-adenosyl-L-methionine + cobalt-factor II <1> (Reversibility: ? <1> [1]) [1]
P S-adenosyl-L-homocysteine + cobalt-factor III

References

[1] Spencer, P.; Stolowich, N.J.; Sumner, L.W.; Scott, A.I.: Definition of the redox states of cobalt-precorrinoids: investigation of the substrate and redox specificity of CbiL from Salmonella typhimurium. Biochemistry, **37**, 14917-14927 (1998)

Precorrin-6A synthase (deacetylating) 2.1.1.152

1 Nomenclature

EC number
2.1.1.152

Systematic name
S-adenosyl-L-methionine:precorrin-5 C^1-methyltransferase (deacetylating)

Recommended name
precorrin-6A synthase (deacetylating)

Synonyms
CobF
precorrin-6X synthase (deacetylating)

2 Source Organism

<1> *Pseudomonas denitrificans* [1]

3 Reaction and Specificity

Catalyzed reaction
S-adenosyl-L-methionine + precorrin-5 + H_2O = S-adenosyl-L-homocysteine + precorrin-6A + acetate (In the aerobic cobalamin biosythesis pathway, four enzymes are involved in the conversion of precorrin-3A to precorrin-6A. The first of the four steps is carried out by EC 1.14.13.83, precorrin-3B synthase (CobG), yielding precorrin-3B as the product. This is followed by three methylation reactions, which introduce a methyl group at C-17 (CobJ), C-11 (CobM) and C-1 (CobF; EC 2.1.1.152) of the macrocycle, giving rise to precorrin-4, precorrin-5 and precorrin-6A, respectively)

Natural substrates and products
S S-adenosyl-L-methionine + precorrin-5 + H_2O <1> (Reversibility: ? <1> [1]) [1]
P S-adenosyl-L-homocysteine + precorrin-6A + acetate

Substrates and products
S S-adenosyl-L-methionine + precorrin-5 + H_2O <1> (Reversibility: ? <1> [1]) [1]
P S-adenosyl-L-homocysteine + precorrin-6A + acetate

References

[1] Debussche, L.; Thibaut, D.; Cameron, B.; Crouzet, J.; Blanche, F.: Biosynthesis of the corrin macrocycle of coenzyme B_{12} in Pseudomonas denitrificans. J. Bacteriol., **175**, 7430-7440 (1993)

[2] Warren, M.J.; Raux, E.; Schubert, H.L.; Escalante-Semerena, J.C.: The biosynthesis of adenosylcobalamin (vitamin B_{12}). Nat. Prod. Rep., **19**, 390-412 (2002)

FSC
www.fsc.org
MIX
Papier | Fördert gute Waldnutzung
FSC® C083411

Kolibri 360 GmbH
Ferdinand-Jühlke-Straße 7
99095 Erfurt, Deutschland
produktsicherheit@kolibri360.de